Introduction to
PHYSICAL GEOGRAPHY

Introduction to
PHYSICAL
GEOGRAPHY

Robert N. Wallen

Mendocino College

WCB Wm. C. Brown Publishers

Book Team

Editor *Jeffrey L. Hahn*
Developmental Editor *Lynne M. Meyers*
Production Editor *Gloria Schiesl*
Designer *Mark Elliot Christianson*
Art Editor *Carla Goldhammer*
Photo Editor *Carrie Burger*
Visuals Processor *Joyce E. Watters*

 Wm. C. Brown Publishers

President *G. Franklin Lewis*
Vice President, Publisher *George Wm. Bergquist*
Vice President, Operations and Production *Beverly Kolz*
National Sales Manager *Virginia S. Moffat*
Group Sales Manager *Vincent R. Di Blasi*
Vice President, Editor in Chief *Edward G. Jaffe*
Marketing Manager *John W. Calhoun*
Advertising Manager *Amy Schmitz*
Managing Editor, Production *Colleen A. Yonda*
Manager of Visuals and Design *Faye M. Schilling*
Production Editorial Manager *Julie A. Kennedy*
Production Editorial Manager *Ann Fuerste*
Publishing Services Manager *Karen J. Slaght*

WCB Group

President and Chief Executive Officer *Mark C. Falb*
Chairman of the Board *Wm. C. Brown*

Cover photo: © Kerrick James Photography

Copyeditor: Andrew L. Alden

Brief Contents

Contents

UNIT I

The Geographic Perspective *9*

1

The Geographic Viewpoint *10*

2

The Earth in Space *25*

UNIT II

The Atmosphere 49

3

The Ocean of Air 50

4

Solar Energy and the Earth's Response 68

5

The Windy Planet 92

6

Clouds, Atmospheric Moisture, and Precipitation 110

7

The Stormy Planet 133

17

Leveling of the Land by Weathering and Mass Wasting 370

The Hydrosphere 387

18

The Work of Water in the Hydrologic Cycle 388

19

Groundwater and the Hydrologic Cycle 406

20

The Work of Ice in the Hydrologic Cycle 424

UNIT VI

The Earth's Systems *483*

23

The Miracle of Design *484*

List of Boxes

List of Tables

Preface

Introduction to Physical Geography is designed for either majors or non-majors in geography. The text is developed in a highly visual manner to aid in understanding geographic relationships and principles. Each chapter acquaints the reader with the major processes, features, and interrelationships of the natural environment.

The theme of energy is used to view the planet Earth in its totality. Energy is traced as it flows from the Sun to the Earth and is transformed into a variety of other forms in the hydrosphere, atmosphere, lithosphere, and biosphere. Energy from the Earth's interior is viewed as the driving force for tectonic plate motion. Careful attention is paid to the role of humans in the modification of the Earth's systems.

The text is structured into four major topics dealing with energy systems of the atmosphere, biosphere, lithosphere, and hydrosphere.

Introduction to Physical Geography approaches the subject utilizing six unique pedagogical approaches:

1. *The Study Area—Integration of Concepts in a Given Place.* The text begins by introducing seven geographic concepts that provide the student with a geographic perspective. The perspective is introduced in the setting of a small valley where the processes and interrelationships of nature are taking place. The valley serves as a focal point, a place where abstract concepts can be related to a place. The valley is revisited occasionally in the text as an example of some of the processes and physical patterns. This approach gives the reader a point of reference to place in proper context the geographic viewpoint before taking on the task of organizing knowledge about the physical environment.

2. *Energy—Integration of the Environment.* Energy is used as a unifying theme to integrate the various divisions of the environment. Frequently, geography texts divide the environment into four major themes—atmosphere, hydrosphere, biosphere, and lithosphere—without integrating these systems. By tracing energy from the Sun and the Earth's interior, the student will be able to understand more clearly the interrelationships between the major elements of the environment.

3. *Balanced Study of the Environment.* The four major systems of the environment are treated with more equality given to each. Plants and animals are treated here not in isolation but as integrated systems in community settings.

4. *Learning Tools.* The text is written for students, and especially the liberal arts freshman and sophomore with little science background. Student objectives introduce each chapter. Contemporary environmental issues give the reader an opportunity to see the applications of the material studied.

5. *Geographic Tools.* Teachers often use topographic maps in the classroom as a teaching tool. In this book, topographic maps are used in each unit to illustrate patterns and processes of nature. This text includes U.S. customary and SI (metric) units side by side.

6. *Illustrated Questions.* Illustrated questions, located at the end of each chapter, are linked to figures in the chapter. This approach is designed to help students better apply their knowledge of new concepts to actual physical environments.

An integrated package of ancillary materials has been designed to support and enhance *Introduction to Physical Geography.* Students will benefit from the skillfully prepared *Student Study Guide* by Miriam Helen Hill. Instructors can incorporate the slides and transparencies reproduced from figures in the text into their lectures and can also make use of the illustrations and photographs found on the Wm. C. Brown Earth Science Videodisc. *Instructor's Manual with Test Item File* and *WCB TestPak,* a computerized testing service, are also available to assist in classroom instruction. We hope you will find *Introduction to Physical Geography* and its ancillaries a valuable teaching tool for you and your students.

Acknowledgments

I am indebted to a great many people who have encouraged me to write about our planet Earth. I continue to be awed by its miracle of design.

I thank the Wm. C. Brown staff who have worked so professionally in the development and production stages. Specifically, I want to thank Ed Jaffe, who appreciated my unique approach to the subject; Lynne Meyers and Jeffrey L. Hahn, who worked so patiently through the development and review stages; Jim Sowder and Joyce Ives for their insights on key manuscript changes; Andrew Alden, who copyedited the text; the production team, who directed the transformation of my raw manuscript into the textbook you are now holding; and Charles Hogue for his excellent photographs.

I extend my gratitude to the members of the review team for their helpful comments and insights:

Charles Zinser
SUNY-Plattsburg

Roland Grant
Eastern Montana College

Miriam Helen Hill
Indiana University-Southeast

Leeland T. Engelhorn
Grossmont College

Peter J. Valora
Ricks College

Keith Runyon
Illinois Valley Community College

Richard A. Crooker
Kutztown University

M. Stanley Dart
Kearney State College

Thomas Terich
Western Washington University

Noel Stirrat
College of Lake County

Diann Keisel
University of Wisconsin-Baraboo

W. Franklin Long
Coastal Carolina Community College

James Feng
Foothill-DeAnza Community College

A special thanks is given to the individuals whose responses to market research helped to develop the initial scope and focus of the text:

Robert Altchul
University of Arizona

Kevin L. Anderson
Augustana College

Robert H. Arnold
Salem State College

Ward Barrett
University of Minnesota–Twin Cities

Susan M. Berta
Indiana State University

Vergean Birkin
Meredith College

David R. Butler
University of Georgia

Anthony O. Clarke
University of Louisville

Lary M. Dilsaver
University of South Alabama

Percy H. Dougherty
Kutztown University of Pennsylvania

Matt Ebiner
El Camino College

Julie Elbert
University of Southern Mississippi

R. E. Faflak
Valley City State University

Gregory E. Faiers
University of Pittsburgh–Johnstown

Patricia L. Fall
Arizona State University

Robert G. Foote
Wayne State College

Robert Franklin
Golden West College

John S. Gaines
King College

Thomas J. Gergel
State University of New York–Oneonta

Robert H. Gray
Diablo Valley College

Barry N. Haack
George Mason University

Jerry Hanson
University of Arkansas–Little Rock

Martha L. Henderson
University of Minnesota–Duluth

Frank N. Himmler
University of North Alabama

Sally P. Horn
University of Tennessee–Knoxville

Solomon Isiorho
Indiana–Purdue University

Brian L. Johnson
Marshalltown Community College

Cecil S. Keen
Mankato State University

Michael Kovalsky
DeKalb College

Garrick B. Lee
Butte College

Robin R. Lyons
University of Hawaii–Leeward Comm. College

Darrel L. McDonald
Stephen F. Austin State University

Harold A. Meeks
University of Vermont

Douglas S. Pease
Grand Canyon University

Robert Phillips
University of Wisconsin–Platteville

Katherine H. Price
DePauw University

R. Douglas Ramsey
Utah State University

Michael Ritter
University of Wisconsin–Stevens Point

Gerald W. Ropka
DePaul University

Roger Sandness
South Dakota State University

George A. Schnell
The College of New Paltz–SUNY

Stan Schumer
Oakland Community College

Suk-Han Shin
Eastern Washington University

Curtis J. Sorenson
University of Kansas

George E. Stetson
Bloomsburg University of Pennsylvania

W. J. Switzer
Southwestern College

Morris Thomas
Lansing Community College

Nicolay P. Timofeeff
State University of New York–Binghamton

Barry Warmerdam
King's River Community College

Charles L. Wax
Mississippi State University

Ted Wilkin
University of Southern Colorado

The review team would have never seen this text if it hadn't been for the inspiration of my father and mother, who taught me to love the outdoors and its beauty as we traveled together on many vacations.

A special thanks to my typist, Nancy Phillips, who labored so faithfully, never missing a deadline, and providing many suggestions for organizing information.

Finally, the steady support and encouragement of my wife, Betty, helped make this project a reality. Her insights provided me with the student perspective reflected in the text.

Introduction

Wild Rose and Twin Peaks, Big Cottonwood Canyon, Wasaton Range, Utah.
© Stephen Trimble

Figure I.1

This volcanic mountain located in the Cascade Range of Washington is a part of a chain of volcanic peaks bordering the Pacific Ocean. How was it formed? What impact does it have on the region's climate, vegetation, soils, and cultural activities? These are some of the questions a geographer might ask. This aerial view was taken during the height of activity on May 18, 1980.

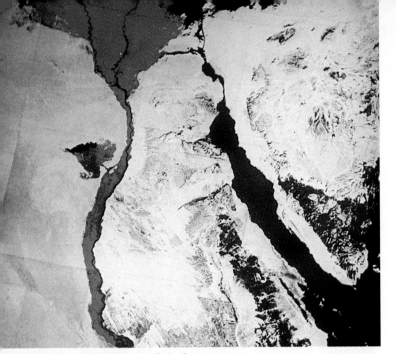

A. The Nile Valley and Delta from space

B. Imperial Valley, California–Mexico Border

C. The Great Pyramids at Giza, Egypt

D. Imperial Valley (surface view)

Figure I.2

World regional geography brings both cultural and physical geography together. The place can be a small valley or the entire planet. The space photos were taken in false color infrared. Anything containing chlorophyll (trees, grass, crops) appears as red. The change in color and pattern near the bottom of photograph B is at the United States–Mexico border. It stands out because of a change in land use pattern and crop condition. Note the offset in crop rows in photograph D. This is the result of a surface rupture along a branch of the San Andreas Fault.

Space photos by NASA.

The aim of geography can be simply stated as the understanding of places. If the place is inhabited, or utilized by people, then people become an extremely important part of it. Geography can be divided into two broad fields of study: physical and cultural geography. *Physical geography* focuses on the physical landscape (fig. I.1). Its task is to describe and explain the components of the natural environment, their distribution, and processes involving their interrelationships. For example, climate is the primary influence on soils and vegetation patterns. Physical geographers define natural regions that are caused by and continuously affected by forces and processes in nature. Physical geography combines climatology, geomorphology, pedology, biogeography, and other scientific fields.

Cultural geography focuses on the interrelationships between people and the natural landscape and is concerned with the distribution of human activity on the surface of the Earth. People move onto the center stage of our study. Cultural geography includes political, economic, urban, and historical components. Both cultural and physical geography are brought together by world regional geography for the purpose of understanding a specific region of Earth. The place of study can be a valley, say the Imperial Valley of California or the Nile River Valley, or the entire planet (fig. I.2).

Not all things are of equal importance; therefore, the geographer selects only items that make up the cultural and physical landscape. The object of geographic study—the place—is no vague concept, but a complex of interacting natural systems as real and well defined as the plants of the botanist, the cultures of the anthropologist, and the civilizations of the historian. It must be studied firsthand in the field, as well as through the research of others.

TWO ESSENTIAL QUESTIONS

Two essential questions must be asked before discovery of geographic knowledge can occur: "Where?", in order to find out how things are arranged; and "Why?", in order to appreciate and comprehend the meaning of the pattern, distribution, and form. The first is a *location* question, and the second is a *process* inquiry. An examination of processes is rewarding because it expands understanding beyond the descriptive level to the functional level of the environment. To answer the questions "where" and "why" requires current information through research, careful observation, data collection, and experimentation to determine the validity of one's understanding. This procedure is called the *scientific method*. A process-oriented inquiry always proceeds in this orderly manner. Throughout this book, subjects will begin with a description of location and natural features and patterns, and then a process inquiry will be made (fig. I.3).

TEXT APPROACH

Geography brings a unique perspective and an interdisciplinary view of the planet. The approach of this text is to examine the natural physical patterns and processes that shape our planet. We will examine how energy provides the driving force for Earth's natural systems. Evaporation of moisture into the atmosphere, followed by condensation and precipitation, are energy systems driven by both the Sun and gravity. Both gravitational and heat energy from the Earth's interior will be viewed as the key driving forces of the geologic systems of the Earth's crust.

TEXT ORGANIZATION

Introduction to Physical Geography focuses on our physical environment: the *atmosphere, biosphere, lithosphere,* and *hydrosphere.* More specifically, weather and climate, water resources, soils, plants, animals, rocks and minerals, and landforms will be viewed as natural elements on a global scale. Unit I, "The Geographic Perspective," will introduce the tools of analysis that will allow you to gain more understanding and insight about your environment. We will visit a small valley, or *study area,* in the Coast Range of northwestern California to illustrate the geographic approach or viewpoint in a microcosm (small-scale) setting. We will then take a macrocosm (large-scale) geographic perspective by examining the Earth's unique place in the solar system.

Unit II, "The Atmosphere," Unit III, "The Biosphere," Unit IV, "The Lithosphere," and Unit V, "The Hydrosphere," examine the Earth's major systems. Unit VI, "The Earth's Systems," looks back at our planet's unique design features that have allowed this quality of life to continue (fig. I.4).

Figure I.3

It is the task of physical geography to describe and explain the distribution of surface features and their interrelationships.

THE ORGANIZER OF NATURE

The text views human influence as more than just a passive consumer of energy and material. Our earliest activities as hunters and food gatherers left the energy systems of the Earth in balance, with input equaling output. About ten thousand years ago, people of various cultures began to cultivate plants and domesticate animals. This reduced the complex biosphere systems to less complex agricultural systems, in order to more efficiently store solar energy in edible forms. Thus we became organizers of our living space as we expanded our energy food base (fig. I.5).

Industrial civilization has caused even greater alteration of the environment in its demand for energy and materials. The transformation and utilization of energy and materials depend on three factors: available resources, technological skill to convert them, and means to safely transport energy and materials. The 1989 Alaskan oil spill is a graphic reminder of our potential to alter the environment. We will give careful attention to human impact on this resource base to gain a better understanding of our role as the organizer of the natural environment.

A

B

C

Figure I.4

Only the planet Earth possesses liquid water, soils, and a biosphere or life zone in the solar system. (A) Hydrosphere and lithosphere—shoreline along a northern California coast. (B) Biosphere—open meadow in a redwood forest. (C) Atmosphere—note the turkey vulture perched on the tree ready for flight.

Space photo by NASA.

A

B

Figure I.5

Early societies made a minimal impact on the Earth's systems. Hunting, fishing, herding, and food gathering left the energy systems in balance. Agriculture reduced complex plant and animal communities to simpler systems. (A) Shipibo Village in the Amazon along the Ucayali River, Peru. (B) Pucalpa, Peru, has grown from 10,000 to 200,000 in just 15 years with virgin forest converted to agriculture and urbanization land use.

Figure I.6

The flows of energy and materials through the Earth's systems originate from both the Sun and Earth's interior. More than 99 percent of the input is solar radiation.

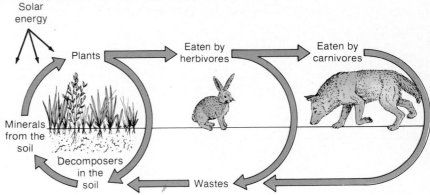

ENERGY: THE COMMON THREAD OF THE EARTH'S SYSTEMS

What is energy? Simply stated, it is the capacity to do work and is used to express many activities and processes. Energy exists in many forms, such as *potential, kinetic* (energy of motion), *chemical, heat, nuclear, radiant* or light, *gravitational,* and *magnetic.* When a car sits parked on a steep hill, we can say it has *potential energy.* If its brakes fail, the potential energy rapidly becomes kinetic energy resulting from the Earth's gravitational pull. Radiation from the Sun is also referred to as energy. When this energy form strikes the Earth's surface, it converts to other forms of energy resulting in such motions as winds, waves, and ocean currents.

Energy can be present also in an inactive chemical form such as gasoline or sugar, latent until the right conditions release its potential energy. All of these examples have one thing in common: motion or potential motion.

Energy and materials flow through the Earth's systems, momentarily going into short-term storage in such forms as plants and long-term storage as fossil fuels, later to be released by humans and nature. Our examination of the natural environment is directed by tracing the flow of energy from both the Sun and the Earth's hot interior core. We will see energy as a thread common to all natural systems (fig. I.6).

Enjoyment of Geography

BOX I.1

Geography is not merely a useful tool to deal with world and local problems. It can also help us enjoy our world around us. A study of places you have traveled to or would like to visit can be an enjoyable experience if you have a knowledge of the geographic perspective. Perhaps we cannot travel to Mexico, but we can go there vicariously through the writings of Carl Sauer, who is well known for his descriptions of the Mexican landscape in its beauty and contrasts.

An Asian philosopher once said, "Had I two loaves of bread, I would sell one and buy white hyacinths to feed my soul." As more and more of our living environmental space is reduced to concrete and plastic, we need to experience the beauty in natural designs and rhythm and the harmony of nature.

Geography can be both bread and hyacinths. It can be used to provide knowledge that gives us greater ability and wisdom to use our resources and support our life. Geography can lead us to fuller understanding of the processes that shape our planet and our lives. Most of our geography curriculum is bread, or materially oriented. It is in demand by governments, industry, agriculture, forestry, and the world traveler.

Although bread is an important factor in the study of geography, so are the hyacinths, or the beauty of the landscape. Through the hyacinths, you can find renewal of your spirit and a new excitement even in your daily travels.

Learn to observe your environment. Hyacinths can be watching a rainbow form in a mist of rain, observing the changing fall colors of a forest, looking for the migratory return of the swallows to their summer home. Remember that a knowledge of geography will provide balance—bread for the table and food for the soul.

It is my aim that you will gain a new perspective of appreciation for this unique and fragile planet. How long it can support the current quality of life will depend not only on the natural systems but on our treatment of these systems. This new perspective of a fragile habitat of integrated systems of air, water, land, and life is the subject matter before you.

Hyacinths on the Ucayali River, Amazon Basin.

OBJECTIVES

Before venturing into Unit I, "The Geographic Perspective," a preview of general objectives for you is essential. After satisfactory completion of this text, you will be able to:

1. View energy as a common thread that integrates the various components of the environment. You will be able to trace the flow of energy from the Sun and the Earth's interior through the Earth's natural systems.

2. Develop an understanding of major processes and forces that shape our natural environment.
3. Develop concepts regarding the relationships of culture to its natural surroundings.
4. Describe in general terms the natural patterns of the environment of the place. These include water bodies, soils, vegetation, climate, and landforms.
5. Develop an appreciation for the beauty, complexity, and uniqueness of the planet Earth, and a realization that its systems are delicately balanced and easily altered by human impact.

The Geographic Perspective

Only the planet Earth is an oasis in space with the right proportions of energy, air, water, and soil to support a diversity of life.

Maligne Canyon, Jasper National Park, Rocky Mountains, Alberta, Canada.

© Wolfgang Kaehler

The Geographic Viewpoint

Objectives

After completing this chapter, you will be able to:

1. Describe the **location** of a place in terms of **site** factors and relative location or **situation** factors;
2. Select the proper map **scale** for geographic study;
3. Define a **region** and recognize regional boundaries and criteria for establishing a region;
4. Recognize the **internal coherence** forces bonding a region or regions together;
5. Identify the **spatial interaction** taking place in a region and recognize the central and marginal zones;
6. Explain the **spatial distribution** of elements being studied in terms of **frequency** or **density** and **pattern;**
7. Recognize **change** as an underlying assumption that the Earth is dynamic and continually changing. Processes are not static but dynamic, cyclic, and variable.

Physical Geography develops our understanding of the natural environment.

© Doug Sherman/Geofile

Figure 1.1

This small valley is the point of departure in our inquiry about our environment.

The viewpoint of each discipline of science and social science can be determined by the questions it asks. Geologists ask questions such as, how was this valley formed? What mountain-building processes are at work? In what stage is the stream erosional process? What is the age of the rock structure?

Geographers look at the same valley and ask some of the above questions but with a broader purpose of understanding of the total environment. First, where is it located? What physical and cultural forces and processes have shaped it, and which are currently functioning in the valley? How does its location affect other places? What is its rate of change? What types of patterns or regions can be observed? How has human activity affected this place? How has the environment shaped human activity? As you can see by these questions, they are interdisciplinary, focusing on spatial interrelationships.

The geographic perspective can best be illustrated in the setting of a real place on the Earth. Since all places vary from one another and possess a unique mix of natural and cultural features, the place selected will not illustrate all places and processes of nature. However, it will bring together, in one location, the geographic viewpoint and subject matter of geography. Here in the *study area,* we will examine seven geographic tools of analysis: (1) location, (2) scale, (3) the regional principle, (4) internal coherence, (5) spatial interaction, (6) spatial distribution, and (7) change.

However, before illustrating these geographic tools in this valley, let us first observe its four major natural systems: the *atmosphere, biosphere, lithosphere,* and *hydrosphere.* This over-view of these four systems will provide you with a mental picture of the natural environment in order to better understand the context of these seven important geographic tools.

THE STUDY AREA VALLEY

Atmosphere and Biosphere

A small, forested valley 3 kilometers (2 miles) long and 800 meters (1/2 mile) wide in the Coast Range of California is our point of beginning in our inquiry. The valley is located about 40 kilometers (25 miles) inland from the Pacific coast and 200 kilometers (125 miles) north of the San Francisco Bay Area.

A cold California ocean current refrigerates the summer onshore breezes. Cool ocean breezes, summer fogs, and winter storms accent the weather. The valley, which we will call the *study area,* is alive, dynamic, and beautiful. It symbolizes the general themes in our study and illustrates the complexities and interrelationships between human activity and the environment (fig. 1.1).

Here in the valley, giant coastal redwoods have stood for centuries, reaching hundreds of feet in height. The ancestors and root systems of these old giants were growing here at the time of Christ. Redwoods ideally need abundant moisture, generally more than 100 centimeters (40 inches), mild temperatures, ranging above freezing and below 30° C (85° F), special soil, slope, sunlight, and drainage requirements. Where the mix of conditions is in the wrong proportions, the trees are either absent, very sparse, or stunted in size (fig. 1.2).

Figure 1.2

The coastal redwoods depend upon a mix of environmental conditions favorable to their growth.

Figure 1.3

The higher slopes are windblown, drier environments where drought-resistant plants establish a foothold.

Figure 1.4
North- and south-facing slopes have contrasts of environmental
conditions producing distinct plant communities.

As the trees of this valley grow to great stature, light is filtered through the branches, creating a shadowy environment on the forest floor that is carpeted with huckleberry plants, sorrel, and a multitude of ferns. Along the streams are a variety of willows and alders preferring shade and moist soils, found growing only where this environment is well defined. As we climb the north-facing slopes of the valley, the redwoods begin to thin out as Douglas firs dominate and compete more successfully. Up on the ridges where high winds rake the slopes, firs are bent, and the more wind- and drought-resistant pines, live oaks, and grasses begin to cover the landscape (fig. 1.3). Looking across on hot, south-facing slopes, large trees have been replaced by sun-loving manzanita, chamise, and live oak, more characteristic of the drier interior Coast Range. Each plant and organism has its own **niche,** or set of environmental requirements. The place of growth represents the favorable location for its needs. A niche can be thought of as the spot best suited for a plant or animal (fig. 1.4).

The forest is quiet and motionless with the exception of a crow or vulture making its presence known. This calm is misleading. Trees are pumping tons of water up into their leaves as

Figure 1.5

Growth of new wood produces an equal amount of fresh oxygen.

solar energy evaporates it into the air. Insects are processing huge quantities of timber through their digestive systems, creating humus through decomposition. As the plants quietly sink their roots into the soil, gases are released to assist in the decomposition process. Solar energy, at the same moment, is being captured in the leaves to be converted and stored as chemical energy. The very oxygen you breathe is being manufactured and replaced by the forest (fig. 1.5). Trees release oxygen as a by-product when solar energy is converted to chemical energy in the green leaves of the plant tissue, commonly referred to as photosynthesis. This process is described in detail in chapter 12.

Looking more carefully at this forest environment, we find evidence of plant and animal consumption by a great variety of animals. Many young seedlings are delicacies for rodents, rabbits, and deer. The larger trees are pruned back continually as high as the deer can reach (fig. 1.6). The coyote, bobcat, mountain lion, and golden eagle prey on the animal population, reducing the numbers of older, weaker animals. The numbers and their activities are too numerous to describe. Each thread of plant and animal activity is woven into the fabric of the biosphere or life zone of the planet.

Figure 1.6

Deer prune trees of lower foliage, creating a neat, gardened effect. Note the small fir growing up through the browsed manzanita.

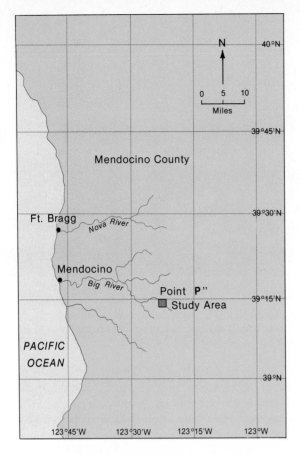

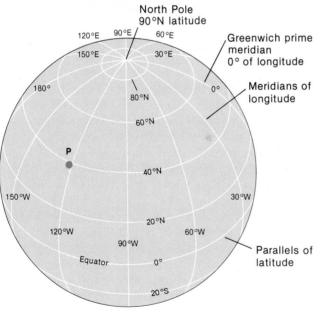

Figure 1.7

The parallels of latitude and meridians of longitude combine to provide a grid system of lines whose intersections define a point's location. Point P, at approximately 39° N latitude and 123° W longitude, is the location of the *study area*. Longitude is measured from the prime meridian or 0° longitude. Latitude is measured from the equator or 0° latitude. Longitude lines run north-south and converge on the north and south poles. Latitude lines run east and west and mark distances north and south of the equator.

The Lithosphere and Hydrosphere

The lithosphere, or Earth's crustal zone, is equally important. Plate tectonic theory holds that the Coast Range was once sea floor. As the Pacific Ocean floor thrust into the North American continent, uplift occurred, and the geologic platform of the biosphere on land was created. Currently, gravitational forces are aiding streams in reducing these soft sedimentary mountains to new flood plains along the lower stream valleys. The valley is narrow and deep in the upper drainage where streams are actively downcutting. Flooding, driven by gravity, is not uncommon, especially where the gradient lowers and the stream loses its energy. Sediments often are deposited around the redwoods, killing all other plants and life forms yet supplying the redwoods with a new layer of soil (see fig. 1.2).

This overview of the valley is our setting in our geographic study. Let us now examine the seven geographic tools in this location by returning to the questions geographers ask. These tools of analysis will give a geographic perspective that can be applied to any place of study.

SEVEN GEOGRAPHIC CONCEPTS

Location

"Where is it" is the starting point in any geographic study. Location fixes the address of the place. The method used depends on the purpose for locating the place. Location can be described mathematically, using a grid system of imaginary lines to fix a position very precisely on a map or on the globe. Key reference points, the North and South Poles and the equator, are defined by the Earth's rotation as explained in chapter 2. The prime meridian, an imaginary line running north and south passing through the Royal Observatory in Greenwich, England, is a key reference line for longitude. The equator, located halfway between the poles, is the chief reference for latitude.

Latitude gives us location in degrees of any point along a meridian north or south of the equator through a range of 90 degrees. **Longitude** is the angular distance measured east and west of the prime meridian through a range of 180 degrees. Both longitude and latitude degrees are subdivided into minutes and seconds. One degree equals 60 minutes of arc ($1° = 60'$). One minute equal 60 seconds of arc ($1' = 60''$).

The mathematical location of the valley, or *study area*, can be described on a globe as a place located 123°22'30" west longitude and 39°15' north latitude. In fact, this only defines a point, not a place. The *study area* spans several fractions of a degree of latitude and longitude (fig. 1.7).

Site

What are the internal aspects of the place? Another aspect of location is **site** or internal aspects. The address says little about the place. All parts of the Earth are unique, and this valley differs from all other places. Site factors, the internal characteristics, forces, and processes of the place, interact to produce a specific environment.

Table 1.1

Physical (natural) and Cultural Landscape Features. Site factors include all internal characteristics of the place, both natural and cultural, that produce a specific environment.

Physical (natural) Landscape	Cultural Landscape
Redwoods	Fences
Oaks	Trails
Deer	Roads
Stream	Houses
Ferns	People
Grasses	Barns
Shrubs	Automobiles
Rocks	Telephone wires and poles
Soils	Bridges
Fog	Blacktop surfaces
Dew	Benches
Atmosphere	
Topography	

Figure 1.8

Site factors: Internal characteristics include both cultural and physical (natural) elements. Can you identify examples of each element?

Consider site factors by comparing the site to a piece of property for sale and asking the following questions: What are the internal aspects, or characteristics, of the property? What does the site include? Accordingly, an inventory is taken of the site. Soils, geology, water resources, plants, animal life, climate, and cultural development on the property are included in the inventory. Although each place on Earth is really a mix of various ingredients, every item is connected to the other by interacting processes.

The inventory represents the subject matter of geography. Table 1.1 gives a brief inventory of the various internal site characteristics. By examining figure 1.8, we can appreciate the general distribution of these factors.

Let us go into the valley and examine a specific site along a forested stream bank of Montgomery Creek. It must be pointed out that our *study area* represents only one point in a mosaic of environmentally different locations. It is unique in the sense that no other place is exactly like it.

The photograph in figure 1.8 shows tall coastal redwoods growing along a stream terrace. The redwoods filter out the light, creating a dark, quiet, fern-covered forest floor. Mosses, lichen, fungus, and rotting logs are randomly spread on the ground in a bed of decomposing leaves. Dampness is everywhere.

Summer coastal fog drifts into the valley at night and lingers most of the morning. Although precipitation may reach 200 centimeters (80 inches), fog drip can double the moisture supply to this forest. Rainfall is very seasonal, coming between October and April, followed by a dry summer. Therefore, the fog flowing into this valley is a key internal site factor providing the needed moisture during the long summer season.

Occasionally a golden eagle or turkey vulture will be seen soaring above the trees. The croaking of a frog or hum of a bee might break the silence. If the wind comes up, the giant trees will sway back and forth, producing a squeaking sound. A large branch or even a whole tree may fall and add to the litter on the ground. (These falling branches are called "widow makers" because they have killed lumberjacks working in the forest.)

A knowledge of soils of the site also helps explain the presence of the trees, especially the coastal redwood. The tall, rapidly growing redwoods grow on the rich, alluvial flood plains and stream terraces. Here, soils are renewed annually by fresh deposits of silt laid down by floods.

In summary, site factors tell the geographer about the internal aspects of the place. The valley possesses a great variety of these elements, which are continually modified by nature and by many years of human occupancy. All of the geographic elements make up the site characteristics.

Relative Location (Situation)

All places on the Earth form interconnecting parts of other regions. No pattern or process stands alone. Relative location, or **situation,** describes how a place is related to another place. Going back to the first question, "Where is the valley?," leads to the next: "How does it relate to other regions?" This is answered in terms of specific criteria such as the main arteries of travel or centers of population.

The valley is located in a rather isolated part of northwestern California, surrounded by rugged terrain, with few roads and low population density. Isolation has been and still is a major factor in shaping the rate of change, which will be discussed as a separate concept. Figure 1.9 illustrates the relative location of the valley.

Scale

Think of **scale** as a comparison, or ratio, of reality to a corresponding photograph or map. Large-scale maps cover small areas, while small-scale maps cover large areas. When "full scale" is used, the map or photograph is said to have a ratio of one to one, written 1:1. If a reduction is made of one-tenth, or one-twenty-four-thousandth, or one-millionth, the scale can be expressed as the ratios 1:10, 1:24,000, and 1:1,000,000, correspondingly. It means one unit on the map represents 10, 24,000, or 1,000,000 units on the land.

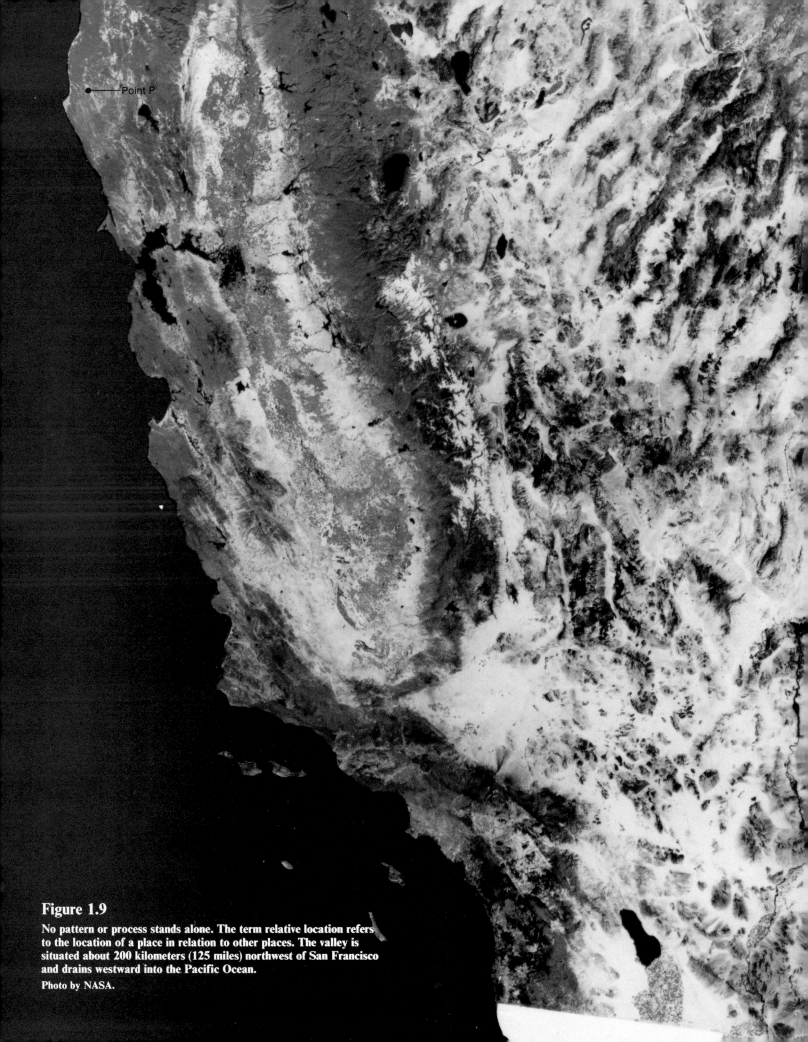

Point P

Figure 1.9

No pattern or process stands alone. The term *relative location* refers to the location of a place in relation to other places. The valley is situated about 200 kilometers (125 miles) northwest of San Francisco and drains westward into the Pacific Ocean.

Photo by NASA.

A. Planet Earth at 35,484 kilometers (22,000 miles) in space

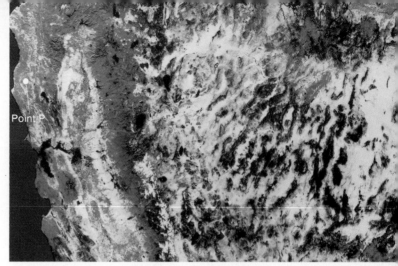

B. 1:19,008.000; inch = 483 kilometers (300 miles)

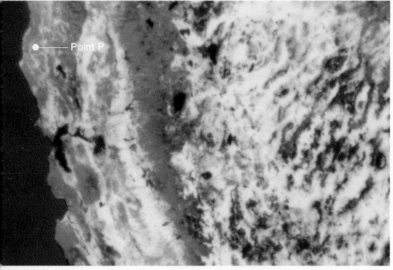

C. 1:5,068,000, inch = 129 kilometers (80 miles)

D. 1:132,000; inch = 3.39 kilometers (2.1 miles)

E. 1:32,000; inch = .81 kilometers (½ mile)

F. Surface view

Figure 1.10

A view of the valley at distances ranging from space to the valley
floor provides changing perceptions and information about the place.
What questions can you ask for each photograph?

Space photos by NASA.

By viewing the *study area* from space and then from various heights and finally on the surface, note how our information changes with scale (fig. 1.10).

A satellite view (fig. 1.10A) from 35,000 kilometers (22,000 miles) in space reveals the relative location of the *study area* on the northwestern coast of California. From this view, we can see its relationship to the major oceans and continents. At this distance, its general latitude and longitude can be determined. By moving in closer, note that we lose certain information such as relative position of the continents and oceans. No longer can we see the curvature of the Earth. However, at the height of 20 kilometers (65,000 feet) above the surface (fig. 1.10C), the *study area* can be viewed in more detail. Internal site factors, and the relative location of the valley to nearby drainage systems, can be mapped and compared to other adjacent systems.

With a closer look, at an altitude of 3,000 meters (10,000 feet) (fig. 1.10E), our attention is refocused on only the valley or the *study area,* and all other areas outside this site are lost to our view. Now, internal site factors dominate our attention. Resources become more prominent. At this scale, we can map detailed landforms, vegetation, soils, and cultural patterns such as roads, fences, houses, and logging operations.

At the surface (fig. 1.10F), we have lost our perspective of general environmental patterns and gained a view of individual plants, rocks, soil units, and wildlife, as well as human activity, at full scale. By changing scales we gain access to new information. To gain a perspective of relative location, or inventory of the site's internal characteristics, a knowledge of the proper scale is of the utmost importance.

The question one asks about a place determines which scale will aid in answering that question. If we are concerned about site factors of the valley, scales such as 1:1,000 will be useful. External relationships (relative location) are answered by small-scale maps (such as 1:1,000,000) where large areas of the Earth can be viewed.

The Region

The Earth's surface is a quilt of patterns determined by culture and nature. Cultural regions include the cultural landscape, including houses, roads, and agriculture. The natural region includes patterns of the physical environment, including forests, soils, climate, and geologic or landform features. Our valley is no exception. It is a geographic region with prescribed boundaries determined by specific criteria. Note in figure 1.11 the variety of regions found in the valley. A region, then, is a part of the Earth that is alike in terms of the specific criteria chosen to set it apart from other regions. Regions can cover only a small area, as in this valley, or span a continent. Therefore, subregions can exist as parts of larger regions. This valley is a subregion of the Pacific Coast Range of California.

The valley is a series of patterns, not isolated, but interwoven into a cohesive whole. Each pattern is linked to the other and held together by the fabric and flow of energy and materials.

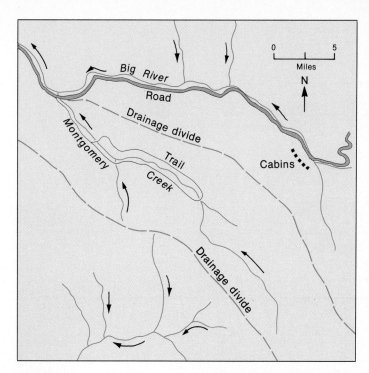

Figure 1.11

A region is a mappable element of the Earth that is similar in terms of specific criteria. These include drainage and vegetation, as well as cultural patterns such as land use.

Internal Coherence

The pattern of one element is related to the pattern of other elements. The valley is a web of internal, cohesive elements interacting to produce bonding relationships between patterns. For example, the coastal low clouds and fog in this valley help to produce the climatic environment outlining the belt of coastal redwoods. Where the fog belt stops, redwoods also become sparse or absent (fig. 1.12). In later chapters, the processes and interrelationships between soils, vegetation, and climatic patterns will be explained.

Landforms and the erosional processes also influence the bonds of the place. Redwoods thrive on stream terraces that are subject to frequent flooding and the resultant silt deposits, which bring new life to the soil while killing competitive species of trees.

The more complex the region is, the greater the potential **internal coherence** it possesses. The bonding forces are more numerous. Fragile environments are often those with fewer complexities and interacting forces and processes (fig. 1.13).

Spatial Interaction

Spatial interaction, or the dynamics within places and between places, is important to recognize when studying a region. Activity occurs on all levels or scales. If the Earth is viewed from space, activity is difficult to detect. Only long-term change such as seasonal color changes are observable.

Figure 1.12

The patterns of a place are interwoven and interrelated. The low clouds and fog belt also represent the prime location for the coastal redwoods in the *study area*.

Photo by Carol Prentice, U.S. Geological Survey.

A

B

Figure 1.13

(A) A beehive system is both fragile and complex. The more complex the system is, the greater the internal coherence and bonding forces. Fragile environments often have fewer interacting forces. (B) Close-up view.

A

B

Figure 1.14

Spatial interaction, the dynamics of places or between places, gives us the mobility dimension of things on the Earth. The Earth, the valley, or even a small soil sample is not static. Change is the norm. These photographs, with Scotia Inn in Humboldt County, California, in the foreground, show the reforestation of an area first clear-cut in 1900 and then repeatedly burned in an attempt to convert the redwood forest to grassland. Photograph A, taken about 1905, shows the inn and the area behind it that was first cut in about 1900. Photograph B, taken in 1965, shows the same area. A dense stand of second-growth redwood and associated species has reclaimed the original clear-cut area.

Photos from the California Conservation Council and the California Redwood Association. Used by permission.

When the Mariner space probes first moved close to Mars, the mystery of the planet's color changes began to clear up. Actual wind storms were observed. Dust was photographed swirling on the surface. Ice was observed, and the planet appeared to be in a dynamic state. Apply this same analysis to the Earth, viewing the valley first from space and then closer to the surface, and activity will become more apparent (figs. 1.10 and 1.14).

All activity is not uniformly spread over the surface of an area. Airflow in the valley is a good example of a phenomenon that is highly variable. As the sea breeze moves onshore, it is channeled up the valley, then moves upslope and spills over the lower divides, flowing with greater velocity in the central channels and through the passes (fig. 1.15). Places in the valley sheltered by landforms or vegetation receive little wind. Some places are centrally located,

Figure 1.15

Central locations are in a greater state of spatial interaction. The shapes and patterns of vegetation in the *study area* reveal exposure not only to light and moisture but also to wind. Trees in the central airflow channels develop branches, all pointing downwind; these branches are referred to as tree flags.

while others are on the margin of activity or removed from the flow of things. Isolation versus central location is important to recognize in geographic analysis. Central location focuses forces, accelerating change and activity. Therefore, the spatial arrangement of things can determine the dynamics of a place.

Spatial Distribution

Places are unevenly populated by the fundamental phenomena that make up the site. If any set of properties is selected to be mapped in the valley, the distribution will be seen to possess two aspects: frequency and pattern. **Frequency** is defined as a number of occurrences of a phenomenon within a region. The most significant economic resources in this valley are the Douglas fir and the coastal redwood. The number of trees per unit area (square kilometer or square mile) describes the frequency, or density. However, fre-

A

B

Figure 1.16

Spatial distribution describes the configuration or pattern characteristics of a place. The density or frequency aspect of a place refers to the number of occurrences of a phenomenon within a region. Trees per acre or board feet per acre, for example, represent the frequency aspect of spatial distribution. (A) High frequency. (B) Low frequency.

quency, or density, does not reveal the fact that redwoods vary greatly in distribution (fig. 1.16). **Pattern** clearly reveals the distribution, or arrangement. In this valley, coastal redwoods dominate the valley floor but begin to thin out above the valley on the steeper slopes. If only frequency is considered, this variation in distribution will be overlooked in the analysis (see fig. 1.14).

Figure 1.17

The present is the momentary interim between the past and the future. Change can be accelerated by human activity.

The spatial distribution of a place continually changes throughout time. The pattern of the past is not the pattern of today. The valley is a classic example of this concept. Logging activities have removed some of the redwood giants of the past. The second-growth forest is very different. Light, temperature, and soil moisture are modified. Nothing is static. Even the rate of change itself is variable.

Change

The first six concepts, (1) location, (2) scale, (3) regions, (4) internal coherence, (5) spatial interaction, and (6) spatial distribution, have an underlying assumption: the Earth is continually changing. Solar energy converted to other forms does the work. Remember, it is the result of energy and materials flowing through the systems. As the redwoods in the valley reach maturity, the microclimate below slowly changes. When trees are shaded, competition for light becomes severe. Only those trees able to compete

continue to grow. When timber fallers clear an area, removing the giants, in effect they suddenly turn on the lights of the forest, allowing energy to flood the forest floor. Thus, environmental change occurs again. Trees that once were almost choked out began a rapid growth. Change is constant for all patterns.

As we focus on a place, we must recognize that its properties are the result of processes occurring over long periods of time. The present is momentarily the state of the evolving physical environment. A valley, forest, prairie, or entire community of organisms is arranged in unique patterns, interrelated and continually in a state of change. Today's technology gives humans a major role in producing change. We are an active force, becoming involved in the processes and patterns of the place (figs. 1.14 and 1.17).

By applying these seven concepts to any region or study area on the Earth, organization and understanding can be brought to the observer. You will be able to view a place with the proper tools, or perspective, of the geographer. New insights will be gained as your geographic perceptions improve.

Learning through Direct Field Experience

Learning through direct field experience is the best way to master the course objectives by applying them to your own locality or place (fig. 1). Your everyday journey to school or work is an opportunity to sharpen your geographic knowledge. Each day's travel can be a self-guided field trip into a cultural and physical landscape that requires your concentration and curiosity.

To be aware means keeping all the senses finely tuned and fully functioning. *Listen* to your environment. Everything has its own sound. Water dripping, rustling leaves during a gentle breeze, birds singing, thunder in the sky, waves pounding on the beach. Don't forget the sounds of human activity.

Go outside your box; escape from the protection of the metal and glass of your car or walls of your house. Urban life shelters us from natural awareness. Human sounds drown out nature. Air pollution reduces visibility. Air conditioning insulates us from real humidities and temperatures of nature.

Touch nature. Rub some leaves to sense the texture, (except poison oak or cactus), feel the cool water of a spring or creek. Examine the texture of the soil. Every object in nature has its own feel and smell.

Smell the air by breathing deeply. How many smells can you identify? Take different plants and detect their odor. Note how the smell of a forest differs from that of a desert after a rain shower. Every plant community has its own odor.

Taste some blackberries or mushrooms from the meadow, and don't forget the refreshing taste of spring water or a cold icy stream coming from a melting snowpack.

See and record the natural elements as you see them. Seeing reveals the dynamics that cannot be pictured on a photographic plate or map. It represents the present. Observe the color of the sky and clouds, the plant forms such as trees, shrubs, and grass that appear so natural. Note the soil colors and texture. Examine the geologic structure on a roadcut or wave-cut beach. Notice the difference between the natural and cultural landscape—those things introduced or manufactured or rearranged by human activity.

Figure 1

Learning through direct field experience.

Figure 2

A geographic perception of your environment will bring new insights, enjoyment, and wiser use of your habitat.

A geographic perception of your environment will bring new insights, enjoyment, and wiser use of your habitat (fig. 2).

SUMMARY

This chapter has presented the geographic perspective and philosophical framework of the text.

1. **Location** of a place is described in terms of **site,** or internal characteristics, forces, and processes of the place producing the specific environment. Relative location, or **situation,** describes how one place is related to another place.
2. **Scale** is the comparison, or ratio, of reality to a corresponding map or photograph. The question you ask about a place determines which scale should be used in answering that question.
3. The Earth's surface is a mosaic of **regions** created by both human activity and nature. A region is a part of the Earth's surface with selected criteria to set it apart from other regions.
4. **Internal coherence.** The pattern of one element is related to another. Cohesive forces bond elements together. The more interaction a region or place has, the greater will be the internal coherence.
5. **Spatial interaction.** Places are dynamic; activity occurs on all levels or scales. Some areas are marginal, out of the mainstream. Central location shapes and focuses forces, accelerating change and activity.
6. **Spatial distribution** has two aspects: frequency and pattern. **Frequency** is simply the number of occurrences of a phenomenon being studied. **Pattern** reveals the distribution or arrangement of the phenomenon.
7. The Earth is in a continual state of **change.** The place is the product of processes occurring over long periods of time. People are an active force capable of speeding up or slowing down the rate of change a place may experience. Change is the result of energy and materials flowing through systems.

ILLUSTRATED STUDY QUESTIONS

1. Which photo provides the best information regarding (see figs. 1.8 and 1.9, p. 16, 17):
   a. site factor?
   b. relative location factors?
2. Using the two photographs of the valley, which one provides you with a small-scale view (see fig. 1.10, p. 18)?
3. Which of the following map scales would give information regarding (see fig. 1.10, p. 18):
   a. the relative location of the valley?
   b. the forest and soil patterns on a square mile, or section of land?
   c. the distribution of houses, stores, churches, and schools in the valley?
4. Describe the internal coherence, or bonding forces, integrating the atmosphere and biosphere (see fig. 1.13, p. 20).
5. The coastal redwoods dominate the region where the photograph was taken in the valley. Describe the general pattern. How does this concept differ from frequency (see figs. 1.12 and 1.16, pp. 20, 22)?
6. How many physical regions can you identify in the photograph (see fig. 1.11, p. 19)?
7. Compare the two photographs taken at exactly the same place and scale, and list all changes in both the physical and cultural landscape (see fig. 1.14, p. 21).

The Earth in Space

2

Objectives

After completing this chapter, you will be able to:

1. Describe the unique position of the Earth in space.
2. Explain five verifications of Earth's shape and size.
3. Describe the planetary aspects of the Earth including **motions, density, mass,** and **volume.**
4. Explain the effects of the motions on the seasons and variation in solar energy.
5. Explain the effects of **rotation** on **tidal action** and the **Coriolis effect.**
6. Explain the relationships between motions and time.

The Earth is a small oasis in space.
NASA

Figure 2.1

The Salton Sea and Imperial Valley, California. Views from space provide a new perspective of the surface of the Earth. Imagery taken in various portions of the electromagnetic spectrum provide information about the Earth's resources. False color infrared penetrates haze and improves resolution. Red indicates reflected infrared energy from green plants.

Photo by NASA.

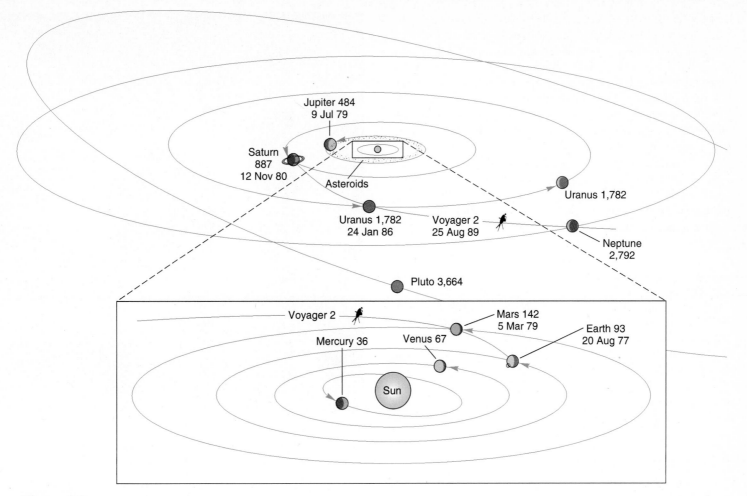

Figure 2.2

The Earth is a member of the solar system, which consists of one star, nine known planets orbited by satellite moons, plus innumerable asteroids, comets, meteoroids, dust particles, and gas. Gravity holds the system together. Voyager 2 was launched on August 20, 1977, and encountered Neptune on August 25, 1989. (Figures show average distances in millions of miles.)

A view of the Earth from space gave us a new perspective and appreciation of this planet we inhabit. Astronaut Michael Collins returned from his journey to the Moon in 1969 with memories of Earth, looking as small as a golf ball, observed through one of the small windows of his spacecraft. He reflected on the significance of this small oasis in the darkness of space (fig. 2.1):

> *I am not a naive man. I don't believe that a glance from 100,000 miles out would cause a prime minister to hurry back to his Parliament with a disarmament plan, but I do think it would plant a seed. New space perspectives have caused environmental concerns on a global scale. Never before have nations been as concerned about survival based on interdependence of its material and energy resources.*

Our space view allows us to have a geographic perspective of the Earth not only as a small golf-ball-sized planet, but also as a member of a solar family of planets and a small speck in the cosmos. What Collins saw in space changed our perspective of the universe and Earth's unique position. From the new perspective, Earth is a bright blue-and-white planet with a yellow companion, the Moon, drifting slowly about the Sun along with eight other known planets and their satellites (fig. 2.2). Beyond this solar system, Collins saw, and we see, the darkness of space speckled by stars and galaxies, reaching to infinity.

THE UNIVERSE IN SPATIAL PERSPECTIVE

An enormous cloud of dust and gas, drawn together under the compressional force of gravity, formed our sun and the planets. This was our beginning, about 4.6 billion years ago.

At the time of our cosmic beginning perhaps 10 to 20 billion years ago it is theorized, all mass and energy were compressed to such a high density that temperatures read trillions of degrees. The universe was then born in a cosmic explosion, the Big Bang. As it expanded, matter and energy condensed out and the elements we know today formed. The haze of hydrogen gas and dust condensed to form stars and collections of stars known as galaxies. Today the universe continues to expand and cool. Radio telescopes detect low-frequency radio waves representing dying embers of the Big Bang. This radiation began its cosmic journey in the much hotter form of light energy. Through expansion this energy has been transformed to a low density hydrogen at 3 degrees Kelvin. This dying ember is one of our best remnants testifying that our universe has evolved from an explosive beginning. We will now look at some key events in time.

Minutes after the Big Bang, hydrogen and helium nuclei formed. Another million years passed before cooling allowed electrons to join atomic nuclei to make atoms. Then another 1,000 million years passed before gravity collected matter together to form stars and clusters of stars, dust and glowing gas known as galaxies.

At the cores of stars where temperatures are high enough to fuse hydrogen into helium, these thermonuclear reactions provide the energy to light the stars and illuminate the universe. As the first stars aged, their interiors compressed and began to manufacture heavier elements such as carbon, oxygen, nitrogen, calcium, and silicon, the stuff the Earth is made of. Eventually this heavier material was ejected during the final dying stage of these first generation stars, the explosive supernova stage.

Interstellar space enriched with these heavier elements eventually condensed into new second generation stars, and the cycle repeated itself many times. Our Sun is one such star that formed over 4.6 billion years ago with its system of planets.

OUR PLACE IN SPACE

Earth is a small planet orbiting in an elliptical path third in order from the Sun. Its orbit requires 365 and one-fourth days to make the journey at an average speed of 29.8 kilometers per second (18.6 miles per second) and an average distance of 150 million kilometers (93 million miles) from the Sun.

The three other inner **terrestrial** planets, Mercury, Venus, and Mars, are Earth-like, with rocky surfaces and high-density cores. Jupiter is first among the large **jovian** planets, which include Saturn, Uranus, and Neptune. Jupiter has more than 300 times the Earth's mass, but the Sun's mass is 1,000 times bigger yet.

The gravitational tether of the Sun locks in orbit not only nine planets but at least 100,000 asteroids, chunks of rocky and metallic debris, millions of comets, and uncountable tiny meteoroids, consisting of ice, metal, or rock. All of the orbiting matter in the solar system represents less than 0.15 percent of the mass of the Sun (fig. 2.3).

Because the Sun is very hot, even elements such as carbon, iron, and calcium are in a gaseous state. At the Sun's gaseous surface temperatures reach about 6000 K; however, in the core where hydrogen is synthesized to helium through nuclear fusion, temperatures reach several million degrees. The sun is a star of ordinary size, yet it has a third of a million times the mass of the Earth, and more than a million Earths could fit inside of the sun's sphere. The Sun's diameter is so immense that 108 Earths could be lined up side by side across the face of the Sun's equator with a diameter of 864,000 miles.

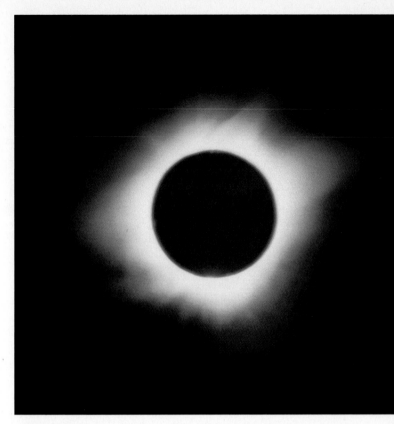

Figure 2.3

The Sun consists primarily of hydrogen gas. The Sun's power system is nuclear fusion, a process that converts hydrogen to helium and produces great quantities of heat, electromagnetic, and magnetic energy. The corona or outer atmosphere of the Sun was photographed during the total eclipse of July 11, 1991. Totality began 7:28 A.M., Kona, Hawaii, lasting 4 minutes, 11 seconds.
Photo by Kathy O'Bryant.

Our nearest celestial neighbor, the Moon, is 384,000 kilometers (239,000 miles) away with a diameter of only 25 percent of the Earth or roughly 3,600 kilometers (2,200 miles). The Moon's orbit would fit well inside the Sun with an orbital radius reaching out only halfway to its photosphere surface.

Light takes about 1 and one-third seconds to reach us from the Moon traveling at the highest possible velocity, the speed of light (300,000 km/s or 186,000 miles/s). This same light reaches us from the sun in 8 minutes and 19 seconds, on average.

Our star was discovered to be part of a large collection of stars known as the Milky Way Galaxy. It consists of billions of stars, hydrogen, dust, and gas. Until the 1920s our galaxy was thought to be the entire universe. Today we know it to be only one of billions of other "island universes" in deep space (box 2.1).

A closer look at our galaxy reveals a pinwheel of rotating stars, gas, and dust bound by the force of gravity. It spans 100,000 light-years across, a light-year being 9.5 trillion kilometers or 5.9 trillion miles. Our sun rotates with the entire galaxy once in 230 million years at 220 kilometers per second. Surrounding the rotating disk is a halo of very old star clusters and dark nebulae, matter that has yet to condense into stellar objects.

Galaxies Form an Enormous Celestial String of Galactic Beads

BOX 2.1 A recently discovered chain of galaxies located in the constellations Pisces, Cetus, and Hercules casts doubt on the current hypothesis that galaxies are randomly distributed and equally spread across the universe.

This galactic string is 100 times more massive than any structure ever observed. It spans 10 percent of the known universe.

Until about 10 years ago, most astronomers believed the largest cluster of stars was a galaxy and that these huge collections of billions of stars were randomly and evenly distributed in the universe. Since then, super clusters have been discovered (fig. 1). However, this latest celestial chain is bigger than anything ever envisioned. The physical process governing this new distribution is unclear, and this grand-scale phenomenon in the universe is unexplained.

Figure 1

This cluster of galaxies in the constellation Coma Berenices has an average distance between galaxies of about a million light-years. This represents a rather richly populated cluster.

Photo by Palomar Observatory Photography. Used by permission.

If we could view our Milky Way from a vantage point 2 million light-years away, it would resemble a typical spiral galaxy such as its sister in the constellation Andromeda. The solar system lies just a few light-years above the plane of the spiral arms near the inner edge of the Orion arm, itself located about 35,000 light-years from the galactic center (fig. 2.4).

The universe appears never to end, no matter where we look with the latest technology, capable of sampling energy ranging from x-rays to infrared and radio waves.

Let us start with the nearest planet to the Sun, Mercury, and make a quick curbside appraisal of each planet in terms of its ability to support life. The following environmental factors will be rated and summarized in table 2.1. These include temperature, air quality, water, mineral resources, and other environmental factors. The information summarized is built on scientific discovery from Earth-based sensors and satellite exploration. All of the planets but Pluto have been visited by spacecraft.

PLANETARY ENVIRONMENTS
Mercury

Mercury, nearest to the Sun, is similar to the Earth with reference to density, size, form, and origin. Its proximity to the Sun and slow rotation rate produce extreme temperatures on the side facing the Sun. A day on Mercury lasts 115.88 Earth days. The amount of energy reaching the planet's surface raises the temperature at noon to over 430° C (800° F), high enough to melt lead and zinc. Because the planet is so small, it lacks the gravitational energy to hold an atmosphere. Even the heavier gases found on Venus, such as carbon dioxide and water in any form, have not been discovered by remote sensors. The only desirable environmental factor might be in its rich crustal mineral resource base.

Venus

Venus so resembles Earth with reference to size, density, and shape that it is often referred to in astronomy circles as our sister planet. This statement is misleading if one draws the conclusion that it is hospitable. The planet has a massive cloud cover, but radar images show widespread volcanoes, and spacecraft studies indicate that surface temperatures are above the 450° C (850° F) mark. Thus Venus is the warmest planet in the solar system. The atmosphere is dense, consisting mainly of carbon dioxide, carbon monoxide, sulfur dioxide, and other deadly gases. The cloud cover prevents harmful shortwave x-ray and ultraviolet radiation from penetrating to the surface, but these same clouds keep the planet's long-wave infrared radiation from escaping. The absence of liquid water at these temperatures has precluded development of any form of life on this planet (fig. 2.5 and box 2.2).

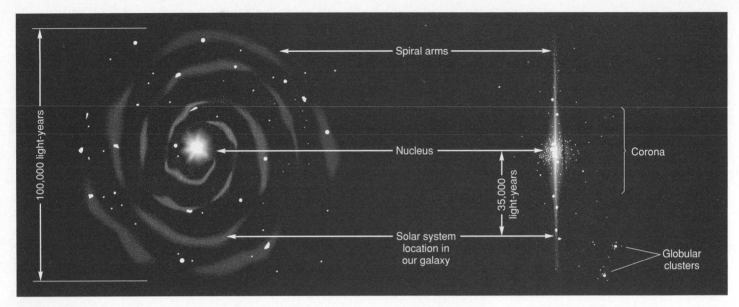

Figure 2.4

Our location in the Orion arm, about 35,000 light-years from the center of the galaxy.

Table 2.1

Environmental Factors of the Solar System

	Temperature	Air Quality	Water	Silicate Minerals	Other
Mercury	Excessive	No atmosphere	None	Rich in silicates	Excessive x-ray and ultraviolet
Venus	Excessive	Toxic	None	Rich in silicates	Excessive "greenhouse effect," very high atmospheric pressure
Earth	Temperate	Excellent, but being impacted	Abundant in liquid form	Rich in silicates	Mild "greenhouse effect"
Mars	Cool	Too thin, low in oxygen	Only in solid form	Rich in silicates	Very low atmospheric pressure
Jupiter	Cold	Lacking oxygen, toxic	Lacking in liquid form	Lacking	Very strong magnetic field, high winds
Saturn	Cold	Lacking oxygen, toxic	Lacking in liquid form	Lacking	High winds
Uranus	Very cold	Lacking oxygen, toxic	Lacking in liquid form	Lacking	Extremes of seasons due to angle of axis
Neptune	Extremely cold	Lacking oxygen, toxic	Lacking in liquid form	Lacking	Very dark due to distance from Sun
Pluto	Extremely cold	No atmosphere	Lacking in liquid form	Unknown	Coldness of deep space

Earth

Earth is located at a very strategic distance where the sun provides just the right energy to sustain life (fig. 2.6). If it were nearer or farther, in the orbital path of either Venus or Mars, temperatures would be very hostile. First and utmost in importance, water has a narrow temperature range in the liquid form. Even if we could take shelter from temperature extremes as the astronauts do, water would not remain in the liquid state. It is liquid water that makes life possible: Most chemical and biological processes require a liquid environment, and most living systems thrive in the temperature range of liquid water. If the oceans were either frozen or turned to vapor, the hydrologic cycle would stop. This means water would cease to recycle from the land and sea to the sky and back to the land again, bringing renewed moisture to the soil.

Our planet's atmospheric composition would also change. Not only would water vapor change but also carbon dioxide, which would have a negative impact on living systems. The Earth's heat budget, or incoming and outgoing energy balance, is regulated by the atmospheric filter, which in turn is controlled by the composition of the atmosphere. Finally, if we were closer to the Sun, harmful radiation dosages might produce genetic defects or totally destroy life. How does Earth fare as a livable planet? Is it an oasis in space?

Figure 2.5

Venus, our sister planet, resembles the Earth in size, density, and shape, but the similarity stops here. The planet has major volcanic ranges and rift valleys. Its dense atmosphere is the warmest in the solar system and deadly to all forms of life. Gases include carbon dioxide, carbon monoxide, and sulfur dioxide, which filter deadly x-rays and ultraviolet radiation, but produce a strong greenhouse effect that maintains a very high temperature.

Photo by NASA.

Figure 2.6

Earth is an oasis in space, nearly three-fourths of its surface covered by water. Its atmosphere filters harmful short wave x-rays and ultraviolet and retains warming infrared. Only here is there an abundance of oxygen and water in its solid, liquid, and gaseous forms. It is alive and dynamic with numerous forms of life, unequaled in the solar system. This view of the crescent Earth and Moon—the first of its kind—was imaged by the Voyager I spacecraft September 18, 1977, 11.66 million kilometers (7.25 million miles) from Earth.

Photo by NASA.

Venus—a Watery Planet?

BOX 2.2 Researchers at the National Aeronautics and Space Administration have developed a theory, known as the "wet" or "moist greenhouse" theory, for Venus.

Currently Venus is a hot desert with active volcanoes and lightning-charged atmosphere. It is hot enough to melt lead and is blanketed by dense clouds, consisting mainly of carbon dioxide and sulfuric acid droplets.

The researchers say Venus once had a hot, near boiling ocean covering major portions of the surface until it boiled away under the intense rays of the Sun.

The scientists, at the Ames Research Center in Mountain View, California, theorize that this planet had an ocean for hundreds of millions of years, much like the Earth. It was only after prolonged exposure to the Sun that the greenhouse effect caused the oceans to evaporate into space.

This knowledge may be useful in trying to forecast our own future and its greenhouse warming trend currently being predicted.

Source: NASA

Figure 2.7

Mars was visited by spacecraft which sent back images of a rocky, desertlike surface. All that is missing is life. Carbon dioxide ice covers the poles, shrinking and expanding during the rhythm of the seasons, which lasts six months. Mars has signs of once having had running streams, but today it has no liquid water.
Photo by NASA.

Earth, our home, is the subject of this text. However, it must be emphasized that only here does abundant water in the liquid form exist accompanied by an oxygen-rich atmosphere with just the right seasoning of carbon dioxide to produce a balance. These ingredients are the key to all life systems.

The only place in the universe known to have a life support system, based on water, carbon dioxide, an oxygen-rich atmosphere, and mild temperatures, is Earth. It is tectonically active with a crustal surface consisting of large continental and oceanic plates that are constantly drifting about and shaken by earthquakes and volcanic activity. It is the only known place with a life zone, or **biosphere,** that is continually evolving as new plants and animals occupy niches once held by now-extinct relatives that could not adapt in earlier environments. The fossil record reveals a dynamic, ever-changing environment, sometimes favoring a species for a period of time and then changing to result in its extinction.

Mars

Mars is the newspaper planet, a title it deserves, because more has been written about the possibility of life here than anywhere else in the solar system. Percival Lowell, working from his Flagstaff, Arizona, observatory in the early part of the twentieth century, mapped the surface of Mars. He believed many of the surface features were the result of intelligent activity. Since his work, NASA space probes, fly-bys, and Earth-based telescope observations have given us a much better understanding of its environment (fig. 2.7). There is no liquid water on the surface of Mars today, but surface stream erosional features indicate that past conditions were once different.

Mars's surface atmosphere is 1/200th the density of the Earth's. It contains 95 percent carbon dioxide with traces of water vapor, nitrogen, and carbon monoxide. The Viking missions sampled the surface soil and detected no life or organic compound. However, pictures from space show a surface eroded by rivers, and active sand dunes exist. Perhaps this planet supported life very early in its history.

Jupiter

Jupiter, giant of the solar system, whose diameter is 143,000 kilometers (88,000 miles), is large enough to allow nearly nine Earths to be lined up side by side across its surface face. In fact it is the closest object in the solar system to becoming a star. It is not merely reflecting light, but radiates more energy than it receives from the Sun. This energy is in the form of invisible infrared radiation generated from gravitational pressure. Even so, temperatures are very low. Water exists only as ice along with such gases as methane,

Figure 2.8

Massive volcano at horizon. Debris is thrown hundreds of kilometers above surface. Jupiter's nearest major moon, Io, is actively erupting volcanic gases of sulfur dioxide and liquid sulfur. Because it is so close to Jupiter and the nearby moon Europa, tidal forces heat up the interior as Io deforms under the gravitational pull. The constant strain on the interior creates heat, just as a piece of metal heats when it is bent back and forth repeatedly. Thus Io's interior is turned to magma or fluid that results in outgassing and volcanic eruptions.

Photo by NASA.

ammonia, hydrogen, and helium. The cloud cover of Jupiter reveals great turbulence in the atmosphere induced by differences in temperature, electrical storms, and a high rotation rate almost three times that of the Earth. In summation, Jupiter's atmosphere is too cold, too windy, and too lethal to breathe. Only the moons of Jupiter resemble Earth in terms of density and geologic conditions. Jupiter's nearest major moon, Io, has erupting volcanoes of sulfur dioxide, but even there, temperatures at the surface are too low and an atmosphere is absent (fig. 2.8).

Saturn

Saturn, a spectacle in the telescope, constantly tipping its rings of orbiting icy gravel, is very similar in composition to Jupiter but differs in size, mass, rings, orbital path, and surface temperatures (fig. 2.9). Saturn's distance from the Sun gives it a temperature even more hostile than Jupiter. Gases in the atmosphere of Saturn are similar to those of Jupiter, consisting mainly of a mixture of methane, ammonia, hydrogen, and helium (fig. 2.9).

Figure 2.9

Saturn with its icy rings is light enough to float if an ocean were large enough to support it. Its atmosphere is cold and toxic with methane, ammonia, hydrogen, and helium. This image of Saturn, taken by the Voyager 1 spacecraft on October 18, 1980, was color-enhanced to increase the visibility of large, bright features in Saturn's north temperate belt. It is believed that these spots might closely resemble gigantic convective storms (similar to, but much larger than thunderstorms in Earth's atmosphere) with upwelling from deep within Saturn's atmosphere.

Photo by NASA.

Uranus, Neptune, and Pluto

Moving on to Uranus and Neptune, we find similar hostile conditions. These planets are just too far from the solar furnace to sustain life. Finally, Pluto, located 5,900 million kilometers (3,670 million miles) from the Sun, is in deep freeze with temperatures among the lowest found anywhere in the universe. There the Sun's energy is not much greater than moonlight.

Our brief survey of the planets points out one major theme. The distance from the Sun or position in the solar system is of paramount importance in determining the nature of the planet's environmental conditions. Therefore, Earth's position in the solar system is a fundamental factor in determining the nature of our life support system.

SHAPE AND SIZE OF THE EARTH

Only a dozen generations ago, most people believed the Earth was flat and the center of the universe. Even after scientific study proved the Earth to be spherical, space technology continues to reveal new dimensions and aspects of our planet in relation to the universe. Each generation contributes additional information. Sir Isaac Newton once said, "If I have seen further, it is by standing on the shoulders of giants." He was referring to Galileo, Kepler, and of course Copernicus. Our understanding will never be final.

A series of observations over the centuries first led to a discovery of the Earth's curvature and then to its sphericity, and finally to the proof that the equator's bulge and other deformations make it less than a perfect sphere.

Submergence on the Horizon

For many years mariners have been aware of the fact that as ships sail out to sea toward the horizon they gradually disappear. The observer's height above the horizon will determine the horizontal distance the observer can see. For example, the crow's nest on a sailing vessel adds several miles to the observer's range. At 6.1 meters (20 feet) above the sea, the height of a deck on a small sailing vessel, the horizon is 9.5 kilometers (5.9 miles) away. Climbing to the crow's nest at 9.1 meters (30 feet), the horizon view extends to 11.6 kilometers (7.2 miles) (table 2.2). This submergence of the horizon can only be explained by a curved surface of the Earth. However, this observation of curvature alone cannot confirm the spherical form of the Earth. We must look beyond ourselves for a better perspective. Studying the Moon can provide more clues.

Lunar Eclipse

The Moon orbits about the center of mass of the Earth-Moon system. This pivot point is located between the Earth's surface and its center. As the Moon orbits about this point, being held by the force of gravity, on predictable occasions it passes into the Earth's shadow. When the Moon passes into the Earth's shadow the circular edge of the shadow is observed on the Moon's face. We call this a lunar eclipse (fig. 2.10). Again this observation alone does

Table 2.2			
Distance of the Horizon			
Height (above sea level)		**Kilometers**	**Statute Miles**
Feet	*Meters*		
1	0.3	2.1	1.3
2	0.6	3.1	1.9
3	0.9	3.7	2.3
4	1.2	4.2	2.6
5	1.5	4.7	2.9
6	1.8	5.2	3.2
7	2.1	5.6	3.5
8	2.4	6.0	3.7
9	2.7	6.4	4.0
10	3.0	6.8	4.2
11	3.4	7.1	4.4
12	3.7	7.4	4.6
13	4.0	7.6	4.7
14	4.3	7.9	4.9
15	4.6	8.3	5.1
16	4.9	8.5	5.3
17	5.2	8.7	5.4
18	5.5	9.0	5.6
19	5.8	9.2	5.7
20	6.1	9.5	5.9
21	6.4	9.7	6.0
22	6.7	10.0	6.2
23	7.0	10.1	6.3
24	7.3	10.5	6.5
25	7.6	10.6	6.6
26	7.9	10.8	6.7
27	8.2	11.0	6.8
28	8.5	11.3	7.0
29	8.8	11.4	7.1
30	9.1	11.6	7.2

Source: From Bowditch, *American Practical Navigator*, U.S. Navy, 1968.

Figure 2.10

A lunar eclipse is an excellent time to view the curvature of the Earth as the Moon moves into the Earth's shadow.

Photo by Gary Bowman.

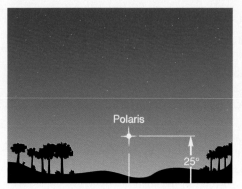

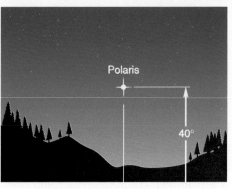

Northern horizon at the equator · Northern horizon at latitude 25° N · Northern horizon at latitude 40° N

Figure 2.11

Because the Earth is spherical, a change in position of one degree of latitude north or south will produce a change in the position of Polaris, the pole star, by the same amount in relation to the horizon. The viewer's latitude is equal to the altitude of Polaris.

not confirm the Earth's true shape. A disk or cylinder could produce the same effect. Viewing not only of the moon but also the North Star, Polaris, at different latitudes reveals the true shape of the Earth.

Changing Altitude of Polaris

The North Star, Polaris, is positioned just 0.7 degree from a point in space directly in line with the Earth's rotational axis. Think of the axis as an imaginary axle on which the Earth spins, running through the north and south poles. If Polaris is viewed from the north pole, it can be seen at the **zenith,** the point directly overhead. By changing your position on the Earth's surface a degree in either a north or south direction, Polaris will shift the same amount from the zenith. Moving this time to the equator, Polaris is seen on the horizon, 90° from the zenith. Therefore, a relationship is established between the altitude or height in degrees of Polaris and one's latitude or position measured in degrees above the equator. Latitude equals the altitude of Polaris (fig. 2.11). The horizon is the line of reference. By changing latitudes, the horizon changes; this can only be true because we live on a curved spherical form. Early mariners used Polaris to determine their north and south position at sea. Of course today's space views leave no doubt about Earth's almost perfect spherical form.

Size of the Earth

Early thinkers also questioned the size of the Earth. In about 200 B.C., Eratosthenes, a Greek librarian in the cultural center of Alexandria, Egypt, calculated the size or circumference of the Earth by using only a rod for observing the Sun's shadow. Eratosthenes knew that for a day or so in summer, the noon Sun appeared directly overhead at Syene, illuminating the bottom of a well, and yet in Alexandria, 5,000 stadia to the north, the noon Sun made an angle of 7° with a vertical post, casting a short shadow. He reasoned that given a curved Earth, light from the distant Sun should strike the different places at different angles. Now instead of using Polaris as previously described, the Sun is the reference point. All celestial objects appear to shift in altitude by the same

degree we move in latitude. If you move one degree north, the Sun shifts one degree south.

By simple geometry, Eratosthenes knew that the 7° 20′ angle between the Sun's rays and the vertical post must be the same angle made by extending lines from the vertical post at Alexandria and the well at Syene to the center of the earth (fig. 2.12). Since this angle is also 1/50 of a circle, then the known distance of 5,000 stadia between the two towns forms an arc of 1/50 of the total circumference. A simple calculation yielded 250,000 stadia as the Earth's circumference, within one-half percent of today's accepted value of 40,000 kilometers (24,800 miles).

Constant Gravity of a Spherical Earth

While exploration of the Earth proceeded during the sixteenth and seventeenth centuries, another form of exploration was going on in the laboratory of Sir Isaac Newton. His discovery of the laws of gravity gave us another verification of the shape of the Earth.

Newton's law of gravitation says that any object and the Earth will have an attraction for each other; they will exert a force on each other that tries to pull them together. The strength of the force is determined by the masses of each body. If either mass is altered, the force is altered by the same proportion. Since the mass of the Sun and Earth are constant, the only variable left is distance. Newton's law also states that the strength of the force depends "inversely on the square of the distance." Gravity constantly pulls the Earth toward the Sun. Our orbital path is a compromise, which is described in the next section of this chapter (fig. 2.13).

Let's examine what this inverse relationship means applying it to Earth and a pendulum clock. Although a pendulum clock keeps accurate time for the place where it is set, the clock will gain or lose a few seconds or minutes per week if it is moved to a location north or south of its former position or to a higher or lower elevation, because the local force of gravity changes the rate of motion of the pendulum.

If the clock is at sea level, it is 6,370 kilometers (4,000 miles) from the center of the Earth's mass. By changing the clock's distance from the center of the Earth, the force of gravity on it

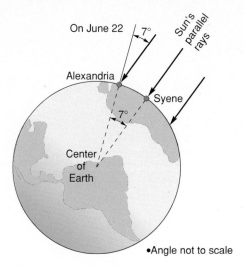

Figure 2.12

Eratosthenes determined the circumference of the Earth using simple geometry. Alexandria and Syene are 1/50th of an arc of the Earth's circumference apart, as he discovered when he measured the angle of the Sun's rays striking the Earth's surface at Alexandria on a day when he knew the noon Sun was vertical at Syene. The distance between the sites was 5,000 Greek stadia or 1/50th of the Earth's circumference.

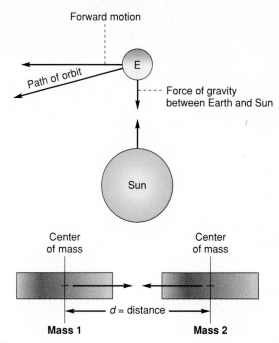

Figure 2.13

The force of gravity is directly proportional to the product of the masses of the attracting bodies. It is inversely proportional to the square of the distance between them. The Earth is continually falling toward the Sun, just as all objects under the Earth's influences are falling toward its center of mass. Its orbital path is a compromise between forward motion and the force of gravity.

will change. The law simply states that if the distance is doubled, for example, the force will be one-fourth. Triple the distance and the force reduces to one-ninth.

This law can be verified by observing the pendulum clock time at various latitudes and elevations. Let's take a journey to Lima, Peru, near the equator. As your plane lands at this capital city, about 3° south of the equator, the pilot announces the local time of 9 A.M. Later in the morning you notice a small shop that is filled with pendulum clocks. You can't resist purchasing one, and the clock stays with you for the rest of your travels. The next morning it's time to visit Cuzco, the ancient Inca capital, at 3,300 meters (11,000 feet). Once in your new hotel room with the clock unpacked and running, you notice it's losing a minute per day. Why? You remember Newton's law of gravitation. Gravity decreases as distance between the center of masses increases. The pendulum responds by moving slower because the gravitational force pulling the pendulum is less farther away from the center of mass of the Earth.

These variations in pendulum activity can only be the result of gravitational variation. Because the Earth is not a perfect sphere, the clock will also run faster in the middle latitudes at sea level than at Lima because the equator is about 10 kilometers (6 miles) farther from the Earth's center than latitude 40° north. The Earth has a bulge. This phenomenon proves the flattened spherical form of the Earth. This is also observable through sensitive navigational and surveying instruments from both land and space platforms. The Earth's diameter is 12,714 kilometers (7,897 miles) through the poles and 12,756 kilometers (7,923 miles) through the equator, or a difference of 42 kilometers (26 miles) greater through the equator. We can thus most accurately describe the Earth's shape as an **oblate spheroid.** This form results from centrifugal effects due to its rotation, like the rotational flattening

of pizza dough when the pizza maker throws it into the air with a spin. As the Earth spins, the motion tends to make objects fly off the sphere, but gravity holds objects in place. The net result produced a bulge over a long period of time.

The equatorial bulge is not a major factor shaping the environment. To the navigator, astronomer, cartographer, astrophysicist, and geophysicist, however, it is a significant characteristic.

Because the motions of artificial satellites are governed by gravity and gravity depends on the size and shape of the planet, careful observation of satellite orbits has given us more precise understanding of the Earth's shape. Satellite orbital patterns combined with surface gravity measurements have led to gravity maps showing regional and local deviations from the oblate spheroid. These *gravity anomalies* are important in geology and geophysics as they provide information about the Earth's interior.

THE EARTH'S MANY MOTIONS IN SPACE

Motion in the universe is constant. Nothing appears at rest. From the smallest atom to the greatest cluster of stars, motion is the norm. Our Earth is in a continuous state of motion, yet to the early observers of ancient times the Earth seemed immobile and vast. The stars, Sun, Moon, and planets appeared to revolve around the Earth, this platform that appeared to hold center stage in the universe.

Figure 2.14

Stonehenge served as an observatory for the prediction of solar eclipses. When the Sun and Moon both appeared in the same arch, an eclipse occurred. The priests of Stonehenge were able to predict this event because of their astronomical knowledge of the motions of the Moon and Sun.

Inset photo by Linda Gandee.

Our ancestors were keen observers who had worked out the celestial positions of the planets, Sun, Moon, and stars to a degree of accuracy that gave them a workable calendar. The Mayan temples of Mexico and Guatemala, the Egyptian pyramids, the Inca observatory at Machu Picchu, Stonehenge on the Salisbury Plain in southeastern England, and numerous lesser constructions all testify to earlier cultures' preoccupation with the motions of the heavenly bodies (fig. 2.14). Their observations were motivated by a number of factors, but foremost was the religious motivation. Priests of these cultures designed their observatories with astronomical orientation and the capabilities of predicting eclipses, seasons, and other significant celestial reference points. Religion and astronomy were inseparable. Celestial bodies were believed to have spiritual power, and those who understood their motion held corresponding power.

Step outside on a clear night and look out at the stars. The beauty of the universe is beyond description. Watch a sunset as the red and golden colors give way to the stars, one by one making their appearance in the evening sky.

Our view of the Earth and its motion in space today begins with the expanding universe from the first Big Bang. Almost every star and galaxy in the sky are receding from us. Compounding this expansion motion, the Sun and Earth participate in the galaxy's rotation with a velocity of about 320 kilometers per second (200

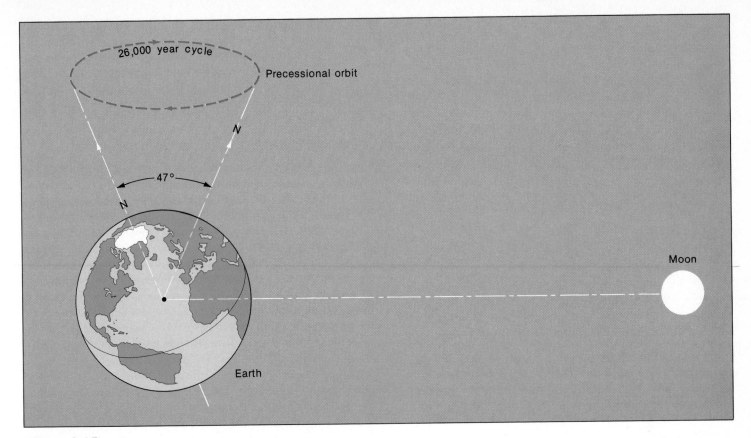

Figure 2.15

The Earth's wobble, or precession. One wobble takes 26,000 years. The Moon's gravitational pull upon the Earth's equatorial bulge causes the Earth to respond like a gyroscope, with a rotation about a vertical axis.

miles per second), making one revolution in 230 million years. In addition to these motions, the Earth is spinning while orbiting about the Sun, completing one **rotation** in 24 hours with respect to the sun and one **revolution** in 365¼ days, averaging 29.8 kilometers (18½ miles) per second in orbit. While you were reading this last sentence the Earth moved about 200 miles in its orbit around the Sun. Finally the Earth's axis is wobbling like a top off balance and taking 26,000 years per wobble. This motion is referred to as **precession** (fig. 2.15).

For the geographer and geographic study, the two most important motions of the Earth are its rotation and revolution. Astronomy defines rotation as a turning or spinning on an axis through a body, and revolution simply means an orbital motion about the center of mass of another body. Let us examine these two principal motions and their influences on the Earth's environment.

Revolution

Just ten generations ago the Earth was viewed by most scholars as the center of the universe, and any notions to the contrary were church heresy. Giordano Bruno, an Italian monk, was burned at the stake for his theories, which held among other things that the Sun was just another star. In 1543 the Polish astronomer and mathematician, Nicolaus Copernicus, released his book *De Orbium Coelestium Revolutionibus,* describing a universe with the Sun at the center and arranging the then-known planets in their proper order: Mercury, Venus, Earth, Mars, Jupiter, and Saturn. He stated:

> *Who in this most beautiful temple would put this lamp at a better place than from where it can illuminate them all? Thus, indeed, the Sun sitting as on a royal throne, leads the surrounding family of stars.*

An interesting proof of the Earth's revolution is the apparent shifting of nearby stars against the stationary background of distant stars as the Earth revolves in its orbital plane. Astronomers refer to this shifting as **stellar parallax** (fig. 2.16).

It is easy to demonstrate this apparent shift or parallax to yourself. Put down your book and sight a nearby object, covering first one eye, then the other. Did you see the object jump back and forth? If not, select a nearer object and repeat the observation. Now hold a pencil at arm's length and repeat the process, then hold it closer and do the same. Note that the apparent displacement of the pencil is greater at a closer distance.

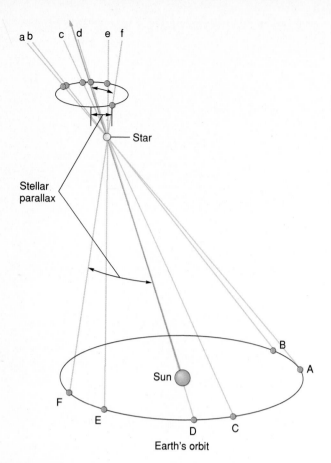

a b c d e f

Star

Stellar parallax

B

A

Sun

F

E

D C

Earth's orbit

Figure 2.16

Stellar parallax is caused by the Earth's change of position in orbit. The apparent motion of a nearby star reflects the Earth's revolution. Most stars are too distant to exhibit stellar parallax.

The nearest stars shift back and forth in exactly the same way because the Earth orbits about the Sun. Nearby stars appear to jump back and forth because the Earth's orbital motion gives us a changing perspective. Stars at greater distances have no apparent shift, so they become the field of reference. This observation can be used to measure distance to the nearest stars because parallax decreases with distance. The greater the distance, the smaller the stellar parallax. Most stars are so far away they have no observable parallax.

In 1619, Johannes Kepler explained that the revolutions of all planets, including Earth, take place in orbits that are elliptical. They speed up their motion when near the Sun and slow when far from it, so that an imaginary line from the Sun to the planet would sweep over equal areas in equal time periods.

The Earth, on its elliptical path, moves closest to the Sun on January 3. This position is referred to as its **perihelion,** and Earth and Sun are 147 million kilometers (91.5 million miles) apart. **Aphelion** is our farthest position in orbit at 152 million kilometers (94.5 million miles) from the Sun. This variation in distance has only a minor influence on the total solar energy reaching the Earth; the ocean and atmospheric circulation offset any variation. The moderating influences of the oceans are discussed more fully in chapter 4.

As it orbits the Sun in its elliptical path, the Earth speeds up and slows down, averaging 29.8 kilometers (18½ miles) per second. If it were to speed up to 42 kilometers (26 miles) per second, Earth would leave the solar system by overcoming the Sun's gravity. Think of the Earth's orbit as resulting from a balance of forces: Forward motion in a straight line is offset by gravity pulling toward the Sun (see fig. 2.13).

The Seasons

As the Earth revolves around the sun, the motion produces seasonal variation in solar energy over the different parts of its surface. The orbit of the earth lies in a plane known as the **plane of the ecliptic.** The Earth's equator is inclined to the ecliptic plane at 23½ degrees. This relationship can be compared to a spinning top with a tilt axis of 23½ degrees with the perpendicular of its plane of orbit or ecliptic plane. It is this permanent tilt and the Earth's revolution around the Sun that cause variation in solar energy, hence the seasons. The Earth's axis remains fixed in this inclined position throughout the orbit. Note in figure 2.17 that the Sun's rays are perpendicular at the Tropic of Cancer, latitude 23½ degrees north, on or about June 21. This latitude is the most northerly perpendicular position of the Sun. This event is called the **summer solstice.** People living in the *study area* valley, described in chapter 1, see the sun 73½ degrees above the southern horizon at noon on this date. For everyone living north of the equator, this date has the longest daylight period of the year. Note in figure 2.17 that the axis has the same inclination six months later, December 21, but now the Sun's rays are only 26.5 degrees above the horizon at noon in the *study area* at the winter solstice (fig. 2.18).

The **winter solstice,** on or about December 21, is the day when the sun is directly overhead on the Tropic of Capricorn, 23½ degrees south of the equator. This is the longest day of the year for those living in the Southern Hemisphere and the shortest day of the year in the Northern Hemisphere. On this date, the sun rises in the southeast 23½ degrees south of the east point and sets 23½ degrees south of the west point. This apparent path or arc of the Sun across the sky is the shortest daily arc of the year observed above the equator. When the combined influences of short days and low angle of the Sun occur during the Northern Hemisphere winter months, the result is less solar energy reaching the Earth in the Northern Hemisphere in terms of duration of daylight and solar energy intensity per unit area (fig. 2.19).

After the winter solstice, the Sun rises and sets farther north each day. This is shown in figure 2.20, a pair of photographs of the sunrise in the *study area* taken on two days in April. As the Sun progresses northward in the sky, daylight in the Northern Hemisphere grows longer until about June 21, the **summer solstice.** At this time the Sun is directly overhead at noon on the Tropic of Cancer.

The Sun moves along the ecliptic plane from the Tropic of Cancer, on June 21, to the equator, on September 21. On this date the sun is directly overhead at noon on the equator, and it rises due east and sets due west. This is the first day of autumn in the Northern Hemisphere and is known as the **autumnal equinox.** Daylight and nighttime both last 12 hours all over the Earth. This event occurs again on March 21, when the Sun returns to the

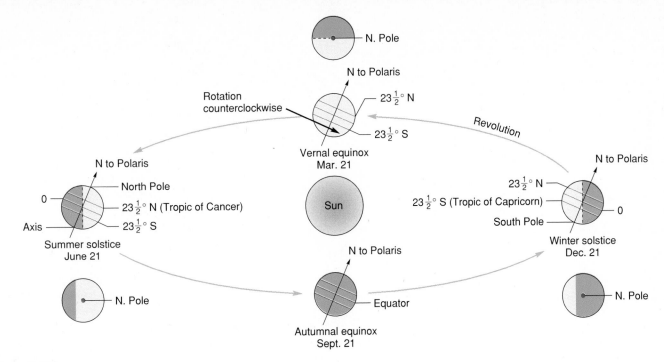

Figure 2.17

The tilt of the Earth's axis, constant the year round, is a chief cause of the changing seasons as the Earth moves in its orbit.

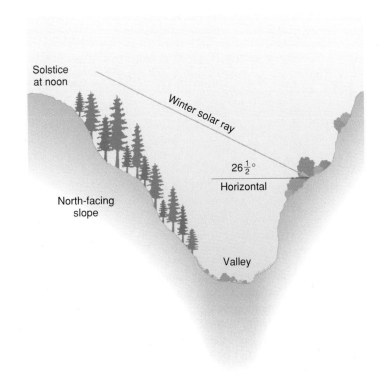

Figure 2.18

The Sun's rays strike the Earth at the lowest angle in the Northern Hemisphere on the winter solstice. This diagram shows the *study area* valley at about 40° N latitude.

equator. This is the first day of spring or the **vernal equinox.** This seasonal cycle equates to approximately 1 degree per day or 365¼ degrees per year. While you completed reading this section the earth moved 1,200 miles (804 km) or 1/1440 of a degree in its seasonal cycle.

Rotation

The second important celestial motion is **rotation.** Rotation, or spinning of the Earth on its axis, produces the daily rhythm of life. This rhythm is synchronized to the spinning earth. Sunrise, sunset, passage of the noonday Sun, and apparent movement of the Moon and stars each night all testify to a spinning Earth. Figure 2.21 shows star trails photographed by simply leaving the camera open for a few minutes as the Earth platform spins. The Earth rotates at a constant rate of 1 degree every 4 minutes or 15 degrees per hour, giving 360 degrees of rotation in 24 hours. If the shutter is left open 4 minutes, a star streak of 1 degree of arc will be made. Rotation gives us not only our daily rhythm of sunrise and sunset but also tidal change, deflection of motion or Coriolis effect, and our time and coordinate grid system.

Rotation and Tidal Action

The work of waves on the beach varies with the tides. Waves erode low on the beach during lowest tide then move upward until high tide, 6 hours later. In some places the difference is 15 meters, or 50 feet. This daily event occurs on time all around the Earth as the Moon and Sun exert their gravitational pulls. Generally, two high tides occur 12 hours apart, separated by two low tides in a 24-hour period (fig. 2.22).

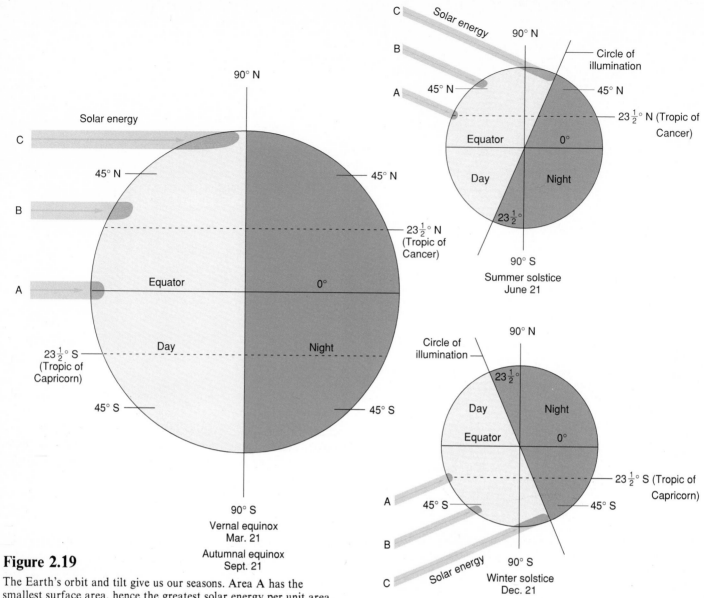

Figure 2.19

The Earth's orbit and tilt give us our seasons. Area A has the smallest surface area, hence the greatest solar energy per unit area. Area C has larger surface areas and thus less solar energy per unit area.

The Coriolis Effect

When any moving object—plane, rocket, bullet, ocean current, or wind system—travels in a straight line over the curved surface of the spinning Earth, it will have two apparent paths. The straight-line path can be observed when using a reference point in space such as a star or planet. When this same path is plotted on the Earth's surface, it takes the form of a curve. The contradiction is really one of perspective. As the object travels it is moving over a surface that varies in velocity with distance from the north or south pole. Although all points rotate at 15 degrees per hour, each latitude has a different velocity. Objects at the equator travel over 1,600 kilometers per hour (1,000 mph) and cover nearly 40,000 kilometers (25,000 miles) in 24 hours, while objects at 40° N latitude cover only 30,700 kilometers (19,200 miles) in 24 hours at

a velocity of 1,280 kilometers per hour (800 mph). If all latitude points on a meridian move at the same velocity, there would be no curved path or **Coriolis effect**.

You can visualize the Coriolis effect by placing any old record on a turntable. Start the record in motion and take a marker pen or piece of chalk and strike a straight-line motion across the record. Note the curved path on the rotating surface. Because the turntable rotates clockwise, the deflection is to the left, just as it is in the southern hemisphere. Counterclockwise rotation, as in the northern hemisphere, will produce a deflection to the right. Deflection is zero at the equator and reaches a maximum at the north and south poles. The effect is also zero when the object is at rest and increases as the velocity of the object increases.

A   B

Figure 2.20

The sun sets farther north each day following the winter solstice, December 22. The photographs were taken on (A) April 15 and (B) April 20, in the vicinity of the *study area*. This northern migration of both sunsets and sunrises, combined with the increasing altitude of the sun, produces longer days and greater solar intensity.

Figure 2.21

Rotation can be seen in the stars. This is a time exposure producing the star streaks due to the Earth's rotation.

Photo by Howard Schaad

A

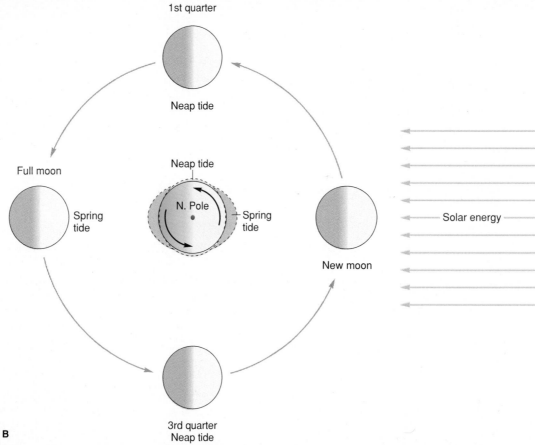

1st quarter

Neap tide

Full moon

Neap tide

N. Pole

Spring
tide

Spring
tide

Solar energy

New moon

3rd quarter
Neap tide

B

Figure 2.22

(A) Grange-Over-Sand, England, an estuary located along the west coast near the Scotland border. (B) Rotation causes changing daily tidal action on the shoreline. The Earth rotates daily into two high and low tide zones. Spring tides result from the combined gravitational pull and alignment of the Sun and Moon. Neap tides occur when the Moon and Sun are aligned at right angles, reducing to a minimum the combined gravitational pull.

Figure 2.23
The Coriolis effect causes objects to be deflected to the right of their forward motion in the Northern Hemisphere. The Coriolis effect increases with latitude. This hypothetical plane flight from Toronto to Miami, essentially a north-to-south path, shows the Coriolis effect due to the rotation of the Earth under the flight path of the plane. The pilot must correct the plane's flight to compensate.

Remember, it is the reference system used that determines the path described. Both the straight-line and curved path are occurring. It is a question of your geographic perspective.

The Coriolis effect will be examined in chapter 5, "The Windy Planet," as it affects the path of wind over the surface of the Earth (fig. 2.23).

TIME IS MOTION

Excuse me, do you have the time? The answer to this question is often given with little thought. We take time for granted until it begins to run out as deadlines approach. What is time? How do we measure it? These are questions we seldom ask.

Our units of time are based on the Earth's two principal motions: rotation and revolution.

The Earth's days are determined by rotation. However, it is not quite that simple. Let's try a simple experiment. Set a post to point directly toward the Sun at noon. During the next 24-hour period the Earth not only rotates on its axis, but it also orbits approximately one degree in its elliptical path around the Sun. Therefore, after one complete rotation on its axis, the post-and-Sun alignment is off by about one degree because the Earth has moved one degree in its orbit. The Earth must continue to rotate nearly four more minutes or one degree (3 minutes, 56 seconds) to bring the post into the aligned position. Thus the "sun day," noon to noon, is nearly four minutes longer than the time to complete one rotation of 360 degrees with reference to a point or star in space (fig. 2.24), which astronomers call a sidereal day.

Because its orbit is an ellipse instead of a perfect circle, the Earth moves faster in its orbit during the time of the year when it is closest to the Sun and slowest at its maximum distance. Thus at certain times of the year the Sun may be as much as 16 minutes ahead or behind in reaching the noon position. Therefore, our time systems are based on "mean solar time." A mean solar day averages out the irregularities of our motion. This is the time our clocks run by.

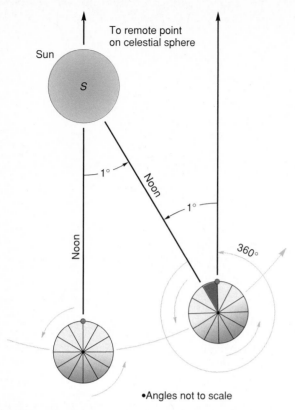

Sun

S

To remote point
on celestial sphere

1°

Noon

1°

Noon

360°

•Angles not to scale

Figure 2.24

Rotation and revolution give us our time equation. Because the Earth revolves 1 degree per day, it must rotate nearly 361 degrees to complete one Sun cycle or solar day. A sidereal day is nearly 4 minutes shorter because its reference point is a point in distant space, not the Sun.

Time Zones

Before 1883 every country had its own local time. When the Sun was on the noon meridian, it was 12:00 noon at that place. One hour before noon, the time was 11:00 A.M. (*ante meridiem*, meaning "before meridian") and one hour after the meridian was 1:00 P.M. (*post meridiem*). This system worked until trains and electric communication brought places closer together in time. No longer could each place operate on its own local time. Therefore, in 1884, the International Meridian Conference in Washington, D.C., divided the world into 24 time zones. Each time zone is centered on meridians 15 degrees apart and includes 7½ degrees on each side of the standard meridians. Everyone living in a given zone operates under the same time assigned to that zone. In the United States, Eastern Standard Time is centered on the standard meridian of 75° W, Central Standard Time on 90° W, Mountain Standard Time on 105° W, and Pacific Standard Time on 120° W. The boundaries of the zones do not follow the meridians too closely. What would be the problem if a time zone cut a city in half? Actually, time zones have been modified to bring economic and political regions together under the same zone. The Hawaiian Islands, for example, are under one zone spanning 30 degrees, overlapping two other time zones. When the Sun is on the meridian at

noon on the Island of Hawaii, the Earth must turn nearly 15 degrees more to bring the Sun to the meridian on Midway Island (fig. 2.25).

International Date Line

Travel halfway around the world from the prime meridian east or west, spanning 12 time zones and 180 degrees of longitude, and you will reach the **International Date Line.** It follows the 180th meridian except for a few jogs to keep islands under the same time and day.

Crossing the International Date Line can add or subtract a day on your calendar. You will turn your calendar a full day back when traveling east toward the prime meridian, or turn it forward if traveling west. If the time is 1 P.M. Tuesday and you are crossing the line going west, it will be 1 P.M. on Wednesday on the western side.

Summing it up, what is time? Time is measured by observing the movements of rotation and revolution of the Earth using the Sun as a reference point. Because the Earth's motion varies from day to day and from season to season, time is measured by averaging out these variations in motion. The time measured by your watch is **mean solar time.**

MATTER AND ENERGY IN THE COSMOS

As we examine the physical environment, it is important not to lose the universal perspective of matter and energy relationships in the universe and the role of **gravity** as a fundamental energy form that fills the universe and yet converts to other forms. Phenomena such as continental plate motion, earthquakes, volcanic eruptions, landslides, glacial action, stream erosion, falling snow and rain, meteorite impact, air movement, atmospheric pressure, falling redwoods, breaking waves on the beach—all are events and phenomena that result directly or indirectly from gravitational energy and keep the Earth alive and dynamic.

SUMMARY

Viewing the earth from space reinforced the Copernican heliocentric view of the Earth as a member of a solar family of nine planets, third in order from the Sun, revolving as a small speck in the cosmos.

Planet Earth belongs to a group of inner **terrestrial planets** high in density, consisting of heavier elements than the large **jovian planets** such as Jupiter.

Because the Sun's mass equals 99.86 percent of the solar system, Earth is tied to its gravitational tether, locked in orbit with its satellite the Moon.

The Sun is a member of billions of other stars in our galaxy, which spans 100,000 light-years in diameter, consisting of a pinwheel of rotating stars, gas, and dust held by the force of gravity.

Earth is strategically located in space at a distance from the Sun to maintain water in the liquid form, and yet keep the atmosphere from freezing or boiling away into space. A curbside appraisal of the other eight planets reveals an unhealthy environment. Temperature, abundance of the right resources, and general life support systems are tipped in our favor only on this planet.

Our space view has not only given us a better perspective of our location in the universe, but also verified our shape and size. These proofs

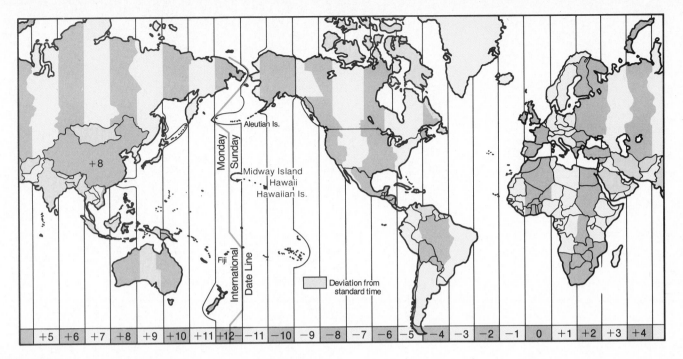

Figure 2.25

The time zones are centered on standard meridians 15 degrees apart. When the Sun is highest over a locality it is solar noon there. Note the difference in solar noon and time-zone noon between Midway Island and the island of Hawaii. Here the islands are all under one time but actually have very different solar time.

include observations of **ship submergence on the horizon, lunar eclipses,** and **changing altitudes of Polaris,** Eratosthenes calculations, and **constant gravity.**

If Earth is perfectly spherical in shape, gravity at any given distance from the center should be the same. But the Earth is slightly oblate, therefore a pendulum clock runs faster in the higher latitudes, where it is closer to the center of the Earth, and slower at the equator.

Motion and gravity are two constants in the universe. The motions of **revolution** and **rotation** are the principal motions that govern the rhythms of life on earth. Revolution determines the length of the seasons as well as the patterns of solar energy from the Sun in each hemisphere. The rotation rate determines our 24-hour day, tidal variation, and reference points for our geographic grid. Winds, ocean currents, and navigation by planes and ships must consider the influence of rotation in the form of the **Coriolis effect.**

Our time zones are also tied to the rotation rate. With a spin of 360 degrees we span 24 hours of time. Thus we have 24 time zones, 15 degrees wide. Another less significant motion, but very real on a millennium time scale is **precession,** taking 26,000 years per cycle.

Finally, matter and energy must be considered as inseparable. All places are a mix. The universe is both matter and energy. Matter is the building block and energy is the glue and driving force.

ILLUSTRATED STUDY QUESTIONS

1. In which time zone is the *study area* located (figure 2.25, p. 47)?
2. The photograph in figure 2.7 on page 32 was taken on the surface of Mars. Describe the environmental features common to the Earth and Mars. Which elements in the photograph are the essentials to life? Which elements are either missing or in short supply on Mars?
3. Polaris can be seen from the *study area* at 40° above the northern horizon. What does this tell us about the latitude of the point of observation (see fig. 2.11, p. 36)?
4. Each of the photographs of the sunrise was taken in the *study area* on the dates indicated. How many degrees did the earth revolve during this time period (see figs. 20.20 and 2.24, pp. 43, 46)?
5. Using figures 2.2 and 2.4, describe the unique location of the Earth in space (pp. 27, 30).
6. List the information Eratosthenes had available to correctly calculate the circumference of the Earth (see fig. 2.12, p. 37).
7. Give one advantage of false color infrared imagery (see fig. 2.1, p. 26).
8. Give the primary ingredients of our Sun. How is it powered (see fig. 2.3, p. 28)?
9. How does Venus resemble our planet (see fig. 2.5, p. 31)?
10. Because the Earth is spherical, a change in position of 10° latitude will produce a change in the position of Polaris, the pole star, how many degrees (see fig. 2.11, p. 36)?
11. Contrast rotation versus revolution motions of the Earth. Give one effect of each motion (see figs. 2.19 and 2.22, pp. 42, 44).
12. What does the effect of stellar parallax reveal about the Earth's motion (see fig. 2.16, p. 40)?
13. When is the Earth closest to the Sun? What effect does this distance have on the seasons (see fig. 2.17, p. 41)?
14. Explain the primary factors determining the seasons (see figs. 2.19 and 2.20, pp. 42, 43).
15. Describe the path of an object such as a plane without correction for the Coriolis effect (see fig. 2.23, p. 45).
16. Describe solar noon and contrast it with mean solar noon (see fig. 2.25, p. 47).

The Atmosphere

UNIT II

The atmosphere is our life support system, enriched by a balanced ratio of oxygen, carbon dioxide, water, and nitrogen, all essential for life. This atmospheric shield filters harmful ultraviolet rays and stores and transports energy over the surface of the Earth.

Santa Lucia Mountains, Wilderness, California.
© David Muench Photography

The Ocean of Air

3

Objectives

After completing this chapter, you will be able to:

1. Explain the difference between **weather** and **climate.**
2. Explain how the atmosphere was formed.
3. Identify the major gases, liquids, and solids in the atmosphere and explain their function.
4. Describe the impact of human activity on the atmosphere's constituent gases and particulates.
5. Identify the primary zones of the atmosphere and describe their major characteristics.
6. Describe the vertical temperature and pressure distribution in the atmosphere.
7. Explain the **gas laws** developed by Boyle and Charles.
8. Explain **adiabatic heating** and **cooling.**
9. Define the wet and dry **adiabatic lapse rates.**
10. Explain latent heat of condensation as it influences lapse rates.
11. Give the principal points of **kinetic molecular theory.**

We live at the bottom of an ocean of air. Clouds over Carribbean.

© Jack Swenson/Tom Stack & Assoc.

We live at the bottom of an ocean of air. The amount of energy from the Sun we receive, and all that we see, smell, taste, and touch, is affected by the quality of this ocean.

The atmosphere is a primary life support system for the planet. All living systems are dependent upon the atmosphere's ingredients. This gaseous envelope that surrounds us filters harmful radiation and helps maintain a tolerable temperature environment at the surface. Without the pressure or weight of this atmosphere, the oceans, lakes, and rivers would boil away. Even the sound of a plane or the bark of a dog would be absent without air to transmit them to our ears.

The atmosphere is never static; energy from the Sun produces continual changes in each of the atmospheric characteristics mentioned above. It is also important to note that these characteristics are interdependent. When a change in temperature occurs, there are usually changes in pressure, density, and perhaps cloudiness. Changes in cloudiness can bring changes in radiant energy reaching the surface. Both air and water respond to solar energy, transporting energy to other locations. This topic is covered in chapter 5, "The Windy Planet."

This chapter considers the following questions:

How does weather differ from climate?
How did our atmosphere form?
What gases make up the atmosphere and what is their function?
What are the major vertical zones of the atmosphere?
How do gases respond to changes in temperature and pressure?
What theories describe the behavior of gases?

These questions focus on how the atmospheric system works. This will prepare you for a better understanding of the geography of our environment.

WEATHER VERSUS CLIMATE

Weather, simply defined, is the state of the atmosphere at a given time and place. Terms or characteristics that describe the weather are temperature, pressure, wind speed, wind direction, humidity, visibility, clouds, and precipitation. **Climate** is a summary of these same characteristics over a period of years. Climatic descriptions deal in averages, and climate is studied by the climatologist and geographer. It sets the environmental conditions for plant and soil development. Weather is the day-to-day state of the atmosphere evaluated by the meteorologist.

ORIGIN AND COMPOSITION OF THE ATMOSPHERE

The Earth's atmosphere is unique in the solar system. Only the Earth is known to have a significant percentage of water vapor and oxygen, so important to the life support systems of living things. It is also unique in that it is probably the only atmosphere that is composed almost entirely of nitrogen (78%) and oxygen (21%). Table 3.1 gives the current composition of the lower atmosphere near sea level.

Table 3.1

Dry Composition of the Lower Atmosphere

Gas	Mass Percent	Characteristics
Nitrogen	78.084	Dilutes oxygen, does not chemically combine easily with other gases.
Oxygen	20.946	Very active with other gases (oxidation).
Argon	0.934	Very inactive.
Carbon dioxide	0.033	Absorbs energy from the Earth and Sun in the infrared; essential in photosynthesis.
Neon *Helium* *Krypton* *Xenon* *Hydrogen* *Methane* *Nitrous oxide*	Combined totals less than 0.003 in decreasing order of percent.	These gases are of minor importance because of their small amounts. They are completely diffused in the homosphere with the major constituents.

Source: From the *Fire Weather—Agricultural Handbook 360,* U.S. Department of Agriculture, U.S. Forest Service.

The origin of this atmosphere is being debated; however, several facts are fairly certain. When the Earth formed, it was probably very hot, making it unlikely to retain any atmosphere. Gas molecules at high temperatures have enough energy to reach speeds great enough to escape the Earth's gravitational field (fig. 3.1).

At the beginning of the Earth's solid existence, its original atmosphere was probably much like those of Jupiter, Saturn, Uranus, and Neptune, with large percentages of hydrogen (H_2), helium (He), methane (CH_4), and ammonia (NH_4) gases. Then as the planet cooled further, these gases were slowly replaced by nitrogen, carbon dioxide, and water vapor escaping from the Earth's interior by outgassing. This process occurs during volcanic eruptions.

Further cooling caused the water vapor to condense and form the oceans. The oceans dissolved most of the carbon dioxide. Oxygen is believed to be the product of primitive plant life capable of photosynthesis, which converted solar energy into chemical forms and released oxygen as by-product.

Some scientists believe another significant source of oxygen was from the water molecule (H_2O) in the vapor form being split apart by high-energy x-ray and ultraviolet radiation, allowing the higher velocity hydrogen to escape from the lower atmosphere and leaving molecular oxygen behind (fig. 3.2).

Most evidence indicates that only minor changes have occurred in the composition of the atmosphere since its formative period. However, scientists are concerned that human activity may now be transforming the composition at a dangerous rate, not only on a local or regional scale but globally. As we examine the composition of the atmosphere, the human impact on the ingredients will be evaluated.

...gravitational energy. ...this gravitational energy, a pressure of about 100,000 pascals (14.7 pounds per square inch) is exerted on the surface. This is enough pressure to maintain the oceans in liquid form. If the Earth's mass were less, it would be a desert without its life support system of water.

Photo by NASA.

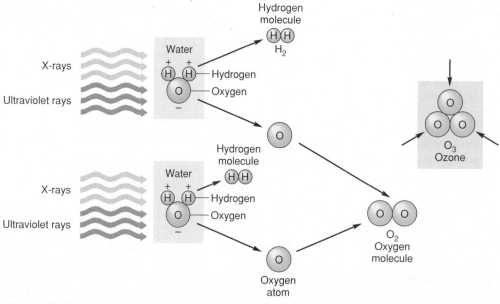

Figure 3.2

The Earth's formative period was a time of cooling and condensation of gases. It is believed that oxygen was formed when the water (H_2O) molecule in vapor form was split apart by high-energy x-ray and ultraviolet radiation. Hydrogen with its greater velocity boiled off the planet's surface, leaving behind oxygen in the atomic (O) and molecular gas forms (O_2) and (O_3) with other heavier gases.

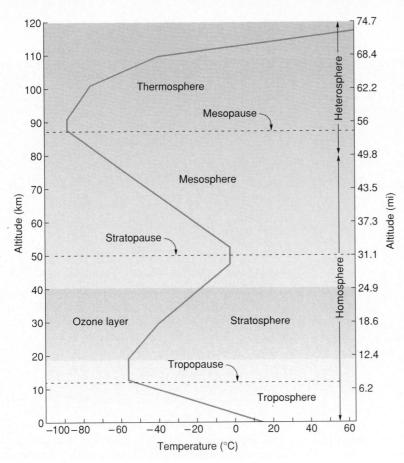

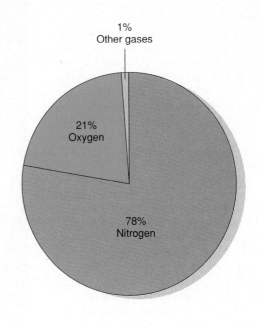

Figure 3.3

The atmosphere is spherically layered into two broad zones, the homosphere and the heterosphere. Below 80 kilometers the homosphere is a blend of heavier gases consisting of 78 percent nitrogen, 21 percent oxygen, 0.9 percent argon, 0.03 percent carbon dioxide, and traces of dust and other gases. Water vapor ranges from near zero to nearly 5 percent.

Composition of the Lower Atmosphere—Homosphere

If you could climb into a hot-air balloon and rise above the Earth taking air samples as you go, you would discover that the ingredients of the atmosphere are very uniform in the lower levels. This lowest layer is the **homosphere**, which reaches an altitude of 80 kilometers (50 miles). Only water vapor shows a significant percentage of variation in this lower layer. Above 80 kilometers, in the **heterosphere**, the various gases tend to be layered according to their molecular mass and density in much the same way two insoluble liquids of different densities will separate in a container with the heavier settling on the bottom.

As we have indicated, 99 percent of the lower atmosphere consists of heavier gases, nitrogen (78%) and oxygen (21%). The remaining 1 percent is divided between argon (0.93%) and carbon dioxide (0.03%) and traces of many other gases both natural and human-introduced (fig. 3.3). In addition to these gases, water vapor ranges from near zero to nearly 5 percent in the humid tropics.

The lower atmosphere also contains a variety of particles consisting of salt, dust, smoke, and industrial by-products in both the solid and liquid state. These solid particles have a significant influence on the clarity of the air by filtering solar energy and reflected light. Because these particles are increasing on a global scale, this filtering effect may be changing the energy budget of the Earth. Some particles serve as nuclei for the condensation of water vapor in cloud formation.

Scientists and policy makers are becoming concerned about the increasing impact of human activity on the composition of the atmosphere. Currently, only the minor constituents appear to be affected, but because of their role in the atmospheric system, these minor constituents can have a severe impact on the quality of our lives. Let us examine the importance of these constituents and evaluate how we are interfering with their function.

Ozone (O_3), a molecular form of oxygen, is found in minute amounts near the surface, usually less than one part per hundred million (0.01 ppm) up to 1 ppm under very smoggy conditions. The greatest concentrations of ozone are found between

The Ozone Hole Is Growing above Antarctica

New observations coming from the Airborne Antarctic Ozone Experiment revealed that chlorofluorocarbons are thinning the ozone in the stratosphere over the continent of Antarctica. Each spring since 1976 the hole has enlarged and decreased ozone has been observed halfway to the equator. The hole is the primary result of a unique weather pattern found over Antarctica, where the extreme cold there in the stratosphere ($-80°$ C) may play a key role in ozone reduction. In the lower stratosphere, ice particles that make up cirrus clouds provide the environment for accelerated chemical reactions (fig. 1).

The lower than normal temperatures observed in the southern hemisphere in recent years may account for the locking up of nitrogen that influences ozone destruction. It may also account for the removal of water vapor to form ice particles, which provide surfaces that accelerate the release of active chlorine, thought to be the culprit in ozone destruction.

Another meteorological connection is in the upper air poleward circulation. Not only is heat transported poleward, but also ozone from the tropical stratosphere where high solar energy manufactures it. Recent observations indicate that these winds have weakened, resulting in southern hemisphere stratospheric cooling by more than $2°$ C between 1980 and 1985, making it colder than ever before (fig. 2).

Directly over Antarctica the stratosphere has cooled as much as $4°$ C since 1980. As ozone declines, ultraviolet radiation is not absorbed as effectively, thus temperatures drop even lower. However, this will eventually create new differences in pressure between the tropics and the high latitudes and stimulate stronger poleward movement of tropical air at higher altitudes, thus restoring the balance.

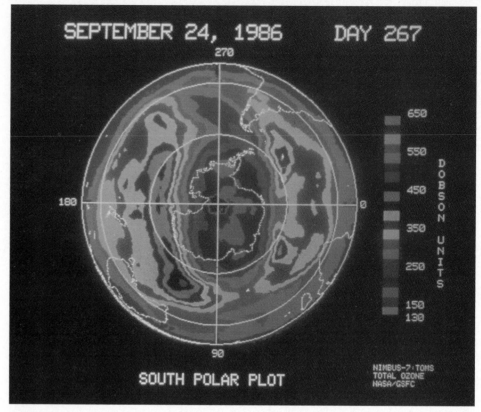

Figure 1

Ozone depletion in the stratosphere over Antarctica is the result of human and natural factors. In this computer-colored image, taken by satellite in September 1986, the black and pink colors over Antarctica show the areas of greatest depletion. The extent of the hole varies with weather patterns.

Photo by NASA.

Here we can see artificial chlorofluorocarbons in concert with nature intensifying a natural phenomenon, unique to the ice continent of Antarctica.

Source: NASA.

---

25 and 40 kilometers altitude. This very important concentration plays a major role in filtering deadly ultraviolet radiation from the Sun. With a weaker ozone layer, our lives would be endangered. Skin cancer would be an increasing disease.

The ozone of the atmosphere is formed two ways. It is thought to be formed naturally where an atom of oxygen (O), a molecule of oxygen (O_2), and a third particle such as nitrogen collide. The end result is ozone (O_3), leaving the nitrogen molecule unchanged. Nitrogen thus acts as a catalyst to help in the chemical process but not enter into it.

In the atmosphere of cities, ozone is a secondary pollutant forming from carbon, nitrogen, and sulfur oxides, with the assistance of sunlight. Solar radiation releases atomic oxygen (O) from primary oxide pollutants, which then combines with molecular oxygen in the atmosphere. Ozone pollution is greatest where the automobile is the primary form of transportation and where there is a high percentage of clear days. Los Angeles, California, and Phoenix, Arizona, are typical climates for the production of ozone.

For those who live in urban and suburban smog belts, the effects are obvious. Ozone shortens your breath, brings tears to your eyes, and corrodes metals, tires, paints, and plasters.

Currently a potential threat to the high-altitude ozone is being debated by scientists. Industrial sources release chlorofluorocarbons, which can drift upward and reach the ozone layer and destroy the ozone molecule. The research is not complete, but along with the international community the Environmental Protection Agency has taken steps to phase out chlorofluorocarbons from industry in the United States (box 3.1).

Figure 2

Upper air circulation poleward is weakening, resulting in less energy and ozone imported from the tropics. These observations were made by this ER-2 (in foreground) which flies at altitudes of 21 kilometers to research the stratospheric chemistry and weather of the Antarctic.

Photo by NASA.

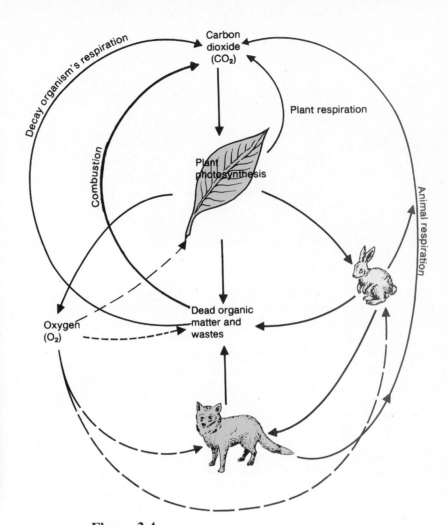

Figure 3.4

Carbon dioxide is essential for all plant life. It is continually released from processes of decay, combustion, respiration, and volcanic activity. The *carbon cycle* can be traced through the Earth's atmosphere, hydrosphere, biosphere, and lithosphere.

Carbon dioxide (CO_2) is a naturally occurring gas in the atmosphere. It is also found dissolved in the oceans and lakes of the world and is continually being produced by plants and animals during respiration and decay. Great quantities are also released during the combustion of fuels and the eruption of volcanoes. Currently over 500 active volcanoes around the world are releasing carbon dioxide and other gases and particles into the atmosphere around the globe. The carbon in the carbon dioxide molecule goes through a natural cycle, which can be traced between the *atmosphere, hydrosphere, biosphere,* and *lithosphere,* known as the **carbon cycle** (fig. 3.4).

Carbon dioxide performs a number of significant environmental roles in each of these major spheres. We shall consider them briefly here for they are described in more detail in chapter

4. In the atmosphere, carbon dioxide is an important absorber of solar energy in certain wavelengths of infrared. Therefore, increased amounts present can raise temperatures on the surface or create a warming trend, or **greenhouse effect.** There is considerable concern among meteorologists and climatologists that we are significantly changing the amount of carbon dioxide in the air by increased energy consumption from fossil fuels. Since the turn of the century, the people of the United States have been doubling their consumption of electrical and fossil-fuel energy every decade. Between now and A.D. 2001, just a few years away, the United States will consume more energy than it has in its entire history.

By A.D. 2001, the annual U.S. demand for energy in all forms will be doubled, and the annual worldwide demand will probably triple. Carbon dioxide emissions from combustion processes in the United States are estimated to have increased ninefold between 1890 and 1965. By 2000, emissions are expected to have increased threefold. This great increase is partially being absorbed by the oceans. Currently we do not know how much or even whether the temperature of the Earth is being affected, but there is evidence that cities having higher carbon dioxide levels tend to

Figure 3.5

Particulate pollution determines the visual clarity of the atmosphere at the pyramids near Cairo, Egypt.

be warmer than surrounding rural regions. Some scientists believe we may be entering a warming trend that could raise global temperatures enough to melt the ice sheets and cause a rise in the sea level of several feet.

Solid particles or **particulates** from both natural and human sources are found suspended in the air. Some eventually fall out while others remain in suspension indefinitely. The particles include dust from the land, salt particles from the ocean, soot from forest and grass fires, industrial exhaust, automobile emissions, pollens, microorganisms, volcanic dust, and meteoric dust. The larger ones fall out more rapidly, while small particles may remain in the air for weeks or months. When the volcano Krakatau in the East Indies erupted in August 1883, particles circled the globe for two years, reflecting and scattering solar rays producing a haze and colorful sunsets around the Earth. It also produced one of the coolest summer periods on record and advanced glaciers in the European Alps.

The Los Angeles basin receives a fallout of industrial particles of over 17 tons per square kilometer (50 tons per square mile) per month. Most of these particles are generated by the automo-

bile engine and abrasion of tires. Even so, Los Angeles is not among the 10 worst American cities in terms of particulates, because most of its pollution is in the gaseous form emitted by automobiles.

Particles produce two problems to our health and environment. First, they can affect the respiratory system if concentrations are at a dangerous level, producing difficulty in breathing. Second, particles may affect the climate by reflecting some of the Sun's energy before reaching the surface. Particulates also reduce visibility. On a smoggy day when particulates are high in Los Angeles, Chicago, or New York, visibilities can drop to less than 3 kilometers (2 miles). The culprit is particulate pollution, while gaseous pollutants such as carbon monoxide, ozone, and other gases do their damage unseen (fig. 3.5).

Particulates may be changing the Earth's climate by raising the atmosphere's **albedo** or reflectivity. If the atmosphere becomes a better reflector, we could enter a cooling trend, hence counteracting the carbon dioxide effect (box 3.2).

Nitrogen gas (N_2) represents the largest percentage, 78 percent by mass, of the lower atmosphere; it does not combine easily with other gases but merely mixes with them. Most forms

Black Soot Discovered in Arctic

BOX 3.2

Particles of black soot similar to those present in polluted urban atmospheres have been discovered in Arctic air. This finding, reported by scientists at the University of California's Lawrence Berkeley Laboratory, raises new concern over the potential impact of fossil-fuel burning on climate.

"The presence of graphitic carbon concentrations in the Arctic can be attributed directly to combustion," said Hal Rosen of LBL's Atmospheric Aerosol Research Group. "Black carbon particles, because they absorb the Sun's radiation very effectively, can contribute significantly to the heating of the atmosphere. Calculations and computer models indicate that these particles could produce a heating effect in the Arctic comparable to that resulting from

a doubling of the carbon dioxide concentration."

Scientists have previously projected that doubling the concentrations of carbon dioxide in the atmosphere may produce a catastrophic "greenhouse effect" by trapping heat on the Earth's surface. Even an average global temperature increase of as little as 2° C might seriously affect distribution of rainfall and could create deserts of agricultural areas. The Arctic is a critically important region because of climatic changes that could result from the melting or receding of the polar ice cap.

The LBL researchers have discovered that concentrations of soot in the air and snow in the remote Arctic region during winter and spring periods are surprisingly high, ap-

proaching those found in American cities. "While a filter might turn black in one day in heavily polluted New York City or in perhaps two or three days in a city with average pollution, it would turn black in the Arctic within a week," said Rosen. "In other words, the amount of soot is only three to four times less than that found in typical urban environments and about ten times less than that found in New York City."

The potential climatic impact of the airborne particulates cannot be conclusively assessed until further studies determine the depth of the sooty air layer, the distance it covers, and the optical properties of the particles.

Source: From A. Kopa, "Energy Newsletter," U.S. Department of Energy, July 1982.

---

of life cannot make use of nitrogen gas; however, very small percentages are converted to the nitrate form by nitrogen-fixing bacteria in the root nodules of plants and transformed into usable forms. Chapter 11 describes this process. Nitrate is also formed in small percentages from the atmosphere when lightning strikes occur.

Oxygen gas (O_2) actively combines with many solids, liquids, and gases in the process of oxidation. Rust on a car is a good example of oxygen and iron combining by oxidation to form iron oxides, or rust. Oxygen gas is involved in the combustion of fossil fuels, the decay of plant matter, weathering of rock minerals, and human and plant metabolism. Every runner knows how important it is in a long-distance race. A pilot flying above 3,000 meters (10,000 feet) knows the importance of oxygen in maintaining an alert mind. Plants and animals require oxygen to burn their fuels and convert them to energy. Solar energy could not be converted to chemical energy without it.

Oxygen currently is in a dynamic equilibrium. The amount in the atmosphere at any one time depends on the balance between plant producers and animal consumers. If either changes, the proportion of oxygen in the atmosphere will be altered. The oxygen fluctuation is similar to water level in a lake, which is a function of incoming water and outflow. If more water enters than leaves, the level rises. The current threat of deforestation on a global scale could upset this dynamic equilibrium.

Composition of the Upper Atmosphere—Heterosphere

Although we do not live in the **heterosphere,** it is a very important shield and filter of deadly solar rays. Until the twentieth century, the atmosphere beyond the clouds was simply "the heavens." Now, with high-altitude balloons, rockets, and spacecraft, our knowledge has greatly expanded.

The blend of ingredients of the atmosphere remains nearly constant up to 80 or 90 kilometers (50–56 miles). Only the densities change. Only 2 millionths of the total atmosphere remains

above this height. However, there are some important variations in minor constituents such as ozone, dust, and water vapor. Dust and water vapor virtually disappear while the ozone molecule becomes split by high-energy radiation from the Sun in a photochemical reaction. A molecule may also gain so much energy that it not only splits, but loses some electrons, or is ionized. In the heterosphere, this **ionization** process is a common occurrence. At 200 kilometers (120 miles), atomic oxygen is even more abundant than molecular oxygen and molecular nitrogen.

Molecular nitrogen is relatively stable, being largely unaffected by ultraviolet radiation; however, it does undergo ionization* in the heterosphere. In fact, a large portion of the atmosphere above 80 kilometers (50 miles) is charged with positive particles and free electrons (fig. 3.6).

This electrically charged zone in the heterosphere is known as the **ionosphere,** and has some unique characteristics. It is useful for radio communication because of its ability to reflect radio waves, making possible long-distance communication. Figure 3.7 shows how radio waves, which travel in straight lines, reflect off the ionosphere. The ionosphere is not static. It varies in height between day and night and can change in reflectivity when the Sun suddenly unleashes a burst of high-energy solar particles, mainly hydrogen nuclei, known as solar wind. A solar storm can last a few minutes, hours, or even weeks, producing fadeouts of communication. The fast-moving particles from the Sun, consisting of ionized hydrogen, helium, and electrons, are directed toward the magnetic poles by the lines of force of the Earth's magnetic field. The particles not only influence radio communication, but also produce the *aurora borealis* (northern lights) and *aurora australis* (southern lights). These colorful displays are produced by atomic collisions between solar wind and atoms in the upper atmosphere.

---

*NOTE: This is a process where atoms gain or lose electrons, causing them to take on positive or negative charges.

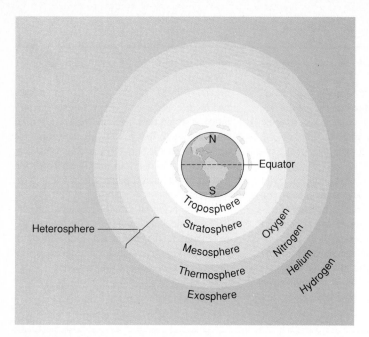

Figure 3.6

Gas composition of the upper atmosphere (heterosphere) is in a spherical shell pattern consisting of oxygen, nitrogen, helium, and hydrogen in ascending order.

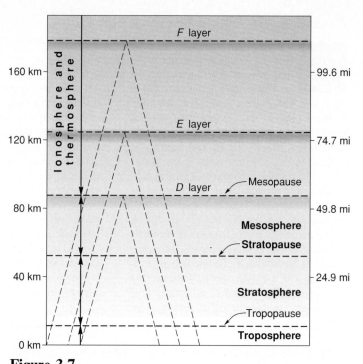

Figure 3.7

Radio waves reflect off the ionosphere, an electrically charged zone in the thermosphere. The D, E, and F layers are highly charged zones in the ionosphere.

THE MAGNETOSPHERE

The surface of the Earth not only has an envelope of atmosphere to shield us but also has a magnetic shield. The **magnetosphere** extends into the outer region of the atmosphere beyond the ionosphere, beginning at a height of 600 to 700 kilometers (350–400 miles). It extends out to 57,000 kilometers (36,000 miles) on the side facing the Sun, but fans out like the tail of a comet hundreds of thousands of kilometers on the side away from the Sun as a result of the solar wind's impact on the magnetic field (fig. 3.8). This magnetic field protects us from high-energy particle radiation from the Sun. Within the magnetosphere, two doughnut-shaped, highly ionized zones circle the Earth. They are known as the Van Allen radiation belts, named after James F. Van Allen, who predicted their existence. The inner belt is centered at 3,200 kilometers (2,000 miles) above the Earth while the outer belt is located at about 16,000 kilometers (10,000 miles). Because the astronauts were protected and spent so little time traveling through this region, there appear to be no harmful effects. However, this is a high-radiation zone and lethal without protection.

HEAT VERSUS TEMPERATURE IN THE ATMOSPHERE

Atoms and molecules are ionized and travel at very high speeds in the thermosphere, which accounts for its high temperatures. However, we must not be misled: it is not hot. At this point we need to draw a distinction between temperature and heat. **Temperature** is a measure of the mean velocity of the molecules and is indicated in degrees. **Heat** is one of the many forms of energy a substance possesses, and is measured in calories (fig. 3.9).

The amount of heat a gas, liquid, or solid possesses is determined by its temperature, the amount of matter present at that temperature, and its specific heat capacity. If two samples of a substance with the same mass and temperature are heated the same amount, then we can say they have received the same amount of heat. If, however, the atmosphere at sea level is heated to 1,400° C (2,500° F), the same temperature as the thermosphere, we can say that the atmosphere at sea level possesses far more energy because more matter is involved. Every substance also possesses its own **specific heat capacity** and has its own precise temperature reaction to a gain or loss of energy. The average rock, for instance, requires one-fifth the calories needed to raise the same mass of ocean water by one degree. If we double the mass of a substance, keeping the temperature constant, the amount of heat is also doubled in the substance. For example, a gallon of water at 70° F. will have twice the calories of a half gallon of water. Heat is a measure of the total energy of molecular motion, while temperature is a measure of the mean velocity of the molecules.

As one can see, then, temperatures can be misleading in dealing with heat or energy values if we are comparing different types of substances or different masses of the same substances.

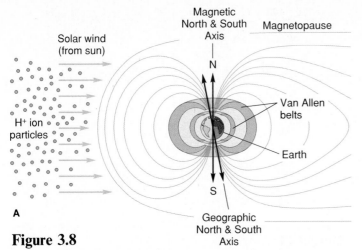

Figure 3.8

(A) Solar wind—high-speed ionized atoms of hydrogen and helium and electrons—is constantly bombarding the ionosphere, producing the auroras as well as interruptions in radio communications. The path of the solar wind is altered by the Earth's magnetic field, sending particles toward the magnetic poles. The Earth's magnetic field fans out like the tail of a comet. The magnetosphere extends beyond the outer atmosphere out to 57,000 kilometers on the side facing the Sun and fans out like the tail of a comet, hundreds of thousands of kilometers on the side away from the Sun. The Van Allen radiation belts are highly ionized particles trapped by the magnetosphere. There are two zones of high concentration located at 3,200 and 16,000 kilometers from the Earth. (B) Halley's comet shaped by solar wind, May 8, 1910.

B: Photo by The Observatories of the Carnegie Institution of Washington. Used by permission.

VERTICAL TEMPERATURE DISTRIBUTION IN THE ATMOSPHERE

As energy bombards the Earth in the form of particles and radiation, the atmosphere is energized in a complex pattern which provides us with another system for subdividing the atmosphere into a series of shells or temperature layers.

Since we have been focusing on the outer atmosphere, let us begin with this zone and proceed toward the surface. A spacecraft on first reaching the outer atmosphere would find extremes of temperatures. On the side facing the Sun, the temperature would be very high. On the shadow side, the temperature would be extremely low.

This outermost temperature zone is called the **exosphere.** At 500 kilometers (300 miles), the density of the atmosphere is virtually a vacuum. The exosphere is a transition zone from the atmosphere to the very thin interplanetary gas, consisting primarily of hydrogen and small amounts of helium. Temperatures have been registered at several thousand of degrees. Matter in this zone is primarily ionized. Here charged particles with very high velocity provide the first shield for the Earth from solar radiation (fig. 3.10).

B

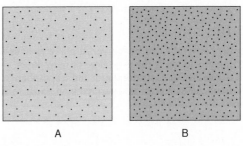

Figure 3.9

Temperature is the average energy of motion of the molecules. Temperature can be the same in both A and B. Heat is the total energy of the molecules. B has more molecules than A and has more energy or heat than A when both have the same temperature.

The **thermosphere** lies below 600 kilometers. There is no sharp temperature boundary between it and the exosphere, but its lower boundary, the **mesopause,** shows a marked minimum in temperature. The thermosphere is composed of highly ionized nitrogen and oxygen and free electrons. Here, too, high-energy x-rays and gamma rays are absorbed.

The mesopause forms a broad boundary not only between the thermosphere and mesosphere, but also between the homosphere and heterosphere. Here temperatures reach −95° C (−140° F), the lowest temperature in the atmosphere.

The **mesosphere** is the zone between 50 and 88 kilometers (30–55 miles) in which temperatures increase downward from their lowest point. The upper mesosphere represents a cool layer between two energy-absorbing layers; above is highly charged, heat-absorbing ionized oxygen and nitrogen in the thermosphere, and below, energy-absorbing ozone reaches a maximum concentration

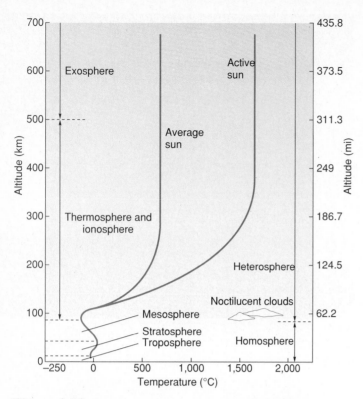

Figure 3.10

The atmosphere is subdivided by temperature layers or spherical shells. Starting at the top is the exosphere or outer layer, then the thermosphere, mesosphere, stratosphere, and troposphere, or sphere of weather. Each zone or layer is separated by a pause or change in temperature.

near the base of the mesosphere, the **stratopause.** In the early twilight hours, very high *noctilucent clouds* (70–90 kilometers or 45–55 miles), resembling thin cirrus, can sometimes be observed in the mesosphere.

The **stratosphere** is a zone beginning with a temperature decline from a peak at the stratopause, its upper boundary, then temperature remains fairly constant down to the **tropopause,** or top of the troposphere. The stratosphere is very dry, with less than 0.02 gram of water vapor per kilogram of air. (At sea level the average kilogram of air holds over 40 grams of water vapor.) As concentrations of ozone decrease below the stratopause, so does temperature, because the **ozone layer** absorbs ultraviolet energy. In the stratosphere the weather is always sunny and stable. Wind patterns are consistent with high velocities. Cirrus clouds occasionally form in the lower stratosphere, but most weather activity occurs below the tropopause.

The **troposphere,** or sphere of weather, is the atmospheric realm of life. It extends 16 to 19 kilometers (10–12 miles) over the equator and 8 kilometers (5 miles) in the polar regions. The region can be characterized by (1) uniform decreases in temperature with height at about 6.5° C per kilometer (3.5° F per 1,000 feet), (2) vertical and horizontal motion of the air due to uneven heating at the surface, and (3) a zone of storms, clouds, winds, precipitation, and other weather phenomena. Near the surface,

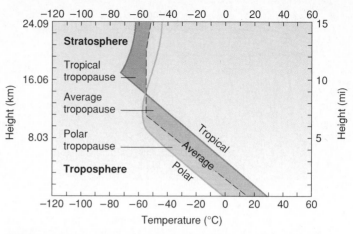

Figure 3.11

The troposphere occupies a narrow zone of temperature where life is possible. This is a region of uniform temperature change with height averaging 6.5° C/km (3 1/2° F/1,000 ft). It also is the scene of atmospheric stirring, storms, and precipitation. The tropopause height decreases with latitude.

temperatures may actually increase with height locally under certain conditions, but a temperature decline with increasing altitude holds true as an average condition on the global scale (fig. 3.11).

ATMOSPHERIC PRESSURE THROUGH THE ATMOSPHERE

The weight of the atmosphere exerts a substantial force on the surface of the Earth, about a million dynes per square centimeter (15 pounds per square inch) at sea level (fig. 3.12). Pressure can be defined as the weight or force on a given surface area. Because this force is spread over the surface of the Earth, the oceans, lakes, and rivers are prevented from evaporating or boiling away. The weight of the air acts like a lid on the waters of the Earth. If you place a glass of water in an airtight container and remove the air with a vacuum pump, soon you will notice small bubbles collecting on the sides of the container. After a few minutes of pumping, larger bubbles will form, and soon the water will begin to boil at room temperature. A thermometer placed in the water will show a steady drop in temperature as the water boils. This occurs because higher energy molecules leave first. When the temperature reaches the freezing point or a little below, the remaining water will freeze under low-pressure conditions. Freeze-dried foods are prepared by a similar method.

The events just described could occur to the waters of the Earth if atmospheric pressure were greatly reduced. Air pressure prevents the water molecule from breaking its liquid bonds and entering the vapor form. Remember, the fall in temperature results from a loss of high-energy molecules from the liquid, which confirms our definition of temperature as the average velocity of molecular motion of a substance. By removing the higher velocity, more energetic molecules, the average speed is lowered, thus reducing pressure until the temperature reaches the freezing point, and the remaining water turns to ice.

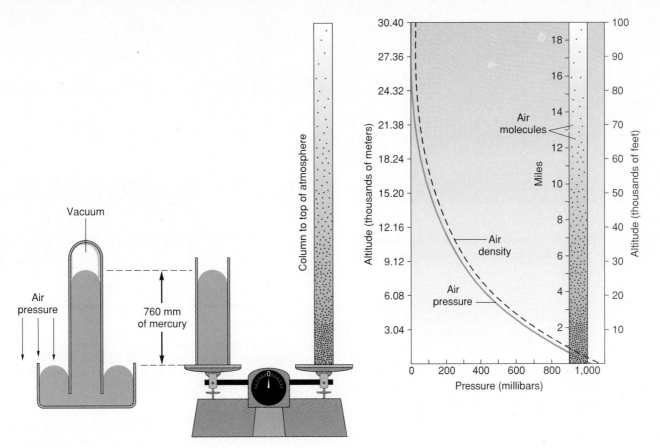

Figure 3.12

Atmospheric pressure is greatest at the bottom of the ocean of air, equaling 1,013 millibars (14.7 lb/in²). This pressure is not felt by your body because it is offset by equal and opposite pressure within. Transport planes are pressurized to offset changes in atmospheric pressure with increasing height. Pressure and density are interrelated: the greater the pressure of a gas, the greater will be the density or amount of a sample of air. When measuring its density at various altitudes, it is apparent that our atmosphere is very thin. Over 90 percent lies below 20 kilometers (12 miles) of altitude. At only 3 kilometers (2 miles) high, atmospheric density and pressure are two-thirds sea level values.

Atmospheric pressure has a profound influence not only on the waters but also on all living things. It is easy to take for granted sea level pressure until we travel in a nonpressurized small plane or hike up a mountain. Lowered pressure means decreased oxygen entering our lungs. At altitudes as low as 2,100 meters (7,000 feet), physiological problems such as headaches, nosebleeds, nausea, and fatigue can occur from lower air pressure and reduced oxygen. In some cases it can be fatal if the person does not quickly return to a lower altitude.

At sea level, atmospheric pressure averages 14.7 pounds per square inch, or 1,013.25 millibars in the metric system. A **millibar** is equivalent to 1,000 **dynes** per square centimeter, one dyne being defined as the force needed to accelerate a 1-gram mass to 1 centimeter per second in 1 second.

Consider the **barometer** as the bathroom scale of the atmosphere. The barometer was first invented by Evangelista Torricelli in 1643. He simply filled a long glass flask with mercury, placed his finger over the open end, and turned it over into a small container of mercury. The mercury did not run out of the tube because it was supported by air pressure (fig. 3.12). The height of the mercury is directly proportional to the air pressure. When the atmospheric pressure increases or decreases, the column rises or lowers accordingly. The column is 76 centimeters or 29.92 inches long at standard sea level pressure. A water barometer could be constructed, but the resulting water column is some 9.8 meters (32 feet) high.

As we will discover in chapter 7, "The Stormy Planet," air pressure measurements are essential in weather forecasting, although the actual pressure changes only slightly horizontally. Vertical changes are quite a different story. For every 275 meters (900 feet) of rise in altitude, pressure lowers by a predictable 1/30th of its value at that given height. By examining a series of locations at different altitudes, it becomes apparent how rapidly pressures fall with increasing altitude (fig. 3.13).

Most of the atmosphere is held close to the Earth's surface by gravity. This causes the lower atmosphere to be more compact or high in density. Note that 50 percent of the Earth's atmospheric mass is below 5 kilometers (3.1 miles), and over 90 percent lies below 20 kilometers (12 miles) from the surface. In our deep ocean of air, our own habitat is a very shallow zone located within 5 kilometers of the surface, and most of humankind lives within 300 meters (1,000 feet) of sea level.

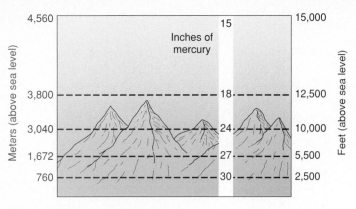

Figure 3.13

Pressure changes with elevation are predictable. If elevation changes 275 meters (900 feet), atmospheric pressure will either rise or fall 1/30th of its value.

THE GAS LAWS AND KINETIC-MOLECULAR THEORY

Matter in the universe exists in three fundamental states. The bulk of the atmosphere is in the gaseous form, consisting of a mixture or blend in the lower atmosphere or homosphere as we have observed. In solids and liquids, with some exception, heating causes expansion while cooling produces contraction as molecular activity decreases. The exact amounts of contraction or expansion depend on the temperature change and the specific heat capacity of the substance.*

The reaction of gases to changes in temperature and pressure, however, can be well described through a few simple formulas expressing laws discovered centuries ago by an English physicist, Robert Boyle, and two French physicists, Jacques Alexandre Cesar Charles and Joseph Gay-Lussac. The discovery of these laws paved the way to a fundamental understanding, at the molecular level, of the behavior of gases. Boyle's law relates pressure and volume while Charles' law relates volume and temperature in all gases.

The Gas Laws

Boyle's law, simply stated, says that at constant temperature the volume of a given amount of gas is inversely proportional to the pressure applied to it. This means when the pressure on a gas is increased, its volume decreases and vice versa (fig. 3.14). Boyle demonstrated that doubling the pressure on a specific quantity of gas, keeping the temperature constant, reduces its volume to half of its original value. This law explains why the troposphere contains most of the atmosphere. Gravity has compressed the atmosphere nearest sea level, producing a greater concentration or density nearest the surface. The weight of the atmosphere has caused the compression and highest density at sea level.

The influence of temperature on a volume of gas was formulated by Charles about 1787. He stated that at constant pres-

*NOTE: The **specific heat capacity** of a substance is the amount of energy necessary to change one gram one degree centigrade.

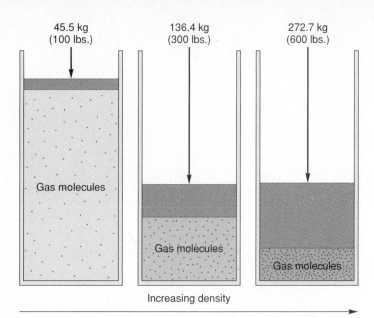

Figure 3.14

According to Boyle's law, the volume of a given amount of gas is inversely proportional to the pressure applied as long as the temperature remains constant. If the pressure doubles, the volume will be reduced to one-half and density will double.

sure the volume of any gas is directly proportional to the absolute temperature (measured in Kelvins or "degrees Kelvin," which equals Celsius temperature plus 273). If the absolute temperature doubles, then the volume doubles. (Note: Absolute temperature is measured in degrees Kelvin. To change a Celsius temperature to Kelvin, add 273.) This relationship is commonly called **Charles' law** (fig. 3.15).

A simple experiment showing this principle is easily done by taking a small balloon attached to a bottle which is then submerged, first in hot water, then in ice water. In the hot water, the gas expands in the balloon as the gas molecules gain more energy and create a greater force on the skin of the balloon. The gas molecules lose energy in the ice water, resulting in the balloon's collapse. In both cases, the number of gas molecules does not change, but the volume does in direct proportion to changes in temperature. This concept, as discussed in more detail in chapter 5, "The Windy Planet," helps to explain why the atmosphere is so dynamic. As the Earth's surface warms and cools, the bottom layer of air expands and contracts. If it expands, the air tends to rise because it has a lower density or more buoyancy. In contrast, cold air contracts and increases in density. Cold air tends to puddle in low places such as valleys during the night as the Earth's surface cools the air in contact with it (fig. 3.16).

Adiabatic Heating and Cooling

When the air expands, it performs some work by spending some of its internal molecular energy of motion. This energy consumption is indicated by a lower temperature. Temperature changes at a very predictable rate known as the **adiabatic lapse rate.** If this air is "dry," that is, not saturated with water vapor, the temperature will lower at a rate of 1° C per 100 meters, or 5.5° F per

Automation for Climatic Observation

The Automated Weather Observing System (AWOS) provides weather and climatic data 24 hours a day, updated every minute, near the *study area* described in chapter 1. The system replaced the human observer with a variety of weather instruments (fig. 1).

The weather sensors feed data to a computer, which collects and analyzes the information automatically. A terminal display plus printout, as well as a voice report, are available to the public by a phone call. Table 1 shows parameters, range, and resolution of the system.

Table 1
Weather Instruments

Instrumentation	Characteristics
Sky vane	Direction 0° to 360° Speed 0 to 200 mph
Precision barometer	Range: 595 to 1,100 millibars
Temperature probe	Range: −55° to 55° C
Dew cell	Range: −35° to 35° C (measures dewpoint temperature)
Visibility sensor	Range: < 0.1 to 20 miles using infrared reflection
Range gauge	Range: 0.01 inch accuracy
Laser ceilometer	Measures cloud height by measuring the time it takes laser light to be reflected by the cloud base back to the transceiver unit. Range: 30 to 12,500 feet

Figure 1

Automated (AWOS) weather station.

1,000 feet of altitude. Upon descent, it will increase by the same amount. This process is described by the **dry adiabatic lapse rate** (fig. 3.17).

The **moist** or **wet adiabatic lapse rate** occurs when the air is completely saturated during ascent. As discussed in more detail in chapter 6, water vapor releases energy when it condenses to a liquid called **latent heat of condensation.** This heat that is released during condensation raises the temperature of the moist parcel of air, thus rising moist air will be considerably warmer than a rising parcel of dry air (fig. 3.18). Table 3.2 gives the values of the moist adiabatic lapse rate at various pressures and temperatures. Note that at colder air temperatures, rates are closer to the dry adiabatic lapse rates because cold air contains less moisture to condense and release heat.

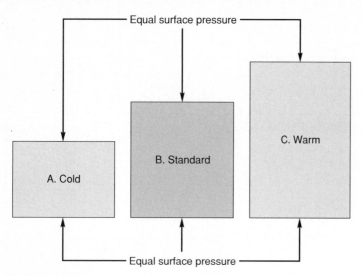

Equal surface pressure

Equal surface pressure

A. Cold

B. Standard

C. Warm

A

B

Figure 3.15

Three columns of air showing how decrease of pressure with height varies with temperature. Left column is colder than average; center column is at average temperature; and right column is warmer than average. Pressure is equal at the tops of the columns. The colder the air, the more rapid is the decrease of pressure with height. According to the gas law developed by Charles, a gas will double in volume if the absolute temperature is doubled, providing the pressure remains constant. The three photographs, looking north over the San Gabriel Valley, California, were taken at (A) 7 A.M., (B) 10 A.M., and (C) 1 P.M. on September 29, 1971. As the atmosphere warms, the air expands, causing the lower layer to expand and become less dense. The temperature at 7 A.M. in the lower layer of poor visibility was 10° C (50° F) while the air temperature above the polluted air was 21° C (70° F). Heating not only causes expansion but mixing and convection as well.

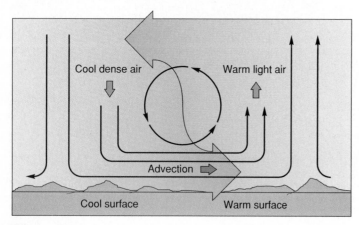

Cool dense air

Warm light air

Advection

Cool surface

Warm surface

Figure 3.16

Cold air contracts and increases in density while warm air expands. These changes in temperature and volume produce air circulation. Warmer air rises and cold air settles in the low spots and valleys. The Earth's surface is the chief regulator of atmospheric temperature in the homosphere.

C

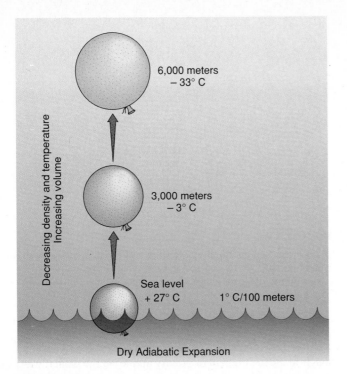

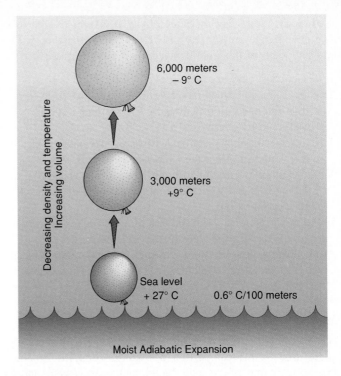

Figure 3.17

The dry adiabatic lapse rate, 1° C/100 m (5.5° F/1,000 ft.), gives the temperature change resulting from air expansion or contraction under nonsaturated conditions. Note how volume, pressure, and temperature change in the parcel of air in the balloon as it ascends from sea level to 6,000 meters. The air cools because of expansion.

Figure 3.18

The moist or wet adiabatic lapse rate (0.6° C/100 m or 3.3° F/1,000 ft.) is less than the dry adiabatic lapse rate because energy is released from the condensation process, warming the air. Therefore the temperature of a parcel of air in the saturated state will fall at a lower lapse rate than a nonsaturated parcel.

Air displaced vertically also experiences pressure changes as well as volume and temperature changes. Compression produces heating, and expansion produces cooling. These are continuing processes in the atmosphere, and moisture modifies the effects of heating and cooling on a parcel of air in motion. The gas laws developed by Boyle and Charles apply to the atmosphere as well as a gas in an enclosed system.

The next time you release the air from a tire or pump up your bicycle tire, note the temperature changes. As you release the air, the air in the tire cools as energy is expended in expansion. Pumping a tire or inflating a ball produces heating in the pump, ball, and tire because we have added heat energy by compressing the gas.

Gases, the least compact of the three states of matter, also represent the most energetic of the three states. Water is a good example. In the solid state (ice), it is a rigid structure, lowest in energy. When it absorbs heat, melting occurs, changing the solid to a liquid, which possesses more energy and mobility. When water absorbs more heat, it becomes a gas. Gases represent the highest level of molecular energy with greatest volume, velocity, and mobility or kinetic energy.

Eighteen grams of water have a volume of 18 milliliters at 4° C. The same amount of water vapor will occupy about 22,400 milliliters in the gaseous state. This is a 1,200-fold increase in

Table 3.2

The Moist Adiabatic Lapse Rate for Selected Temperatures, in ° C per 1,000 Meters

Pressure (mb)	Temperature (° C)				
	−40	*−20*	*0*	*20*	*40*
1,000	9.5	8.6	6.4	4.3	3.0
800	9.4	8.3	6.0	3.9	2.8
600	9.3	7.9	5.4	3.5	2.6
400	9.1	7.3	4.6	3.0	2.4
200	8.6	6.0	3.4	2.5	2.0

volume. We can conclude from this difference in volume that (1) gas molecules are farther apart than in liquids and solids, (2) gas molecules are capable of being significantly compressed into smaller spaces, (3) the space occupied by gas molecules is rather empty of matter.

The behavior of gases as we have described has been summarized in a theory known as the **kinetic molecular theory** based on the motions of molecules. Only at extremely high temperature and high pressure do gases deviate from this theory. The principal points of the kinetic molecular theory are these:

1. Gases consist of tiny molecules.
2. The distance between molecules is large compared to the size of the molecules themselves. The volume of a gas consists mostly of empty space.
3. Gas molecules have negligible attraction for each other under standard atmospheric densities and pressures.
4. Gas molecules move in straight lines in all directions, colliding frequently with each other and with the walls of their container.
5. No energy is lost by the collision of a molecule with another molecule or with the walls of the container.
6. The average kinetic energy for molecules is the same for all gases at the same temperature, and its value increases with temperature.

Keeping these key points in mind, let us briefly review the facts supporting the theory. Point number 1 is based on the size of atoms and molecules scientists have been able to measure. Points 2 and 3 are based on the comparison of volumes occupied by equal amounts of matter in the solid, liquid, and gaseous state, and on the observable fact that a gas expands to fill the size of any container. Point 4 is supported by the fact that gases exert a pressure, and continue to expand into larger containers as well as diffuse from one vessel into another.

Diffusion is shown by the ability of two or more gases to spread out into one another when they are mixed. The diffusion principle also supports the assumption that gas molecules have no significant attraction for each other. To illustrate this principle, figure 3.19 shows two containers, one containing air and the other ammonia vapor. When the stopcock between the containers is opened, the two gases will diffuse into each other, and after a while, both containers will contain ammonia and air.

To illustrate points 4, 5, and 6 that gases exert a force by collision on the walls of a container and that this force changes with temperature, one can place pressure gauges at different locations on an enclosed volume of air. Billions of molecules continue to move about, colliding with each other and with the walls of the container, and yet pressure or temperature changes do not occur. Therefore, we can assume that collision produces no energy loss in the gas system. Energy will be transferred between molecules in collisions, but the total energy will remain constant. The average **kinetic energy** or energy of motion for molecules remains constant. It can be expressed by the formula $KE = 1/2\ MV^2$ where M is the mass of the molecule, V the velocity, and KE is the energy of motion. This means that molecules of different gases, although they have the same energy at constant temperature, do not have the same velocity. Remember, each type of atom has different mass, therefore, to have the KE remain constant, the equation can only be valid if velocities increase as masses decrease. Hydrogen molecules, then, have 4 times the velocity of oxygen molecules because oxygen molecules are 16 times as massive. This fact explains why hydrogen is not found in the lower atmosphere. At sea-level temperatures and pressure, hydrogen molecules are traveling fast enough to escape the pull of gravity and leave the planet, whereas oxygen is kept by gravity near the surface.

Thus an understanding of our atmosphere at the molecular level, expressed by the gas laws and the kinetic molecular theory of gases, is of primary importance in understanding the large-scale weather and climatic events that are considered in the following chapters.

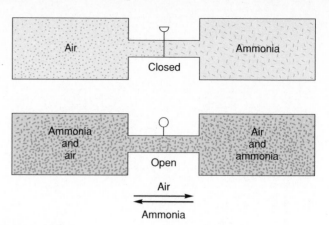

Figure 3.19

Diffusion can be illustrated by allowing two different gases, such as air and ammonia, to be exposed to each other. In time they will become totally blended, and both containers will contain a mixture of the gases.

SUMMARY

The atmosphere is a primary life support system for the biosphere. The two key gases are oxygen and carbon dioxide because they perform important life-giving functions.

It is believed that the Earth's original atmosphere was more like that of Jupiter or Saturn with large percentages of hydrogen, helium, ammonia, and methane gases. Early changes in the Earth's environment caused a replacement of these gases with nitrogen, oxygen, carbon dioxide, and water vapor at percentages much like they are today.

The atmosphere can be divided in two major zones based on temperature and gas composition. The lower zone, below 80 kilometers (50 miles), is the **homosphere.** The homosphere consists of a mixture of nitrogen (78%), oxygen (21%), argon (0.93%), carbon dioxide (0.03%), and traces of many other gases. Water vapor varies from near zero to nearly 5 percent in the more humid areas.

Currently scientists are concerned that human activity is causing changes in the composition of some of the ingredients such as **ozone,** which shields us from ultraviolet radiation. Carbon dioxide concentration also is rising, which could result in a changed global climate by reducing the energy leaving the Earth through the **greenhouse effect.** There is also concern about increased solid particles in the air. They affect the lungs, the atmospheric visibility, and possibly the reflectivity of the planet, which may result in a global cooling trend. It is unclear what the global temperature trend will be.

The upper atmosphere or **heterosphere** (above 80 kilometers or 50 miles), composed of molecular and ionized gases, is our protective shield and filter. A large portion is in a charged state of positive particles and free electrons in the area known as the **ionosphere.** Because of its electrical properties, the ionosphere is capable of reflecting radio waves, making possible long-distance global radio communication. The ionosphere is also the zone of colorful displays known as the **aurora borealis** (northern lights), and the **aurora australis** (southern lights).

The magnetosphere is the outermost region of the Earth's environment beyond the exosphere. It extends out to 57,000 kilometers on the side facing the Sun and fans out like the tail of a comet hundreds of thousands of kilometers on the side away from the Sun. This magnetic zone shields us from high-energy particles from the Sun. The Van Allen belts are two zones where many high-energy particles are held by the Earth's magnetic field.

Vertical temperature patterns show several maximums and minimums. In the thermosphere, temperatures reach several thousand degrees and then fall to their lowest levels at the mesopause, rise to the **stratopause,** then decline in the **stratosphere.** Temperatures then rise through the troposphere at a consistent rate of 1° C/100 m or 3.5° F/1,000 ft from the tropopause to the surface.

Atmospheric pressure or weight of the air creates a force of 1,013 millibars (15 pounds per square inch) at sea level. This weight acts like a lid on the oceans, lakes, and rivers and prevents them from boiling away. Most of the atmosphere lies very close to the surface. Over 90 percent lies below 20 kilometers (12 miles). For every 275 meter or 900 feet rise in altitude, pressure drops by 1/30th of its value.

The gas laws and **kinetic molecular theory** explain in a predictable fashion how gases respond to changes in temperature, pressure, and volume. **Boyle's law** explains pressure-volume relationships while **Charles' law** defines the relationship between volume and temperature. Kinetic molecular theory formulates and summarizes the primary facts about the behavior of gases at the atomic level. Thus weather on a global scale can be better understood, as presented in the next chapters.

ILLUSTRATED STUDY QUESTIONS

1. Explain the statement, "We live at the bottom of an ocean of air" (see fig. 3.1, p. 52).
2. Name the major constituents of the atmosphere and give one important characteristic of each (see table 3.1 and figs. 3.3 and 3.4, pp. 51, 53, 55).
3. In which atmospheric zone does the planet's weather and climate primarily occur (see fig. 3.11, p. 60)?
4. Describe the lapse rate above the tropopause for the next 16 kilometers (10 miles) (see fig. 3.11, p. 60).
5. What does the barometer measure (see fig. 3.12, p. 61)?
6. How does the ozone layer protect life (see box 3.1, p. 54)?
7. Describe the Van Allen radiation belt in terms of shape and chemistry (see fig. 3.8, p. 59).

Solar Energy and the Earth's Response

4

Objectives

After completing this chapter, you will be able to:

1. Describe the **electromagnetic spectrum.**
2. Explain two theories of electromagnetic energy.
3. Describe energy transfer by **radiation, conduction, convection,** and **evapotranspiration.**
4. Explain the factors that determine the distribution of solar energy over the Earth's surface.
5. Explain the role of the atmosphere in **scattering, filtering,** and **reflecting insolation.**
6. Explain the influence of surface features on the daily pattern of temperatures.
7. Explain the factors that contribute to the **nocturnal inversion.**
8. Explain the three major characteristics that cause the oceans to moderate global temperatures.
9. Describe global patterns of temperature.

Solar energy warms and energizes the planet.
© David Muench Photography

Figure 4.1

The Sun's energy is the source of energy and driving force for all natural systems. Conversion processes transform this energy into global circulation systems of the oceans and atmosphere.

Photo by NASA

The Sun's energy is the fundamental driving force of nearly all natural systems. Once solar energy reaches the Earth, a series of conversion processes transform this energy into large-scale atmospheric and oceanic circulation, chemical changes, and biological activity, and some is stored as fossil fuel (fig. 4.1). This energy is also used to evaporate surface waters and moisture from plants. Green plants use solar energy to manufacture sugars and release some heat in the process. The fossil fuels—coal, petroleum, and natural gas—are stored forms of solar energy captured by green plants during earlier geologic periods.

Not all places on the Earth receive the same amount of energy on any given day, month, and year. The amount really depends on the location, the time and season, and conditions in the atmosphere.

THE SUN AND ITS ENERGY

The Sun, our nearest star, has been consuming hydrogen since it condensed into a stellar form, about 5 billion years ago. When the nuclear burning began, helium started to accumulate as the "ash" in its interior core. Today, after steady burning, the concentrations of hydrogen at the Sun's center have been depleted by half and helium has more than doubled. In the lifetime of the Sun, only about 5 percent of its fuel supply has been converted to helium. Astronomers project the Sun to be a middle-aged star with about 6 billion more years left.

Although the Sun appears to be about the size of the Moon, it is many times larger. In fact, you could line up about 108 Earths side by side across the Sun's diameter of 1,382,000 kilometers (864,000 miles). It has a volume 1.3 million times that of the Earth. In fact, 99.86 percent of all the mass in the solar system is in the Sun (fig. 4.2).

Because of the Sun's mass, the Earth is held in a fixed orbit by gravitational energy and energized by solar luminous power. If the Sun were smaller or larger, warmer or colder, closer or farther away, conditions as we experience them today would be quite different. Let us examine the energy that leaves the Sun and energizes the Earth.

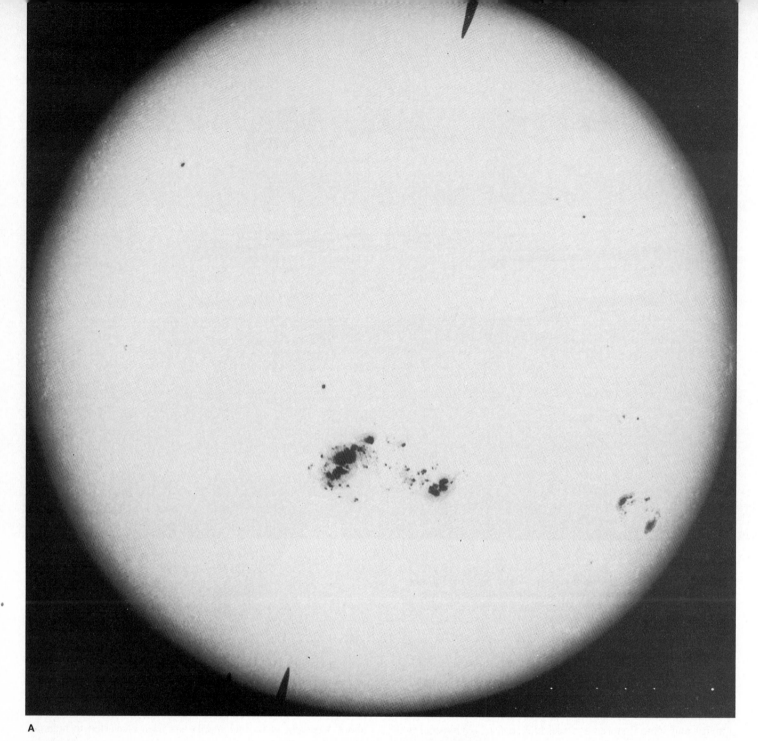

A

Figure 4.2

(A) The Sun holds 99.86 percent of the
mass of the solar system. Over one million
Earths could fit inside its volume. (B) The
largest sunspot, in lower center, equals five
Earths in diameter.

Photo by The Observatories of the Carnegie
Institution of Washington. Used by permission.

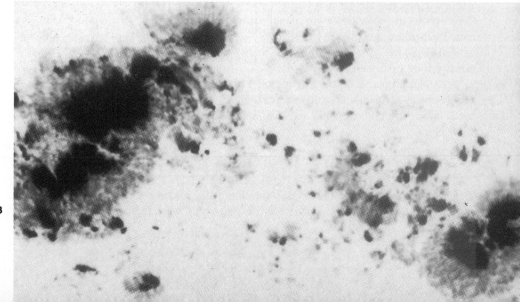

B

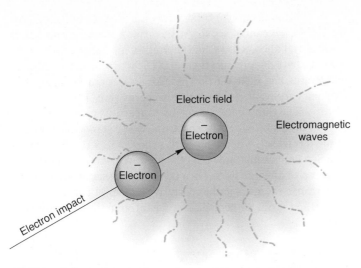

Figure 4.3

Electromagnetic energy has both electrical and magnetic properties. Surrounding each electron is an electric field similar to a magnetic field around a magnetic object. If the electrons are excited or vibrated, perhaps by collisions with other electrons, these vibrations are sent outward in all directions as electromagnetic energy. The frequency of vibration determines the nature of the wave form and its wavelength.

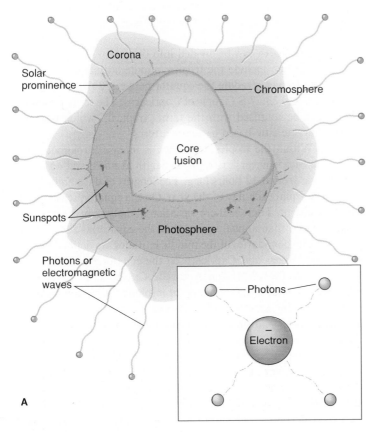

Figure 4.4

(A) Sometimes electromagnetic energy exhibits characteristics of particles, called *photons.* Gravitational fields attract these particles as if they possessed mass. Light behaves as particles or as *electromagnetic waves,* but never both at the same time. The sun is photographed in violet light. (B) The jetlike structures are solar prominences rising great distances above the solar surface.

THE NATURE OF LIGHT

Most of the information we possess about the Sun, Moon, stars, and planets has come to us in the form of **electromagnetic energy,** primarily as luminous energy or light. According to James Clerk Maxwell, the nineteenth-century English physicist, this form of energy can be described as vibrations of electrical force capable of traveling through a vacuum.

What is a vibrating electrical force? If you take some rabbit fur and rub a rubber rod on it, electrons will collect on the rod and build up an electric charge on the surface that is capable of producing a spark and attracting small objects with the opposite charge. Surrounding the rubber rod is an electric field or field of electric force, similar to a field of magnetic force around a magnetized object. If this field is vibrated by vibrating the electrons on the rod, the vibrations will be transmitted outward in all directions (fig. 4.3). Electrons impacting other electrons can cause this vibration. It is the same as if you and a friend held a rope tightly between you, and one of you gave it a snap or series of jiggles. The result is a train of waves moving from the person transmitting the motion to the receiver. The hand receiving the wave motion vibrates at the same rate or frequency.

This is exactly how light is transmitted from the Sun's three atmospheric zones: *photosphere, chromosphere,* and *corona.* The vibrating electrons send out waves into space in the visible and also nonvisible frequency range. When these waves reach your eye or some other material, they stimulate the electrons in the atoms of your retina or some other substance to vibrate at the same rate.

Sometimes, however, electromagnetic energy exhibits properties more closely resembling streams of particles called **photons.** We can observe gravitational fields pulling on these particles as if they possessed mass. Light exhibits either a wave or a particle nature in an experimental setting, but never both at one time. The wave theory and particle concept complement one another and provide information in a predictable manner about the source of light and the medium it travels through or is reflected from (fig. 4.4).

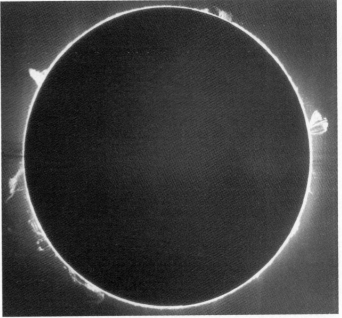

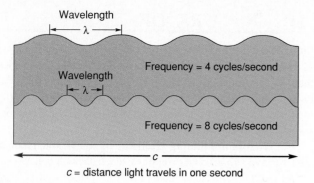

Wavelength
λ

Wavelength
λ

Frequency = 4 cycles/second

Frequency = 8 cycles/second

c

c = distance light travels in one second

Figure 4.5

Electromagnetic energy travels at the rate of 299,783 km/s (186,283 mi/s) in a vacuum. Its wavelength is inversely related to its frequency.

The wave theory of light states that energy from the Sun travels in wave form. Wavelength is the distance between successive crests in a wave train. These vibrating electric fields produce a vibrating magnetic field that is interwoven inseparably with the electric waves. For this reason, the energy is called **electromagnetic radiation** instead of just electric waves. Light energy, then, possesses electric and magnetic properties.

All electromagnetic waves travel at the same velocity in the same medium. In the vacuum of space, the velocity of light is 299,793 kilometers (186,283 miles) per second. The various forms of electromagnetic energy differ from one another in terms of wavelength (fig. 4.5). Each wavelength is like a key that unlocks or stimulates different elements in the environment. Your eye is sensitive to visible wavelengths, while your skin feels longer wavelength, invisible infrared radiation.

THE SUN'S ELECTROMAGNETIC SPECTRUM

The **electromagnetic spectrum** is a graphic representation of electromagnetic energy in all wavelengths. The electromagnetic spectrum of the Sun can be divided into two primary categories by wavelength or by **frequencies.** Think of a frequency as representing the number of wave crests passing a reference point every second, measured in hertz (Hz) or cycles per second. High-frequency energy has short wavelengths and low-frequency energy has long wavelengths. The two divisions of the electromagnetic spectrum are **longwave** and **shortwave radiation.** Visible light falls in the shortwave range. Our eyes are sensitive to wavelengths between 0.0004 and 0.0007 millimeter, or the range of visible light. When wavelengths of 0.0007 mm strike the retina of your eye, you see red. At the other end of the visible region, you see violet when wavelengths of 0.0004 mm are received. Hence, the rainbow of colors or color spectrum is determined by the exact wavelength of visible light (fig. 4.6).

If the electromagnetic energy is a shorter wavelength than the visible, we cannot see it. At this end of the spectrum are **ultraviolet** radiation, **x-rays,** and shortest of all, **gamma rays.** Longwave radiation—greater than 0.0007 mm—includes **infrared, microwaves,** and **radio waves.** These long waves are considered low-frequency forms of electromagnetic energy (box 4.1).

The Sun's energy happens to reach its greatest intensity in the visible portion of the spectrum. Some stars emit primarily in the ultraviolet, while others are strongest in x-ray frequencies. The Sun emits a mixture of light of all wavelengths in a blend that we perceive as white light. On a percentage basis, about 9 percent of the Sun's energy is in the shortwave region, 41 percent is in the visible, and about 50 percent is in the infrared through radio portion.

INSOLATION ON THE EARTH

The Sun's energy reaches Earth at a value of 2 calories per square centimeter per minute. One calorie per square centimeter is also the definition of one **langley.** So the Sun has a **solar constant** at the outer edge of the Earth's atmosphere of 2 langleys per minute.

Insolation denotes this energy received from the Sun, whereas radiation is energy transmitted by the Sun. Let's focus now on the one two-billionth of the Sun's total radiant energy we receive, or insolation. The insolation on any given day is a function of latitude, the time of day and season, and the clarity of the atmosphere.

The Earth can be divided into three latitudinal zones of insolation, which include the tropics, middle latitudes, and polar zones. The **tropics** straddle the equator between 23½° N (Tropic of Cancer) and 23½° S (Tropic of Capricorn). The term "tropics" should not be confused with the term "tropical," which is a climatic characteristic based on temperatures.

Within the tropics, insolation is intense throughout the year. The Sun is never far from the zenith at noon, and daylight periods are around 12 hours long. The lengths of days and nights are nearly equal each day of the year. On the equator, seasons are virtually absent. There is more variation in daily temperature, although small, than annual variation in the equatorial zone.

The boundaries of the middle latitudes are defined by the Tropics of Cancer and Capricorn on the equatorial side and the Arctic Circle (66½° N) and Antarctic Circle (66½° S) toward the poles. Here seasonal variation in insolation is caused by both variations in the length in daylight and angle of the Sun's rays. Insolation at 40° latitude varies from a little over 300 langleys in January to over 900 langleys in June (fig. 4.7).

Poleward of the Arctic and Antarctic Circles lies the land of extremes in variation in insolation and length of daylight. At the North and South Poles, one will experience six months of daylight then six months of darkness lasting from equinox to equinox period. Note that at 90° north and south latitudes, langleys per day range from zero in January to over 1,000 in June, which represents the widest possible annual range in insolation.

If we compare the insolation during the summers and winters in each hemisphere, the Southern Hemisphere receives slightly more langleys during the summer, and slightly less during

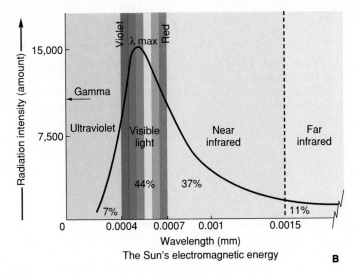

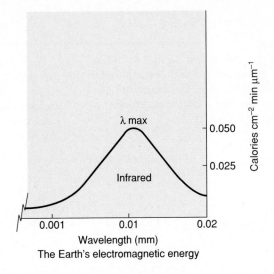

A The Sun's electromagnetic energy

B The Earth's electromagnetic energy

Figure 4.6

(A) The electromagnetic spectrum can be divided into gamma, ultraviolet, visible, near infrared, far infrared, microwave, short radio, and long radio waves. Vision is possible because our eyes are sensitive to wavelengths between 0.0004 mm and 0.0007 mm. All waves have the same velocity in a vacuum but vary in other aspects related to frequency. (B) The Earth emits energy only in the infrared, peaking at 0.01 mm wavelength and 0.05 calorie cm^{-2} min μm^{-1} of energy.

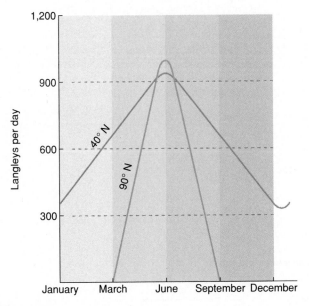

Figure 4.7

Insolation at 40° N latitude varies from over 300 langleys per day in January to over 900 langleys per day in June. The variation is caused by both variation in the length of day and insolation from season to season. The widest annual range occurs at the poles. In this graph the north pole varies by 1,000 langleys between June and March, and June and September.

the winter, than the Northern Hemisphere. This can be explained by considering our orbital pattern. The elliptical orbit of the Earth takes us some 5 million kilometers (3 million miles) closer to the Sun when the Southern Hemisphere is in its summer season and the same distance further away during the winter season.

Solar Energy: Scattered, Reflected, and Filtered

Solar energy penetrating the troposphere encounters atmospheric **scatterers, reflectors,** and **filters.** Each atmospheric constituent influences specific wavelengths or frequencies of the electromagnetic spectrum.

Scattering

Some light is scattered in all directions by dust particles, water droplets, and ice crystals. This scattered energy is sent both back into space and toward the Earth.

Have you ever wondered why the sky is blue and sunsets are red? On the Moon, the astronauts saw only a black sky and a brilliant white Sun because the Moon has no atmosphere. Air molecules scatter over 50 percent of the blue wavelengths while nearly all of the red waves are transmitted through the atmosphere. Wherever you look in the sky, the air there is scattering blue light toward you (and in all other directions). Thus the sky appears blue.

When the Sun is near the horizon, only the reds can effectively penetrate this longer denser path of atmosphere. Dust, smoke, haze, and smog can produce very red sunsets because of scattering of the shorter blue and violet rays, leaving primarily oranges and reds.

Reflection

The next time you look at yourself in the mirror, take a closer look. You're not what you see. Most mirrors only reflect about 80 percent of the light they receive, so you are only seeing about four-fifths of your true image brightness. The proportion of reflected energy is called the **albedo** and is expressed as a decimal or a percent; that is, 0.80 or 80 percent reflectance.

View from the Top

"A view from the top" or 20 kilometers (65,000 feet) provides NASA high-altitude aircraft photographic and digital data in support of a variety of research projects.

The ER-2 high-altitude aircraft fly above the troposphere in an undisturbed atmosphere, collecting data on the Earth's surface. Thousands of square kilometers can be covered in just a few hours flying at a rate of 11.1 kilometers (6.9 miles) per minute.

Most recent photography has been in the color infrared film range because it penetrates haze and is especially sensitive to infrared reflected by vegetation and other surface features. With multiple cameras, natural color, black and white, and color infrared film can be

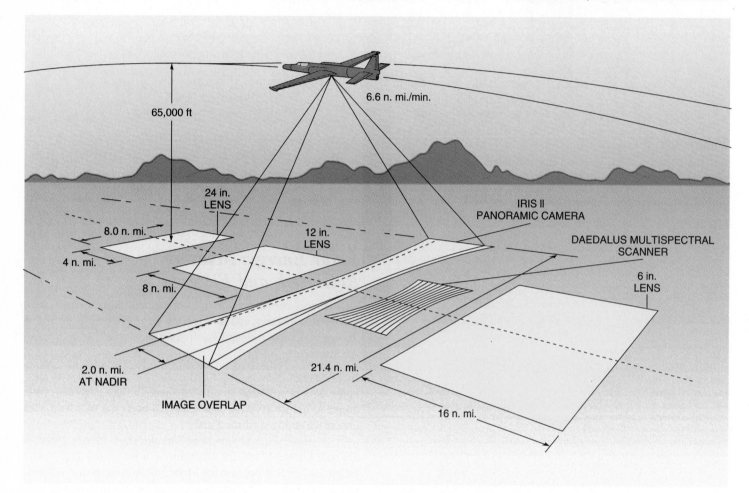

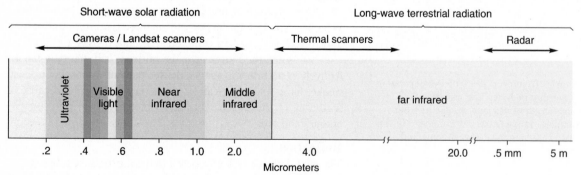

Figure 1

Sensors and coverage of the ER-2 aircraft. The ER-2 is designed for sustained flight at very high altitudes and at consistent speeds, allowing collection of virtually distortion-free data. On a single flight of a few hours, they can collect data on thousands of square kilometers of land. Sensitivities of the ER-2's are shown along with wavelengths of the electromagnetic spectrum, in micrometers. One micrometer equals 1 one-millionth of a meter.

used simultaneously over a given area to provide multiple images.

In addition to these traditional tools, a multispectral scanner is employed to provide data in a variety of wavelengths (fig. 1). The Daedalus Thematic Mapper Simulator (TMS) or multispectral scanner senses the Earth in 11 wavelengths ranging from the visible to infrared. The TMS measures both reflected sun-light and heat energy from the Earth's surface. It then converts this information to a digital format. Computers then reduce the data to a variety of maps that portray the subtle variations of the Earth's surface (fig. 2).

Applications include water surface temperature mapping, land cover classification, and land-use mapping. The advantage here is in the wider spectral sensitivity not provided by color infrared or visible photography.

NASA's high flyers also carry atmosphere sensors or "sniffers" to detect and measure the concentrations of chemicals in the troposphere and stratosphere. These include ozone, methane, freon, ammonia, and water vapor. Mapping of the ozone hole over Antarctica was made possible by this instrumentation.

Source: NASA.

A

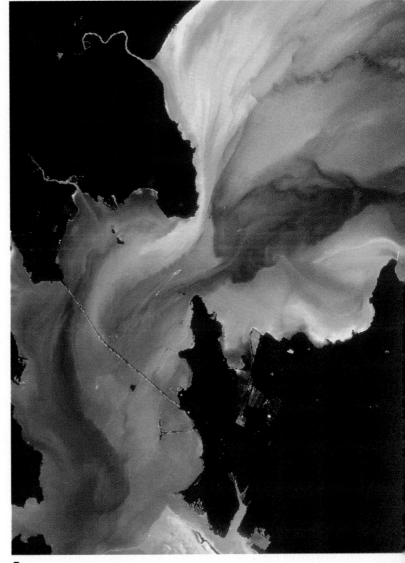

B

Figure 2

These two images illustrate multiple applications of a single set of data. The water-enhancement image (A) uses three TMS channels, or frequencies, patterns, and distribution of suspended sediment as clear ocean water mixes with outfall of the Sacramento River. The color infrared image (B) uses visible and infrared waveband data to point up vegetation and land-use features. Color infrared images are useful for distinguishing between basic land cover types (vegetation, water, bare earth, and urban settlement) as well as for distinguishing type, maturity, or condition of specified areas of vegetation.

Photos by NASA.

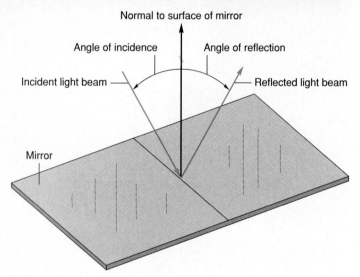

Figure 4.8

Reflection is not selective: all wavelengths are altered in direction according to the *law of reflectivity*. Clouds, dust, sand, snow, and ice all contribute to the reflectivity of the Earth's atmosphere. The law of reflection states that the angle of incidence equals the angle of reflection. The angle is measured from the normal to the surface, or line perpendicular to it.

Photo by Gary Bowman.

Reflection is unlike scattering in that there is no wavelength selectivity; the entire light beam is turned back into space by particles of dust, water, ice, or salt (fig. 4.8). Clouds are the chief reflectors, aside from zones of heavy air pollution or regions of dust storms and forest fires. Volcanic eruptions can also increase the Earth's reflectivity. Under certain conditions clouds can reflect as much as 80 percent of the incoming energy.

On a global scale, clouds reflect about 25 percent of the total solar energy reaching the atmosphere and absorb less than 5 percent.

How good is the Earth's reflectivity? What are the mechanisms for reflecting or returning light to space? As we have indicated, light is reflected from clouds and the surface. These together produce an albedo of about 33 percent for Earth and atmosphere as a whole over the entire surface. Weather satellites

Table 4.1	
Typical Albedo or Reflectivity of Various Surfaces	
Surface or Object	**Albedo (percent)**
Fresh snow	75 to 95
Clouds (thick)	60 to 90
Clouds (thin)	30 to 50
Venus	78
Ice	30 to 40
Sand	15 to 45
Earth and atmosphere	30
Mars	17
Grassy field	10 to 30
Dry, plowed field	5 to 20
Water (daily average)	10
Forest	3 to 10
Moon	7

Source: After C. Donald Ahrens, *Meteorology Today,* 4th ed., p. 67. St. Paul, Minn.: West Publishing Co., 1991.

reveal considerable albedo variation over the Earth's surface, with a variation of over 80 percent from the surface itself.

Vegetation is a good absorber, while sand, snow, and ice have some of the highest albedos. The albedo of water when the Sun's angle is greater than 40° above the horizon is less than 4 percent. Considering the fact that over two-thirds of the globe is covered by water, it is interesting that Earth's surface as a whole has such a high albedo. Primarily clouds and the atmosphere account for this high reflectivity (table 4.1).

Filtering and the Greenhouse Effect

Although the atmosphere contains only small percentages of water vapor and carbon dioxide, these two gases have a major influence on the heat balance of the atmosphere. Without these gases the Earth's night-time temperatures would plunge to very low levels. These two gases are almost transparent in the visible wavelengths, but are nearly opaque to infrared radiation in specific bands on the electromagnetic spectrum. This characteristic—the **greenhouse effect**—prevents the Earth from rapidly losing its heat by returning most of the reradiated energy back to the surface to be absorbed and reradiated again, perhaps many times, before being finally released to space (fig. 4.9). The lunar surface plunges to −129° C (−200° F) at night because there is no atmosphere to absorb energy.

Clouds also have a filtering and reflecting influence. Clouds consist of water droplets that reflect shortwave and visible radiation, and absorb and emit infrared radiation. Overcast skies reflect visible radiation and restrain the loss of infrared from the surface.

In summary, the atmosphere works much like a greenhouse. Shortwave radiation is easily transmitted through the atmosphere and the clear glass panes of a greenhouse. However, like the atmosphere the glass is highly opaque to infrared waves. The air in a greenhouse remains warmer than the outside atmosphere because infrared energy emitted by plants and soil is trapped by the glass and is reradiated to the surrounding interior greenhouse

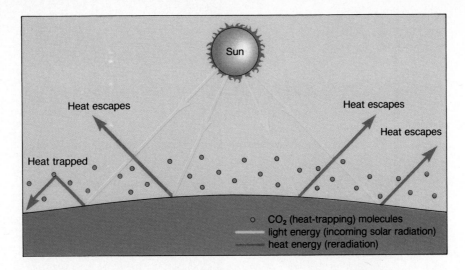

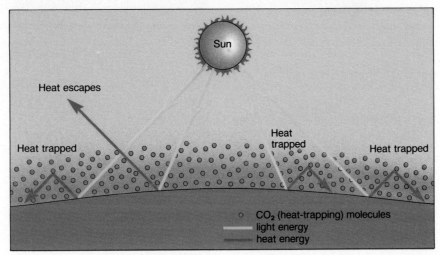

Figure 4.9

The *greenhouse effect* is operating on several levels. (A) The Earth's atmosphere is a giant greenhouse trapping infrared energy and reradiating it back to the surface before being dissipated away into space. (B) On a local and regional scale, clouds reduce energy losses and moderate moist climates. Therefore, the cloudy zones of the Earth are regions of fewer temperature extremes resulting from the greenhouse effect.

environment before being slowly lost to the outside by conduction and radiation. The same **greenhouse effect** is experienced when you enter a car on a cold but sunny day: the car's interior is warm.

ENERGY TRANSFER

Because the Earth's surface is covered by vegetation, soil, rocks, water, and atmosphere, energy is transferred, not only by **radiation,** but by several other important methods. These include **conduction, convection,** and **latent heat** of **evaporation** and **condensation.** Without energy transfer, the tropics would become excessively hot and the middle and high latitudes would grow very cold. Energy transfer evens out temperature by redistributing energy.

Conduction

If you place a metal rod in a fire it will only take a minute before your unprotected fingers sense the heat. Heat travels from the rod to your hand by **conduction** (fig. 4.10). Heat energy always flows from warmer to cooler objects. The greater the difference in temperature and the greater the substance's heat conductivity, the greater the rate of flow (table 4.2). As the Sun warms the ground and water surface, the air and cooler zones below the surface and water are warmed. On the average, nearly 10 percent of the Earth's absorbed energy is conducted to the layer of air in contact with the ground, but it does not stop here. Energy sets water and air in motion.

Metal rod heats by conduction

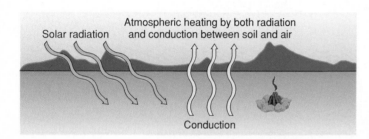

Solar radiation

Atmospheric heating by both radiation and conduction between soil and air

Conduction

Figure 4.10

Energy transfer by conduction is a series of chain reactions. When atoms are energized, their energy passes on to less energetic atoms they come in contact with. Energy always flows down energy slopes or from warmer to cooler objects. Nearly 10 percent of the Earth's absorbed energy is conducted from the ground to the air layer in contact with it.

Table 4.2

Heat Conductivities of Common Substances

Material	Heat Conductivity (cal/s/cm/° C)
Still air	0.0000614 (at 20° C)
Dry soil	0.0006
Water	0.00143
Snow	0.0015 (density 0.5 g/cm³)
Wet soil	0.0050
Ice	0.0053 (at 0° C)
Sandstone	0.0062
Granite	0.0065
Iron	0.161
Copper	0.918

Source: After C. Donald Ahrens, *Meteorology Today,* 4th ed., p. 56. St. Paul, Minn.: West Publishing Co., 1991.

Figure 4.11

Convection mixes the energy of a fluid throughout the system by circulation of the fluid. When the atmosphere and oceans receive greater amounts of energy in one location, this energy moves to cooler locations. The redtail hawk soars effortlessly on a rising convective air current generated by surface heating.

Convection

When the air and water layer are warmed, they expand and circulate by **convection,** a stirring or mixing process common to all fluids. The stirring causes energy to be blended throughout the system (fig. 4.11).

Energy from the Earth's surface enters the atmosphere and oceans, warming them through conduction then radiation. Convection mixes the air thoroughly up to the tropopause.

Latent Heat Transfer by Evaporation, Condensation, and Fusion

When a solid, liquid, or gas changes state, energy is transferred. **Evaporation** occurs when liquid water is converted to vapor by an input of 540 calories of energy per gram of water converted. This same energy is released when vapor is converted to liquid in **condensation.** Plants contribute water in a process known as **transpi-**

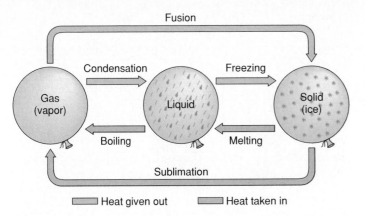

Figure 4.12

Energy is transferred from one location to another when water changes state between solid, liquid, and gas. For instance, to convert water from liquid to vapor requires 540 calories per gram. This same energy is released into the environment when condensation of this gram occurs.

ration, taking water out of the soil and releasing it as vapor in the atmosphere. Evaporation plus transpiration combined is known as **evapotranspiration.**

When water turns to ice, 80 calories is released per gram. Thus **freezing** adds energy to the environment. **Melting** absorbs this energy of 80 calories per gram of ice. **Fusion** releases energy from the gaseous form of matter when it changes to the solid form. **Sublimation** is the reverse. Over 600 calories is required to convert one gram of ice to a vapor. When an air mass picks up moisture over a body of water, energy is also picked up and temporarily stored in the vapor. Clouds are sites where **latent heat of condensation** is being released. Thus the atmosphere and oceans serve as a storehouse and transporter of solar energy in the latent form (fig. 4.12).

THE ENERGY BUDGET

The clouds of the atmosphere not only release heat but also absorb, scatter, and reflect solar energy. Therefore it is necessary to consider atmospheric influence on the energy reaching us from the Sun.

Let's start at an altitude of 160 kilometers (100 miles) in the thermosphere, where nearly 100 percent of the atmosphere lies below, and let us assign a 100 unit value to a sample of solar energy. As this theoretical sample streams toward the Earth, it spends itself in a variety of ways. Nearly all x-rays and gamma rays are absorbed in the top 80 kilometers (50 miles) of atmosphere. Then in the lower mesosphere and upper stratosphere, ultraviolet is absorbed by the ozone layer, reaching maximum absorption between 30 and 60 kilometers (20 and 40 miles) above the surface in the ozone layer described in chapter 3. Visible energy has an "open window" until it reaches the lower atmosphere or homosphere; that is, only the shortwave gamma rays, x rays, and ultraviolet radiation are significantly absorbed in the upper atmosphere.

Figure 4.13 summarizes solar energy coming in and reaching the Earth's surface on an average year for the entire planet. Because the global average temperature is quite constant,

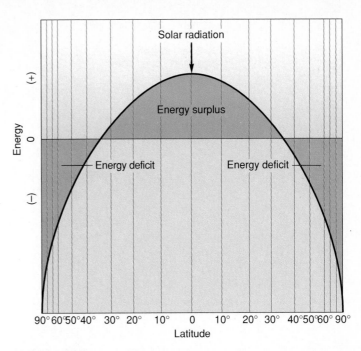

Figure 4.13

The Earth receives energy at the surface in varying amounts. The low latitudes are zones of surplus that supply energy to the higher latitudes by convection and transfer to latent energy.

the amount of energy arriving from the Sun and departing from the Earth must be equal. As energy passes through the atmosphere, 50 percent is absorbed at the surface directly from the Sun as shortwave visual radiation. Dust and clouds absorb 18 percent. Reflection, absorption, and scattering by the surface and atmosphere account for the remaining 32 percent (fig. 4.14).

Thus latent heat transfer, conduction, convection, and evapotranspiration assist in dissipating surface energy not already reradiated and reflected by the surface. Therefore, the system is in balance. Without an atmosphere, energy calculations would be greatly simplified. There would be no atmospheric reradiation, greenhouse effect, evaporation, condensation, convection, or latent energy exchange. Insolation at the surface produces a wide range of temperature responses not only seasonally but also daily. These daily rhythms of temperature respond to a variety of environmental factors at each location.

THE DAILY RANGE IN TEMPERATURE

Daily temperatures are recorded by thermometers and *thermographs,* which trace the temperature on graph paper fixed to a slowly moving drum. The slowly turning thermograph records the daily pattern of temperature. If we could examine enough thermographs taken from stations around the world, it would become apparent that no two locations have the same pattern. However, all locations under clear skies and calm conditions will show one maximum and one minimum in a 24-hour period, resulting from

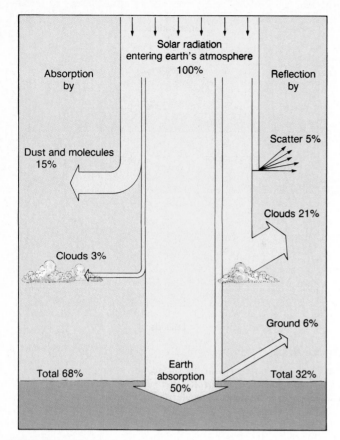

Figure 4.14

The energy budget of the atmosphere is always in balance. Environmental factors such as clouds, dust, reflectivity, and water vapor produce a complex flow pattern. The 68 percent absorbed by the atmosphere and surface, plus 32 percent reflected, equals 100 percent insolation. This entire amount is returned to space in the long run.

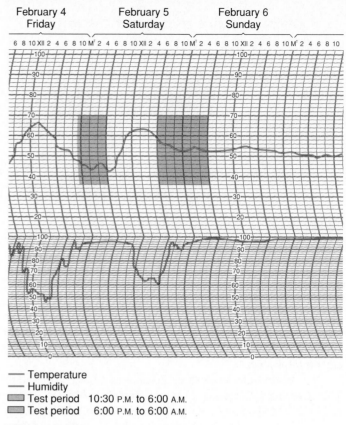

— Temperature
— Humidity
▨ Test period 10:30 P.M. to 6:00 A.M.
▨ Test period 6:00 P.M. to 6:00 A.M.

Figure 4.15

On the clear night of February 4–5, 1966, rapid cooling with nocturnal winds produced the fluctuations in the trace. Clouds moved in on February 6, 1966, and reduced the difference in temperature.

daily variation in insolation as the Earth turns on its axis. For most middle-latitude locations, the maximum temperature is reached 2 to 4 hours after the Sun has reached its zenith while minimum temperatures occur at about sunrise or a little after (fig. 4.15).

Let's follow a temperature trace made during a typical winter period in southern California from 6 A.M. on February 4, 1966, to February 6 at 6 A.M. We can learn about the general state of the atmosphere and surface environmental influences by interpreting the record (fig. 4.15).

The morning and evening of February 4 was clear, and water vapor content of the air was relatively low. Therefore, the Earth efficiently reradiated absorbed energy back to the atmosphere and space in infrared wavelengths. This caused the surface to cool to 44° F at 6 A.M. As the Sun's rays warmed the surface, the air in contact with the surface rapidly warmed to a maximum 65° F by 2:15 P.M. Remember that the Sun is highest at 12 noon, or very close to it, yet there is a lag in time between maximum energy input and maximum temperature. The surface is still gaining more energy than it is losing until later in the afternoon. After the maximum temperature is reached, energy losses are greater than gains so the temperature declines. If the Earth could

lose and gain energy at the same rate, our nights would be unbelievably cold and days would not warm up very much. Our atmosphere is a heat recovery system that reduces heat loss.

Following the night cooling trend on February 4 to sunrise at 7 A.M. on February 5, note that the rate of decline in temperature is slower than its earlier daytime rise. Note also that the slope of the trace is much greater before the maximum temperature at 2:15 P.M. This phenomenon can be explained by the fact that the atmosphere sends infrared energy back to the surface, recycling the lost energy by adding to declining insolation in the afternoon. Shortwave visible light is transmitted easily through the daytime air, allowing energy to rapidly warm the surface. Also note that the decline in temperature in the afternoon and evening is irregular.

In the early morning hours between 2 and 4 A.M. on February 5, the temperature even climbed 4° F, then fell three degrees, reaching a minimum at 6 A.M. Before dawn the wind stirred the air, raising the temperature by moving warmer air into the environment.

Topographic Influences

A rise in temperature during the early morning before sunrise is not uncommon on clear nights at this location. This can be ex-

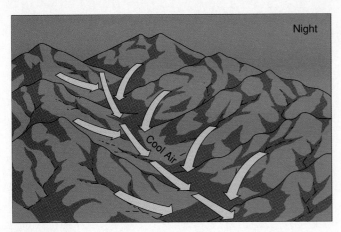

Night

Cool Air

Figure 4.16

Cold air drainage off the slopes causes mixing of cold and warm air, thus raising the mean temperatures in the lower layers.

plained by considering the terrain influences. The station is located in a valley surrounded by hills that rise 150 to 200 meters (500–700 feet) above the valley floor. As the surface cools, a layer of cool air builds up on the surface. Once the cool air layer reaches a certain depth, it begins to slide off the slopes following the drainage patterns down the valley. The net effect is to mix the air where drainage is good because air flow produces turbulence and mixes warm and cool layers (fig. 4.16). Where air drainage is poor, cold air puddles on the surface.

Cloud Cover Influences

On February 5 the Sun rapidly warmed the air to a maximum of 62° F at noon. At this time, clouds moved into the area and reflected insolation back into space, keeping the temperature 3° F below February 4 and causing the maximum to be reached earlier. During the next 12 hours, the temperature fell only 10° F, with a minimum recorded at 52° F. February 6 showed a maximum of only 54° F. During this time, low clouds covered the valley and the Earth's heat could not effectively escape. Here is a good example of the influence of cloud cover.

Not only is there a daily rhythm of temperature, but also an annual cycle that reveals the influences of the surface environment as well as insolation.

THE ANNUAL CYCLE OF TEMPERATURE

Daily and annual patterns of temperature at stations located around the world at the same latitude show great variations in temperature, although each station receives the same potential insolation. Figure 4.17 shows five stations located at latitude 40° N and their annual cycles of monthly mean temperatures. There are many environmental factors that produce such temperature variation. We have discussed the effects of topography and cloudiness, but other temperature controls include water vapor, smoke, smog, elevation, air mass influences, percentage of carbon dioxide in the air, storms, ocean influences, prevailing winds, vegetation cover, and cultural features. These are all considered in the discussion of global climate in chapter 8.

Although each of these five stations varies widely in mean monthly temperatures, several common temperature traits are visible. Note in figure 4.17 that all show maximum mean monthly temperatures late in July and early August and minimum mean monthly temperatures in late January and early February. Except for equatorial locations, this pattern is typical around the globe. Although potential insolation—that is, solar energy at the top of the atmosphere—reaches 900 langleys per day in June, the Earth continues to warm, reaching its peak in July and August, because cooling is slower than the heat gained. In December, at the winter solstice, potential insolation drops to nearly 300 langleys per day. Temperatures continue to fall, with January and February being the coolest months, while insolation increases to nearly 400 langleys per day during this time. The time lag reflects the fact that outgoing energy is greater than insolation. The mean temperature varies from station to station depending on the many environmental local temperature controls.

LAND VERSUS WATER EFFECTS ON HEATING

When equal amounts of solar energy are absorbed by the continents and oceans, great contrasts in heating occur. Under similar insolation (solar angle, duration of daylight, transparency), the daily and annual temperature ranges at marine locations are much less than locations in the interior of a continent or even a short distance from the coast. Note in figure 4.18 that Iceland and Ireland have only 10 to 15° C annual ranges in air temperature. However, in the same latitudes Siberia experiences 60° C annual ranges in mean monthly temperature. January's mean temperature for Iceland and nearby Ireland is just above freezing, while Siberia during the same winter period experiences mean January temperatures of −30° C to −40° C (−22° F to −40° F) and lower, and yet warms to 10–15° C (50–60° F) during July.

There are four primary reasons for these temperature differences. First and most important, water distributes and mixes incoming energy through a deep layer. Its transparency allows solar rays to penetrate through a deeper zone. Fish can make use of daylight more than 100 meters (several hundred feet) below the surface. Second, because water is a fluid, surface heat is circulated and transported vertically and globally by convection currents to cooler regions. Third, water has a higher **specific heat** than land. The same amount of heat applied to land and water can raise the temperature of the land five degrees for every one degree the water is warmed (table 4.3). Fourth, water surfaces have higher potential evaporation. Evaporation causes cooling, so water surfaces can remain cooler. These four factors prevent energy from building at the surface of water in contrast to land.

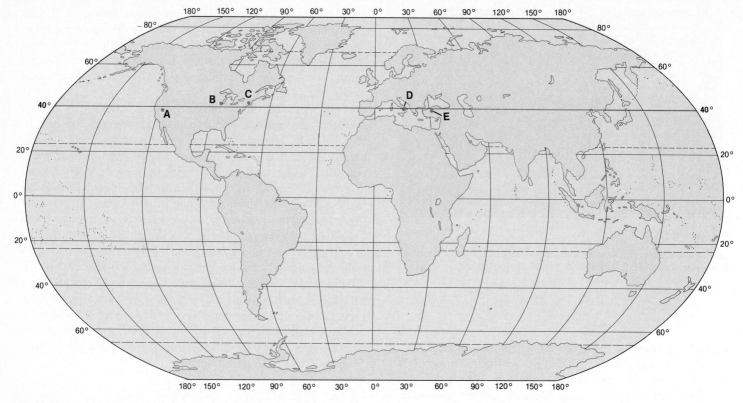

Figure 4.17

Five stations located at 40° N latitude show several common temperature traits, such as similar maximum and minimum periods. They also illustrate considerable difference in temperature throughout the year, although each receives the same potential insolation. The temperatures of a place depend not only on insolation but also on local temperature controls of elevation, slope exposure, vegetation, cloudiness, oceanic versus continental influences, prevailing winds, and human influences. Note the following July and January temperatures: (A) Reno, Nevada—July 70.5° F, January 32° F; (B) Chicago, Illinois—July 73.5° F, January 25° F; (C) New Haven, Connecticut—July 72.5° F, January 28.5° F; (D) Naples, Italy—July 76.5° F, January 48° F; and (E) Ankara, Turkey—July 73° F, January 31.5° F.

If we contrast land surfaces with water bodies, land surfaces heat up rapidly and cool just as quickly (fig. 4.19). Because land surfaces are opaque, light stops at the surface and heat transmission then occurs by conduction. Have you ever burned your feet on a hot sandy beach? There is quick relief by simply scraping off the top inch or two down to where the sand is actually cool. Sand is a poor heat conductor, and the Earth's surface in general is also a poor conductor of energy. This is why caves maintain temperatures close to the annual average at their locations while outside summer temperatures exceed 38° C (100° F) and winter minimums frequently fall below freezing. The root cellar and basement take advantage of this same characteristic.

Near our *study area* described in chapter 1, mean monthly temperatures along the coast vary only 5° C to 7° C (9° F to 13° F) annually between July and January. At the same latitude, 32 kilometers (20 miles) inland, mean monthly temperatures vary 32° C to 36° C (90° F to 97° F) between January and July.

Oceans have a strong moderating influence on coastal locations with strong onshore prevailing winds. Looking at temperatures on the east and west coasts of North America at the same latitudes, for example, note how more moderate the January and July temperatures are on the west coast. These locations are under stronger marine influences because the prevailing westerlies are onshore. East coasts experience westerly winds out of the interior, reflecting continental temperature extremes. San Francisco has a January mean monthly temperature around 10° C (50° F) while Boston at a similar latitude experiences mean January temperatures of 0° C (32° F).

Table 4.3
Specific Heat of Various Substances

Substance	Specific Heat (cal/g × °C)
Water	1.00
Wet mud	0.60
Ice (0° C)	0.50
Sandy clay	0.33
Dry air (sea level)	0.24
Quartz sand	0.19
Granite	0.19

Source: After C. Donald Ahrens, *Meteorology Today,* 4th ed., St. Paul, Minn.: West Publishing Co., 1991, p. 53.

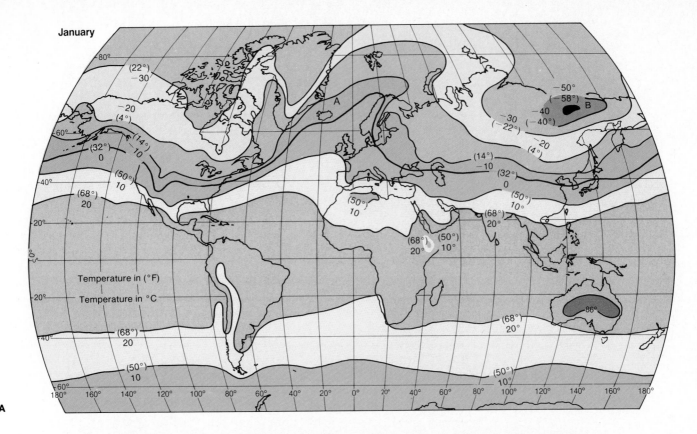

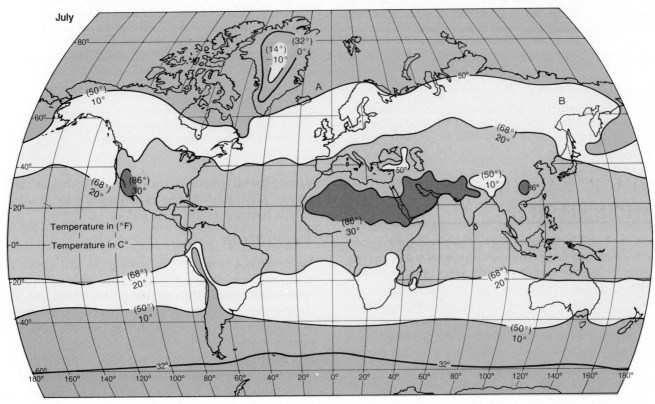

Figure 4.18

Global January and July temperature patterns reveal the influence of land and sea. Compare Iceland (A) with a Siberian location (B) at the same latitude and one can see the moderating influence of Iceland's marine location.

Figure 4.19

Land and sea locations differ in temperature due to four primary factors: (1) water is transparent, (2) it is also fluid, making energy distribution by convectional mixing more efficient, (3) water has a higher specific heat, and (4) water has a greater potential surface of evaporation. The fog and low clouds are dispersing first over the warmer land surface, which heats more rapidly.

SURFACE FACTORS AND TEMPERATURES

A view of the Earth from space shows a blue planet partially covered with clouds and large continents of various colors, ranging from white at the polar latitudes to deep green in the coniferous forest belts of the upper middle latitudes and tropical rainforests of the low latitudes. Each color absorbs and reflects different amounts of energy. The albedo of the Earth is greatest where colors are light. Although more energy is absorbed in the darker forests, grasslands, and ocean regions, it is important to realize that insolation energy is not all converted to sensible heat, or heat causing a temperature change. Some energy is used in transpiration, which causes dark vegetation surfaces to be cooler than bare ground with the same albedo (fig. 4.20).

Other factors change the effect of insolation on temperature change, even on surfaces having the same color. White snow fields and white sand dunes, for example, have different responses to insolation, although each might have similar reflectivity. Energy striking snow is reflected or converted to latent heat in the melting process, while energy striking a sand dune is reflected or consumed to raise the temperature of the surface.

Vegetation

Vegetation is a moderating influence on air and surface temperatures for several reasons. Photosynthesis, explained in chapter 13, extracts a portion of the solar energy. Vegetation also intercepts incoming light before it reaches the ground and prevents rapid loss of outgoing infrared energy reradiated from the surface. Also, a cover of vegetation exchanges heat in a deeper air layer than the thin ground surface. Evapotranspiration, or the combined water loss from plants and soils by evaporation, produces cooling in the same way perspiration cools the human body. Evaporation of 1 gram of water removes 540 calories from the air and surface that otherwise would heat the area (fig. 4.21).

The type of vegetation and its density can cause considerable variation in moderating the temperature. A young forest has a different surface temperature than a mature one. A young forest permits more light to reach the ground surface. A deciduous

Figure 4.20

The colors of the Earth's surface at Simpson's Gap near Alice Springs, Northern Territories, Australia, have a strong influence on temperature patterns. Light sandy surfaces reflect most incoming energy and remain cooler, while dark surfaces tend to absorb energy and heat up. Transpiration by green plants and evaporation from water bodies cool surfaces even when the albedo is the same as a bare surface of soil.

forest in the summer has considerable temperature variation during the day at the surface, due to the shadows from the trees. During the winter, the same forest shows much less variation in temperature because the foliage is absent, allowing sunlight to penetrate more completely to the surface during the entire day.

We can consider the top of a canopy of a forest to be the effective air contact surface. At the top, temperatures are much higher during the day and cooler at night, resembling an open clearing or meadow.

At night, trees in a dense timber stand absorb the infrared radiation from the ground and recycle it back to the surface. Temperatures under a forest canopy may be as much as 6° C (10° F) warmer at night, and cooler during the day than an adjacent open meadow. Have you ever noticed how much more morning dew falls in a meadow on a clear night than in the adjacent forest? This increase in dew is the result of lower evening temperatures in the clearing. As the air cools, its water holding capacity decreases until saturation occurs. This is called the **dew point,** and at that temperature, when the air is saturated, water vapor condenses as dew or frost, depending on the temperature. Usually the temperature will decline more slowly after the dew point is reached because latent heat of condensation or freezing is released to the air.

Figure 4.21

Ferns and moss carpet a rock surface on Cocos Island, Costa Rica. Vegetation moderates temperature of the Earth's surface by shading, transpiration, reradiation, and consumption through photosynthesis.
Photo by Charles L. Hogue.

Topographic Influences on Temperature

North- versus South-Facing Slopes

Topography or surface morphology plays a major role in the local surface temperature. Variations in a slope's steepness and orientation, or aspect with reference to the Sun, affect the angle at which sunlight strikes the surface. Slopes perpendicular to the Sun receive more energy per unit area than those forming a small angle. A south-facing slope in the Northern Hemisphere will have a much higher daytime temperature than a slope facing north at the same latitude (fig. 4.22). Midsummer temperatures in direct sunlight on south-facing slopes can reach as high as 63° C (145° F).

Figure 4.22

A south-facing slope will have a much higher average annual temperature than a north-facing slope. The Golden Gate Bridge connects heavily populated San Francisco Peninsula with the sparsely populated Marin Headlands. In this false-color infrared aerial photograph, the south-facing slopes north of the bridge are lighter red in contrast with cooler north-facing slopes with denser shrubs, in darker red, where insolation is less.

Photo by NASA.

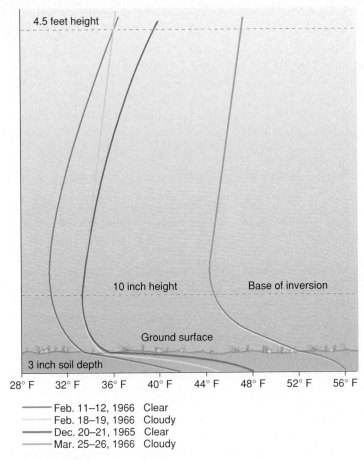

28° F	32° F	36° F	40° F	44° F	48° F	52° F	56° F

——— Feb. 11–12, 1966   Clear
——— Feb. 18–19, 1966   Cloudy
——— Dec. 20–21, 1965   Clear
——— Mar. 25–26, 1966   Cloudy

Figure 4.23

The lowest minimum temperatures occur just above the surface where a heat sink occurs due to reradiation to the surface and space. The temperature curves were measured on two clear nights and two cloudy nights during the winter and spring of 1965–66 at Nogales High School Weather Station, Walnut, California.

East- versus West-Facing Slopes

East-facing slopes reach their maximum temperatures earlier in the day than west-facing slopes. In general the highest surface temperatures occur on southwest slopes, which receive early afternoon sunlight at the warmest time of the day.

Nocturnal Temperatures and Inversions

On calm clear nights, the air in contact with the ground cools rapidly. The air a meter or so (several feet) above the surface cools more slowly than the surface because of lack of contact. Even with downward reradiation from the atmosphere, more rapid cooling takes place in the layer in contact with the ground. This layer undergoes additional cooling by heat exchanges through convection and evapotranspiration with air layers at higher levels. These factors cause the lowest minimum temperatures to be just 8 to 25 centimeters (3–10 inches) above the ground surface (fig. 4.23). The nocturnal inversion starts above this height, where an increase in temperature takes place with increasing height. Figure 4.24 illustrates readings taken at the base of San Jose Hill and three

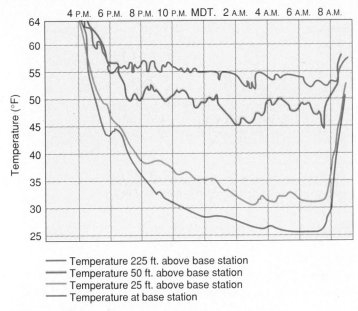

——— Temperature 225 ft. above base station
——— Temperature 50 ft. above base station
——— Temperature 25 ft. above base station
——— Temperature at base station

Figure 4.24

On calm, clear nights temperatures will increase with height. Also, the duration of minimum temperatures decreases with elevation.

different heights above the base. During the night hours, the temperatures were without exception warmer as elevation increased. Figure 4.24 also illustrates that the duration of minimum temperature shortened with increasing height. At the highest level, the lowest temperature lasted only a few minutes whereas a much colder low temperature lasted several hours on the surface.

Figure 4.25 shows that the air is first cooled next to the ground, forming a shallow layer that deepens during the night, reaching a maximum depth at sunrise.

A strong relationship exists between strength of the nocturnal temperature inversion and the maximum temperature of the preceding day. There is also a strong positive correlation between calm air, high pressure, and low humidity and the inversion. Under these conditions, the Earth is most efficient in reradiating infrared energy back into the atmosphere and convecting heat away from the surface. Both water vapor and clouds tend to reduce temperature contrasts between the surface and upper layers by redistributing the heat more uniformly in the lower layers.

Cold Air Drainage and the Inversion

As we have noted earlier, topography plays a key role in the channeling of cool air drainage. Topography also influences the formation and intensity of inversion development. The orientation or aspect of the slope appears to be of no importance at night, but the degree of slope, relative differences in elevation or relief, and morphology are significant factors.

Valleys serve as cold air reservoirs which collect cold air drainage from the surrounding hills. Table 4.4 shows minimum surface temperatures on clear nights in a cross section of the San Jose Creek drainage system in southern California. Note that the warmest temperatures were located at the highest altitude; however, low clouds moved in during the night of March 25–26 at 3

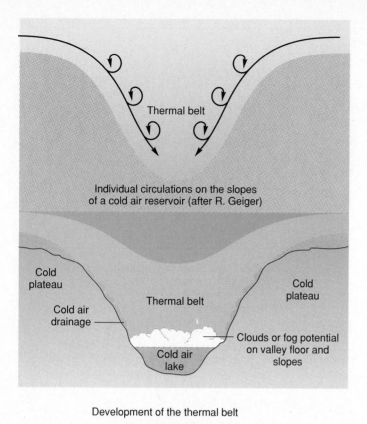

Development of the thermal belt

Cold ☐ ☐ ☐ Warm

Figure 4.25

(A) Cold air puddles on plateaus and valley floors creating cold air lakes. (B) Highest minimum temperatures occur in the thermal belt on the slopes, where air flow is unrestricted.

A.M., raising the minimum temperature. On clear nights, when energy loss is most efficient, there was an average of 6° F (3.3° C) difference in minimum temperatures between the ridges and valley in the region.

Cold Air Drainage and Vegetation

When looking at the nocturnal inversion near the surface in rolling terrain, several important relationships can be observed from data collected on 17 nights (fig. 4.26 and table 4.5). Readings proved in some cases to be high on both low and high ground in certain terrain and vegetation configurations, but the general rule—increasing height produces an increased temperature—held true. Cold air flows more like a thick syrup and tends to puddle and stagnate in low or flat spots. The highest minimum temperatures occurred in areas of maximum wind velocities. Note that station 5 had a 39.1° F average while station 8 had a 32.9° F average, yet station 8 was only 14 feet higher than station 5 and just a few yards away. The vegetation and increased humidity at station 5 in the creek bottom prevented heat loss and intrusion of cold air drainage. At station 5 the vegetation sheltered the station from outside cold air intrusion and heat loss from within. Station 8 was in an open field where puddling of cold air from the drainage system occurred.

Table 4.4

Minimum Surface Temperatures (°F) on Clear Nights in a Cross Section of the San Jose Creek Drainage System

Station A: Crest of the San Jose Hills, 305 meters (1,000 feet).

Station B: Nogales High School weather station, 144 meters (472 feet).

Station C: San Jose creek bottom adjacent to Nogales St., 130 meters (425 feet).

Station D: Crest of the Puente Hills, 305 meters (1,000 feet).

Date	A	B	C	D
March 25–26, 1966*	50	51	51	52
March 26–27, 1966	53	48	47	55
April 14–15, 1966	57	49	44	51
Average	53.0	49.3	47.3	52.6

*Low clouds moved in at 3 A.M. to raise minimum temperatures in lower elevations.

Downslope Winds and Minimum Temperatures

Downslope winds, shown in figure 4.26, create the warmest locations in areas of highest wind velocity in the more exposed creek drainage sites at stations 13, 14, 7, and 15. Several points were noted during the wind observations. First, valley constrictions decreased wind by producing a damming effect and puddling of cold air. Second, air velocity and temperatures increased as air exited through the narrow gaps, much like water passing through a hose nozzle. In our small valley of chapter 1, tree falls are more common where the valley opens up from a constriction gap. Third, wind velocity varied with the height of the drainage system. Gravity moves colder air faster off a steeper and higher slope. Fourth, lower velocities occurred where vegetation and rugged terrain produced surface friction. Wind velocities were observed at various heights above the ground. The highest wind velocities occurred at the 24-foot height, and here temperatures were near the maximum of 46° F observed at the 15-foot height. Here we can see the mixing influence of wind. Colder surface air is mixed with higher warmer air, raising the average temperature.

Summary of Surface Influences on Minimum Temperatures, Inversions, and Downslope Air Drainage

From our discussion of surface factors and their influence on minimum temperature, we can conclude the following:

1. The nocturnal inversion is modified by hilly terrain by causing pooling of cold air and turbulent mixing of the air.
2. Nights with strong inversions follow warm, clear days with high pressure and low humidity.
3. Cloud cover raises minimum temperatures at the surface and destroys or weakens the inversion.
4. Winds weaken or destroy the inversion by mixing the air.

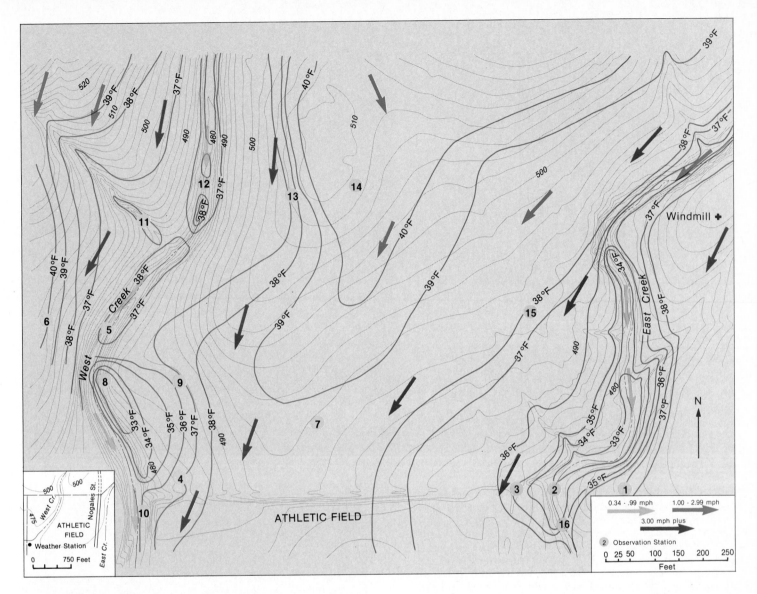

Figure 4.26

Mean minimum temperatures and downslope winds show a strong correlation. Highest minimum temperatures occur where air can flow unrestricted. Lowest minimum temperatures develop where air flow is restricted, producing a damming effect.

5. The amount of water vapor and carbon dioxide in the atmosphere influences the nocturnal inversion by decreasing outgoing infrared radiation, thus enhancing the greenhouse effect.

6. The concept of increased height producing an increased temperature during inversion conditions applies only to areas with large-scale relief or broad, open valleys. On the smallest scale involving only a few feet difference in elevation, high elevations do not always have warmer temperatures. Minimum temperature patterns primarily reflect the influences of surface morphology and vegetation controls.

7. The size and shape of the air drainage system determine the volume, velocity, and consistency of downslope air movement. Friction of an irregular surface slows this flow.

8. Terrain with the greatest slope has the greatest downslope wind velocity because gravity is the driving force.

9. Nocturnal warm areas are confined to zones of best drainage and exposure to the downslope wind.

10. Cold areas occur where air drainage is weak, resulting from a dish-shaped terrain. This causes pooling of cold air.

WORLD PATTERNS OF TEMPERATURE

Temperature patterns on a local or global scale can be illustrated with **isotherms,** lines connecting points of equal temperature value. Figure 4.27 shows global mean monthly temperature values expressed in isotherms for January and July.

Solar Energy and the Earth's Response   **89**

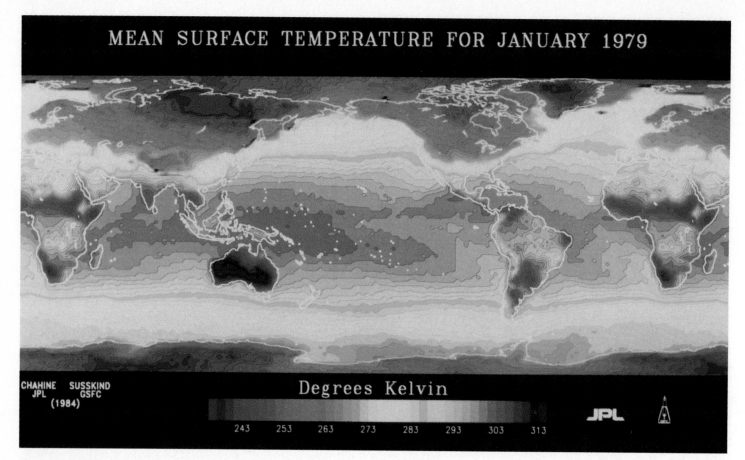

MEAN SURFACE TEMPERATURE FOR JANUARY 1979

CHAHINE   SUSSKIND
   JPL   GSFC
      (1984)

Degrees Kelvin

243   253   263   273   283   293   303   313

JPL

Figure 4.27

Global temperature patterns for January and July strongly reflect the controls of latitude and continental effects for each season. NASA's Jet Propulsion Laboratory and Goddard Space Flight Center processed this view using National Oceanic and Atmospheric Administration satellite data. This view shows surface temperatures on a typical January day in 1979. Higher temperature levels are indicated by deeper shades of red and cooler temperature levels are indicated by deeper shades of blue. Bright yellow bands indicate moderate temperatures. Notice the moderate temperatures on South America's western coast.

Photo by NASA.

Temperature patterns are primarily controlled by inso-lation, the presence of oceans, snow cover, vegetation, ocean and wind currents, altitude, and transparency of the atmosphere. Let us examine global temperature patterns, keeping in mind these environmental controls.

First, note that isotherms have an east–west orientation in all seasons. Temperatures decrease toward the poles. Because the Southern Hemisphere is primarily water, summer and winter isotherm patterns there show less annual fluctuation than in the Northern Hemisphere. Second, in the Northern Hemisphere the location of large land masses in the subarctic and Arctic results in wide ranges in annual temperature. The coldest places with the widest annual ranges in temperature are in the interiors of Asia, North America, and Greenland, while the ice-packed North Pole is slightly warmer, with less annual range in temperature. Third, equatorial isotherms show little change in value and position during

Table 4.5

Average of Observed Minimum Temperatures at the Sites in Figure 4.26

Station	Number of Observations	Average Minimum Surface Temp. at Stations
1	6	37.5° F
2	10	33.7
3	6	35.0
4	5	36.6
5	10	39.1
6	6	40.2
7	7	38.4
8	8	32.9
9	5	36.8
10	7	34.2
11	5	36.2
12	5	37.7
13	7	37.9
14	6	40.7
15	8	37.7
16	6	34.9

the year. This results from relatively uniform insolation throughout the year. Select any isotherm crossing the middle latitudes and note how it shifts poleward in the summer, especially over land, and equatorward in the winter. This shift is greatest on the eastern coasts, where the prevailing westerlies come from the interior.

Fourth, highlands and ice-cap regions of Antarctica and Greenland are always cold, with little range in temperature throughout the year. Because temperatures decline with altitude, temperatures will always be lower in the mountains and plateaus. The plateau highlands of Tibet and the Andes are two examples of arctic conditions in the middle and low latitudes.

SUMMARY

From nuclear energy in the Sun comes **electromagnetic radiation,** which warms and illuminates our planet in a variety of forms.

Electromagnetic energy ranging from gamma to radio waves is emitted from the photosphere, chromosphere, and corona of the Sun's atmosphere.

Not all places on the Earth receive the same amount of **insolation.** The amount depends on location (geographic factors), the time and season, and clarity of the atmosphere. As sunlight penetrates the atmosphere, it is **filtered, reflected,** and **scattered.** The degree of influence of each of these factors determines the energy budget at the surface.

Because the Earth has a dynamic atmosphere and ocean, energy reaching one location is transferred through **convection** by winds and water, and to a very limited extent by conduction through the soil. Energy is also exchanged by **evapotranspiration,** a process which includes both **evaporation** and **transpiration.** Interior land areas heat and cool rapidly in contrast with coastal locations.

The horizontal and vertical temperature patterns reflect surface and atmospheric conditions as well as the latitude and season of the year. Temperatures reach their highest and lowest annual levels after the maximum and minimum periods of insolation. The lag in temperature results from the **greenhouse effect** and the Earth's ability to store and transfer energy.

Surface conditions such as water versus land, rugged hilly terrain versus plains, desert vegetation versus rainforest, as well as color, all produce variations in temperatures over the surface of the Earth.

Thus global temperature patterns are a response to a wide variety of controls ranging from surface and atmospheric factors to the planetary position of the Earth in space.

ILLUSTRATED STUDY QUESTIONS

1. Describe the fusion process in the core of the Sun and trace the flow of energy from the Sun's core to the Earth (see figs. 4.1 and 4.4, pp. 69, 71).
2. Describe the electromagnetic spectrum and percentages of energy of visual, ultraviolet, and infrared radiated by the Sun (see fig. 4.6, p. 73).
3. Contrast heating of land and water by solar energy (see figs. 4.18, 4.19, and 4.27, pp. 83, 84, 90).
4. Describe the essential factors necessary for the nocturnal inversion and cold air drainage (see fig. 4.15, p. 80).
5. Give three important roles the oceans play in moderating the Earth's temperatures (see figs. 4.18, 4.19, and 4.27, pp. 83, 84, 90).
6. Which continent has the greatest annual range in temperatures (see figs. 4.18 and 4.27, pp. 83, 90)?
7. Which hemisphere has the greatest annual range in temperatures (see figs. 4.18 and 4.27, pp. 83, 90)?
8. Describe the Earth's albedo over land and water (see table 4.1, p. 76).
9. Explain the role of carbon dioxide and water vapor in absorption of solar energy (see fig. 4.9, p. 77).
10. Describe the greenhouse effect on both a local and global level (see fig. 4.9, p. 77).

The Windy Planet

5

Objectives

After completing this chapter, you will be able to:

1. Explain the causes of wind.
2. Define high and low pressure.
3. Explain the forces on a parcel of air that determine the direction and velocity of the air.
4. Describe the major surface pressures and wind systems of the Earth.
5. Describe upper air global circulation and relate this motion to **Rossby waves** and the **polar jet stream.**
6. Relate mean precipitation and temperature patterns to Rossby waves in the upper atmosphere.
7. Explain the effects of ocean currents and land–sea temperature gradients on the position of upper air waves.
8. Explain the **sea breeze, land breeze, valley and mountain breezes,** and **foehn wind system.**
9. Explain the effects of local wind systems on human activity.
10. Describe the **Asian monsoon** wind system.

Solar energy stirs the atmosphere.
© Doug Sherman/Geofile

Figure 5.1

Uneven heating of the Earth's surface causes the air to circulate vertically and horizontally in a series of convectional systems. Photo taken in infrared wavelengths, June 4, 1975. Warmest temperatures are black and dark shades, while white represents cold temperatures such as snow and high clouds consisting of ice crystals. Near the center and to the north and east a tropical storm is forming off the coast of Mexico.

Photo by NOAA.

Motion is the normal state of the atmosphere, with the Sun's energy as the driving force. When solar energy warms the air, activity increases, and the atmosphere stirs and churns restlessly over the surface of the Earth. This is an example of solar energy being transformed into wind energy through convection. If the Earth were uniformly heated, air circulation would be similar to that of the Sun's interior, where hot gases rise and cooler gases sink to be reheated in a cyclic fashion. However, because there is a relative surplus of insolation in the lower latitudes and a poleward global cooling, the atmosphere circulates poleward carrying heat to these cooler zones. In sum, the Earth's wind systems move by solar power and transfer energy from lower to higher latitudes (fig. 5.1).

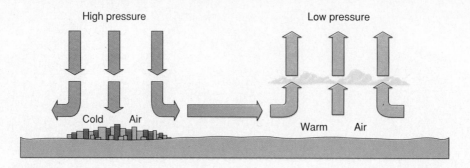

Figure 5.2

Winds are caused by differences in pressure. Air always moves from zones of higher pressure into zones of lower pressure until pressure differences are reduced or neutralized.

WINDS AND PRESSURE

Winds are fundamentally caused by differences in heating over the Earth's surface. These produce contrasts in atmospheric density, hence pressure. Remember, **density** is the amount of matter in a given space. More matter means more force on a given area, or **pressure.** Low-pressure regions are zones where the air exerts a lesser force than 1,013 millibars, the standard pressure at sea level.

Air always moves into zones of lower pressure to equalize the pressure over the Earth's surface (fig. 5.2). The sea breeze, discussed in chapter 4, results from differences in heating between land and water, which creates a pressure difference causing air to move onshore.

High pressure develops where the air is in a compressional subsiding state. When air subsides, it creates a greater force on the surface than the standard pressure. Think of the atmosphere as a fluid of gas, expanding and contracting, rising and sinking, always moving. The dynamics of the fluid both produce and result from changing pressures.

FORCES ON A PARCEL OF AIR
The Pressure Gradient Force

Pressure patterns are plotted on maps as **isobars**, or lines that connect points of equal pressure. The National Weather Service and all international weather observers map global weather patterns using millibars as the standard unit. A millibar is equal to a force of 1,000 dynes per square centimeter.

Pressure maps show not only the isobaric patterns but also potential wind velocities and direction. On figure 5.3, note that the isobars are not uniformly spaced. The distance between isobars tells the rate of change in pressure, or **pressure gradient.** Isobars are read like contour lines on a topographic map. Where the lines are closer together, the barometric slope is steeper, hence the rate of change is greater. Just as water flows faster down a steeper gravity slope, wind velocities are also greater down a steeper **barometric slope.** Air always moves down the pressure gradient from higher to lower pressure.

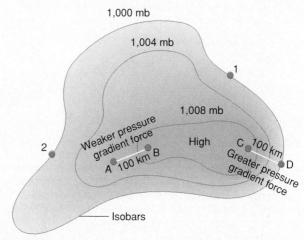

Figure 5.3

Isobars connect points of equal pressure. Points 1 and 2 have equal pressure. The distance or spacing between isobars reveals rates of change in pressure. If the spacing between lines decreases, the pressure difference will be greater. The *pressure gradient* is greater between C and D while the horizontal distances are the same between A and B and C and D. The difference in pressure between C and D is over 8 millibars, while the difference in pressure between A and B is 3 millibars. The pressure gradient is 3 millibars per 100 kilometers between A and B. Winds will be stronger where gradients are steeper.

Other Forces on a Parcel of Air

The driving force for water is gravitational energy, while air is driven primarily by the pressure gradient force. Just as in the case of gravity driven water flowing in a creek, there are several reactionary forces opposing the pressure gradient force. These include friction of the channel or surface, Coriolis effects resulting from the Earth's rotation, and centripetal effects as the air changes direction.

If we isolate a parcel of air, we can analyze the factors that cause its ultimate path. This path will reflect the results of all forces acting on the parcel. Figure 5.4 shows a parcel P located between two isobars well above the surface so as to avoid friction with the ground. Let's examine the forces on this parcel of air.

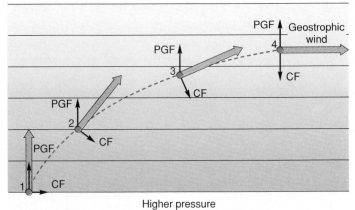

Lower pressure

PGF Geostrophic wind

PGF

CF

PGF

CF

PGF

CF

PGF

CF

Higher pressure

PGF Pressure gradient force
CF Coriolis force
⇨ Wind

Figure 5.4

A parcel of air travels a path that is determined by the series of forces acting simultaneously on the parcel. The Coriolis effect is the result of the Earth's rotation. The parcel P moves forward and to the right of its path until point 4, when the Coriolis effect and the pressure gradient forces are equal and opposite. At this point, neglecting friction, the parcel P will move parallel to the isobars. These conditions occur about 1 kilometer (3,500 feet) above the surface. This wind is termed a *geostrophic wind.*

The Coriolis Effect

If we could take snapshots of this blob of air as it starts from rest in figure 5.4, the following sequence would be observed. At point 1, the parcel P is accelerated by the pressure gradient force down the barometric slope. At point 2, as mentioned in chapter 2, the Earth's rotation produces the Coriolis effect, which causes any moving object to move to the right of its path in the Northern Hemisphere and to the left in the Southern Hemisphere, regardless of its direction of motion. The Coriolis effect is zero when the object is at rest and at the equator. It reaches a maximum at the poles. Coriolis effects also increase as the velocity of the object increases. At points 2 and 3 our snapshots show the Coriolis effect deflecting the parcel's path more and more until at point 4 the pressure gradient force and the Coriolis effect are equal and opposite. The parcel now moving in the isobaric channels parallel to the isobars is called a **geostrophic wind.**

If the Earth did not rotate, we would remove the Coriolis force, leaving only the pressure gradient force. Therefore, air would flow at right angles to the isobars down the barometric slope at velocities proportional to the steepness of the barometric slope. On the smallest scale, for instance when you use a straw or vacuum pump, air is primarily influenced by the pressure gradient force. Only over an extended distance of several kilometers does the Coriolis effect have a significant influence.

The Coriolis effect is an apparent force. Objects are deflected providing the Earth's surface is the reference system used to determine direction. However, if a point in space is used, the

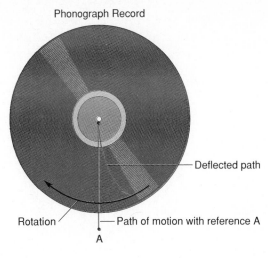

Phonograph Record

Deflected path

Rotation Path of motion with reference A

A

Figure 5.5

The Coriolis effect can be demonstrated by placing a phonograph record in motion and then making a straight-line motion with a piece of chalk across the surface. As a result of these two motions, you will see on the record a curved path. From your observation point, the chalk motion was in a straight line. However, from the perspective of the record surface a curved path is observed. On the Earth, the latitude reference system, being fixed to the turning platform, reveals a curved path. If celestial reference points are used, a straight-line motion is observed. Point A is a stationary point in space.

object will have no deflection. Figure 5.5 illustrates this concept. You can demonstrate the Coriolis effect by placing an old phonograph record you wish to discard on a turntable and set it spinning. Next, take a piece of chalk and make a mark, using a straight-line motion, on the moving record. You will see that no matter how fast or slow you move it, the chalk always produces a curved mark on the record. From your perspective the motion with your hand is a straight line. On the surface of the record the line is an arc. Thus, the deflection effect is real when the Earth's surface is our reference system. The turntable, which turns clockwise, is a model of the Southern Hemisphere turning toward the east around the South Pole. The Coriolis deflection is to the left. In the Northern Hemisphere, which rotates counterclockwise, the deflection is to the right.

Centripetal Force

When air flows around the center of a high- or low-pressure system, **centripetal force** enters the picture. This force is directed toward the center of the low-pressure system and works either in concert with the pressure gradient force in a low-pressure system or the direct opposition to it in a high-pressure system. All of us experience centripetal force when we round a curve. Centripetal force only comes into play when a curved path is taken. The force keeps the car on the road through friction between the tires and the pavement. The strength of the force varies with the radius of the curve and the magnitude of the wind velocity (fig. 5.6). Thus, as air flows in the isobaric channels around centers of high or low pressure,

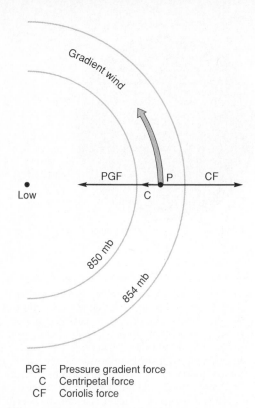

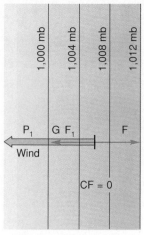

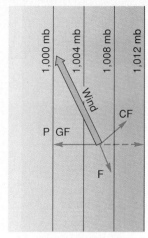

PGF   Pressure gradient force
C   Centripetal force
CF   Coriolis force

Figure 5.6

The centripetal force (C) on a parcel (P) of air pulls the parcel toward the center of the curve in concert with the pressure gradient force PGF. The Coriolis force is slightly less than the opposing forces, thus the parcel stays in a curved path. This wind, neglecting friction, is termed the gradient wind.

Winds at the equator have no Coriolis forces

Winds at 30° N cross the isobars toward the low pressure

Figure 5.7

Friction plays an important role not only in slowing the forward motion on a parcel of air, but also in reducing the Coriolis effect. Because the Coriolis effect decreases with decreased velocity, the pressure gradient force dominates and the parcel moves directly toward the direction of the low pressure. PGF is now balanced by the sum of friction F and Coriolis force CF. Notice that no Coriolis force is present at the equator. Thus, a parcel of air moves at right angles to the isobars under the driving force of the pressure gradient force.

centripetal force must be added to the influences on the parcel of air to explain the curved path of the wind above the friction layer. This wind is termed **gradient wind.**

Friction

When air or water moves over the Earth's surface, its velocity is reduced and its direction is changed by surface drag or **friction.** Rugged, mountainous country creates more friction than a lake or ocean surface. A parcel of air moving over a rugged surface will have lower velocity and less Coriolis effect at the lower velocity caused by frictional drag. If the Coriolis effect is weaker, the pressure gradient force is able to direct the air on a path across the isobars, toward a center of low pressure (fig. 5.7).

The combination of forces creates a counterclockwise spiral for low-pressure systems in the Northern Hemisphere and a clockwise rotation south of the equator. High-pressure systems rotate clockwise in the Northern Hemisphere and counterclockwise in the Southern Hemisphere.

Vertical Forces

The discussion until now has focused on the horizontal forces upon a parcel of air. Let us now consider the vertical forces influencing a parcel of air, which include gravity and buoyancy (fig. 5.8).

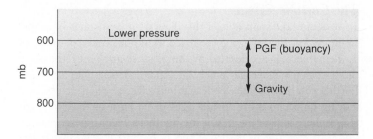

Figure 5.8

Gravity and buoyancy determine the vertical motion of the air. If the density of the parcel of air is lower than the surrounding air, the parcel will rise as buoyancy dominates. When it has a higher density, the parcel will sink as gravity dominates. If the vertical pressure gradient force (PGF) equals the gravity force, the air is in hydrostatic equilibrium.

Gravity is continually pulling air downward toward the center of the Earth. If it didn't, the Earth would soon be breathless and barren of life as is the Moon, where surface gravity is only one-sixth of the Earth's. Gravity is also countered by **buoyancy.** If the parcel of air has a higher density than the surrounding air, gravity dominates and its motion is downward. Rising air currents over a campfire illustrate the buoyancy effect. Heating of the air expands its volume and lowers the density, hence its pressure, according to the gas laws. When the air rises, buoyancy is the stronger of the two vertical forces. When the air is stagnant, all forces are in balance.

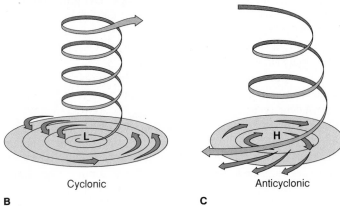

A

Cyclonic

Anticyclonic

B   C

Figure 5.9

(A) Hurricane John and cyclonic motion. (B) Cyclonic motion can be described as air that is ascending in a spiral motion under low-pressure conditions. (C) Anticyclonic patterns are the opposite of cyclonic motion. The air descends under high pressure in the opposite direction of cyclonic motion. For both cyclones and anticyclones, the direction of rotation is the reverse south of the equator due to the Coriolis effect.

Photo by NOAA.

Next, let's turn our attention to the many patterns of pressure and winds that result from the many forces on a parcel of air.

CYCLONES AND ANTICYCLONES

As air circulates over the Earth's surface, seldom does it maintain a constant course. The forces just described are continually changing, causing the air to alter its velocity and direction. Two major circular patterns of wind and pressure dominate the global circulation. The meteorologist calls these **cyclonic and anticyclonic systems** (fig. 5.9). Cyclonic systems are low-pressure regions of ascending, spiraling air currents. Anticyclonic systems are high-pressure spirals of descending or subsiding air. In the Northern Hemisphere, cyclones rotate counterclockwise and anticyclones

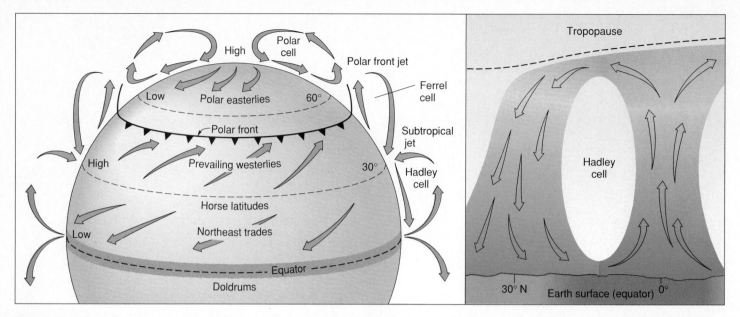

Figure 5.10

Global surface pressure and wind systems result from differences in heating over the Earth's surface. The wind and pressure belts are arranged in a latitudinal pattern. Low-pressure belts straddle the equator and also a broad belt of the upper middle latitudes (50° to 65° north and south). High-pressure patterns cap the poles and ring the Earth at latitudes 20 to 40° north and south. The major wind belts are sandwiched between the zones of low and high pressure, feeding and relieving these pressure differences. The winds always blow from highs to lows down the barometric slopes. Moving from the equator poleward one first finds the trade winds, which feed the doldrums or intertropical convergent zone. Then the anticyclonic subtropical high pumps air out into the westerlies, which flow toward the cyclonic belt of the polar front. At the polar front, moist tropical air confronts dry polar air, resulting in a series of wave cyclones and anticyclones. At the poles, air is under high pressure and moving equatorward in the belt of the polar easterlies. Global winds drive corresponding ocean currents.

have a clockwise spiral. The Coriolis effect reverses the direction in the Southern Hemisphere. Cyclonic systems range in size from more than 1,000 kilometers across to small dust devils spiraling across a hot, dry field or parking lot on a summer day. Hurricanes, tornadoes, and the middle-latitude wave cyclone are included in the cyclonic systems.

GLOBAL SURFACE PRESSURE AND WIND SYSTEMS

The Sun's energy heats the Earth's atmosphere and surface unevenly to produce a series of cyclonic and anticyclonic eddies and global wind patterns. Potential insolation is strongest at the equator, where a permanent zone of low pressure forms there (fig. 5.10). This equatorial low-pressure zone is called the **intertropical convergent zone** or **doldrums.** The air here rises and flows poleward in convectional circulation.

Air that rises along the equator drifts poleward to settle or subside in zones of high pressure located near 30° latitude and over the poles. Of the subsiding anticyclonic air at 30° latitude, a portion returns to the intertropical convergent zone by way of the trade winds. The rest flows poleward in the **westerlies,** forming an anticyclonic motion around semipermanent high-pressure cells.

Another belt of low pressure known as the **subpolar cyclonic low,** centered between 50° and 65° latitude, represents another convergent zone. This low is best developed in the Southern Hemisphere, where land masses are absent. Pressure here averages 984 millibars. At the poles are high-pressure caps. Cold descending air from the top of the troposphere subsides on the ice-covered polar zone and moves toward the equator near the subpolar lows in the wind belts of the **polar easterlies.**

The subpolar low-pressure belt is a zone where polar air of the polar easterlies mixes with warmer humid subtropical air of the westerlies. Along this convergent zone, the polar front, cold and warm air collide to produce a whole series of cyclonic storm systems explained in chapter 6.

The seasonal variations in insolation cause a seasonal shift in surface temperatures. This is especially true in the Northern Hemisphere because large land masses are located at latitudes where considerable seasonal variation in solar energy occurs. Recall that land areas heat and cool rapidly in contrast to bodies of water. This temperature characteristic produces a wide range in temperature, pressure, and wind patterns between summer and winter seasons in the Northern and Southern Hemispheres.

If we examine global circulation by looking at the profile of motion in figure 5.10, it is apparent that there are three circulation cells on each side of the equator, extending vertically to the tropopause. The atmospheric cell nearest the equator is called the **Hadley cell,** and the driving force is solar energy. As the Hadley cell rotates, we can compare it to a driving gear which is interconnected with the poleward cells, the **Ferrel** and **polar cells.** If the Hadley cell stopped rotating or lost power, the other cells would come to rest because they receive their energy from its motion.

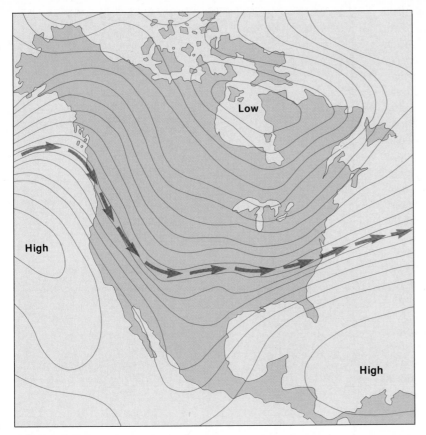

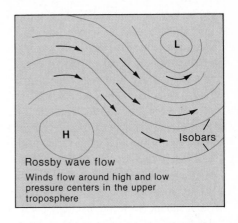

Rossby wave flow

Winds flow around high and low pressure centers in the upper troposphere

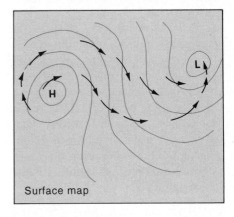

Surface map

Figure 5.11

Upper air motion is in wavelike meandering patterns that circle the globe in the tropics, subtropics, and upper middle latitudes. The long waves in the upper middle latitudes are known as Rossby waves. On the surface, winds cross the isobars and converge on the low-pressure center. Friction causes this inward spiral toward the low and outward spiral from the high-pressure center. Higher in the troposphere the winds follow the isobars in the *jet stream.* The stippling shows where the jet stream is strongest.

Thus, the atmosphere is constantly stirring in a broadly systematic manner. Motion is both horizontal and vertical around centers of high and low pressure. Rotation and surface conditions, in conjunction with changes in the seasons, produce continual changes in the patterns just described.

UPPER AIR CIRCULATION— ROSSBY WAVES

Circulation in the upper atmosphere is responsible for daily weather patterns. Forecasters watch the upper air motion closely to predict future weather. The atmosphere begins to flow smoothly and becomes less turbulent with increasing altitude. Near the top of the troposphere and into the stratosphere, air motion is more consistent and wavelike. Long waves known as **Rossby waves,** shown in figure 5.11, meander around the poles. Air motion in these waves can reach speeds in excess of 300 kilometers (200 miles) per hour. This high-velocity wind in the upper middle latitudes is called the **polar jet stream.** It resembles a stream of water flowing with variable velocity. It has short blasts of high-velocity flow along its course.

The shape and position of the wave play a major role in determining the weather path. Both temperature and precipitation patterns, as well as cyclonic and anticyclonic development, are under the influence of these waves. The oscillating wave systems with the westerly jet stream encircle the polar region and extend deep into the stratosphere, particularly during the winter season. Not only do the westerlies and accompanying Rossby waves intensify during the winter, but they also shift toward the equator. When this happens, polar air moves to lower latitudes along the wave troughs or zones of lower pressure, and warm tropical air shifts poleward along the west side of the higher pressure ridges. If the wave pattern is persistent, weather conditions also remain consistent. Both droughts and above-average precipitation can be attributed to static upper air waves and accompanying wind patterns.

Weather Forecasting and Rossby Waves

Examine figure 5.12 at positions A and B to understand the significance of the upper air flow to the weather conditions. If the general pattern remains constant for a period of several days, a predictable temperature and precipitation pattern develops for the surface location. In location A of the trough, southward air flow is introducing polar air masses, lowering temperatures in its path. In location B, beneath the ridge, warm air moves north from lower, warmer latitudes. Because of these waves, great quantities of energy and moisture are transported poleward to high-latitude regions. When a trough remains stationary for an extended period of time, above-average precipitation can result at this location. Precipitation is already enhanced by the general upward motions characterized by this portion of the trough in the jet stream axis zone. Clockwise rotation or **negative vorticity** to the rear of the trough encourages subsidence and lower-than-normal precipitation. Counterclockwise rotation or **positive vorticity** in the upper level trough enhances ascent, instability, and high precipitation. Figure 5.13 shows precipitation patterns, mean cyclonic tracks, and the jet stream axis.

The intensity of wind velocity and amplitude of the wave increase as the fall and winter months approach. An increased thermal gradient between the equator and the pole produces a corresponding pressure gradient, resulting in higher wind velocities in the jet stream circulation. The jet stream wind surges result from a variety of causes, but all are associated with steep temperature gradients. In fact, we see that every middle-latitude cyclonic system has an associated polar jet stream in the Rossby wave system. The polar jet stream is the storm track of the middle-latitude storm system, controlling both middle-latitude temperatures and precipitation patterns.

Temperatures and Precipitation and Rossby Waves

Now let's apply this upper air flow to actual temperature and precipitation patterns on the North American continent. Figure 5.14 shows a typical trough and ridge pattern that persists especially during the summer season. Summers are periods of low amplitude, smaller waves, and lower wind velocities, located at latitudes primarily over northern Canada.

Typically, winter periods show intensification of westerly circulation, high amplitudes, and long waves, with a decrease in frequency or number of Rossby waves (fig. 5.15). It is believed that increased temperature gradients between land and sea and also between higher and lower latitudes induce these changes. Figure 5.15 shows that a trough is situated off the Asian coast and a ridge is wedged over western North America. These ridge and trough systems show a rather persistent pattern because of underlying surface conditions that contribute to their existence.

If we examine the western Pacific winter trough where precipitation is high, the trough appears to be strongly influenced by the warm, underlying waters adjacent to cold, continental air from the interior of the Asian continent. This strong temperature gradient encourages cyclonic activity. The eastern Pacific ridge, which extends over western North America, is enhanced by the cold ocean current along the west coast of the continent and the cold surface of the Coast Ranges and Rocky Mountain system. The cold ocean current reduces precipitation potential by cooling the air, thus stabilizing it.

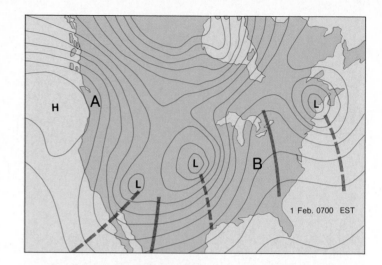

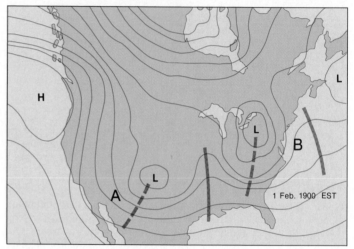

1 Feb. 0700 EST

1 Feb. 1900 EST

Figure 5.12

A Rossby wave or long wave consists of high-pressure, anticyclonic ridges and low-pressure troughs. These two upper troposphere (500 mb) charts, 12 hours apart, illustrate short waves moving through the long-wave (solid line) pattern. Short waves are indistinct in the long-wave ridge position in the Gulf of Alaska, at left. Short-wave troughs (dashed lines) tend to deepen in the long-wave trough position, which extends into northern Mexico. Short-wave ridges (solid lines) are indistinct in the long-wave trough but develop as they move out of the trough, as did the one that moved from the Southwest and northern Mexico into the Mississippi Valley. Letters A and B show movement of the short waves in a 12-hour period.

The trough along the northeastern coast of North America results from the same factors encouraging the eastern Pacific trough. Warm ocean currents and the mixing of warm, humid tropical air with cold, dry continental air from the North American interior contribute to the results. Although daily patterns may deviate from these mean positions, these wave positions are highly probable during the winter months and become the basis for weather prediction on both a short-term and long-term basis, extending for periods of 14 days or more. Chapters 8 and 9 discuss how wave patterns correlate with climatic zones in the middle-latitude belt of the westerlies.

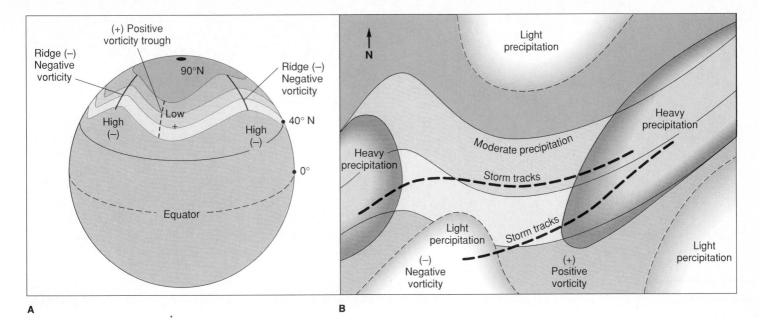

A   **B**

Figure 5.13

(A) Precipitation patterns, cyclonic tracks, and the jet stream axis
are predictable patterns in a Rossby wave. (B) Positive vorticity
(rotation) and precipitation patterns occur on the eastern side of the
trough.

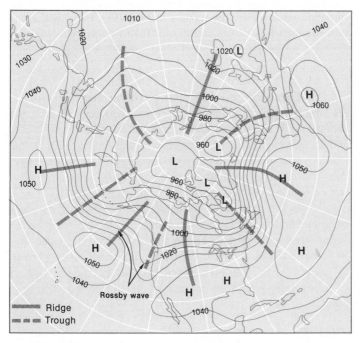

Figure 5.14

Summer periods produce smaller wave amplitudes, lower wind
velocities, and less cyclonic precipitation. The wave pattern also
shifts considerably poleward. Winter results in larger waves, higher
winds, and more cyclonic precipitation extending toward the equator.

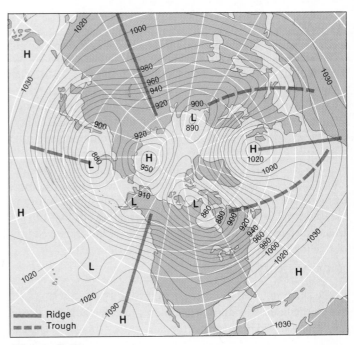

Figure 5.15

During the winter months, increased temperature gradients between
land and sea and between higher and lower latitudes are believed to
be contributing factors to the larger amplitude wave patterns and
persistent location centered on 40° latitude.

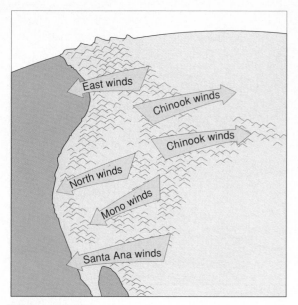

Figure 5.16

Foehn winds are known by different names in different parts of the mountainous West. In each case, air is flowing from a high-pressure area on the windward side of the mountains to a low-pressure or trough area on the leeward side. A strong foehn may flow down the leeward side of the mountains, bringing warm and extremely dry air to lower elevations.

LOCAL WINDS

Foehn Winds

Between a ridge and trough of the Rossby wave, subsidence with compressional heating can offset lower polar outbreaks of cold air. The result is a warm, dry wind called a **foehn** (fig. 5.16). This foehn, called a warm **chinook** wind east of the Rockies, results from this type of upper air flow. These foehn winds descend around a wave of high pressure and clockwise rotation (negative vorticity) and blow downslope across the Great Plains, often warming the air by 17° C to 22° C (30–40° F) in just a few hours.

The **Santa Ana** winds of southern California represent another case of descending currents in upper-air Rossby waves. Compressional heating can bring 32° C (90° F) temperatures to the region in December and January, drying the chaparral to extreme fire conditions. Many of the great chaparral and forest fires of this region have been fanned by these foehn winds.

The Sea Breeze

In the summer, along most coasts, the Sun warms the land and the overlying air to higher temperatures than the adjacent ocean (see chapter 4). Heating over the land causes the air to expand and rise, and cool ocean air begins to move onshore toward the lower pressure. This type of flow is very localized and shallow, usually extending only a few hundred meters deep into the atmosphere. Light onshore winds begin in midmorning as inland temperatures rise, creating convectional circulation (fig. 5.17).

These light midmorning winds begin just offshore and extend inland 30 to 50 kilometers (20 to 30 miles). They reach their maximum velocity in the late afternoon. After sunset, the sea breeze begins to taper off as the land begins to cool, which stops the convectional process.

Along the Pacific coast, the cold ocean current chills the sea breeze, producing a sizable temperature gradient between shoreline locations and the warm interior valleys. Because the surface marine air is cooler and denser than the overlying warmer drier air, an inversion is formed (as described in chapter 4), and air pollution remains in this cool marine layer. The cold ocean current, adjacent to the coast, intensifies the inversion (fig. 5.18).

The sea breeze is best developed during the summer, when the largest temperature gradient develops between the coast and the interior. It is not uncommon to see a 22° C (40° F) difference in daily maximum temperatures between the Pacific coast and the interior valleys. Once this marine layer is heated in the late afternoon to the same temperature as the overlying drier air, the inversion is destroyed and mixing of the air occurs.

In the *study area* of chapter 1, the sea breeze is a key environmental factor. Each afternoon during the summer, the tall coastal redwoods are refreshed by the cool marine air, laden with moisture in the form of fog. Westerly wind speeds reach 24 to 40 kilometers (15 to 25 miles) per hour until sunset. Then as the land cools down, the great trees cease to sway and return to their stoic positions of vertical grandeur.

The Land Breeze

The rising of the full Moon after sunset can signal the land breeze. Cool air gains momentum from pressure gradient forces. The result is gentle winds back to the sea, through the channels of the drainage systems. This diurnal summer rhythm is nearly as punctual as the daily tides.

At night the land cools more rapidly than the sea, and pressure patterns reverse with corresponding wind reversals (fig. 5.19). The land breeze is usually less developed, very shallow, and extends only a short distance offshore because temperature differ-

A

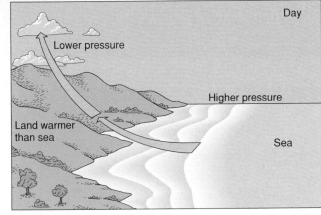

B

Figure 5.17

(A) Panama Beach, Florida. As the air warms over the land it rises, then cools, and the cloud pattern forms. (B) The sea breeze occurs locally along the shorelines of the world. Bodies of water produce a local microclimate and create a surface pressure gradient. Onshore air movement develops in the afternoon when land temperatures warm up and produce convectional currents. After sunset, winds subside and may reverse to an offshore drift when the land temperatures are lower than the water. This offshore and onshore movement is a very shallow surface wind, cyclic with the diurnal heating and cooling of land and water.

ences between land and sea are less during the night than during the day. Like the sea breeze, it is best developed during the summer when stronger differences in temperature develop between land and sea. Clear skies and low humidity promote radiation exchange and temperature contrasts.

Valley and Mountain Breezes

Valley and mountain breezes arise from contrasts in temperature between the valley floor and mountain slopes. As the Sun warms the valley floor and exposed mountain slopes, a "chimney effect" is created, and air rises up the valley and mountain slopes, reaching maximum velocity in the late afternoon. With evening, temperatures begin to fall on the shady valley floor and on the shady slopes. A cool air layer, in contrast with the surface, builds up and then drains off under the pull of gravity, as described in chapter 4. The nocturnal, gravity-driven winds follow the path of least resistance downslope (fig. 5.20). Because air is compressible, adiabatic heating, accompanied by turbulent mixing of cooler surface air, results in raising the temperature several degrees in the more wind-exposed zones. This warming influence is of paramount importance where frost is a problem for agriculture. In many mountainous or hilly regions, the growing season is extended later into

A

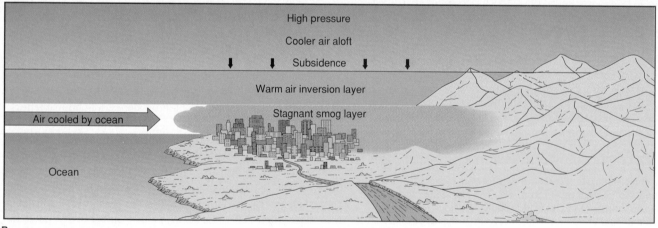

B

Figure 5.18

(A) Los Angeles basin viewed from the San Gabriel Mountains at 3,000 meters (10,000 feet) elevation. (B) A strong inversion can develop along the coast if the atmosphere is in a state of subsidence and high pressure.

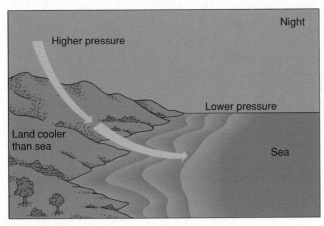

Figure 5.19

The land breeze is the reverse of the sea breeze. Air moves in a shallow layer offshore during the night. Weaker than the onshore sea breeze, it is best developed when strongest contrasts in temperature occur between land and sea. The sea must be warmer than the land for the wind to fully develop.

autumn and begins earlier in the spring where these winds at the mouths of canyons maintain warmer nighttime temperature conditions (box 5.1).

In some mountainous regions where humidity is high, upslope valley winds and increased cloudiness occur in the late afternoon. If moisture and wind conditions are right, precipitation may follow. Then, just as suddenly as the rain began, clearing skies occur toward evening. When the uplift stops, the upslope valley wind process reverses after sunset and the downslope gravity-driven wind follows. Solar energy is the key to upslope movement. Gravity dominates once the Sun sets.

There are a number of local names for these gravity-driven, cool downslope winds. Along the European Adriatic coast, a wind known as the **bora** brings cold winter air downslope from an interior plateau region. A similar wind known as the **mistral** occurs in the winter over southern France and blows southward across the Mediterranean Sea. Where cold air drains off the fiord glaciers of Norway and Greenland, similar winds are generated. Sometimes these winds can have considerable force if barometric

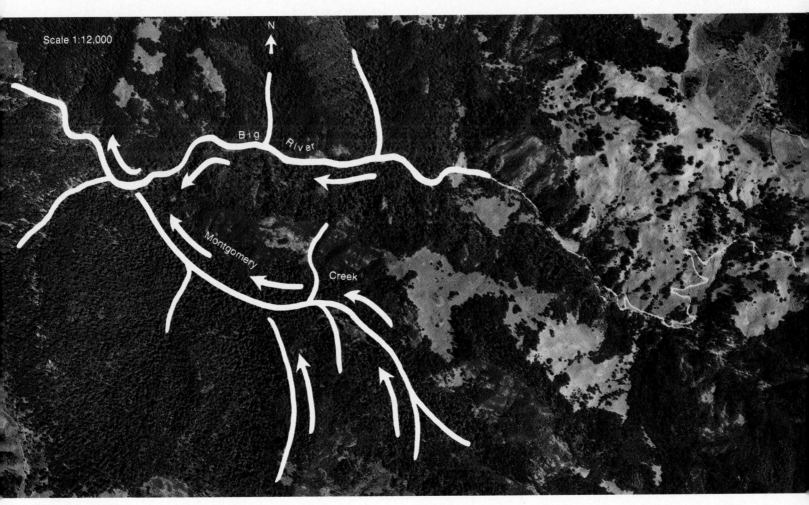

Figure 5.20

Nighttime air drainage in the *study area* of chapter 1. Valley and mountain breezes are caused by sharp contrasts in temperature between valley floors and mountain slopes. Heating on the slopes creates increased convection. At night, air in contact with the surface cools rapidly and drains off the slopes and down the valley.

Turbulent mixing and adiabatic heating raise the temperature of the moving air, especially in the more fluid zones. The larger stream drainage to the north has a larger air drainage than the tributary Montgomery Creek.

Photo by NASA

Impact of Downslope Winds on Land Use in Hilly Terrain

BOX 5.1 A knowledge of surface influences on air drainage and minimum temperatures has aided agriculturists, foresters, and industry planners in determining the best sites for frost-sensitive crops, timing for controlled burning, spraying, and location of manufacturing plant sites (fig. 1). For example, knowing where cold air drifts can direct the meteorologist in the strategic placement of wind machines to make maximum use of downslope winds and reduce cool air puddling.

Downslope winds must also be considered in the positioning of industries that release noxious fumes at night. Since these winds are confined to the surface layers of the atmosphere, industrial gases will not rise but be transported downslope. Therefore, planners can apply this knowledge in developing healthful zoning patterns for future development.

Controlled burning and selective herbicidal spraying of forests are also important applications of this knowledge of downslope air drift.

By knowing the potential for cool air drainage and low-level temperature inversions in a region, we can predict the severity of air pollution for the area. When an inversion develops, pollutants become trapped in the cool layer in the valley. The lower the inversion lid in the basin, the greater will be the concentration of pollutants. The top of the inversion is at the height where temperatures cease to climb and begin to decrease (fig. 2).

Figure 1

Controlled burns require a knowledge of atmospheric winds. This fire was ignited from the helicopter seen in the smoke.
Photo by Gary Bowman.

A

B

Figure 2

Inversion height is an excellent indicator of the severity of air pollution a location will have. The inversion height in A is 150 meters at 7 A.M. By noon (B), the inversion had deepened reducing the density of the pollutants, but reducing visibility in a deeper layer. Heating of the cooler air on the surface expands this layer. (C) Successive plots of temperature against height on a clear day show that early in the morning a shallow layer of air is heated, and gradually the warmed layer becomes deeper and deeper, reaching its maximum depth about midafternoon.

C

forces are working in conjunction with gravity. At the mouths of canyons, these winds have exceeded hurricane force of 120 kilometers per hour (75 mph).

MONSOON CIRCULATION

Summer Monsoon

In the middle latitudes of the Northern Hemisphere, large land masses warm up in the summer and create low-pressure conditions in the interior. This is especially noticeable in southern Asia, where there is a great onshore flow of tropical marine air from the Indian and Pacific Oceans known as the **summer monsoon** (fig. 5.21).

When this moist, unstable air reaches the warm Asian continent, increased convection occurs and heavy rains fall. The heaviest rainfall is along the windward facing slopes. The winds and rain begin in May and continue until about October. During the fall, continental temperatures begin to decline and pressure begins to rise. October is a transition period; the air becomes calm and clear and the onshore movement comes to a halt. The summer monsoon can be compared to the sea breeze, with similar causes that differ in scale and magnitude.

Winter Monsoon

October and November mark the time when the interior of Asia becomes cooler, and the pressure gradient force steepens over the interior land mass. This difference in pressure causes an offshore flow of cool, dry air mainly toward the equator and then across the tropical oceans to the warm Australian continent. Warm tropical air invades the northern coast of Australia, resulting in the summer monsoon for this land south of the equator. Air that was once dry becomes moist as it moves over the warm tropical ocean. This **winter monsoon** lasts until late April or early May when the pressure gradient disappears momentarily, then the cycle repeats itself. The traditional societies of India, Pakistan, and Southeast Asia gear their planting and their religious festivals and other cultural activities around these seasonal shifts in the wind and precipitation. If the winds are late, it can mean famine.

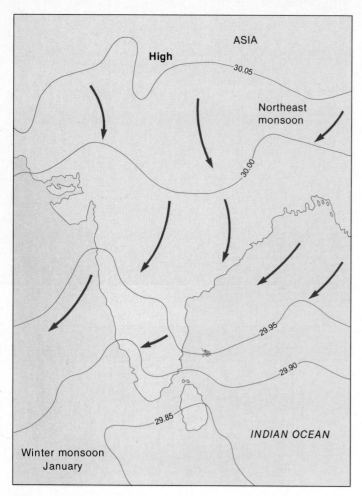

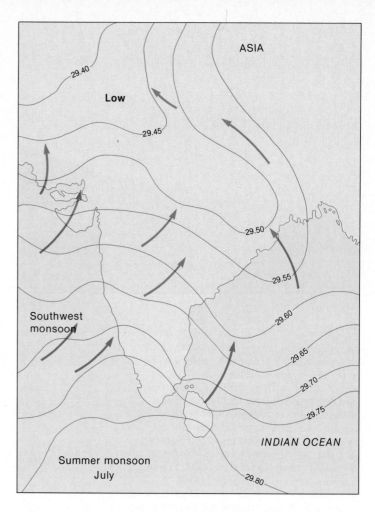

Figure 5.21

Monsoon circulation results for large-scale seasonal heating and cooling of the continents. In the summer the Asian land mass warms up creating low pressure in the interior. As the pressure differential grows between land and sea, moist tropical air moves onshore and brings heavy convectional showers lasting from May through

October over the Indian subcontinent. As summer shifts to the Southern Hemisphere, pressure patterns reverse, causing the air to move out of the higher pressure interior of Asia and into the warm interior of Australia, bringing moisture to the northern coast of Australia.

Even the centuries of trade between India and the east coast of Africa are tied to the monsoonal wind patterns. In the early days of wind-powered vessels, Indian merchants set sail southwestward toward Africa where they traded their wares for African products. The Indians were able to return when the summer monsoon began and carried them homeward to India's west coast.

Our own continent has a modified monsoon wind, but it is less developed than the system experienced by Asia. The North American continent does not have the extreme shift of summer and winter temperatures and pressures experienced by Asia. However, summers east of the Rockies are warm and humid. Thundershowers resulting from an influx of tropical air from the Gulf of Mexico over a warm land mass create instability, rain, hail, and tornadoes as described in chapter 7.

During the winter, high pressure builds up over Canada causing cold, dry air to build up. It then moves southward, sometimes as far as southern Florida and Mexico, bringing freezing

temperatures into these regions. A cold outbreak at these southern latitudes can destroy tropical agriculture and cause many people to become ill with respiratory diseases, where freezing temperatures are unusual and unexpected.

In the Southern Hemisphere, the Amazon rainforest of Peru and Brazil can suddenly be invaded by cool Antarctic air, dropping maximum temperatures to below 20° C (68° F), with heavy rains and high southerly winds known locally as afrio, or cool winds.

South of the equator, only Australia and the east coast of Africa have a significant monsoon pattern. While Asia is cooling off, Australia and the southern part of Africa are warming. The winter monsoon winds of Asia move south across the warm Indian and Pacific Oceans toward these continents. After picking up moisture from the warm oceans, these winds spread rain and high humidity into northeastern Australia and eastern Africa. When the monsoon winds shift, north desert air moves out of Australia's interior and is replaced by the forces of high pressure, producing several months of dry weather along the northern coast. A similar wind shifts along the southwestern coast of Africa, blowing from the southwest toward southwest Asia and India.

SUMMARY

The atmosphere is dynamic and constantly changing, and solar energy is the fundamental driving force. As solar energy warms the surface and overlying air, convectional processes transport energy from the lower latitudes poleward. Thus, differences in temperature cause differences in pressure. Air always moves from regions of high pressure to zones of low pressure. When pressure gradient forces increase, winds also increase in velocity.

Pressure patterns are described on weather maps by using isobars, or lines that connect points of equal pressure, with the millibar as the standard unit of measurement. The spacing and distribution of the isobars gives the weather forecaster information about wind velocity and direction.

The direction of air flow is influenced by a number of forces and effects working in concert to produce the resultant path. The forces or influences include the **pressure gradient force, Coriolis effect, centripetal force, friction, gravity,** and **buoyancy.**

If air circulation is centered around a low-pressure circular system, the motion is described as **cyclonic and positive vorticity.** Air motion around a high-pressure cell is **anticyclonic or negative vorticity.**

A global view of **cyclonic and anticyclonic pressure and wind patterns** reveals a belt of low pressure along the equator known as the **intertropical convergence zone, doldrums,** or **equatorial trough.** The northeast trades and southeast trades converge on this low-pressure belt. Air that has drifted poleward in the upper atmosphere begins sinking at 30° latitude, producing the **subtropical highs of anticyclonic motion.** The westerlies blow from these highs toward the **subpolar lows,** located at about 60° latitude. The **polar front** forms the boundary between the convergence of the polar easterlies and the westerlies. It is along this boundary that Rossby waves loop our globe and form the track of the polar jet stream.

Upper air circulation is wavelike around the poles. The wave motions, known as Rossby waves, form ridges of high pressure followed by troughs of low pressure. The mean position of these waves is an im-

portant factor in determining future weather forecasts. Surface features, such as water bodies, ocean currents, mountain systems, and shoreline configuration, influence the position of these waves.

Local winds are caused by local pressure gradients, induced by differences in heating between adjacent locations. These winds include the sea breeze, land breeze, and valley and mountain breezes. On a larger scale, Asia, Africa, and Australia experience the summer and winter monsoons. These wind systems create sharp contrasts in seasonal precipitation and temperature. The summer tropical monsoon brings heavy rains and high temperatures to southeastern Asia and India, while the winter monsoon introduces dry, cool air from the Asian continental interior. As the winter monsoon winds of Asia move across the ocean to Australia and the southeastern coast of Africa, moisture is collected and heavy summer precipitation falls on these continents.

ILLUSTRATED STUDY QUESTIONS

1. Trace the flow of energy from the Sun to conversion to wind energy (fig. 5.1, p. 93).
2. Using isobars, draw both strong and weak gradients on the map on page 96 (fig. 5.3, p. 96).
3. Describe the major locations of the world's high- and low-pressure belts (fig. 5.10, p. 98).
4. Explain the cause of the Coriolis effect (figs. 5.4 and 5.5, p. 95).
5. Contrast cyclonic and anticyclonic motion (fig. 5.9, p. 97).
6. Define and locate the horse latitudes and doldrums. What do they both have in common (fig. 5.10, p. 98)?
7. Name and describe the forces on a parcel of air (figs. 5.6, 5.7, and 5.8, p. 96).
8. Explain the causes of the monsoon wind (fig. 5.21, p. 108).
9. Describe the sea breeze, mountain and valley, and chinook winds (figs. 5.17, 5.18, 5.19, and 5.21, pp. 103–6).
10. What important role does the polar jet stream play in advection of warm and cold air (fig. 5.11, p. 99)?
11. Locate the westerlies, trade winds, and the polar easterlies (fig. 5.10, p. 98).
12. Describe the Rossby wave pattern and its relationship to the polar jet stream, precipitation, and the chinook wind (Figs. 5.11, 5.12, 5.13, and 5.14, pp. 99–101).

Clouds, Atmospheric Moisture, and Precipitation

6

Objectives

After completing this chapter, you will be able to:

1. Describe the major cloud types and the meteorological significance of each in terms of meterological events.
2. Explain the sequential steps of cloud and precipitation formation.
3. Compare and contrast the **solute and curvature effects** governing the growth of a droplet.
4. Explain the dynamic forces on a falling drop of rain.
5. Explain the processes leading to fog formation.
6. Describe the various types of **hydrometeors** and atmospheric conditions leading to their formation.

Cumulus cloud buildup.
© Shattil/Rozinski/Tom Stack & Assoc.

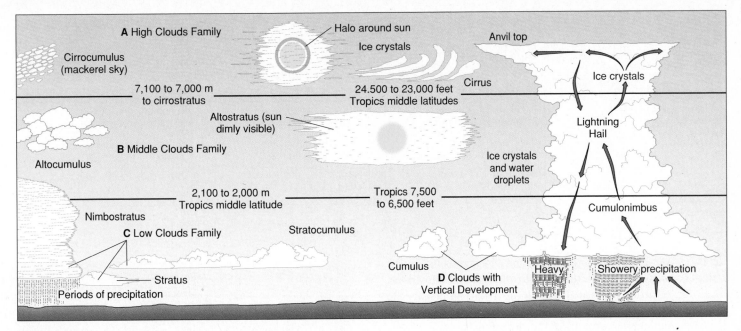

Figure 6.1

Clouds are named according to height and form. Cloud sheets or layers are described as stratus. A heaped or globular cloud is referred to as cumulus, streaked forms are called cirrus, and precipitation clouds are known as nimbus.

Insolation not only warms the surface but converts water into the energy-rich vapor form. When vapor cools and condenses to form a cloud, a visible signpost is present giving us clues to a dynamic atmosphere.

Clouds add a dynamic visual element to the atmosphere. Without them there would be no lightning, rainbows, or snow, fog, or mist. A cloud is not only a visible collection of tiny water droplets or ice crystals, but also a heat source. Each droplet and ice crystal releases energy during the formation process. Think of clouds as warm spots where energy in the latent form is being released—it is actually warmer inside a cloud. Clouds also scatter and reflect visual energy as well as absorb infrared from both the Sun and Earth, hence they regulate the Earth's heat budget. Clouds can provide information regarding atmospheric stability and instability.

This chapter begins with a discussion of clouds and their significance. Next, our focus will be on six steps leading to precipitation. Finally we will examine the **hydrometeors,** which include **rain, snow, hail, mist,** and **fog.**

CLOUD CLASSIFICATION AND SIGNIFICANCE

When you are outside and observing the sky, clouds can appear in an infinite variety of forms and colors of the visual spectrum and tones of pure white to black. At sunrise and sunset, the cloudless sky can also take on a brilliance of colors which include red, orange, yellow, green, and deep blue. Air pollution, both natural and human, can add special color effects to produce some of the most unusual color patterns and atmospheric glows. Forest fires or volcanic eruptions can help create a beautiful red sunset from the injected smoke and volcanic dust. This same particulate material can also contribute to cloud development by servicing as a surface for condensation.

Consider clouds to represent visually complex meterological events or processes. The color, shape or form, height, and rate of change give important clues to the trained meteorologist about the coming weather. Therefore, we can think of clouds as natural billboards advertising coming meterological events. A primary objective of this chapter is to sharpen your understanding of these signposts so that you will be able to explain their meterological significance.

The first published classification of clouds was published by Jean-Baptiste de Lamarck, a French naturalist, in 1802. Then in 1803 an English pharmacist and naturalist, Luke Howard, developed a more complete cloud classification which is the basis for the present system. He introduced Latin, the scientific language of the day, to describe or classify a system of related phenomena. Cloud sheets were called **stratus** (from *stratum,* layer); a heap or globular cloud **cumulus** (Latin for pile); a streak cloud cirrus (Latin for hair); and a cloud producing precipitation nimbus (Latin for rainstorm) (fig. 6.1).

Stratocumulus was added by Kaemit, a German meteorologist. This term describes a transition cloud form which is changing from a layered to a globular shape. Renow, director of

Table 6.1

Heights of Cloud Groups (étages) in Various Latitudes

Cloud Group	Tropical Region	Temperate Region	Polar Region
High	6,000 to 18,000 meters (20,000 to 60,000 feet)	5,000 to 13,000 meters (16,000 to 43,000 feet)	3,000 to 8,000 meters (10,000 to 26,000 feet)
Middle	2,000 to 8,000 meters (6,500 to 26,000 feet)	2,000 to 7,000 meters (6,500 to 23,000 feet)	2,000 to 4,000 meters (6,500 to 13,000 feet)
Low	Surface to 2,000 meters (0 to 6,500 feet)	Surface to 2,000 meters (0 to 6,500 feet)	Surface to 2,000 meters (0 to 6,500 feet)

Source: NOAA

the observatories at Parc Saint-Maur and Montsauris in France, introduced the terms **cirrocumulus, cirrostratus,** and **altostratus.** These terms further delineated cloud forms at various heights. He also introduced clouds of medium level between stratus clouds of low-level and high-level clouds of the **cirrus family.**

Observations of clouds from the vantage points of planes, balloons, satellites, and the ground reveal that clouds can be found over a wide range of altitudes. Cloud heights vary from sea level to the tropopause and even into the stratosphere. This upper boundary ranges from 18 kilometers (60,000 feet) at the equator to 7,500 meters (25,000 feet) at the poles. The atmospheric levels, where clouds appear, can be divided into three levels, or "étages," high, middle, and low. Specific cloud types or genera are commonly found at each level.

The highest clouds belong to the cirrus family. These include cirrus, cirrocumulus, and cirrostratus. In the middle level is the alto family, which include altocumulus as well as altostratus. The lowest level stratus family include stratocumulus and stratus. Several special cases of clouds that develop vertically are represented by nimbostratus, cumulus, and cumulonimbus. These clouds can be found extending into several levels. Cumulus and cumulonimbus can have their base in the low level with their tops sometimes reaching into the middle or even high levels.

Table 6.1 gives the approximate heights of the levels in the tropical, temperate, and polar regions. Note that the levels overlap and their upper limits vary with region or latitude.

Cloud Classification

Let's now examine the ten genera of clouds in terms of appearance, ingredients, or physical makeup and significance regarding weather phenomena.

1. **Cirrus** (Ci). These are high-level clouds in the form of delicate white patches or narrow fibrous bands with a silky sheen. Cirrus are composed exclusively of small ice crystals of low density, accounting for their transparency. Before sunrise and after sunset, cirrus color the sky in bright red, orange, and yellow. They remain visible much longer than other clouds because of their height. The cirrus cloud is common in fair weather, but it can be a forerunner of a storm. The atmosphere is stable when they are present (fig. 6.2).

Figure 6.2

These cirrus (Ci) clouds seen above the Himalayan peaks in Nepal crest at 7,250 meters (26,800 feet). They are composed exclusively of small ice crystals and have a fibrous, silky sheen.

Photo by Sally McGregor.

Figure 6.3

Cirrocumulus (Cc) are high clouds consisting of white patches, sheets, or ripples. They are composed of ice crystals and sometimes supercooled water droplets. This satellite view of India reveals cirrocumulus over the Indian Ocean. Cumulus clouds build up over land where convection is greater.

Photo by NASA.

2. **Cirrocumulus** (Cc). White patches, sheets, or ripples consisting of small grainlike elements characterize these clouds. The ripples and sheets can resemble the sand ripples along the beach or sand dunes. These clouds also are composed of ice crystals and sometimes supercooled water droplets which quickly turn to ice. The air is unstable when these clouds are present (fig. 6.3).

3. **Cirrostratus** (Cs). A thin, whitish, transparent veil which does not blur the outline of the Sun and Moon. Haloes are common around these celestial bodies. The sky can sometimes have a milky diffuse pattern with enough sunlight penetrating the clouds to produce shadows on the ground. These clouds follow cirrus as a storm approaches (fig. 6.4).

Figure 6.4

Cirrostratus (Cs) create a milky white sky with the Sun and Moon still able to penetrate this thin veil. Often a ring will form around the Sun and Moon caused by light refraction through the ice crystals that make up the cloud. The contrail of a jet aircraft is also a cirrus cloud formed of ice crystals.

Figure 6.5

Altocumulus (Ac) at Caborca Sonora, Mexico, form in the middle heights with globular patches, sheets, or layers. These cloud forms have distinct masses or rolls and consist primarily of water droplets.

4. **Altocumulus** (Ac). White or gray globular patches, sheets, or layers of clouds with distinct masses or rolls, sometimes partly fibrous or diffuse. Altocumulus clouds are usually composed of water droplets; however, ice crystals can form to produce a fuzzy-looking cloud. These clouds also represent instability in the middle level (fig. 6.5).
5. **Altostratus** (As). Following the cirrostratus and at a lower level, the altostratus clouds of middle height nearly always have a layered look covering great horizontal extent. These clouds can effectively filter the Sun, totally blocking its penetration. Altostratus do not produce haloes, which are associated exclusively with cirrostratus (fig. 6.6).
6. **Nimbostratus** (Ns). Dark, nearly uniform, clouds of precipitation, including rain or snow, usually of lower levels.

Precipitation is steady, often lasting for hours. Beneath the main cloud mass there is usually a low, ragged series of clouds called scudi fractocumulus or fractostratus. Nimbostratus may occur without rain when precipitation falls but does not reach the ground because of evaporation. This effect is called a virga sky. Typically, nimbostratus evolves from a layer of altostratus which grows thicker (fig. 6.7).
7. **Stratocumulus** (Sc). A patchy or layered cloud sometimes composed of globular masses or rolls. They appear soft and gray with some dark patches. Rolls are often so close that their edges are joined together forming a wavy appearance over the whole sky. Stratocumulus can change into stratus or cumulus depending on the state of the atmosphere's

Figure 6.6

Altostratus (As) clouds have a layered look, cover a great horizontal extent, and filter the direct rays of the Sun and Moon.

Figure 6.7

Nimbostratus (Ns) bring rain to a Peruvian valley in the Andes. They typically evolve from altostratus.

Clouds, Atmospheric Moisture, and Precipitation   *115*

Figure 6.8

Stratocumulus (Sc), viewed over the *study area,* appear patchy or layered in long rolls or globular masses.

stability. Stratocumulus usually consist of water droplets but sometimes ice crystals if temperatures are well below freezing (fig. 6.8).

8. **Stratus** (St). A low, uniform, layered cloud commonly associated with west coast summer weather of the middle latitudes. Stratus on the ground is called fog. It gives the sky a hazy appearance with occasional mist and drizzle. Stratus can easily be mistaken for nimbostratus when there is no precipitation. Often it is present under stable atmospheric conditions (fig. 6.9).

9. **Cumulus** (Cu). Tops are dome-shaped while their bases are flat. They can evolve into stratocumulus or altocumulus by spreading out. Cumulus clouds are formed from water droplets. If they grow vertically, cumulus clouds may

develop into towering cumulonimbus composed of ice crystals above the freezing line. These clouds develop under conditions of vertical air movement and instability. They are commonly found on the windward slopes of mountains (fig. 6.10).

10. **Cumulonimbus** (Cb). Massive clouds of great vertical development often resembling domes or towers. An anvil-shaped head can form at the top of the cloud resulting from strong horizontal jet stream flow at the top of the troposphere. Vertical wind velocities can exceed 300 kilometers (200 miles) per hour during this stage. The base of the cloud is usually horizontal and layered with very low ragged clouds below it. These are referred to as fractostratus and fractocumulus. Cumulonimbus clouds

Figure 6.9

Stratus (St) along the coast of northern California. When stratus rests on the surface it is called fog. Notice that the cloud is breaking up over the warmer land surface. This cloud dominates the cold ocean current areas of the world.

Figure 6.10

Cumulus (Cu) clouds indicate vertical air movement and instability in the atmosphere. Tops are dome-shaped and bottoms are flat.

Clouds, Atmospheric Moisture, and Precipitation   *117*

Figure 6.11

Cumulonimbus (Cb) clouds result from atmospheric turbulence and build massive domes. At the top of the troposphere the jet stream produces a shearing action and flattens the top into anvil-shaped heads. These giants of the atmosphere are often associated with hurricanes, tropical storms, tornadoes, and heavy orographic precipitation. This cumulonimbus was generated by a major fire in northern California. The top of the cloud, flattened at the tropopause, consists of ice crystals.

Photo by Gary Bowman.

are associated with violent weather such as tornadoes, hurricanes, and thunderstorms, described in chapter 7 (fig. 6.11 and box 6.1).

Atmospheric Conditions and Clouds

The form or morphology of the cloud is the chief clue to the state of the atmosphere. We can describe it as having either a stable or unstable state or in a process of changing from one to the other condition.

Atmospheric stability is a state of lowest energy. The atmosphere is in equilibrium. It resists change. Instability is an atmospheric condition when the system has potential for motion away from its point of origin. For example, a rock will tend to move downstream under the forces of gravity and stream current. Similarly, when the atmosphere is set in motion by differences in pressure it is considered unstable, lacking equilibrium, thus moving to a new location.

A Stable State and Clouds

Because the air in a stable state will resist upward motion, the air will spread out horizontally and the cloud form will also be horizontal or layered. Stratiform clouds such as cirrostratus, altostratus, nimbostratus, and stratus form in stable atmospheric conditions. Stability can occur when the air aloft is warmed or the surface air cooled by contact with a cool surface of land or water. Advection of air over a cool surface can also result in a stable condition as well as radiation cooling. These conditions can produce

Mean Cloud Cover and Altitude for July 1979

BOX 6.1 This photo shows global cloud cover as viewed by a National Oceanic and Atmospheric Administration satellite on a typical July day in 1979. The data were assembled by NASA's Jet Propulsion Laboratory in collaboration with the NASA Goddard Space Flight Center. Red indicates extensive clouds, green indicates moderate cloud cover, and blue minimal to no cloud cover. Note the heavy band of cloud cover around the equatorial belt and in the western Pacific Rim and North American Gulf and Caribbean areas (fig. 1).

Figure 1

Mean cloud cover and altitude for July 1979.

Photo by NASA.

---

persistent fog or stratus clouds. Summers along the Pacific coast of Peru, Chile, and California are typically foggy, resulting from a cold ocean current just offshore. Winter radiation fog in the Central Valley of California is very persistent during the rainy season. It is caused by a high dewpoint temperature and cooling near the surface on calm, clear nights. Both fog conditions result from very stable atmospheres.

An Unstable State and Clouds

The unstable state of the atmosphere is often observable in cumuloform clouds. Convection, topographic influences, and frontal uplift are the major mechanisms for formation of clouds and the triggering of instability. These are also the mechanisms leading to precipitation, to be discussed in the next section.

Clouds are the sources of precipitation. Without them the planet would be a desert. Clouds do not guarantee precipitation, but they are essential before it can occur. Most clouds never produce a drop of rain or flake of snow. Those that do follow a series of steps.

STEPS TO PRECIPITATION

For precipitation to occur certain conditions must be present. These include six events or processes: (1) **dirty moist air,** (2) **ascent,** (3) **cooling,** (4) **saturation,** (5) **condensation,** and (6) **droplet growth.** These steps occur both sequentially and simultaneously. To have the seventh step, precipitation, all of the previous steps must occur. Ascent and cooling are two steps that occur simultaneously, with saturation following as a result.

Precipitation is the culmination of events beginning with tiny cloud droplets, which are so light they remain in suspension.

Figure 6.12

Precipitation results from a series of steps. Notice the streamers or fall streaks of ice crystals falling from this altocumulus cloud. Before they reach the ground, evaporation returns these crystals to water vapor. This picture exemplifies many of the steps to precipitation.

How do they grow larger and heavier to finally fall? This is our focus to be summarized in seven steps or events (fig. 6.12).

"Dirty" *Moist Air—Step One*

"Dirty" moist air, the first step in the precipitation process, is essential for cloud formation. The air must be "dirty" as well as saturated before any condensation can occur. "Dirty" air contains many small microscopic particles originating primarily from ocean salt particles in aerosol form. Volcanic activity, desert storms, combustion, or various kinds of air pollutants are also secondary sources of "dirty" air. These small particles, also known as **hygroscopic nuclei,** become the microscopic surface for condensation. Pure water

vapor must have a surface on which to cling. Hygroscopic nuclei are "water lovers." For example, salt crystals attract water to their surfaces at relative humidities as low as 76 percent. In regions of severe air pollution, clouds and fog often form when the relative humidity is near this percentage.

Because of the molecular structure of water, it can dissolve more substances and adhere to more particles than any other liquid. For this reason it carries the title of "universal solvent." Figure 6.13 shows its molecular geometry and bonding characteristic that allows it to electrically attract salt in solution. The water molecule has negatively charged oxygen on one side and two positively charged hydrogen atoms on the other (fig. 6.13). This

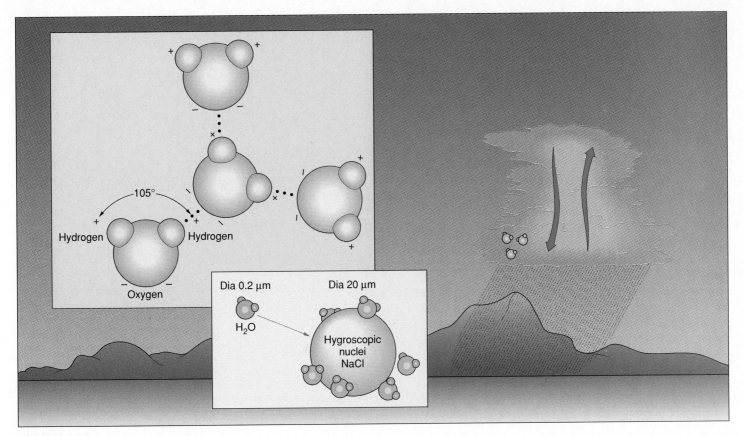

Figure 6.13

Condensation nuclei or hygroscopic (water-seeking) nuclei are the particle surfaces for condensation. The molecular geometry is the key to water's solvent and cohesive properties. Water molecules bond electrically to hygroscopic nuclei such as salt (NaCl), dust, or smoke particles. The water molecule is composed of oxygen, with a negative charge, and hydrogen, with a positive charge. Thus water molecules are dipolar or charged at each pole, attracting other molecules to form hexagonal crystals of ice.

geometric form produces a *dipolar* molecular structure, a molecule with two oppositely charged poles. Thus water molecules can hold, by electrical forces, a variety of charged atoms or ions in solution. The dipolar molecule gives water its internal cohesive bonding characteristic that causes water to form large droplets, and also gives it bonding adhesion to other surfaces such as hygroscopic nuclei or any surface where water is in the liquid state.

Once the "dirty" air attracts water vapor to condense on these particles, five additional conditions are necessary to build larger droplets that fall as a form of precipitation. For precipitation to occur, the droplets must grow in size until they can no longer be suspended. Ascent helps to accomplish this growth.

Ascent—Step Two

Convection Ascent

Convection starts where the Earth's surface is warmer than the surroundings. The air in contact with this warm surface forms a bubble of air, or **thermal,** and breaks away, rising as an invisible warm parcel of air. If it cools to the condensation point, the bubble takes the form of a cumulus cloud. Growth of this cloud is dependent on energy, the driving force of convection. If the atmosphere is unstable with a steep environmental lapse rate and abundant

moisture supply, growth potential is enhanced. (The **environmental lapse rate** is the rate of change of temperature with altitude in an air mass.)

As the cloud grows in size, it creates its own microclimate. First, it shades the ground. This shading effect can reduce convection by cooling the surface. Without a continual supply of rising air, the droplets begin to evaporate. The crisp silver lining of the cloud becomes jagged and fuzzy, finally dissolving. Often one will see a cumulus cloud building, gradually disappearing, then another reforming in its place.

If the day is warm and humid, by midafternoon the sky will be full of cumulus clouds. The cloud bases will all be at the same height above the surface where the dewpoint temperature is reached.

Let's now apply our knowledge about the effects of moisture on instability and cloud development as described in chapter 3. Suppose the environmental lapse rate in the lowest layer of 1,000 meters is greater than the dry adiabatic lapse rate. (Recall that this is the rate of change of air temperature of a parcel of air as it ascends or descends.) We can describe the atmosphere as absolutely unstable. Now, if a parcel of air with a dewpoint temperature of 35° C starts to rise and the air temperature is 40° C, the parcel will cool at the dry adiabatic rate of 1° C per 100 meters until it reaches saturation at the dewpoint temperature of 35° C. As long as the temperature in the cloud is warmer than the air surrounding the cloud, the buoyancy effect will keep the air rising freely (fig. 6.14). The **dewpoint** temperature is the temperature of the air when it can hold no more moisture or is saturated.

When the dewpoint temperature is reached, the air gains heat energy from the latent heat of condensation as the water vapor condenses. The dewpoint temperature is reached at the clouds' base where the wet adiabatic lapse rate (0.6° C per 100 meters) takes over. Ascent will continue until the temperature in the rising air parcel equals the temperature outside the rising parcel of air. Equal air temperatures indicate equilibrium and stability. This will be the case at the top of rising columns of air in the cloud.

An afternoon sky full of flat-based cumulus clouds with no vertical growth indicates fair weather and atmospheric stability. If the clouds continue to build into towering cumulonimbus, the air is in a state of absolute instability.

Seldom do cumulonimbus clouds punch through to the stable stratosphere, where temperatures actually increase. Thus anvil-shaped heads cap the tops of cumulonimbus clouds at the tropopause. Spreading results from the stable conditions there and high upper-level horizontal winds.

Topographic Uplift

If we lived on a smooth-surfaced planet the atmosphere would still lift by convection, but because the Earth has a rough surface, air can rise in a forced ascent over a land barrier. We refer to this type of ascent as **orographic uplift.** When an air mass approaches a chain of mountains like the Sierra Nevada of California, the Cascade Range of Oregon and Washington, or the Andes in South America, the air must rise over the barrier. This lifting can produce clouds if the air has the right dewpoint temperature. First, the air will rise and cool at the dry adiabatic lapse rate and then, when the air is saturated, follow the wet or moist rate. Condensation produces a cloud that takes on a lenticular shape forming at the crest of the mountain. As the air descends down the mountain slope, it is drier since it loses its moisture to condensation during the cloud formation process. If the air is rich in moisture, large cumulus congestus or even cumulonimbus can form with resulting heavy precipitation. This state is called conditional instability. The conditions are met when the rising air becomes warmer than the air outside the ascent zone, hence instability and cloud growth are the result.

This type of cloud buildup usually occurs in the afternoon. Both convection and orographic uplift work in concert when interior valleys heat up. Heating creates a thermal uplift or convection. This interior heating draws humid marine air onshore and up over the mountain barriers, cooling and condensing during ascent on the upslope side. The result can be afternoon thundershowers that follows the six steps to precipitation.

Widespread Ascent

Mountain barriers are visible obstructions to air flow and forced ascent. However, a denser cooler air mass can produce the same effect. A cyclonic storm system (discussed in chapter 7) is one of the most common causes of widespread ascent of the air. Associ-

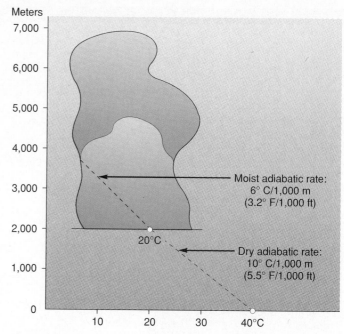

Figure 6.14

A rising column of air cools at a predictable rate of 10° C per 1,000 meters (5.5° F per 1,000 feet), the dry adiabatic lapse rate, before saturation is reached. Cooling results from expansion. The air loses energy during the expansion process. When air reaches saturation, condensation occurs, and further ascent causes cooling at the moist adiabatic lapse rate of 6° C per 1,000 meters (3.2° F per 1,000 feet). The moist rate is lower because cooling is offset by heat liberated during the condensation process.

ated with these systems are weather fronts that form air-mass boundaries where sharp contrasts of temperature, humidity, and density occur. For example, along a cold front where warmer moist air is forced to ascend, cumulus cloud development will run for hundreds of kilometers along the frontal zone spanning more than 100 kilometers in width. It is along a cold front boundary that the violent cumulonimbus cloud can form, and in some places spawn tornadoes.

The warm front boundary is a zone where warm air gently glides up over a wedge of cooler air, often resulting in clouds that cover thousands of square kilometers. The clouds that form are usually layered with the forerunners consisting of cirrus up to 1,000 kilometers (700 miles) ahead of the surface boundary of the warm front. Middle-layer clouds of altostratus follow cirrus. Then the lower level nimbostratus, or stratus, arrive as the warm front approaches. Therefore, clouds are visual manifestations of the atmosphere as it stirs and circulates in both stable and unstable conditions.

Ascent results in adiabatic cooling. This is energy loss resulting from expansion during ascent.

Cooling—Step Three

As air rises, expands, and cools, the expansion process causes the air to do some work or give up some energy. A rising air mass cools at the rate of 1° C per 100 meters (5.5° F per 1,000 feet), the dry

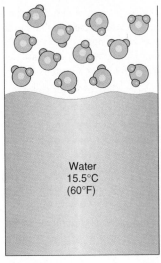
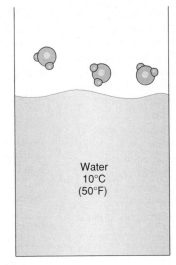

Vapor pressure
higher over
warmer water

Water
15.5°C
(60°F)

Water
10°C
(50°F)

Equal volumes

Figure 6.15

Vapor pressure is the index used to measure water holding capacity. Warm air has a higher vapor pressure than cold air. Saturation vapor pressure occurs when there are as many molecules leaving the liquid state as entering the vapor form. As the air temperature rises or falls the vapor pressure will also change. Warm air has a greater vapor holding capacity, so the vapor pressure is higher.

adiabatic lapse rate. When condensation occurs, cooling will be at the moist adiabatic rate of around 0.6° C per 100 meters (3.2° F per 1,000 feet). Both cooling and ascent are inseparable in terms of related events, and both are necessary to produce precipitation.

Cooling can also result from contact with a cooler surface or heat sink. When this happens, the air rapidly reaches saturation and condensation will occur. This type of cooling produces condensation but does not lead to precipitation. Cooling by contact with a cooler surface will produce dew, frost, fog, and stratus, which are all very important influences on the Earth's water and temperature budgets. However, the atmosphere in contact with the cooler surface becomes stable and ascent stops.

Let's examine more closely the saturation process leading to condensation and finally, precipitation.

Saturation—Step Four

What is **saturation**? Think of it as the state of the atmosphere when it can hold no more water vapor. Why does warm air hold more water vapor than cold air? Part of the answer to this question lies in the fact that at higher temperatures molecules have more energy and molecular activity. This is why evaporation rates increase as the temperature rises. The other part of the answer lies in the capacity of the atmosphere to hold water, in other words how full the atmosphere is. The index used to measure water holding capacity is the **vapor pressure** of the atmosphere. This is the pressure exerted by vapor in the atmosphere. When saturation

occurs over a water surface there are as many molecules leaving the liquid surface as entering it. This is when vapor pressure is at its maximum for that temperature. The air is experiencing no net gain or loss of water molecules; it is in equilibrium. If the temperature of the air rises, more molecules will be able to escape the liquid surface and a new equilibrium level of saturation will be reached at a higher vapor pressure. It is important to note that although warm air can hold more water vapor than cold air, the actual amount it does hold is a function of the moisture available to the air mass. Air masses over oceans are more likely to reach saturation than continental locations, which have less moisture to contribute from the soil (fig. 6.15 and box 6.2).

Condensation—Step Five

Once the air is saturated or has reached a point where no more moisture can be held in the vapor form, condensation can occur. This is the cloud formation stage. The condensation process is simply the conversion of a gas to a liquid. In the atmosphere, water vapor converts to water in the liquid or solid form when the air is cooled and saturation reaches approximately 100 percent relative humidity. Remember, **relative humidity** describes, in percent, the amount of moisture in the vapor form the atmosphere is holding at that temperature compared to its potential capacity. A 50 percent relative humidity means the air is 50 percent saturated at that temperature. The abundance of hygroscopic nuclei can affect this percent. Supersaturation can occur if hygroscopic particles are sparse in the air mass. Condensation can also result at values less than 100 percent when large amounts of the particles are present; this is why "dirty" air is necessary.

The condensation process is an energy exchange process. When one gram of water vapor condenses into the liquid form at 20° C, approximately 600 calories of heat—latent heat of condensation—is released into the atmosphere (fig. 6.16). This latent form of energy is a significant heat source and a major factor in energizing and accelerating the air upward in a cloud. This additional energy aided by the condensation process is called **latent heat of condensation** because the energy is not expressed as sensible heat or measurable until water condenses and a temperature change occurs. Table 6.2 lists four heat transfer methods.

The droplets that form during the condensation process can remain small and suspended in the air without falling as precipitation or grow very rapidly to fall as precipitation.

Droplet Growth—Step Six

Once the droplets are formed, the rate of growth is a compromise between two opposing processes, the curvature effect and the solute effect.

The **curvature effect** is greatest when the droplet is small. The smaller the droplet, the greater will be its resistance to growth. Surface tension produced by electrical charges in the water molecule produces the strongest bonds on a tightly curved surface. The smaller the surface area and radius of the droplet, the greater the cohesive forces will be. Droplets with a radius of less than 0.0005 millimeter will show virtually no growth until the relative humidity reaches or exceeds 100 percent. However, real atmospheric

Relative Humidity—"Relative or Not"

BOX 6.2

Humidity can be described as the amount of moisture in the atmospheric air sample. Most weather reports refer to the water content of the air in terms of **relative humidity,** or percent of saturation. This is the water vapor present in an air mass expressed as a percentage of the total amount that would be present were the air saturated at that temperature. It can be determined simply by dividing the actual vapor pressure exerted by the water molecules in the atmosphere, at a given temperature, by the pressure exerted by the water molecules in saturated air at that temperature. Then, multiply by 100. This concept can be expressed by the equation $RH = E/E_s$. Let RH represent relative humidity where E is the actual vapor pressure and E_s is the saturation vapor pressure. The ratio of E to E_s is known as the **saturation ratio.** When E and E_s are the same, the air is saturated and the relative humidity is 100 percent. If E is one-half of E_s, the relative humidity is 50 percent, and so on (fig. 1).

Vapor pressure, or pressure exerted by the water vapor present in the air, is another way to describe atmospheric moisture or humidity. It can also be described in terms of **specific humidity,** or grams of water vapor per kilogram of air being sampled. Specific humidity can be expressed by the simple formula:

$$\text{Specific humidity} = \frac{\text{Mass of water vapor}}{\text{Total mass of air}}$$

Mixing ratio is a useful way of describing humidity because it measures the mass of water vapor in a parcel and compares it to the mass of dry air in that parcel. We can calculate the mixing ratio by the following formula:

$$\text{Mixing ratio} = \frac{\text{Mass of water vapor}}{\text{Mass of dry air}}$$

Finally, the **absolute humidity** can be determined by extracting water vapor from an air sample and comparing it to the volume of air sampled. Expressed mathematically:

$$\text{Absolute humidity} = \frac{\text{Mass of water vapor}}{\text{Volume of air}}$$

Each of these expressions of humidity is useful in understanding the nature of the moisture content.

Figure 1

Different measurements of humidity.

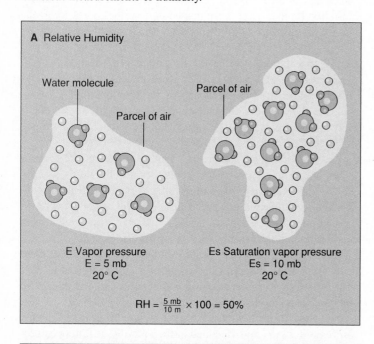

A Relative Humidity

Water molecule

Parcel of air

Parcel of air

E Vapor pressure
E = 5 mb
20° C

Es Saturation vapor pressure
Es = 10 mb
20° C

$$RH = \frac{5 \text{ mb}}{10 \text{ m}} \times 100 = 50\%$$

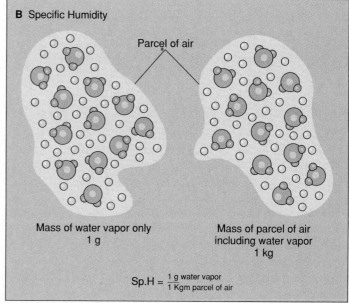

B Specific Humidity

Parcel of air

Mass of water vapor only
1 g

Mass of parcel of air
including water vapor
1 kg

$$\text{Sp.H} = \frac{1 \text{ g water vapor}}{1 \text{ Kgm parcel of air}}$$

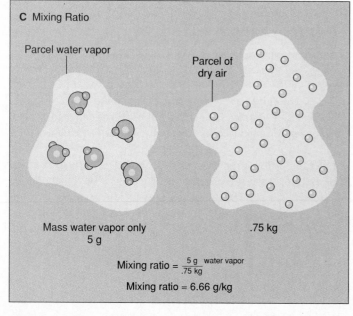

C Mixing Ratio

Parcel water vapor

Parcel of dry air

Mass water vapor only
5 g

.75 kg

$$\text{Mixing ratio} = \frac{5 \text{ g}}{.75 \text{ kg}} \text{ water vapor}$$

$$\text{Mixing ratio} = 6.66 \text{ g/kg}$$

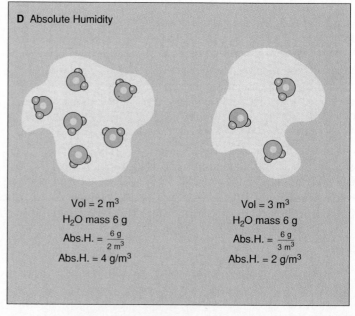

D Absolute Humidity

Vol = 2 m³
H₂O mass 6 g
$$\text{Abs.H.} = \frac{6 \text{ g}}{2 \text{ m}^3}$$
$$\text{Abs.H.} = 4 \text{ g/m}^3$$

Vol = 3 m³
H₂O mass 6 g
$$\text{Abs.H.} = \frac{6 \text{ g}}{3 \text{ m}^3}$$
$$\text{Abs.H.} = 2 \text{ g/m}^3$$

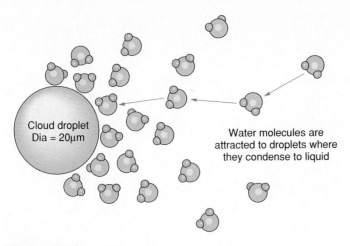

Figure 6.16

Condensation is simply the conversion of water vapor to liquid form. When this process occurs, latent heat of condensation is released into the air at the rate of approximately 600 calories per gram of water.

Table 6.2	
Four Ways the Earth and Atmosphere Transfer Heat	
Method of Energy Transfer	**Characteristics**
Radiation	Wave form electromagnetic energy ranging from the shortest gamma through the longest radio wave. It warms all absorbing surfaces. This form of energy travels at the maximum speed of light in a vacuum.
Conduction	Direct heat transfer through matter from a point of higher temperature to one of lower temperature. More energetic molecules excite adjacent molecules by contact, sending heat energy to lower energy areas.
Convection	The mass movement of a gas or liquid as a result of differences of temperature hence differences in density within the medium.
Latent heat energy	Energy is transferred when water or any liquid evaporates. This is called latent heat of evaporation or energy necessary for evaporation. When this same vapor condenses in the atmosphere or on the surface, this same heat is given off as latent heat of condensation. Latent heat is also given off when sublimation occurs.

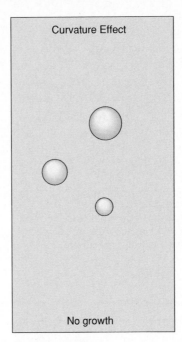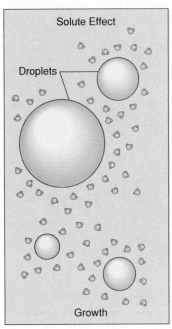

Curvature Effect — No growth

Solute Effect — Droplets — Growth

Figure 6.17

Droplet growth is a function of two opposing processes. The *curvature effect* opposes droplet growth while the *solute effect* promotes rapid growth. While the droplet is very small the curvature effect dominates; however, if the air is rich in condensation nuclei and droplets are colliding and fusing together, the solute effect will dominate and promote rapid droplet growth.

This is the principle behind artificial cloud seeding; condensation nuclei are introduced into a saturated environment where droplets are small. The solute effect then increases and droplet growth occurs as a result of this natural cloud seeding process.

Crystallization

Where temperatures are below freezing, water vapor or liquid turns to ice. An ice particle can form two ways: water droplets can freeze or water vapor can change directly into ice crystals. This latter process is called **sublimation** and is the major process by which water is transformed into the solid state in the atmosphere. Pure water ice always melts when the temperature reaches 0° C (32° F) at standard sea-level pressure, but it is curious that water may be supercooled below 0° C, remaining in the liquid state in clouds with temperatures as low as −40° C (−40° F). As temperatures approach 0° C, the chances of crystallization become less and less probable because the internal bonding forces are overcome by molecular activity of the water molecule. Its energy of motion prevents crystallization.

The geometry of ice crystallization is always a hexagon or six-sided figure (fig. 6.18). Photographs of snowflakes or ice crystals found in nature show infinite variations. They take many forms including flat plates, columns, prisms, cups, needles, and dendrites—all variations of the hexagon form. The ultimate shape a crystal takes is determined by two factors: the molecular bonding structure of water, and the atmospheric environment. Since the molecular structure is a constant, and causes all flakes to have a

conditions rarely exceed 100 percent relative humidity by more than a few tenths of a percent. Therefore, very small droplets remain very small.

The **solute effect** created by "dirty air" causes droplets to grow rapidly because hygroscopic substances attract water vapor to the small droplets containing the nuclei or solute particles. If the air is rich in nuclei, droplet growth will occur rapidly and the curvature effect will be overcome by the solute effect (fig. 6.17).

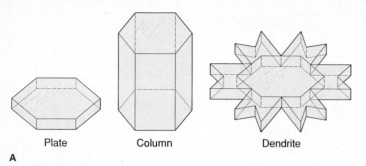

Plate   Column   Dendrite

A

Figure 6.18

(A) The geometry of ice results in a six-sided figure or hexagon. However, within this configuration rarely will two snowflakes or ice crystals be identical. (B) Crystal growth will take the form of flat plates, columns, prisms, cups, needles, and dendrites, but all are derived from the hexagonal form.

Photo by NOAA.

Table 6.3
Ice-Crystal Shapes That Form at Various Temperatures

Environmental °C	Temperature °F	Crystal Form
0 to −4	32 to 25	Thin plates
−4 to −10	25 to 14	Columns
−10 to −12	14 to 10	Plates
−12 to −16	10 to 3	Dendrites
−16 to −22	3 to −8	Plates
−22 to −50	−8 to −58	Hollow columns

Source: NOAA

B

hexagon structure, the major factors determining the crystal variation in shape and size are the temperature and the moisture available in the environment.

Pioneer research conducted by B. J. Mason, at Imperial College in London, determined the influences of the environment by growing crystals in a variety of temperatures (0° C to −50° C) and humidities ranging from minimal saturation to 300 percent supersaturation (table 6.3). The warmest temperatures produced thin plates while the coldest formed hollow prisms or cups. Temperature determines the geometry or form within the framework of the hexagon molecular structure, but the rate of growth and size is strictly a function of moisture availability, not temperature.

The Wegener/Bergeron/Findeisen Theory of Ice Crystal Growth

When the air is saturated, the water vapor exerts a pressure, but that pressure depends on the temperature (fig. 6.19). For normal sea-level atmospheric temperatures (10° C to 30° C), water vapor pressures range from 2.1 to 32 millimeters of mercury (2.8 to 43 millibars). At 100° C, the value rises to 760 millimeters of mer-

cury (1,013 millibars), standard atmospheric pressure, the pressure inside a bubble of boiling water. This is why boiling water bubbles: Vapor pressures equal atmospheric values. At the other end of the graph between −10° C and 10° C, note that the vapor pressure is lower over an ice surface than over water. This difference favors the growth of ice crystals at the expense of the supercooled water droplets, as the German meteorologist, Alfred L. Wegener pointed out in 1911. Thus, today we have an explanation for the growth of snow crystals in clouds with mixed ice particles and supercooled droplets. When this happens, the numbers of water droplets decline as snow crystals increase.

We have traced the steps of droplet and crystal growth to a size where gravity pulls the hydrometeor back toward the Earth's surface as **precipitation.**

Precipitation—Step Seven

What causes rain or snow to fall from one cloud but not from another? Why are some showers heavy and others light? Why do most clouds fail to precipitate? Condensation only ensures that

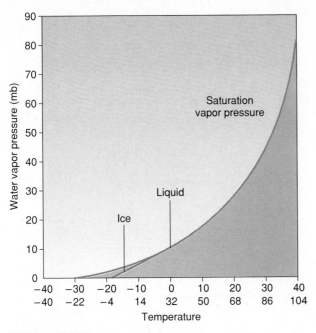

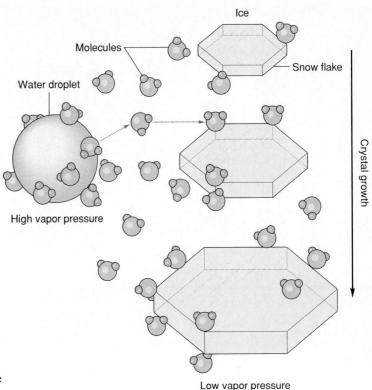

Figure 6.19

The vapor pressure over various surfaces such as ice, water, or soil will be influenced by the surface. Note vapor pressure at temperatures between −10° C and 10° C will be lower over the ice than over the water surface. Thus water present will tend to leave the liquid form and sublimate to ice crystals if the temperatures are lower than 10° C.

clouds will form or dew will cover the ground. Yet the tiny cloud droplets or ice crystals are often so small that they remain suspended. It is only after they gain some mass that they fall from the clouds. If the droplets or ice crystals are very small they may evaporate before leaving the cloud.

Let's explore the anatomy of a cloud and examine further the dynamics of droplet growth processes which lead to precipitation.

Remember, if the cloud has a mixture of water droplets and snowflakes, the vapor-pressure difference between the water and ice will cause the water droplet to evaporate and the ice crystal to grow into a snowflake. When the particle becomes large enough, it will fall as snow. Whether or not it reaches the ground will depend on several factors. These include the thickness of the cloud, height of the cloud above the surface, and the temperature and moisture content of the air.

Once water droplets are formed from either melted snowflakes or through condensation, their final fate is determined by their growth rate. **Coalescence,** the process of droplet combination by collision, seems to be the most likely method of rapid droplet growth. (The dipolar electrical charge causes droplets to adhere or produce coalescence.)

The key to coalescence is droplet size and motion. If all droplets are moving at the same speed in the same direction, no impacts will occur. However, if droplets are moving in various di-

rections and at different velocities, the frequency of impacts greatly increases (fig. 6.20).

Forces on a Drop of Rain

Forces are at work on all matter but gravity is the primary driving force in precipitation. There are three forces acting on a drop of water or snowflake in air that determine its direction and velocity: (1) gravity, which pulls it downward, (2) buoyancy reflecting upward motion of the air, and (3) the drag force of friction, which only comes into play when the particle moves in any given direction. If the gravitational and buoyancy forces are equal, it will not move and friction will not come into play. Friction is a reactionary force. When the falling drop or flake reaches a velocity where the drag of friction equals the difference between gravity and buoyancy, it reaches a terminal or maximum velocity (fig. 6.21). Small droplets have very low terminal velocities, according to table 6.4. This is why cloud droplets often do not produce precipitation. They are just too small; they are held up by buoyancy and drift in the wind supported by the atmosphere, never really coming to rest.

In a typical rain cloud, droplets are not moving at the same speed and they vary greatly in size. Their size may range from 20 microns to 2,000 microns in diameter or have a size ratio of 100 to 1. (A micron, or micrometer, is one-thousandth of a millimeter.) If you could place yourself on a 2,000-micron drop you would see the 20-micron droplets moving upward toward you at

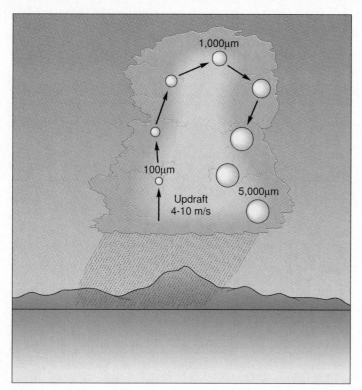

A

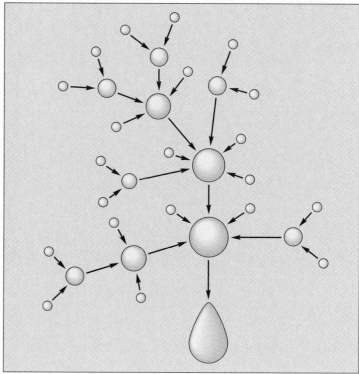

B

Figure 6.20

(A) The key to droplet growth is turbulent motion. (B) Droplets traveling at various rates collide and combine or coalesce into larger droplets. Gravity then takes over and pulls the particle to the surface as rain, snow, sleet, or hail.

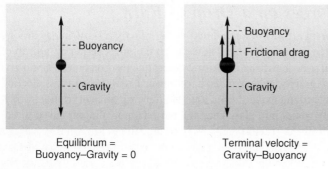

Figure 6.21

The terminal velocity of a falling drop is reached when frictional drag equals the difference between gravity and buoyancy forces. When gravity and buoyancy are equal, the droplet remains in suspension.

Table 6.4

Terminal Velocity of Different Size Particles Involved in Condensation and Precipitation Processes

Diameter (microns)	Terminal Velocity		Type of Particle
	m/s	*ft/s*	
0.2	0.0000001	0.0000003	Condensation nuclei
20	0.01	0.03	Typical cloud droplet
100	0.27	0.9	Large cloud droplet
200	0.70	2.3	Large cloud droplet or drizzle
1,000	4.0	13.1	Small raindrop
2,000	6.5	21.4	Typical raindrop
5,000	9.0	29.5	Large raindrop

Source: NOAA

4 meters per second (13 feet per second). Instead of colliding with you, the 20-micron droplets will flow in a streamline path around the large drop, like water moving around a rock in a stream. As the ratio between drop and droplet size decreases, the rate of collisions increases. The highest collision rates seem to occur when the droplet ratio falls between 0.6 and 0.7. Then as the droplet approaches the size of the larger drop, collision rates drop rapidly

to zero. Air flow patterns of two large drops interfere with one another and prevent collisions. Thus, there is a critical size for the greatest frequency of collisions.

When impact does occur, the drops either combine or bounce apart. If two large drops of nearly the same size collide, they can break apart. Therefore, collision will be determined by three factors. These include (1) relative sizes of the drop and droplet, which effects (2) the difference in their falling velocities and (3) impact angle. It is also believed that atmospheric electrical field conditions produce attractive forces. Rain generated by coalescence of drops is referred to as "warm rain" because it was first recognized in tropical clouds where freezing temperatures are rarely realized. Coalescence also occurs in supercooled clouds, where temperatures are well below freezing in the cloud.

In review, remember that precipitation is a process requiring six important preliminary stages or steps. If any of these steps are missing, precipitation will not occur. Let us examine six forms of hydrometeors, or precipitation and condensation, ranging in size from fog droplets to hailstones.

DESCRIPTION OF HYDROMETEORS

The various kinds of hydrometeors form distinctive classes described in the 1956 International Cloud Atlas, prepared by the World Meterological Organization. Each type is a product of the atmospheric environment. Hydrometeors, as the term suggests, are water-formed meteors ranging in size from large hailstones over 5 centimeters (2 inches) in diameter to the small droplets of mist suspended in air that make up the cloud. Table 6.5 gives the typical characteristics of various types of hydrometeors.

Rain and Drizzle

Precipitation of liquid water—rain—is the principal form watering the Earth, because the Earth's mean temperature is above the freezing point. Raindrops are greater than 0.5 millimeter (0.02 inch) in diameter; smaller drops are called **drizzle.**

Snow

The second most common form of hydrometeor is **snow,** which falls on both the highlands and high latitudes of the Earth (fig. 6.22). Snow consists of ice crystals resulting from the transformation of water vapor into the solid form directly, or passing through the liquid state. Snow pellets are white, opaque grains of ice (2–5 millimeters or 0.1–0.2 inch) that are brittle and easily crushed. They usually occur in showers mixed with snowflakes or raindrops when surface temperatures are near freezing. Snow grains are very small white and opaque grains of ice less than 1 millimeter (0.04 inch) in diameter. They fall from fog or stratus clouds.

Ice pellets or **sleet** are frozen raindrops which have a diameter of 5 millimeters (0.2 inch) or less. They form when rain falls through a colder layer of air and freezes. Sleet can also form if snow partially melts then freezes again, encasing the snowflake in a thin coat of ice.

Hail

Hail is a unique and less common form of precipitation than rain, snow or sleet. **Hail** is small balls of ice (hailstones) with diameters ranging from 5 to 50 millimeters (0.2 to 2 inches) or more. They consist almost entirely of transparent layers of ice at least 1 millimeter (0.04 inch) thick. Hail is generally associated with tur-

Table 6.5	
Hydrometeor Types	
Type	**Characteristics**
Rain	Liquid. Diameter equal to or greater than 0.5 mm with a spherical or oval form.
Drizzle	Diameter less than 0.5 mm and spherical in shape.
Snow	Crystal structure ranging from thin plates to hollow columns. All with a hexagonal crystal pattern.
Ice Pellets	Frozen rain with a diameter of 5 mm or less.
Hail	Pieces of ice larger than 5 mm. Some have been measured larger than a baseball.
Mist	Suspended fine water droplets.
Fog	Suspended very fine water droplets.

Source: NOAA

bulent thunderstorms and cumulonimbus clouds. The largest hailstone on record fell in Potter, Nebraska, measuring nearly 14 centimeters (5½ inches) in diameter. Hail can completely destroy a crop and kill animals in its path. The Nebraska hailstone weighed 1½ pounds. The hailstones in that event fell so hard they buried themselves in the soil.

Hailstones that have been collected and preserved for study show a variety of shapes ranging from cones to spheres. Many, but not all, are composed of concentric shells which alternate from clear to opaque ice (fig. 6.23).

Hail is usually absent at polar and tropical latitudes; however, hail is a common event in the high Andes of Peru in the late afternoons. Oddly, in thunderstorm regions there may be considerable hail from one storm system with associated thunder and lightning and little, if any, from a comparable storm system.

Mist

Mist is a suspension of fine water droplets and wet hygroscopic particles. With mist, the air does not have the clammy feeling of a fog, and visibility is much better. Both mist and fog can be important sources of moisture for vegetation; they contribute over 50 percent of the available water supply to the soil and ground cover in fog vegetation belts along the west coasts of continents where cold ocean currents exist.

Fog

Fog is a suspension of very small water droplets in the air which reduces ground-level visibility to one kilometer (0.6 mile). Fog in the air gives it a raw, clammy feeling, and when air pollutants are mixed with it the mixture becomes **smog,** a word coined by combining "smoke" and "fog." Ice fogs form when the dewpoint temperature is below freezing, causing a suspension of small ice crystals that produce haloes around luminous objects.

Fogs can be classified according to the processes involved in their formation. They can be produced by frontal activity or as one of several types of air-mass fogs, described next.

Figure 6.22

Snowflakes are nature's second most abundant form of precipitation after raindrops. Crystallization occurs when water vapor is transformed into ice below the freezing point. The crystal size and shape are governed by temperature and available moisture.

Advection Fogs

Advection suggests horizontal movement. **Advection fogs** occur when warm, moist air moves over a cooler surface that causes both a lowering of the air temperature and a rise in relative humidity, resulting in condensation. The coastal stratus and fog are common along the east and west coasts of North America in the summer or winter: Foggy summer air moving over a west coast cool ocean current can come in with the sea breeze, like clockwork. As the foggy air mass drifts inland during the day, it warms up and "burns off" in the afternoon. Then the fog returns in the late afternoon or early evening (fig. 6.24).

Steam Fog

Steam fogs, also called sea smoke, develop when a dry, cool air mass drifts over a warmer moist surface. This condition produces a rapid evaporation of moisture from the wet surface and saturation and condensation of the cool air mass. This type of fog can

Figure 6.23

Hail consists of small balls of ice that are layered like an onion with concentric shells.

Photo by NOAA.

Figure 6.24
Advection fog occurs when warm moist air moves horizontally over a cool surface. Coastal stratus and fog along the west coast of North America during the summer months is a classic example of this type of fog formation.
Photo by NOAA.

be observed in arctic regions and over open, unfrozen lakes in the middle latitudes during the winter when a cold and dry continental air mass moves into the region.

Radiation Fog
The result of cooling of the air by the moist ground surface, **radiation fogs** form when the skies are clear and calm, allowing cold air to puddle and lie stagnant over wet soil. As the Earth reradiates energy at night, the surface layer of air falls in temperature and increases in relative humidity until saturation and condensation occur. The process of evaporation reverses this sequence in the morning. Fogs of this type—also called ground fog or tule fog, after the tule reeds in marshes—are often very shallow but dense; one may be able to see the Moon and stars overhead and yet only a few yards horizontally. These fogs are common during the winter after heavy rain, which contributes to the moisture content in the air at the surface. Radiation fogs can create extremely hazardous driving and airport landing conditions.

Upslope Fog

Upslope fogs are common on the Great Plains, especially on upper slopes where upslope movement aids condensation through adiabatic cooling. The Great Plains slope gently toward the Rocky Mountains. As moist marine air from the Gulf of Mexico moves upslope, the air cools to the dewpoint temperature and fog results.

SUMMARY

Atmospheric stirrings of the planet are revealed by the clouds. Each cloud form is an expression of the state of the atmosphere. These include the **cirrus** family of high clouds, middle-level **alto** clouds, low-level **stratus,** and **cumulus** of vertical development. The color, shape, height, and rate of change give signals to the trained meteorologist about the weather.

To produce clouds and then precipitation, a body of air must go through a series of steps or stages leading to precipitation: (1) dirty, moist air, (2) ascent, (3) cooling, (4) saturation, (5) condensation, (6) droplet growth, and (7) precipitation.

Once droplets form, grow, and begin to fall as precipitation, three forces act on the drop to determine its direction and velocity. These are gravity, buoyancy, and frictional drag. The terminal velocity is determined by the interplay of these forces.

Hydrometeors include **rain, drizzle, snow, sleet, hail, mist,** and **fog.** Each hydrometeor form is the result of a combination of factors which include temperature, moisture supply, wind, and evaporation rates.

ILLUSTRATED STUDY QUESTIONS

1. Classify the clouds in figure 6.1, p. 111, according to cumulo or stratus forms.
2. Describe the general cloud heights for the clouds in figure 6.1, p. 111.
3. Describe the physical properties of the clouds in figure 6.2, p. 112.
4. Explain the cause of the ring that forms around high-altitude cirrus shown in figure 6.4, p. 113.
5. Explain the effect the clouds, shown in figure 6.6, p. 115, have on the heat budget of the Earth.
6. What do these clouds tell us about atmospheric stability (fig. 6.10, p. 117)?
7. What atmospheric conditions are necessary for the fog in figure 6.24, p. 131?
8. Describe the role of cooling of a rising column of air before it reaches saturation (fig. 6.14, p. 122).
9. Contrast vapor pressure over warm and cold water (fig. 6.15, p. 123).
10. Contrast the curvature effect and solute effect on droplet growth (fig. 6.17, p. 125).
11. Which has high vapor pressure; water or ice (fig. 6.19, p. 127)?
12. What determines the terminal velocity of a droplet (fig. 6.21, p. 128)?

The Stormy Planet

7

Objectives

After completing this chapter, you will be able to:

1. Explain the primary elements involved in the formation of the **wave cyclone.**
2. Define a **cold** and **warm front** and explain the processes leading to the formation of each.
3. Describe the life history of a **middle-latitude wave cyclone.**
4. Describe the interrelationships between **Rossby waves, air masses,** and the **middle-latitude wave cyclone.**
5. Define an **air mass** and factors influencing its life history.
6. Compare and contrast **tornadoes** with the **tropical cyclone.**
7. Explain the key elements necessary for the formation of **hurricanes** and **tornadoes.**
8. Define an **easterly wave** and its relationship to the **tropical cyclone.**

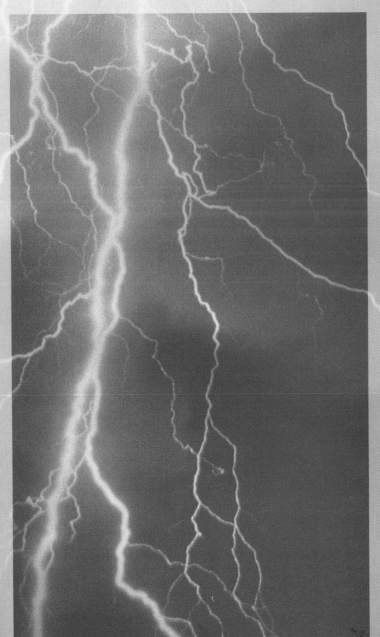

Solar energy stirs the atmosphere to explode.
© Gary Milburn/Tom Stack & Assoc.

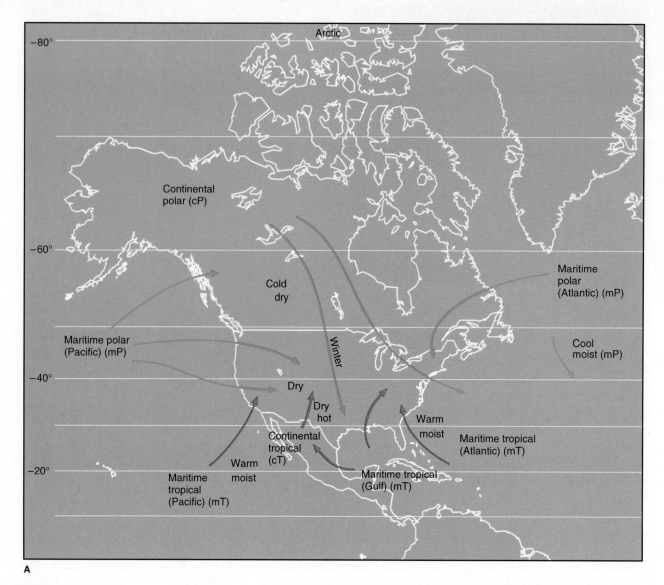

A

Figure 7.1

(A) An air mass source region is the birthplace of air masses. When air remains stagnant for several weeks in a source region, the air takes on the temperature and moisture characteristics of the surface. Source regions are generally located in zones of persistent high pressure where the air stagnates long enough to take on the properties of the surface. (B) In the NOAA satellite photograph a continental polar air mass (cP) is moving southeast over the southern United States, Gulf of Mexico waters, and Atlantic Ocean. Clouds form, indicating instability, over the warmer waters.

Photo by NOAA.

Each storm system is really an energy system powered by the Sun's radiation and the Earth's gravity. For every drop of rain that falls under the force of gravity, it took solar energy to evaporate the water from the oceans and continents and carry it up into the atmosphere. This same solar energy is stored in a latent form to be unleashed in storm systems.

Previous chapters focused on the key elements that make up weather; specifically, the last chapter focused on clouds and the precipitation process. Of course, daily weather is really the product of larger atmospheric systems. These include air masses and storm systems that form when the atmosphere discharges its energy.

An understanding of air masses and storm systems as a composite of weather elements is the final ingredient to incorporate in this examination of climatic belts of the Earth that shape its environmental realms.

AIR MASSES

An **air mass** is a large body of air that takes on homogeneous properties of temperature and moisture from the surface. Air masses are continental in size. They are associated with **source regions,** generally major high-pressure areas located over both land and

B

water. The nature of these air masses is of primary importance in determining the characteristics of storms and climatic patterns.

Air masses are also classified according to their low-level surface temperature relative to the ground over which they travel. If the air mass is cooled from the surface, then the body of air is classified as warm (w); heating from below gives the air mass a cold (k) designation. When very dry air descends from aloft, causing heating by compression, it is called a superior (s) or subsidence air mass. Because air masses move horizontally as well, they not only influence the region of immigration but also become modified as they travel along. Thus, each air mass has a life history.

Life History of an Air Mass

There are three aspects to the life history of an air mass, the (1) source region, (2) properties of the region traveled over, and (3) age of the air mass.

The source region determines the atmospheric properties of the air mass. When air remains for a long period of time (weeks) over a portion of the Earth's surface, it takes on the temperature and humidity characteristics of that surface. If the surface is cold and dry, the air will become cold and dry. Such a surface must be large—perhaps a million square kilometers (several hundred thousand square miles) in extent—to qualify as a source region (fig. 7.1).

A survey of major source regions reveals four principal types of air masses, first classified by T. Bergeron in 1928. His classification system (table 7.1) groups air masses, according to geographic conditions, into four principal types: **maritime polar** (mP), **continental polar** (cP), **maritime tropical** (mT), and **continental tropical** (cT). "Tropical" indicates a source region in the warm lower latitudes. The term "polar" is a higher latitude designation. "Continental" is in contrast to "maritime," which indicates that the source region is oceanic while continental denotes

Table 7.1		
Air Mass Types		
Source Region	**Polar (P)**	**Tropical (T)**
Land, Continental (c)	Cold, dry, stable (cP)	Hot, dry, stable aloft, unstable at surface (cT)
Water, Maritime (m)	Cool, moist, unstable (mP)	Warm, moist, usually unstable but stable over cold ocean current (mT)

land masses of continental size. Some meteorologists further classify polar air masses into arctic (A), which originate over a snow or ice surface, and polar (P), which originate over cold lowland or water. Tropical air masses have also been called equatorial (E) when the source region is located over water in the equatorial belt of the doldrums.

When an air mass leaves its source region, modification begins immediately. The air is no longer in balance with the surface; it loses or gains energy from the surface with which it is in contact. Atmospheric stability results when air moves over a cooler surface, while instability occurs when cooler air comes in contact with a warmer land area. The changes that occur are closely related to the amount of time spent with the modifying influences. This determines the age or survival time of the air mass. The lower layers especially undergo transitions in temperature and humidity while the upper layers remain less affected unless instability occurs. If the (1) source region, (2) properties of the region traversed, and (3) age are known, then the weather a given region experiences can be forecast and described. For example, in the spring, air that reaches Michigan from central Canada will be quite different from air reaching Michigan from the warm, humid Gulf of Mexico. The cool, dry Canadian air is classified as continental polar (cP) whereas the air mass from the Gulf of Mexico is classified as maritime tropical (mT). Each will bring a new set of atmospheric conditions which include changes not only in temperature and humidity, but also in winds, clouds, and precipitation.

Air Masses and Rossby Waves

Middle-latitude storm systems form in convergent zones of contrasting air masses where potential energy is converted to kinetic energy. The fronts are boundary lines between air masses where convergence occurs.

A satellite view of upper-air Rossby waves provides us with a clue to the development of middle latitude storms. Recall from chapter 5, "The Windy Planet," that upper-air waves form a snakelike meandering pattern in the upper middle latitudes consisting of high-pressure ridges and low-pressure troughs. The troughs often result in outbreaks of maritime or continental polar air while the ridges of high pressure can introduce warm, moist maritime or Continental tropical air into the region. Storm sys-

tems usually develop along the eastern side of the low-pressure trough where a ripple in the **polar front** develops and evolves into a **wave cyclone storm system** (fig. 7.2). Frontal development or *frontogenesis* occurs where sharp contrasts exist in temperature and moisture of two converging air masses.

As the amplitude of the wave grows larger, the pressure continues to fall at the trough and isobars become more compacted, indicating increasing winds and sharper contrasts in temperature both vertically and horizontally through the air mass. Ascent of the air around the center of the system also increases.

Wave cyclones are born along the eastern side of upper-air troughs. The geographic position of the trough governs which air masses converge along the frontal boundary. If the trough is located over either the Atlantic or Pacific Ocean, the two air masses will be maritime tropical (mT) originating in the mid-oceanic high-pressure region, and maritime polar (mP) from the Gulf of Alaska or Icelandic and Greenland waters. When the trough dips over a continental region, the air masses will be modified by continental conditions. Thus, continental polar (cP) air may converge on either maritime tropical (mT) or continental tropical (cT) air (see fig. 7.1).

The most violent North American weather and storm activity results when continental polar air (cP) meets maritime tropical (mT) Gulf air. Warm surface temperatures can add fuel to the system if the invading air is cooler. The result is instability and clouds of vertical development. This is especially prevalent in the late spring when the land is warming up. If warm, moist air (mT) traveling north invades a cold surface, stability will result and cyclonic development will be weakened. Stratus clouds and fog are often characteristic of this type of air mass invasion.

Wave cyclones travel in families or series connected together in various stages of development along troughs of the polar front. Hence the meandering Rossby wave form dictates the storm track and site on the wave where frontogenesis will develop (fig. 7.3).

Let us examine the primary elements involved in the development and life cycle of a wave cyclone.

LIFE CYCLE OF THE MIDDLE-LATITUDE WAVE CYCLONE

A pioneer Norwegian meteorologist, V. Bjerknes, presented the first model of a wave cyclone in 1919. His model has been continually refined, but it remains the standard for meteorologists.

The Bjerknes model, illustrated in figures 7.4, 7.5, and 7.6, shows a cold front and warm front originating along the global polar fronts and boundaries between air masses. Cold fronts develop where cooler air invades a warmer air mass. When warmer air overrides a cooler air mass a warm front is created. Each stage represents an increasing magnitude with time. Each front radiates outward from a center of lowest pressure like spokes on a wheel. As the system evolves, the spokes pivot around the low in a counterclockwise fashion in the Northern Hemisphere and clockwise in the Southern Hemisphere. Eventually, the trailing cold front

Figure 7.2

A wave cyclone often has its origin on the eastern side of an upper-air wave of low pressure where a ripple in the Rossby wave develops into a storm system. In this NOAA satellite photograph maritime polar (mP) air moves southeast invading maritime tropical (mT) air along the west coast of North America, creating a band of clouds and unstable air. (Photo taken at 21.15 Greenwich Time.)

Photo by NOAA.

Figure 7.3

Cyclonic systems travel in families or series, each in a different stage of development, and travel in an easterly direction across the middle latitudes in the zone of the polar front and jet stream. This infrared photograph of two cyclonic systems was taken on October 25, 1975. The lighter the tones, the colder the temperature. Here the white tones are high clouds and the tops of cumulus clouds in the tropics at 20° N latitude. Dense white areas also occur over the Mississippi Valley and the Pacific Northwest.

Photo by NOAA.

Figure 7.4

The V. Bjerknes wave cyclone model consists of a warm and cold front or boundaries between opposing or contrasting air masses. The entire cyclonic system is a three-dimensional dynamic process.

Notice the changes in this system by comparing figure 7.4 with figure 7.2, which was taken three hours later. (Photo taken at 18.15 Greenwich time.)

Photo by NOAA.

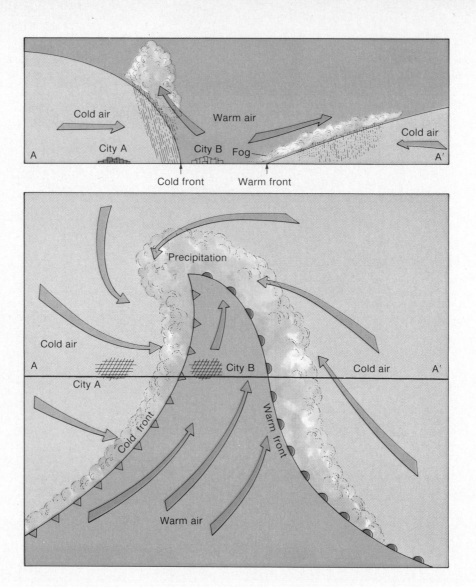

Figure 7.5

A cross section of a warm and cold front reveals a widespread horizontal and vertical distribution of layered clouds. At the leading edge, announcing the coming of the warm front are cirrus, followed by altostratus, stratus, or nimbostratus. The cold front is dominated by cumuloform clouds ranging from cumulus to cumulonimbus. The cold front produces a set of predictable events. Cumulus cloud forms increase near the frontal boundary. The slope of the cold front is much steeper than the warm front and it moves more rapidly. This produces a snowplow effect or churning action along the boundary often causing cumulonimbus clouds, hail, lightning, and thunder.

overtakes the warm front and the system is said to be occluded. An analogy of this process is a wave approaching the beach. First, the wave grows in size and crests as its bottom drags on the ocean floor, and then spills onto the beach with its form destroyed. Although the atmospheric wave is generated by totally different processes, both wave examples evolve, then dissolve, when all of the potential energy has been converted to kinetic energy and maximum entropy has been reached in the system.

To better understand this energy system, map surgery is necessary.

By taking a cross section from southwest to northeast through any cyclone system we can observe, in three dimensions, the front's air masses and precipitation patterns (fig. 7.5).

The Cold Front

The more energetic cold front brings up the rear of the cyclonic system. The **cold front** is a temperature border when cooler air is invading warmer air. The advance of the cold front produces a set of predictable events. Winds increase and are generally out of the south or southwest as the cold front approaches. Pressure continues to fall as winds bring in warmer, more humid air resulting in increasing cumulus cloud development near the frontal boundary. Near the frontal zone, which is about 80 kilometers (50 miles) wide, showers often become heavy, forming a line of cumulus and cumulonimbus clouds, followed by clearing skies as the front passes. Winds then shift from southerly or southwesterly to northerly or

Table 7.2
Atmospheric Changes Resulting from Frontal Passage

Weather Element	Before Passing	While Passing	After Passing
Typical Weather Associated with a Cold Front			
Winds	South-southwest	Gusty, shifting	West-northwest
Temperature	Warm	Sudden drop	Colder
Pressure	Falling steadily	Sharp rise	Rising steadily
Clouds	Increasing Ci, Cs, then either Tcu or Cb	Tcu or Cb	Often Cu
Precipitation	Short period of showers	Heavy showers of rain or snow sometimes with hail, thunder, and lightning	Decreasing intensity of showers, then clearing
Visibility	Fair to poor in haze	Poor, followed by improving	Good except in showers
Dew point	High; remains steady	Sharp drop	Lowering
Typical Weather Associated with a Warm Front			
Winds	South-southeast	Variable	South-southwest
Temperature	Cool-cold	Steady rise	Warmer
Pressure	Usually falling	Leveling off	Slight rise, followed by fall
Clouds	In this order: Ci, Cs, As, Ns, St and fog; occasionally Cb in summer	Stratus-type	Clearing with scattered Sc; occasionally Cb in summer
Precipitation	Light to moderate rain, snow, sleet, or drizzle	Drizzle	Usually none; sometimes light rain or showers
Visibility	Poor	Poor, but improving	Fair in haze
Dew point	Steady	Slow rise	Rise, then steady

Source: D. Ahrens, *Meteorology Today,* 3rd ed. St. Paul, Minn: West Publishing Co., Table (14.2, 14.3) p. 3.62, p. 3.64.

northwesterly, bringing lower temperatures and humidity as well as scattered fair weather cumulus clouds. The slope of the cold front is much steeper than the warm front. The advance of the steeper cold front produces a snowplow effect; the cooler invading air causes a churning, mixing motion of the warmer moist air being overrun.

If the invading cooler air ceases to advance, the front takes the name **stationary front.** Once it moves forward again it becomes a cold front.

The Warm Front

Tracing the cross section at the **warm front** boundary (fig. 7.5), we can observe cold air in a retreat poleward, and warmer, lighter air is gliding up over the top of the colder, denser air mass. The slope of the warm front is gentler (1 part vertical to 250 parts horizontal) than that of the cold front (1:75). Because of these slope differences, clouds and precipitation characteristics are also quite distinct. Along the warm sector, clouds are layered, or stratiform, and precipitation is usually light and steady.

The cross section of a warm front also shows a profile of cloud development with potential precipitation patterns from each cloud type. As the warm front advances, high cirrus appear first followed by altostratus, and then low-level stratus clouds follow extending over large areas. Winds are generally southerly to southwesterly and humidity is high. Precipitation falling continuously ahead of the frontal surface will be light to moderate, increasing as the front approaches. When the warm front passes, atmospheric changes are less marked than the passage of a cold front. Table 7.2 summarizes the changes that occur with frontal passages. It must be pointed out that many fronts produce no precipitation. In fact, most pass without much fanfare, although there are changes in temperature, humidity, winds, and clouds as frontal passage occurs.

Occlusion

Because the warm front has less energy and velocity, it is overtaken by the more active and fast-moving cold front. At that point the warm air between the two fronts is lifted off the surface. This lifting process is called occlusion, marking the beginning of the end of the system or the dissolving stage. As mentioned earlier, this can be compared to a wave breaking on the beach and dissolving (fig. 7.6).

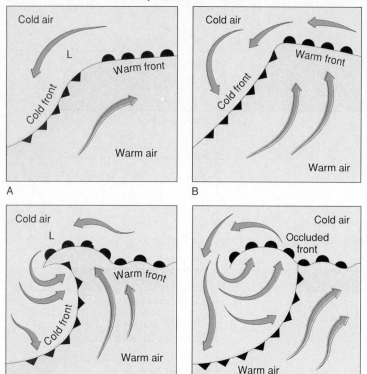

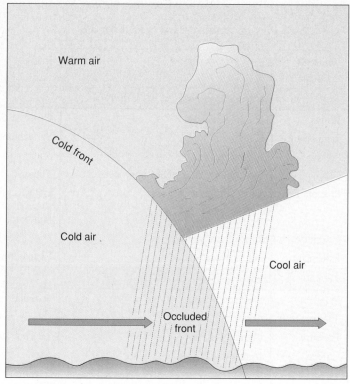

Figure 7.6

Occlusion occurs when the cold front overtakes the warm front. As this event occurs, the warm air mass located between the two frontal boundaries is lifted off the surface. This event marks the beginning of the dissolving stage.

CATASTROPHIC CYCLONIC STORMS: TORNADOES AND HURRICANES

In earlier chapters, the atmosphere was described as a heat engine that is energized by the Sun. Electromagnetic energy reaching the Earth is converted to other forms such as heat and motion to produce a variety of physical phenomena. Without this energy there would be no winds, ocean currents, evaporation, precipitation, or any other related events.

When this heat engine becomes excessively loaded with potential energy, the atmospheric system has a release valve that takes two major cyclonic forms: the **tornado** of the middle latitudes and the **tropical hurricane** born in the trade winds. These two atmospheric systems behave like a pressure release valve or the spillway on a dam. The tornado and hurricane have several things in common. First, they are both cyclonic low-pressure storm systems releasing great quantities of energy. Second, there are no frontal components directly associated with them. Fronts may or may not be present as they drift across the land and sea. Third, cumulonimbus clouds, thunder, lightning, and high wind are common to both. However, these two storms differ in terms of formation, region of origin, air masses, size, energy, and life expectancy or meteorological history.

The Tornado

The Midwest is a wonderful place in the springtime as the signs of winter disappear and the dormant plants spring back to life. However, when the skies become covered with gray thunderheads or cumulonimbus clouds, it's time to look toward the storm cellar or at least seek the National Weather Service advisories.

The tornado is a killer, striking without warning like a sneak bomber attack, leveling everything in its path. In the period from 1916 to the present, an average of 230 people per year were killed by tornadoes, along with many millions of dollars in property damage.

The unique feature of this storm is the funnel cloud which extends earthward from the base of a cumulonimbus cloud (fig. 7.7). The funnel shape can vary from a ropelike structure to a cone or vertical cylinder. The funnel cloud moves across the countryside averaging 50 to 65 kilometers per hour (30–40 mph), cutting a path typically about 15 kilometers (10 miles) in length. Observers recorded a tornado that covered 164 kilometers (102 miles) near

Figure 7.7
Tornado funnels take many forms from ropelike structures to cones, as shown here, and vertical cyclones.

Photo by the National Weather Service.

Hutchinson, Kansas, on May 8, 1927. Because the funnel is usually only a few hundred feet or yards in width, structures may be completely demolished in one block and untouched across the street. Complete destruction of structures results from two factors: extremely high winds and severe pressure drops to as low as 600 millibars. Because of these two factors, objects as large as freight cars are lifted off their tracks and houses torn off their foundations. When the pressure falls so low over such a short distance, the pressure gradient force is very great. This accounts for the very high winds and explosion of structures. Straws have been found buried in heavy planks. Laboratory tests have demonstrated that winds of 500 to 650 kilometers per hour (300–400 mph) are necessary to give a light object such penetrating power.

A drop in pressure of just 68 millibars, a fraction of the strongest drops recorded, can be equated to one pound per square inch of pressure. This means 144 pounds of force on every square foot from the air inside a structure pushing outward. Therefore, a roof measuring 20 feet by 40 feet with an area of 800 square feet would experience an upward thrust of 115,200 pounds. This is more than enough to lift most structures off their foundations (fig. 7.8).

One observer near Lincoln, Nebraska, told *Weatherwise* in April 1943 about the passage of a tornado directly overhead: "We looked up into what appeared to be an enormous hollow cylinder, bright inside with lightning flashes, but black as blackest night all around. The noise was like ten million bees, plus a roar that begs description."

Figure 7.8

When a tornado strikes, damage results from both very high winds (in excess of 600 kilometers or 400 miles an hour) and severe sudden pressure drops, which cause objects to explode and boxcars and houses to be lifted off the ground.

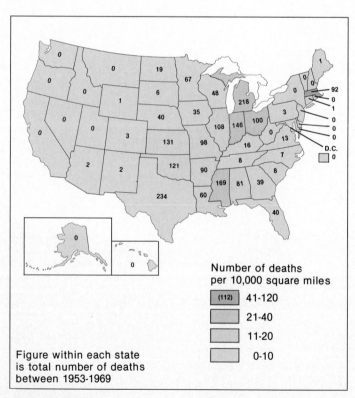

Number of deaths per 10,000 square miles

(112) 41-120

21-40

11-20

0-10

Figure within each state is total number of deaths between 1953-1969

Figure 7.9

Tornadoes occur more frequently on the Great Plains of North America than any other place on Earth. They reach their peak during the month of May when warm and cold air masses clash along the polar front.

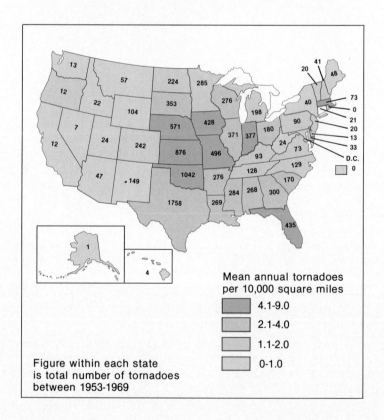

Mean annual tornadoes per 10,000 square miles

4.1-9.0

2.1-4.0

1.1-2.0

0-1.0

Figure within each state is total number of tornadoes between 1953-1969

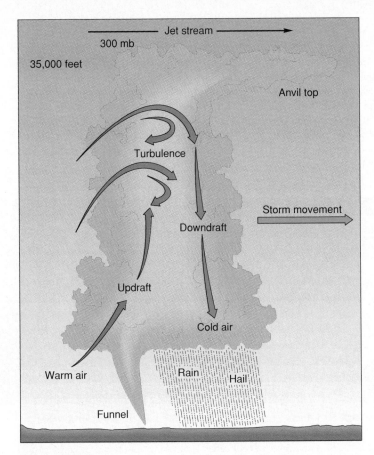

Figure 7.10

Three primary factors contribute to the formation of a tornado:
(1) upper-level jet stream low pressure associated with a wave
cyclone, (2) small low-pressure ripples within a larger wave system,

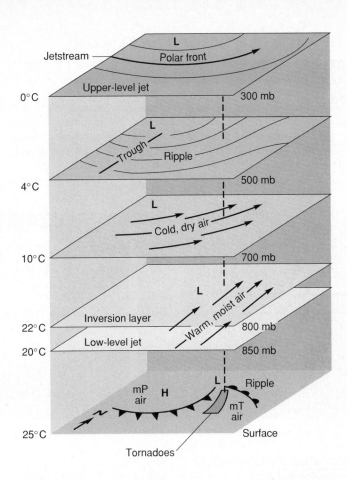

and (3) contrasts in temperature between surface and upper levels,
producing very steep lapse rates. The inversion layer of 800 millibars
acts like a lid on the moist lower layer.

Where and When

Tornadoes are best known as a North American phenomenon. Only
on the Great Plains and eastern United States do they occur with
any frequency. Iowa, Arkansas, Kansas, Mississippi, Alabama, and
Missouri lead the world in frequency of tornadoes, over 2 per year.
They do occur in all states and most countries of the world, but
very infrequently, and they tend to be small and short-lived.

Tornadoes can form at any time and in most latitudes,
but the Great Plains of North America is the "tornado alley" of
the world (fig. 7.9). Spring is tornado season. During the month
of February, frequencies begin to increase and reach a maximum
in late April and early May.

The center of maximum frequency lies over the central
Gulf states in February, then shifts eastward to the southwest Atlantic states during March and April. When May arrives, the center
shifts to the southern Plains states. The northern Plains and Great
Lakes area eastward to New York becomes the zone of maximum
frequency in early summer. This changing center reflects the increasing penetration of warm moist air northward. Tornadoes occur

when cool, dry air moving south meets unstable maritime, tropical
air. Therefore, as the polar front battle line shifts northward with
the advancement of summer, tornadoes also shift poleward.

Because winters are dominated by cool air out of Canada,
fewer encounters occur between warm and cold systems. The entire
continent is under the influence of the continental polar air mass.
Therefore, tornado frequencies reach their lowest values in December and January. The zone favored for winter tornadoes falls
mainly in Missouri, Arkansas, and Mississippi. It is here that the
upper-air charts often show a well-developed Rossby wave low-
pressure pattern and jet stream in the upper troposphere, oriented
from southwest to northeast. There is also evidence that smaller
low-pressure patterns or ripples within the large system contribute
to the development of tornadoes (fig. 7.10).

A tremendous release of energy results from the active
convection of moisture-rich, unstable air (box 7.1). In summary,
three factors contribute to a tornado's formation: first, a polar jet
stream with positive vorticity; second, a smaller low-pressure ripple
within the upper-level trough; and third, a steep lapse rate caused

Life Cycle of a Cumulus Cloud

BOX 7.1

Convection can produce four stages in the life cycle of a cumulus cloud. Stage one is fair weather cumulus clouds, characteristic of the trade wind circulation and conditions following a middle-latitude wave cyclone. If energy is added to the atmosphere, then fair weather cumulus will grow, resulting in cumulus congestus or swelling cumulus, which resembles large cauliflower clusters. This is stage two. Stage three represents a violent upward thrust where the cauliflower head changes to an anvil head with the icy, streaky anvil top pointing downwind. This is the cumulonimbus associated with violent thunderstorms, tornadoes, hurricanes, and severe wave cyclones. At the onset of precipitation, strong downdrafts develop. Falling rain and hail reverses rising air columns by frictional drag, and eventually the cloud dissolves. This is the fourth or dissipating stage of the thunderstorm. As one cell or cumulonimbus cloud disappears, another is born with a life expectancy of one to two hours, repeating the stages just described (fig. 1).

Cumulus stage 1

Mature stage 3

Dissipating stage 4

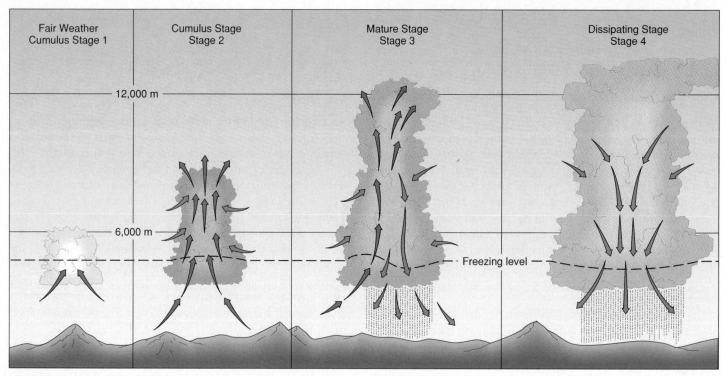

Figure 1

Cumulus cloud development associated with convection occurs in four primary stages. In stage 1, the fair weather cumulus form. Stage 2, the cumulus stage, results in cumulus congestus. In stage 3, the mature stage, cumulus congestus grows to towering cumulonimbus often associated with thunder, lightening, hail, and violent updrafts and downdrafts. Stage 4 is the dissipating stage triggered by precipitation. Frictional drag of falling rain or hail or snow eventually slows the ascent and finally causes strong downdrafts and the death of the cloud.

Photos by Bruce Carroll (left) and NOAA (middle).

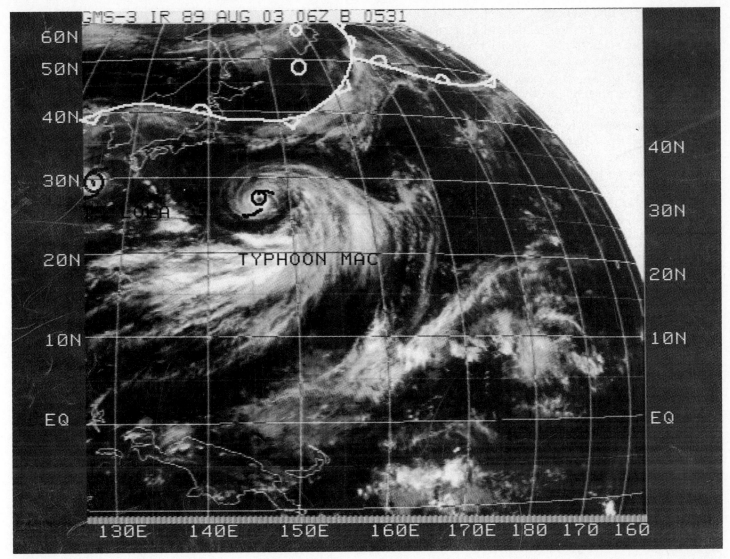

Figure 7.11

Hurricanes and tropical cyclones spawn in the trade winds and follow devious paths that arc poleward. They can stand still, back up, or meander widely. Winds in the system exceed 120 kilometers per hour (75 miles per hour) and often reach speeds in excess of 160 kilometers per hour (100 miles per hour). This infrared photograph taken August 3, 1989, shows Typhoon Mac and tropical storm Lola off the coasts of Japan and China.
Photo by NOAA.

by invading cool air aloft with warmer moist air near the surface. When all three of these factors occur simultaneously, the result is likely to be a tornado.

Hurricanes—Tropical Cyclones

Almost every year, the United States is battered by hurricanes along the Gulf Coast and eastern seaboard. These tropical cyclones are among the most powerful natural systems on the Earth. One hour of hurricane activity equals all the electrical energy generated in the United States in one year. Since 1900, hurricanes have claimed the lives of over 12,000 people in the United States. The hurricane that struck the coast of Bangladesh on November 12, 1970, claimed the lives of over a quarter-million people, most of whom drowned in high tidal waters created by the storm surge. This was the greatest single natural disaster of the twentieth century.

The word *hurricane* is probably derived from the Mayan storm god "Huraken," the Quiche god of thunder and lightning. A hurricane is defined as a tropical cyclone accompanied by sustained winds of 120 kilometers per hour (75 mph) or higher. They go by such names as "typhoon" in the western North Pacific (fig. 7.11), "hurricanes" in the Atlantic and Gulf of Mexico, and "tropical cyclones" over most of the South Pacific and Indian Oceans. In the Philippines, these storms are called "baguios" and in Australia "cyclones." The terms differ with the location, but the events are the same.

Easterly Waves and Genesis

A hurricane starts out as a tropical disturbance or squall in the low-latitude trade winds where the Coriolis force is still active, but very weak. The Coriolis effect influences rotation or deflection of air into a wave pattern known as an **easterly wave** (fig. 7.12). The storm is born in the trough of this easterly air flow where maritime equatorial air is convectively unstable. There are no fronts; the entire system develops where easterly tropical waves deepen into hairpin form and then isobars close to form concentric rings around a center of low pressure.

Hurricane season in the Northern Hemisphere occurs primarily from June through October. In the Southern Hemisphere, December through April are peak months. The primary routes hurricanes follow in each source region are located between 5° and 20° latitudes, in the warm, moist trade winds (fig. 7.13). Once they have developed, their path can deviate considerably. In the Atlantic Ocean, hurricanes often turn northward and slam into the Gulf and east coasts of the United States. Once they move over land or over cold water, they lose energy until they are reclassified as tropical storms.

The Hurricane's Energy System

The hurricane is a thermal engine that takes energy from the sea in the form of sensible heat and latent energy of the evaporation-condensation cycle. The results are wind, rain, thunder, lightning, and more heat from the condensation process.

Warm ocean water in the tropics plays a key role in providing energy for the storm. The heat stored in the top layer of ocean water is the primary source of energy for the storm. If a hurricane moves over cold water or land, it tends to weaken but revitalizes if it moves back over warm water. The energy absorbed through evaporation is released into the atmosphere as latent heat of condensation (nearly 600 calories per gram of water). As long as the evaporation and condensation cycle are strong, the system is well fueled.

Air temperatures both inside and outside the interior of a hurricane are very similar. The pressure difference, however, is considerable. Pressures in the core's center in some cases is equivalent to those pressures usually found at 1,500 meters (5,000 feet) altitude. Thus the energy for expansion of the atmosphere in the system must come primarily from the ocean surface as latent heat, which is released during the condensation process.

Pressure, Wind, and Ocean Current Influences

When we examine global pressure patterns, oceanic high-pressure cells dominate the middle latitudes and control the path of hurricanes. The winds on the low-latitude side of these great subtropical high-pressure cells move westward and toward the equator over warm ocean water where hurricanes form. Here, too, warm ocean currents move northward and contribute to the unstable moist air. On the eastern side of the global ocean basins, cold ocean currents cool the descending dry warm air from the subtropical high. The result is a stable inversion. As a result of these oceanic and

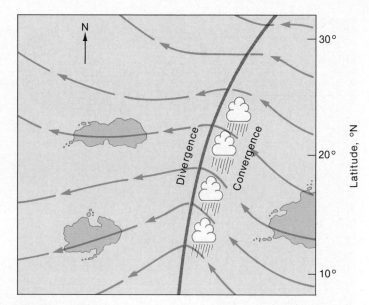

Figure 7.12

Hurricanes are born in the trough of an easterly wave. Hurricanes are thermal engines that received their energy from the ocean through the evaporation-condensation energy cycle. Great quantities of latent heat are released during the condensation portion of the cycle fueling winds, lightning, and thunder in cumulonimbus clouds.

atmospheric differences, hurricanes are born primarily in the southwestern ocean basins of the trade winds where warmer ocean temperatures are likely to occur. A cross section of this heat engine reveals its secrets (fig. 7.14).

Anatomy of a Hurricane

In the fully developed storm, a warm core or eye develops in the upper altitudes. At 9,000 meters (30,000 feet), the air may be 10° C warmer in the core. It is believed that sinking air, heating at the dry adiabatic lapse rate, in the eye region, produces the higher temperatures. While this air is sinking in the center, warm moist air spirals toward the center at lower levels outside the eye. The convection of air near the eye reaches the upper troposphere, releasing great quantities of energy from cumulonimbus clouds in the form of latent heat of condensation. Above clouds in the upper troposphere, there is a reversal of wind direction and the air moves outward away from the center in a counterclockwise motion. The eye is a most unusual feature. Not only is the air warm and dry, the air is also calm and clear for a radius of up to 80 kilometers (50 miles). The system's life and death is determined by the path of the eye.

Life History of a Hurricane

Hurricane tracks follow a path around the large mid-oceanic high, moving first in the trade winds, then veering northward and back eastward in the westerlies (fig. 7.15). In the trade winds, the center travels about 15 kilometers per hour (10 mph). When the center

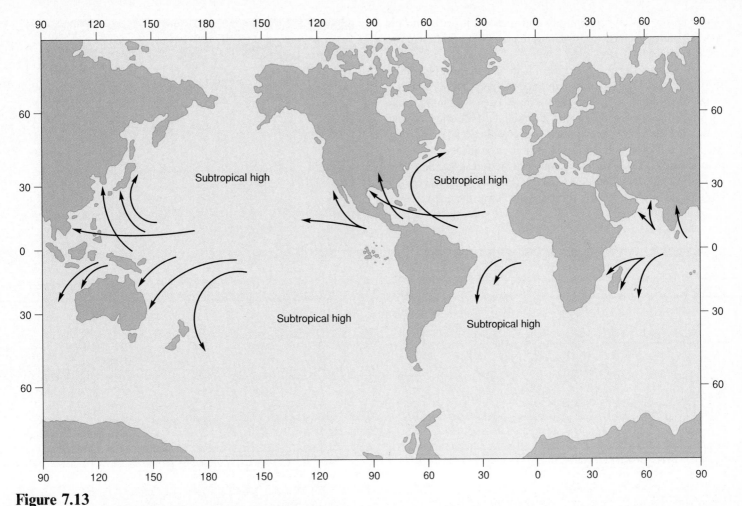

Figure 7.13

Hurricanes or tropical cyclones spawn on the western sides of the major ocean basins of the world. Other factors that contribute to their development include warm ocean currents and convectively unstable air.

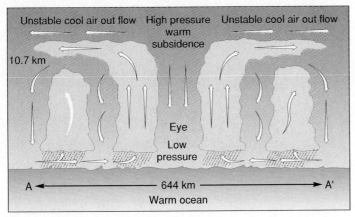

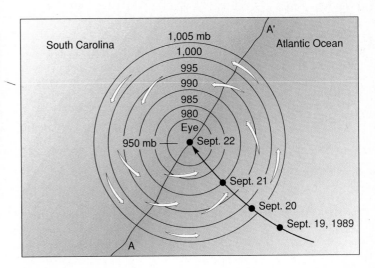

Figure 7.14

Anatomy of Hurricane Hugo, September 1989. A cross section of a hurricane or tropical cyclone reveals several unique features. These include a warm dry core, or *eye,* in the air aloft, moist unstable air surrounding the warm core and high winds spiraling toward the center at lower levels, and cumulonimbus clouds that release great quantities of moisture, sometimes in excess of 25 centimeters (10 inches) in a 24-hour period.

moves northward, the system picks up speed as it drifts into the zone of the subtropical jet stream, often reaching 80 kilometers (50 miles) per hour (fig. 7.16).

Once a storm experiences a landfall and moves inland, it slackens in speed because it loses its energy supply, the warm waters of the ocean (fig. 7.17). The system may degenerate into a rainy squall, or it can even merge with a middle-latitude wave cyclone to produce excessive precipitation over a widespread area, often resulting in floods. As hurricane Hugo moved on shore at 4:30 A.M., September 22, 1989, its winds pounded the forest, leveling trees and stripping vegetation of its foliage (fig. 7.18). However, this onshore assault lost speed a short distance from the coast.

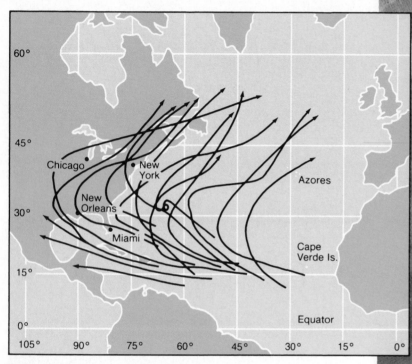

Figure 7.15

Although hurricanes follow unpredictable paths from day to day, their general track forms a loop around the mid-oceanic high. They first develop in easterly waves of the trade wind belt and then drift poleward into the zone of the subtropical jet stream and westerlies. If the storm moves onshore, it loses velocity because it ceases to be connected to its energy system, the warm ocean waters.

Figure 7.16

Hurricane Hugo followed an arc around the southwestern side of the subtropical mid-oceanic high. It then drifted northwest into the belt of the westerlies, triggering heavy rains in the southeastern United States. In this infrared photograph taken at 6:01 P.M. EDT on September 21, 1989, Hugo's eye is clearly visible off the South Carolina coast.

Photo by NOAA.

Figure 7.17

As a hurricane passes over a cold water surface or land area it dies from lack of energy. Hurricane Hugo suffered such a death as it moved onshore on the morning (4:30 A.M.) of September 22, 1989, when this photograph was taken. It has become poorly organized without an eye.

Photo by NOAA.

Figure 7.18

Hurricane Hugo moved onshore with great fury, leveling portions of the Francis Marion National Forest. The primary damage was east of Weatherbee Road running northeast to southwest.

Photo by Walter Salazar, U.S. Department of Agriculture, U.S. Forest Service, Southern Region.

SUMMARY

Because our atmosphere is in a dynamic state resulting from uneven heating over the surface, the air stirs and mixes, equalizing energy over the Earth's surface and through the thin layer of atmosphere. This layer takes on the characteristics of the surface it rests on and becomes a unique air mass.

The nature of the advancing **air mass** is determined by its life history. This history is governed by the characteristics of the source region, properties of the region the air mass comes in contact with in travel, and the age or time spent in traveling.

Atmospheric stirring results in the **wave cyclone** of the middle latitudes, first scientifically described by a Norwegian meteorologist, V. Bjerknes. The cyclonic system is the result of contrasting air masses coming together on a collision course along the **polar front.** This process releases energy in the form of winds, precipitation, and overturning of the atmosphere along the frontal boundary. It is along this line that warm and cold fronts develop.

Each front radiates like spokes from the hub of lowest pressure. The **cold front** is the boundary between polar and tropical air where polar air is merging equatorward. **Warm front** conditions result from polar air retreating poleward beneath tropical air masses.

The **tornado** and **hurricane** are also major stirrings of the atmosphere. Tornadoes result from a rapid drop in pressure where contrasting air masses meet along the cold front of the wave cyclone. Wind speeds reach 600 kilometers (400 miles) per hour in a funnel cloud which can take a ropelike form or cylindrical shape beneath a cumulonimbus cloud. Tornadoes are most common in a widespread zone extending across the Great Plains and eastern United States. Although they are usually only a few hundred yards in diameter, destruction is total as structures explode under the strain of a sudden fall in pressure and high winds.

The **tropical cyclone,** known as a **hurricane** in the Atlantic, typhoon in the Pacific, and **baguio** in the Philippines is a tropical phenomena. The storm begins as a tropical disturbance in a trough of low pressure or easterly wave. When wind speeds reach 120 kilometers per hour (75 mph), the tropical cyclone is classified as a hurricane or typhoon. The hurricane is another example of a heat engine in the atmosphere taking its fuel from the sea. Latent energy of condensation and sensible heat are extracted in the trade wind belt. The life history of the hurricane is determined by its path. They generally travel around the large midoceanic high by moving first westward, then veering north and back eastward. If their path veers over cold water or dry land, the system rapidly loses its energy. The major destruction and loss of life occurs from flooding due to higher tides (storm surge) and rain-swollen rivers as well as high winds.

ILLUSTRATED STUDY QUESTIONS

1. Describe the nature of the front generating the cumulonimbus clouds in this photograph (fig. 7.5, p. 140).
2. Identify the stability of the atmosphere in these photographs (fig. 7.10 & 16, pp. 145, 150).
3. If you were caught in the eye of the hurricane shown in figures 7.14 & 7.15, pp. 149, 150, what would the weather conditions be?
4. Give the life history of a hurricane (fig. 7.15, p. 150).
5. Which cloud type in these photographs has the potential for a tornado (box 7.1, fig. 1, p. 146)?
6. Describe the air masses shown on this map that help produce tornado weather (fig. 7.1, p. 134).
7. Describe the key factors in the life history of an air mass using continental polar (Cp) as an example (fig. 7.1, and tables 7.1 and 7.2, pp. 134, 136).

Climate: Controls, Classification, and Design

Objectives

After completing this chapter you will be able to:

1. Explain the primary factors that produce the global climates described by the Köppen climatic system.
2. Explain the primary factors that make up the **climate** of a place.
3. Use the Köppen climatic system to describe global climatic patterns.
4. Describe 11 major climatic patterns developed by the Köppen climatic classification system.
5. Explain the role of climate as an environmental influence affecting the cultural patterns of a place.
6. Describe the influence of climate on house design in tribal and traditional society.
7. Give several examples of the impact of climate on human activity.

Mali, Dogon Country, Banani Village. Cliffside above dwellings of Tellen tribe reflect climatic design.
© Wolfgang Kaehler

When the term **climate** is used to describe the nature of a place, it means the long-term average weather conditions. Accurate climatic descriptions are generalizations about the atmospheric conditions of a place based upon weather data collected for several decades. **Weather** can fluctuate daily and seasonally, but climatic descriptions remain constant. This does not mean that climates do not change, but the changes are slow and occur over centuries or thousands of years.

In this chapter, we will first look at the physical bases of climatic patterns. Then we will consider methods of climatic classification, followed by a look at the impact of climate on human activity.

CLIMATIC CONTROLS: THE PHYSICAL BASES FOR CLIMATIC DIFFERENCES

The Earth's weather and climate are controlled by terrestrial and atmospheric controls. Terrestrial controls include **latitude, distribution of land and water, ocean currents,** and **mountains.** Lesser terrestrial controls also include **vegetation** and **land-use patterns** such as agriculture, cities, and other human works. Surface conditions, such as snow and ice and soil color and texture, are also factors influencing the Earth's climate.

Atmospheric influences or controls include the **general circulation of the atmosphere** and **cyclonic storms.** The climate of a place is the result of a blend of both the above *climatic controls* and *meteorological elements* such as **insolation, temperature, evaporation, precipitation, humidity, winds, pressure,** and **air masses.** Let's examine each of the principal controls of climate that influence the major global climatic patterns.

Latitude and Solar Energy

Because we live on a spherical planet, solar energy reaches the Earth and atmosphere on any given day at angles ranging from perpendicular (90°) to horizontal (0°). The **tropics,** located between 23½° N (Tropic of Cancer) and 23½° S (Tropic of Capricorn), represent a zone where the Sun's rays are perpendicular twice during the year. Poleward of these boundaries, the Sun's rays reach the Earth at a lower and lower angle with increasing latitude. Seasons are the norm outside the tropics because of major changes in insolation during the year. The isotherms of global mean annual temperatures run generally east-west and decline in value poleward. **Latitude,** then, controls the amount of energy reaching the atmosphere and surface. Recall from chapter 4 that by knowing the latitude and time of the year, one can compute insolation expressed in langleys for any location (fig. 8.1).

Distribution of Land and Water

Although the isotherms generally run east-west over the oceans, on land they shift poleward in the summer and equatorward during the winter, especially in the middle latitudes (fig. 8.2). Recall from chapter 4 that land surfaces are poor storehouses for energy; con-

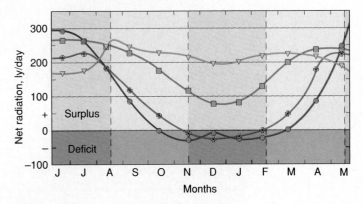

Place	Lat	Annual total kly/yr
●— Anchorage, Alaska	61° N	32
✳— Hamburg, Germany	54° N	35
■— Cherrapunji, India	25° N	64
▽— Manaus, Brazil	3° S	73

Figure 8.1

On any given day, each latitude of the Earth receives a predictable percentage of the total solar energy reaching the Earth during the year.

tinents heat up rapidly, but also cool off when energy supplies decline during the winter. Oceans, on the other hand, moderate the climate of a place by preventing rapid heating during the summer and sudden cooling in the winter season of lower insolation. The coldest spot in the Northern Hemisphere is located on the Asian land mass, not at the Oceanic North Pole. Land locations also have the widest annual ranges in temperature.

The impact of land and water patterns is primarily on regional climatic patterns. Climates can be distinguished on the basis of either marine or continental characteristics. **Marine,** or coastal climates, are characterized by smaller annual ranges in temperature, and delayed maximum summer and minimum winter temperatures. Maximums occur usually in August or September instead of July, while minimums often arrive in late January or February instead of January. They also have more fog, greater cloudiness, and higher humidity.

Continental climates have larger annual ranges in temperature, hotter summers and cooler winters. Precipitation usually reaches a peak in the summer while marine climates tend to have a maximum in the winter.

Therefore, isotherms show greater seasonal shifts over land in contrast to ocean surfaces. Because the Northern Hemisphere is the land hemisphere, we can also see the greatest mean annual range in temperature or seasonal fluctuation in the middle latitudes.

If the Earth were uniformly covered by either land or water, the Southern Hemisphere would have the greatest mean annual range in temperature, because it is closer to the Sun during its summer than during the hemisphere's winter. Fortunately, the

A

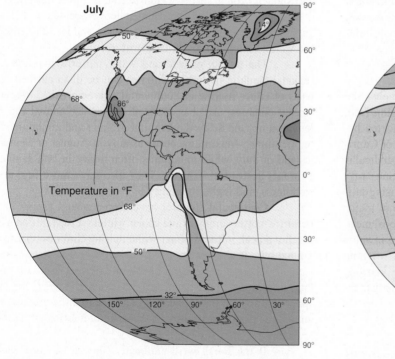

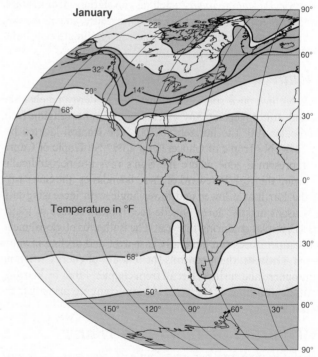

B

Figure 8.2

(A) The cloud band at 10° N latitude is the July location of the equatorial trough of low pressure or intertropical convergent zone. In January it shifts to 10° S latitude. Both temperature and cloud patterns shift seasonally. (B) Isotherms shift poleward in the summer and equatorward in the winter as a result of the Earth's response to seasonal changes in insolation over the continents and oceans.

Photo by NOAA.

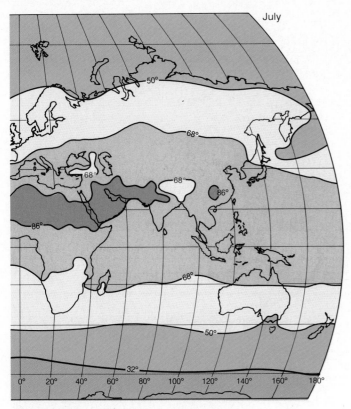

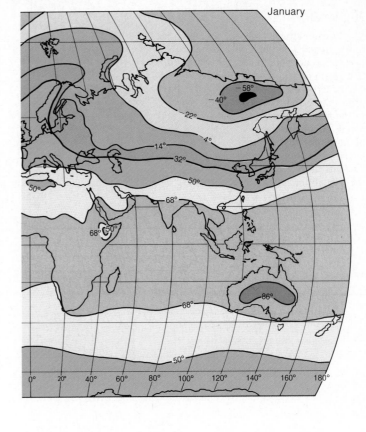

Figure 8.3

The greatest seasonal range in mean temperature is located in the upper middle latitudes of the continents. Notice the greatest range (°F) is located in Siberia, U.S.S.R.

variation in distance from the Sun, caused by our elliptical orbit, has minimal influence on global temperatures because the moderating effects of the oceans that dominate the Southern Hemisphere counter the greater seasonal insolation variation (fig. 8.3).

Ocean Currents

Oceans, when viewed more closely, not only produce marine climates, but also influence the redistribution of energy and affect the stability of the atmosphere and patterns of precipitation. Ocean currents are driven by wind systems. Friction between air and water is the driving force. The ocean contains a complex circulation system of warm and cold currents, shown in figure 8.4, that transports solar energy received in the lower latitudes to higher latitudes. Thus the ocean currents serve a vital function by balancing the uneven distribution of energy on the planet.

Not only do warm and cool currents redistribute the Earth's energy, but they also affect the moisture supply and stability of the atmosphere. Warm currents contribute to instability, storm development, and moisture content through a deep layer of the atmosphere. Major storms are usually born in warm water regions.

Cold oceans and currents influence stable atmospheric conditions by cooling the overlying air. The cooled air holds less moisture, and it becomes sluggish and often associated with foggy atmospheric inversions on the surface layer of air. Coastal locations influenced by cold currents are usually areas of lower rainfall and less frequent precipitation from cyclonic activity because of the stabilizing influence of cool water. Cold water puts a damper on the energy systems of the atmosphere by drawing the energy out of the system. Hurricanes, wave cyclones, and convectional squalls quickly dissipate over cold water. The major arid regions of the Earth are also adjacent to cold currents where they border the coast.

Mountains

If the Earth had no mountains or valleys, the atmosphere would flow more smoothly over the surface and precipitation patterns would be quite different. Mountains exert a major influence on local or regional climatic patterns both because they alter the flow of air and because they receive solar energy at various angles on an uneven surface. Mountains induce updrafts and the **orographic effect** on their windward sides, resulting in higher precipitation on these slopes. Leeward sides lie in the **rain shadow** or zone of lower precipitation. A classic example can be found on the Olympic Peninsula in the state of Washington. Rainfall exceeds 500 centimeters (200 inches) annually in the mountains of Olympic National Park, and yet just 25 kilometers (15 miles) downwind at Port Townsend, Washington, rainfall averages less than 50 centimeters

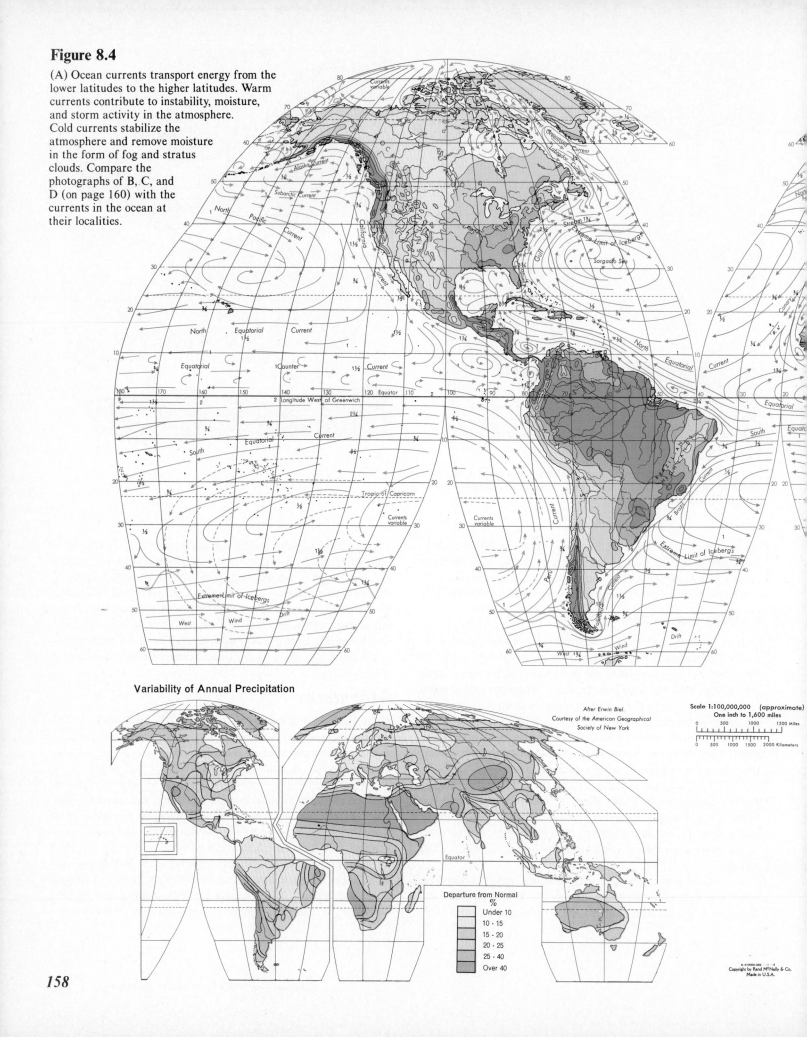

Figure 8.4

(A) Ocean currents transport energy from the lower latitudes to the higher latitudes. Warm currents contribute to instability, moisture, and storm activity in the atmosphere. Cold currents stabilize the atmosphere and remove moisture in the form of fog and stratus clouds. Compare the photographs of B, C, and D (on page 160) with the currents in the ocean at their localities.

Variability of Annual Precipitation

After Erwin Biel.
Courtesy of the American Geographical
Society of New York

Scale 1:100,000,000 (approximate)
One inch to 1,600 miles

Departure from Normal
%
Under 10
10 - 15
15 - 20
20 - 25
25 - 40
Over 40

A-510000-669
Copyright by Rand McNally & Co.
Made in U.S.A.

158

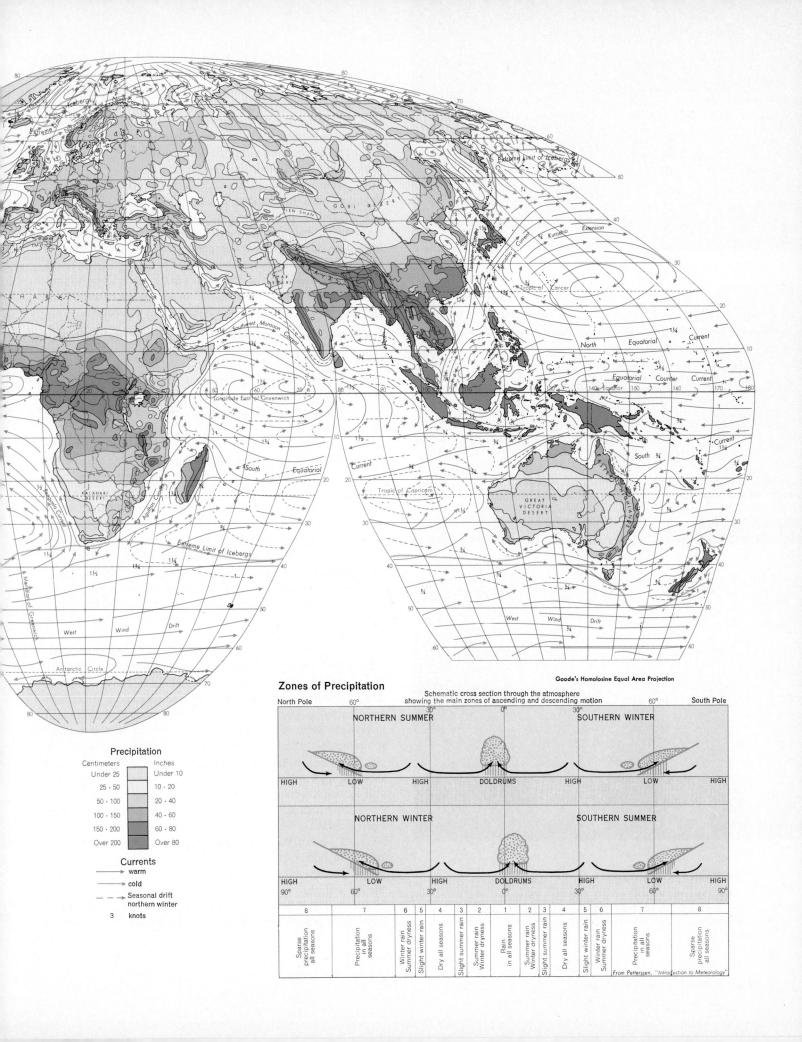

Goode's Homolosine Equal Area Projection

Zones of Precipitation

Schematic cross section through the atmosphere
showing the main zones of ascending and descending motion

North Pole   60°   30°   0°   60°   South Pole

NORTHERN SUMMER   SOUTHERN WINTER

HIGH   LOW   HIGH   DOLDRUMS   HIGH   LOW   HIGH

NORTHERN WINTER   SOUTHERN SUMMER

HIGH   LOW   HIGH   DOLDRUMS   HIGH   LOW   HIGH

90°   60°   30°   0°   30°   60°   90°

8	7	6	5	4	3	2	1	2	3	4	5	6	7	8
Sparse precipitation all seasons	Precipitation in all seasons	Winter rain Summer dryness	Slight winter rain	Dry all seasons	Slight summer rain	Summer rain Winter dryness	Rain in all seasons	Summer rain Winter dryness	Slight summer rain	Dry all seasons	Slight winter rain	Winter rain Summer dryness	Precipitation in all seasons	Sparse precipitation all seasons

From Petterssen, "Introduction to Meteorology"

Precipitation

Centimeters	Inches
Under 25	Under 10
25 - 50	10 - 20
50 - 100	20 - 40
100 - 150	40 - 60
150 - 200	60 - 80
Over 200	Over 80

Currents

→ warm
→ cold
- - → Seasonal drift northern winter

3   knots

B

C

D

Figure 8.4 *Continued*

(B) Fog along the Peruvian coast near the equator.
(C) Cumulus clouds, unstable air, Cape Canaveral, Florida.
(D) Stratus and fog in the coastal valleys of California in the *study area.*

(20 inches) annually (fig. 8.5). Pacific storms drop their moisture on the windward slopes and then continue eastward into the rain shadow zone of Port Townsend, where the air is in a state of descent down the leeward slopes of the Olympic Mountains. This leeward location is mild and dry in both summer and winter as a result of the barrier effect of the mountains. The same rain shadow effect can be observed along the leeward slopes of the Southern Alps or on South Island, New Zealand.

On a microclimate scale, mountains produce wide ranges in temperature. By changing elevations on a mountain slope, one can experience climatic variation. Arctic climates are located in the tropics of both South America and Africa on the slopes of the highest mountains. It is also important to recall the influence of slope orientation. A south-facing slope in the Northern Hemisphere will be warmer and drier than a north-facing slope. In summary, mountains control the flow of air and produce temperature and precipitation variations on varying slopes and heights.

General Circulation of the Atmosphere and the Flow of Energy

Just as ocean currents transport energy from the lower latitudes to the higher latitude regions, winds produce the same effect. The transport process equalizes to some extent the uneven patterns of global heating and moisture.

The equatorial tropics is a zone of high precipitation. As the air circulates, equatorial moisture and energy are transported poleward on the western side of the oceanic high-pressure cells. Cooler drier air returns toward the equator on the eastern side of these cells. At the equatorial intertropical convergent zone, where the oceanic whorls of high pressure converge, the result is low pressure, cloudiness, and abundant rainfall.

In the summer this **convergent zone** shifts poleward, producing equatorial westerlies along the equator. Then in the fall the convergent zone moves back toward the equator. Thus, winds, solar heating, and pressure patterns complete a latitudinal cyclic pattern as the Earth completes its annual journey around the Sun (fig. 8.6).

The monsoon winds of the Indian and Pacific Oceans, described in chapter 5, also complete a cyclic pattern by seasonally shifting with the Sun. As the Asian continent heats up in the spring, moist tropical air on the western side of the oceanic whorl of high pressure moves strongly onshore, bringing with it moisture and monsoonal winds. This movement is reinforced by the summer subtropical jet stream, which shifts poleward, drawing in surface air. Then, as the Asian continent cools down in November, the subtropical jet stream shifts south of the Himalayas, introducing cool continental polar air into the tropics and across the equator into both Africa and Australia. Again, in the spring the pattern repeats itself.

While tropical air masses are converging on Asia, another pattern of middle-latitude convergence is occurring on the poleward side of the mid-oceanic highs. Cold, dry air accumulates over the ice and snow-covered surfaces of the polar region, poised to invade the middle latitudes while in the regions of the mid-oceanic high, moist tropical air develops and advances poleward toward

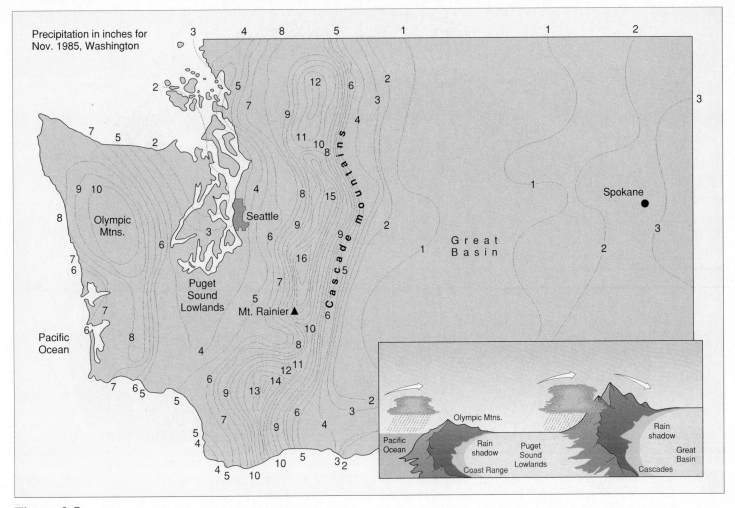

Figure 8.5

Topographic barriers induce updrafts and the orographic effect on the windward slopes while a rain shadow forms on the leeward sides. The Olympic Peninsula in northwestern Washington is a classic example of the phenomenon: Rainfall exceeds 500 centimeters (200 inches) annually on the windward slopes, but falls short of 50 centimeters (20 inches) annually in the rain shadow zone.

the cold polar air. The battle results in cyclonic eddies of the wave cyclone, accompanied by cloudiness, precipitation, and winds as described in chapter 7. Thus we can divide the Earth into two major zones of air-mass convergence that represent the primary regions of precipitation. These shift with the Sun on a seasonal rhythm, bringing wet periods to the zone under their influence. In contrast, the quiet calm parts of the atmosphere are the staging ground for the Earth's major air masses. These areas are located in regions of high pressure or zones of divergence. They are located in the polar lands and subtropical high pressures of the middle latitudes centered on 30° latitude. The Earth's major deserts and region of low precipitation are associated with these two latitudinal zones.

Cyclonic Storms

Cyclonic activity is a climatic control that brings rapid weather changes (fig. 8.7). As pointed out in chapter 7, it is a dynamic influence caused by the release of latent energy of condensation. When the atmospheric system becomes excessively loaded, the safety valve of the cyclonic storm comes into play. These eddies of the air are not randomly distributed but can be predictably charted in the zones of convergence, both in the tropics and the middle latitudes. We can thank the tropical cyclone (typhoon), subtropical tornado, and wave cyclone for the precipitation they bring. Even the most destructive hurricane blesses the land it rakes with life-giving moisture. The rainfall budget would be in a deficit state some years in the tropics and in the subtropical middle latitudes if cyclones ceased to spin. Frontal precipitation of the wave cyclone is the primary supplier of moisture in the middle latitudes. On the west coasts of the major continental land masses the wave cyclone invades the land, supplying over 90 percent of the precipitation. Thus, storms of cyclonic origin represent a major control on global patterns of precipitation and climate.

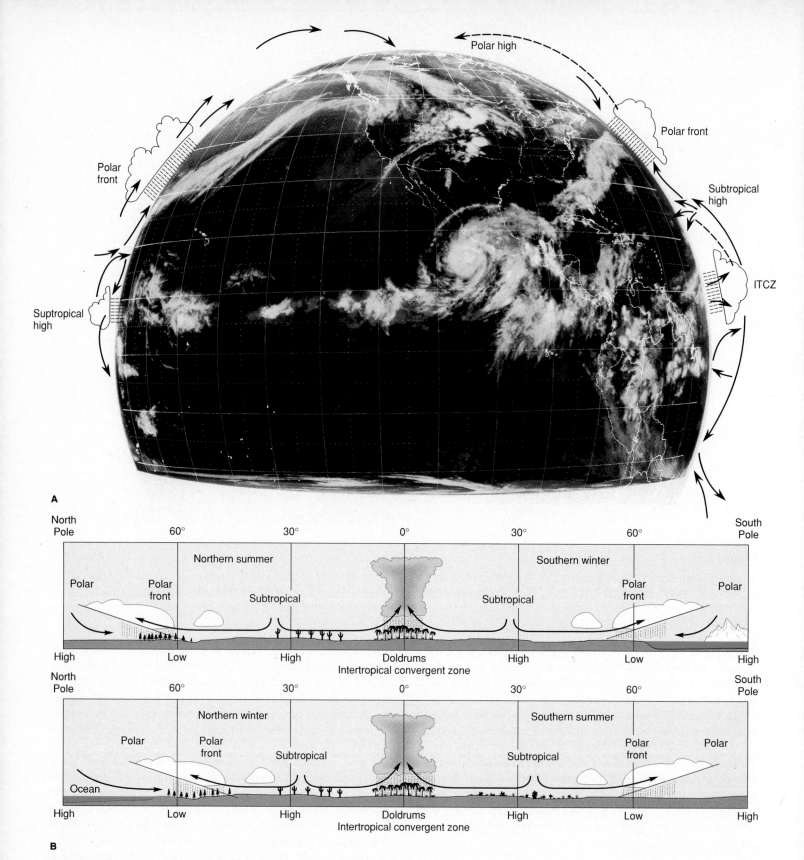

Figure 8.6

General circulation of the atmosphere. The flow of air over the Earth's surface redistributes heat energy. Convectively unstable air rises in the intertropical convergent zone (ITCZ), moves poleward, subsides at the subtropical high, converges with polar air at the polar front, and sinks again at the polar high. Notice the bands of clouds at the two latitudes of convergence, 0° to 10° and 40° to 60° N. In the top portion of B, the cross section of general circulation represents a June condition, and the bottom portion represents a December condition.

Photo and diagram by NOAA.

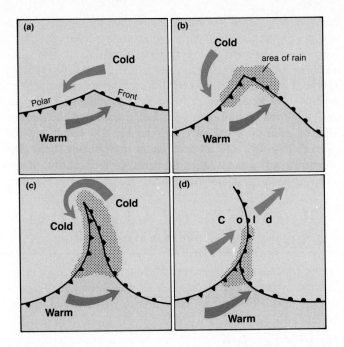

Figure 8.7

Cyclones mix the atmosphere along the convergent zones of the low and upper middle latitudes and release latent energy and moisture.

CLIMATIC CLASSIFICATION

Climatic patterns emerge as temperature, precipitation, wind, and evaporation, together with other elements of weather, give a distinct atmospheric environment.

Many attempts have been made to define climatic boundaries using a variety of schemes. Most systems have stressed precipitation and temperature as the key elements; others have introduced evaporation and vegetation patterns.

The problem in classifying climate or any other natural system lies in the complexity of the natural system and lack of available data. For example, we could use the traditional Greek divisions of global climate into temperate, torrid, and frigid zones; yet some of the warmest and the coldest temperatures ever recorded near sea level have occurred in the temperate zone. The truly temperate climates are located along the shores of the tropical islands in the torrid zones. Climatic classification on a global scale also becomes a problem when the Earth's surface changes in elevation. One can ski in Hawaii, hike a glacier on the equator in Peru, or experience frostbite on the slopes of Kilimanjaro in equatorial Tanzania. All of these chilly experiences are located in the tropics above approximately 3,600 meters (12,000 feet). Because of these and many other examples, no one system is the answer to climatic classification. However, because the Köppen climatic system is widely accepted, we will describe climates with boundaries based primarily on temperature and precipitation.

GLOBAL CLIMATIC ENVIRONMENTS

The Earth's climates are viewed under the Köppen system as within the boundaries of 11 major climatic types. Each is a mix of varying climatic controls and elements. Where a blend of these elements occurs in different parts of the globe, a climate as defined can be mapped as a pattern on the Earth.

The primary purpose of any classification system is to provide an objective method that permits an analysis of the data being observed. From the standpoint of cause and effect, a genetic classification system is most desirable because processes are interpreted. However, such a classification system is yet to be developed. The classification system described in Appendix I, originally developed by W. Köppen in 1918, defines the major climatic types on the basis of primarily climatic elements—temperature and precipitation—without regard to causes.

The geographic distribution of each type of climate is predictable if the variables of temperature and precipitation are known. Köppen constructed an ideal continent to illustrate the predictable patterns of the 11 major climatic types. In figure 8.8, the continent resembles an ice-cream cone which tapers to a point in the Southern Hemisphere, conforming to the general pattern of the continents. The latitudinal range extends from 80° N to about 60° S. A bulge in the cone in the Northern Hemisphere at 30° N is in conformance with the actual continental land masses of the Northern Hemisphere. The climatic types of the oceans are also shown but with less variation and detail because fewer data are available on oceanic zones, and also because oceans tend to produce more uniform environments due to the more uniform surface temperatures.

We can summarize the general distribution by noting that there are five major belts labeled A, B, C, D, and E. The 11 climatic types, described in detail in chapter 9, are subdivisions of these five. Note that E climates (Tundra and Ice Cap) occupy the polar latitudes and extend equatorward to roughly 60° latitude. Farther equatorward, but only in the Northern Hemisphere, are the D climates: continental types with wide seasonal variation in temperature. They extend southerly to 40° N on the eastern sides on continents but cannot be found south of the equator, because the continents in the Southern Hemisphere taper in the upper latitudes to produce strong marine influences and milder climates in the same equivalent latitudes. The temperate rainy C climates of the humid middle latitudes are next, located between 20° and 40° in both hemispheres. The C climates extend more poleward on the western sides of continents where warm ocean currents and southwesterly winds carry marine moderating influence to higher latitudes.

Arid B climates, or deserts and steppe, are centered on 30° latitude and extend poleward on the interior and equatorward of the west coasts where cold ocean currents parallel the coasts and permanent high-pressure cells encourage atmospheric stability.

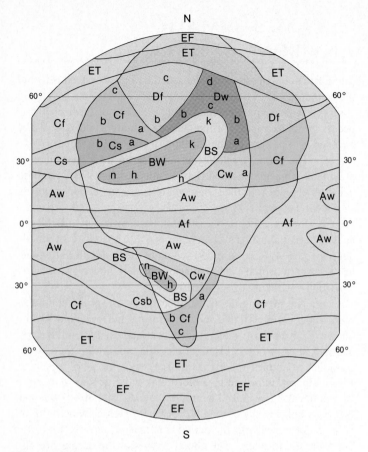

Figure 8.8

Global climates on the hypothetical continent. Refer to appendix A for abbreviations of Köppen's climatic system.

As the equator is approached the B climates merge with more temperate rainy C climates. Finally, the tropical rainy A climates emerge between the Tropics of Cancer and Capricorn. Here temperature variation is minimal throughout the year, and seasons experienced by the middle latitudes do not exist. Solar energy on a day-to-day basis shows little variation. Note how these global patterns are latitudinally oriented, especially over the oceans, and are found on both sides of the equator at approximately the same latitudes, with the exception of the continental D climates.

In terms of global surface area over land and water, the A climates cover the largest area followed by the C climates, and together they cover more than half the Earth's surface. If the Earth were entirely water, these percentages would be even higher. The B and D climates combined cover nearly 50 percent of all land masses, while the A and C climates extend over three-fourths of the ocean surface and about one-third of the land area.

In the Köppen system, numerical limits for each climate are set by using vegetation patterns as a general indicator. Plants reflect both temperature and moisture conditions when data are not available. Temperature and precipitation are the two meteorological elements selected for the precise classification of a given location. A region of high monthly rainfall and warm tempera-

tures can be inferred to have high humidity, considerable cloudiness, and storm activity under low-pressure conditions most of the year. As we use temperature and precipitation data, it is important to never view them in isolation or independently. The same amount of precipitation may have a very different impact on two separate locations. For instance, a north-facing slope in the *study area* can have the same rainfall as a south-facing slope, but the temperatures on each slope will vary considerably. The result is a cool, moist climate with a forest cover on the north-facing shady slope, in contrast with a sunny, drier, chaparral-covered southward slope receiving an equal amount of rainfall.

THE IMPACT OF CLIMATE ON DESIGN FOR LIVING

An understanding of the impact of climate is basic to an understanding of the nature of a place. Climate is the chief environmental influence affecting both human activity and cultural patterns.

House Design

The climate of a place has a strong influence on house design and settlement patterns, especially in tribal and traditional societies. Examples of house design can be observed in Amazonian Shipibo culture in their bamboo houses on raised platforms. In these humid tropics, houses are built on poles not only to gain a better view but to allow air to freely circulate and cool the home environment. This design also keeps insects, small animals, and especially floodwater from invading their living space (fig. 8.9).

At the other end of the precipitation spectrum, the desert dwellers of both North America and the Middle East successfully designed their own microclimate with ubiquitous mud. Thick adobe walls and recessed small windows trap the cool night air inside and prevent the midday heat from invading their living space (fig. 8.10).

Many settlements reflect in their design the prevailing winds. Roofs are sloped to allow the air to move with little resistance over the top. Buildings are also strategically spaced to produce wind protection.

In primitive and pre-industrial society, builders could not ignore climate as an influence on design. That is not true today; for example, a commonly expressed view by builders today is that there is no area in the United States that does not require air conditioning. This indicates that we are totally ignoring climate in design. Through ingenious heating and cooling systems we can make our buildings bearable, although at an expense that may exceed the cost of the building frame. The comfort created has often led to unforeseen problems in dwellings, which can include increased allergies, insulation toxicity from formaldehyde materials, as well as an overcontrolled environment resulting in high energy consumption.

Early cultures lacked the technology to make major changes in the climate environment or ignore its influences in their design. They had to create buildings with limited materials and tools that responded successfully to the climate.

Figure 8.9

The Shipibo house reflects a successful design to live in harmony with its climatic environment. Notice the thatched roof and no walls. What design benefits do these features have in a tropical humid climate?

Figure 8.10

Adobe homes fit desert climates. Thick walls and recessed windows insulate and shield the occupant from temperature extremes.

Climate as an environmental force on house design is operating on various scales. This scale can range from a need for no shelter to areas where a maximum amount of protection is needed. The severity of the climate limits the variation in structures.

Because tribal and traditional societies are technologically limited and materials are often in short supply, their attitude toward climate is one of working in harmony. Their structures make use of local materials, wasting little, in an economy of scarcity, and along with site selection reflect considerable awareness of the local microclimate. Before construction, a site study is conducted under a wide range of weather conditions. There are many examples of taking the environment seriously in house design on our own continent. For example, the southern Pomos of our *study area* in

chapter 1 located their wooden structures with the door opening downwind to reduce the inflow of cool marine air off the Pacific Ocean. The Pueblo people of the southwestern United States constructed their stone and adobe houses where they could maximize solar heating during winter (fig. 8.11).

Houses of the early European settlers of Australia, the United States, and Mexico reflect successful solutions similar to tribal society. Consider the well-built, snug houses of New England, oriented to give protection from snow, wind, and rain, with their undercover links to barn and storehouse; also the cool, well-ventilated plantations of the South and similar Australian farmhouses. Peru and its eighteenth-century monasteries are constructed with thick-walled, patio-centered courtyards, designed to

Figure 8.11

The Pueblo Indians constructed their homes to maximize solar heating during the low winter sun. This photograph was taken in July looking east at 9:30 A.M. Winter afternoon low sun warms this site at Mesa Verde, Arizona.

A

B

Figure 8.12

Principal house types associated with major climatic environments.
(A) The lower floor was built of stone by Incas around A.D. 1200.
The second floor was constructed by descendants using adobe, a good
insulator for frosty nights at high elevation. (B) Stone house under
construction in Scotland. These granite stones retain heat in a cool
marine climate. (C) Mud plaster and adobe in the highlands of
Nepal are an available material and insulate the house for cool
winters and warm summers.

C: Photo by Sally McGregor.

C

provide both indoor and outdoor living space in tropical environment. Although cultures may be widely separated in time and
space, the similarities of solutions to climatic stresses in similar
environments are amazing.

The climatic variables of temperature, humidity, wind,
rain or snow, radiation, and light combine in various degrees to
produce a given environment. Figure 8.12 shows the principal types
of structures associated with variation in climatic environments.
Note how the materials used, architectural style, and site factors
are strongly influenced by climate.

Human Physiology and Climate

Climate also has an impact on the human physiology. Your fin-
gernails grow 20 percent faster in the summer. Ultraviolet rays
produce increased pigmentation, darkening the skin, during the
summer. Warmer temperatures cause your heart rate to increase
to keep body temperature at 37° C (98.6° F) by circulating your
blood coolant faster. In addition, increased temperature cause you
to perspire and produce a cooling effect due to evaporation. This
same evaporation can be a chilling experience if the wind is blowing.
Wind chill is not psychological; the wind increases evaporative
cooling, making body surfaces cooler.

The Natural Environment

In chapters 10 to 13, the influence of climate on vegetation patterns and soil-forming processes is described. Climate also influences weathering and erosion, resulting in a variety of landforms
characteristic of specific climates. Desert climates, for example,
produce a variety of arid landforms not typically found in humid
climates of the tropics.

A

B

Figure 8.13

Climate controls land use, especially agricultural patterns. Within each climate we have a spectrum of choices. (A) Market selection in Tashkent, USSR, where a warm summer plus irrigation produces a variety of subtropical and middle-latitude fruits and vegetables. (B) Shipibo tribal people of Peru market tropical products in nearby Pucalpa along the Ucalali River.

A: Photo by Eugene Klug.

Agriculture

Within the wide spectrum of climates, we can observe that crops, like natural vegetation, have specific limitations or boundaries that shape their distribution and growth. Each climate produces boundaries and limits upon human and biological activity (fig. 8.13). When these boundaries are ignored, disaster follows. The Dust Bowl of the 1930s is a reminder of this principle.

Climate, then, is a strong force that produces limits and regulates both human activity and natural systems either in a subtle or very direct manner. Therefore, a knowledge of climatic controls described in this chapter is of prime importance in understanding the geography of a place.

SUMMARY

Climate is the long-term, average weather conditions of a place. Each climate is uniquely different because the climatic elements and controls vary over the surface of the Earth to produce various blends or combinations. Climatic elements such as *insolation, precipitation, pressure,* and *wind* are interrelated to and modified by **latitudinal position, land,** and **water** as well as **vegetation, terrain,** and **atmospheric and oceanic circulation.**

Today the most widely accepted Köppen classification system uses temperature and precipitation data to subdivide the surface of the Earth into 11 major climatic zones and a number of smaller subunits. The letters A, B, C, D, and E are used to divide the earth into five major climatic zones. Humid climates, supporting forests, are designated by A, C, and D. The letter A covers the tropical forests and savanna, and C stands for the middle-latitude savanna and forests. The snowy forested zones of the upper middle latitudes of the Northern Hemisphere are classified as D climates. Zones of scant precipitation are designated by the letters B and E, with B representing warm arid lands and E cold polar areas.

In the Köppen system, boundaries reflect vegetation patterns, although the actual boundary limits are determined by formulas using the variables of temperature and precipitation. It is also important to note that climates vary not only with latitude but also elevation. A mountain slope can have a variety of climatic zones within a small horizontal and vertical distance.

A knowledge of climate is of paramount importance in understanding both the cultural and physical patterns of the Earth, nation, city, or village. The climate of a place plays a major role in shaping the design of house types, building materials, and tools of construction. Considerations of the environment vary with technological knowledge. Because tribal and traditional societies have few options in design, environmental factors are taken seriously, and the house type and location, building materials, and tools are reflected in their cultural patterns. Climate also has a major impact on human and plant physiology. Although technology has allowed us to modify climate, agricultural patterns are still under the controls of climate.

ILLUSTRATED STUDY QUESTIONS

1. Using figure 8.5, p. 161, describe the effect of this climatic control.
2. Describe the influence of climate on the design of each house (figs. 8.9, 8.10, 8.11, and 8.12, pp. 165, 166, 167).
3. Explain the climatic control of the warm ocean current parallel to the coast in this photograph (fig. 8.4b, p. 160).
4. Why is fog a dominant feature of the coastal Peruvian scene in figure 8.4b, p. 160?
5. Which location has the least annual variation of insolation (fig. 8.1, p. 155)?
6. Describe the location with the greatest annual range in mean temperature (fig. 8.3, p. 157).
7. Explain the effect of a mountain barrier on precipitation patterns and total annual amounts (fig. 8.5, p. 161).
8. Describe the seasonal shift in the intertropical convergent zone (ITCZ) (fig. 8.6, p. 162).
9. Give one example of the role of climate on land use (fig. 8.13, p. 168).
10. Why are continental "D" climates missing in the Southern Hemisphere (fig. 8.8, p. 164)?
11. Which climatic controls influence wide annual ranges in temperature in the upper middle latitudes (figs. 8.1 and 8.3, pp. 155, 157)?
12. Which climatic control influences the net insolation of a place? Which place in figure 8.1 has the widest range? The least? (figs. 8.1 and 8.2, pp. 155, 156)
13. Give one effect of a cold and warm ocean current (fig. 8.4, p. 158).
14. Explain the role of a cyclonic system as a climatic control (fig. 8.7, p. 163).

Climatic Patterns

9

Objectives

After completing this chapter you will be able to:

1. Describe the general patterns of the Earth's major climatic zones.
2. Describe each of the following climates in terms of location, temperature, precipitation, and major climatic controls.
   a. **Tropical Rainforest** (Af)
   b. **Tropical Monsoon** (Am)
   c. **Tropical Savanna** (Aw)
   d. **Hot Desert** (BWh)
   e. **Hot Steppe** (BSh)
   f. **Cool Desert** (BWk)
   g. **Cool Steppe** (BSk)
   h. **Mediterranean** (Csa) (Csb)
   i. **Subtropical Monsoon** (Cwa)
   j. **Humid Subtropical** (Cfa)
   k. **Marine West Coast** (Cfb)
   l. **Snow Forest** (Dfa, Dfb, Dfc, Ds, Dw)
   m. **Polar Ice Cap** (EF) and **Tundra** (ET)
   n. **Highland Polar** (EH)

Antarctica, King George Island.
© Anna E. Zucherman/Tom Stack & Assoc.

The geographic patterns of climate are a response to the primary energy systems driven by the Sun. This survey of global climates will begin in the low latitudes of surplus energy, then proceed poleward to regions of greater temperature variation and areas of deficient insolation.

THE TROPICAL RAINY CLIMATES (A)

The climates of the tropical world are separated from all other zones by having no winter. All mean monthly temperatures are above 18° C (64.4° F), typically between 21° C (70° F) and 27° C (80° F). The annual range in temperature is small, usually less than 5.5° C (10° F) while the daily range sometimes exceeds 11° C (20° F). The high angle of the Sun each month results in abundant energy, which is exported to the higher latitudes primarily by wind and ocean currents.

The small annual and daily range in temperature is due to small annual variation in insolation in the tropics. The length of days and nights throughout the year remains nearly the same in these latitudes. Changes in monthly temperatures result more often from change in cloud cover. During the cloudy rainy season, mean temperatures often drop a few degrees as the clouds reflect more solar energy back into space. The warm ocean, above 27° C (80° F), is also a contributing factor to the minimal temperature range.

The tropical rainy climates, as the name implies, are wet, rarely with less than 88.9 centimeters (35 inches) annually. Precipitation is mostly the result of convectional processes, while tropical cyclones are significant rainfall producers, especially in the belt of the trade winds. At many coastal locations and islands located in the trades, orographic precipitation is a major contributor to annual rainfall on the windward slopes. In the Amazon Basin, where convection is prominent, precipitation falls almost daily, usually in the late afternoon or evening.

The amount and patterns of precipitation vary much more than annual variations in temperature in tropical climates. Therefore, precipitation patterns are the basis for subdividing the **Tropical Rainy** (A) climates into three distinct subregions: the **Tropical Rainforest** (Af), **Tropical Monsoon** (Am), and **Tropical Savanna** (Aw).

The Tropical Rainforest rainfall even during the driest month exceeds 6.1 centimeters (2.4 inches). If it is less than this amount, the climate is classified as Tropical Savanna (Aw). In coastal Southeast Asia, the vegetation can resemble the tropical rainforest type even where rainfall is below this mean monthly minimum. This results from large annual monsoon rainfall and winds coming during summer. Because moisture remains sufficient to support a rainforest, the climate is classified as Tropical Monsoon (Am). Figure 9.1 gives global patterns and divisions for the A climates.

Tropical Rainforest (Af)

The major regions of the Tropical Rainforest climate are mainly located near or on the equator in the doldrums or intertropical convergent zone and on the windward slopes in the trade wind belts. Typical regions include the equatorial Amazon Basin in South America, the Congo Basin in Africa, the Guinea coast of West Africa, the East Indies, and many islands of the tropics in the Indian, Pacific, and Atlantic Oceans (fig. 9.2).

Tables 9.1 and 9.2 give a summary of temperature and precipitation for Hilo, Hawaii, and the Volcanic Observatory at the summit of Kilauea, 1,243 meters (4,078 feet).

Hilo is located on the windward side of the island of Hawaii with a mean annual temperature of 23.7° C (74.7° F) and rainfall of 348 centimeters (137 inches). Typical of most locations on the Hawaiian Islands, precipitation is greatest in the cooler months of January and February. The rainfall pattern is closely related to the mountains and exposure to the prevailing northeast trade winds. On the windward slopes rainfall increases from sea level to about 1,800 meter (6,000 feet) and then declines toward the summit. On the island of Hawaii, the zone of maximum rainfall lies between 750 meters (2,500 feet) and 900 meters (3,000 feet) elevation on the windward slope, averaging 250 to 750 centimeters (100–300 inches) annually, making it one of the Earth's wettest spots. Precipitation turns to snow above about 2,700 meters (9,000 feet), making winter sports a reality on this island during the cool season. Leeward of Hilo, rainfall declines to less than 50 centimeters (20 inches). Here we can see the rain-shadow and orographic effects at work when the northeast trades reach their maximum in the cooler season. The summer months of June through September experience convectional thunderstorms and variable rainfall as the intertropical convergent zone (doldrums) shift poleward. Note, the range in mean annual temperature is less than 5.5° C (10° F) while precipitation exceeds 40.1 centimeters (15.08 inches) in March but drops to 18.8 centimeters (7.4 inches) in June. At the Volcanic Observatory we see the same weather patterns but temperatures and rainfall are much lower.

The chief control of temperature and rainfall at the Volcanic Observatory is elevation. Here the mean annual temperature is 16° C (63° F) and rainfall is 246.15 centimeters (96.91 inches) annually. Thus in just a few miles one can leave one climate and enter another by changing one important control, elevation.

Tropical Monsoon (Am)

The Tropical Monsoon climate occurs along the lowland tropical coastlines where monsoon winds blow onshore during the summer period. Although there is a distinct dry season where at least one month receives less than 6.1 cm (2.4 inches) of rain, the annual rainfall totals exceed 100 centimeters (40 inches). Coastal Southeast Asia, Sri Lanka, and the northern coasts of South America and Australia are the primary sites for this tropical climate.

(Af) Amazon rainforest, Pulcalpa, Peru, latitude 7° S, longitude 73° W

(Am) Kakadu National Park, Northern Territory, Australia, latitude 12° S, longitude 131° E

(Aw) Darwin, Northern Territory, Australia, latitude 12° S, longitude 132° E

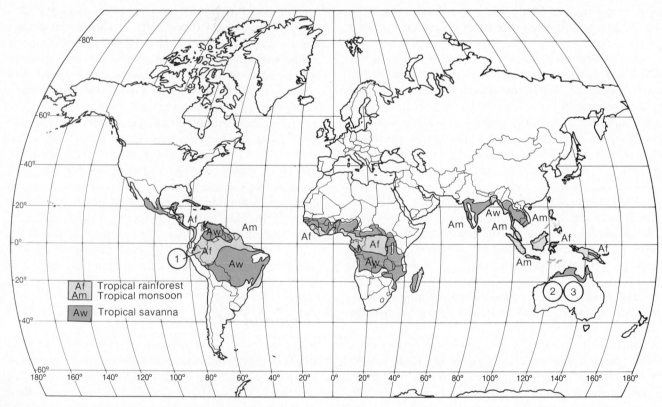

Figure 9.1

Global locations for Tropical Rainy (A) climates.

Top left: Photo by Dr. Charles Hogue, Curator of Entomology, Los Angeles County Museum of Natural History.

Figure 9.2

Tropical Rainforest (Af). Daily rainfall in the form of
thundershowers maintains a lush tropical rainforest in the Amazon
Basin, Peru.

Table 9.1

Temperature for the Island of Hawaii

Station	Length of Record	January Average	Average	Maximum	Minimum
Hilo	33 yrs	69.9° F	74.7° F	85° F	50° F
Volcanic Observatory	26 yrs	57.8° F	63.0° F	84° F	40° F

Source: Climate and Man. Yearbook of Agriculture, 1941, USDA.

Table 9.2

Average Precipitation for the Island of Hawaii

Station	Length of Record	Jan.	Feb.	Mar.	Apr.	May	June	July	Aug.	Sept.	Oct.	Nov.	Dec.	Annual
Hilo	40 yrs	11.77	10.82	15.08	13.06	9.21	7.40	9.89	11.81	10.84	10.97	13.80	12.47	137.12
Volcanic Observatory	26 yrs	12.27	7.78	10.76	8.66	5.92	4.18	6.18	7.23	6.68	6.59	9.94	10.72	96.91

Source: Climate and Man. Yearbook of Agriculture, 1941, USDA.

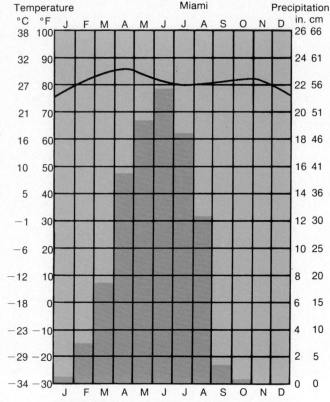

City: Rangoon, Burma
Latitude: 16°46'N   Am
Altitude: (5.5m) 18 ft.
Yearly precipitation: 252 cm (99.2 in.)

Climate name: Savanna (monsoon type)
Other cities with similar climates:
Bombay, Calcutta, Darwin,
Miami

Figure 9.3

Tropical Monsoon climate (Am), Rangoon, Burma. Monsoon winds and rains reach their peak between May and October as moist tropical air moves onshore from the equator, dumping over 2.5 meters (99 inches) of precipitation during this period.

Rangoon, Burma, is located in the path of the southeast monsoon winds at 16.46° N latitude. The maximum monthly rainfall peaks at 55 centimeters (almost 22 inches) during the month of July when the southeast winds are strongest, bringing in moist tropical air from the Indian Ocean (fig. 9.3) In sharp contrast, December and January experience less than 1 cm (0.4 inch) of rain. During the dry season, winds are offshore coming from the arid interior of the Asian continent. These same winds move southeast across the equator toward Australia, bringing heavy monsoon rains from November through March.

Tropical Savanna (Aw)

Like the Tropical Monsoon climate, rainfall in Tropical Savanna climate is very seasonal, coming in the period of higher sun. Total annual rainfall is smaller than either the Af or Am types of climate.

The Tropical Savanna climate, located poleward of the Tropical Rainforest, is in a transition zone from wet to dry cli-

mates. The Tropical Savanna is located in northern and eastern South America, northern Australia, southeast Asia, southern India, and in two belts on both sides of the Tropical Rainforest in equatorial Africa.

The strong seasonal rainfall pattern is the result of a shift in the intertropical convergent zone poleward during the summer period, bringing convectional showers and cloudy skies. The dry season is a warm, sunny period under the influence of the subtropical high-pressure belts that dominate the arid regions. The maximum temperatures occur before the period of highest sun when the skies are clear. Cuiaba, Brazil (latitude 16° S) has its maximum mean monthly temperature of 28.3° C (82.6° F) in October just before the summer rainy period, which occurs from November through March.

The climate is dominated by three air mass systems—maritime tropical (mT), maritime equatorial (mE), and continental tropical (cT) air—characterized by upper-level subsidence and stability. The dry season is under the influence of the continental tropical (cT) air mass that dominates the tropical deserts of the world. As the Sun shifts poleward, maritime equatorial (mE) and maritime tropical (mT) zones shift poleward, bringing unstable moist air into the higher latitudes of the tropics. Because the Sun crosses the zenith twice each year in the belt of the tropics, stations such as Cuiaba have two maximums of rainfall. The first occurs during October and the second in April. Both maximums correspond closely to the highest altitude of the Sun and dominance of maritime tropical (mT) and maritime equatorial (mE) air. This double maximum of annual variation of precipitation is found between about 20° latitude north and south. Closer to the equator, the two maximums become more distinct; with increasing distance from the equator a single maximum emerges and total annual rainfall declines.

THE DRY CLIMATES (B)

The arid regions of the world are extremely varied, but they all experience greater potential evaporation than precipitation. Rainfall can vary from nearly zero to over 63.5 centimeters (25 inches) per year and still be considered an arid region. The key determining factor is available soil moisture. Therefore, precipitation and temperature viewed separately from evaporation cannot explain the dry lands such as the Sahara Desert or great deserts of the Middle East and Australia. Both the daily and annual range in temperature are among the most varied in the world in the dry climates (fig. 9.4).

The dry climates are widely distributed; nearly one-fifth of the land surface of the Earth is made of dry lands. Most deserts are located in the middle latitudes centered on 30° in the regions of the subtropical high-pressure belts. Deserts are also located in the interior of the continents and on the leeward sides of mountains in rain shadow zones. Some of the driest deserts are also found along the west coasts of continents where cool ocean currents moderate temperatures and discourage precipitation. The major dry regions in the Northern Hemisphere are located in North Africa and the Middle East and continue into the Asian interior of the

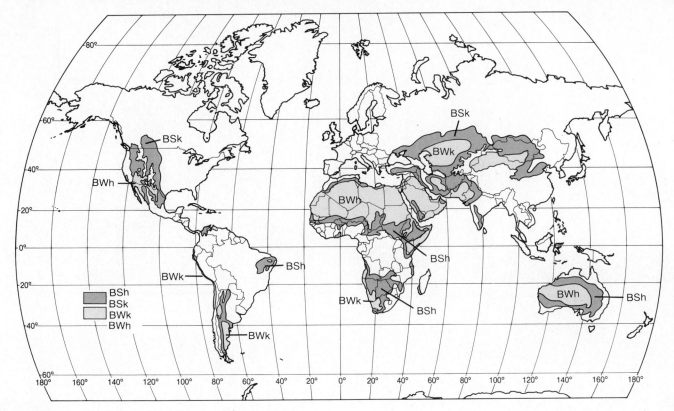

Figure 9.4

The Dry climates (B). Rainfall can vary from none to over 60 centimeters (25 inches) per year and still qualify as a B climate. Considerations of temperature, precipitation, evaporation, cool ocean currents, and high-pressure patterns are important factors in determining the distribution of climatic patterns.

southern Soviet Union and western China and Mongolia. The western half of the United States and northern Mexico also have large dry regions at comparable latitudes.

In the Southern Hemisphere, deserts and semiarid lands are also associated with the belts of subtropical high pressure near 30° S. Dry lands make up most of the Union of South Africa, southwest Africa, and Australia. South America has limited dry lands along the Peruvian and Chilean coasts, from 5° to 30° S, and in southern Argentina from 30° S to nearly 50° S.

The dry climates designated by Köppen with the letter B are defined in terms of combinations of mean annual temperature and precipitation. The effectiveness of precipitation on vegetation is strongly governed by evaporation. Evaporation data are sparse, but its effects can be taken into account by considering both precipitation and temperature jointly. Köppen defined a dry climate as a place where the annual precipitation (in inches) is equal to, but not greater than the value determined by the following formula:

$$r = 0.44t - 8.5$$

where r represents the annual rainfall and t stands for mean annual temperature in degrees Fahrenheit. If the annual temperature is above 10° C (50° F), then the climate would be considered dry if the annual precipitation did not exceed 34.3 centimeters (13.5 inches). If the mean annual temperature is less than 10° C (50° F) for this amount of precipitation, the climate would be considered moist.

Hot Desert (BWh) and Steppe (BSh)

The BWh and BSh climates are located near 20° to 25° latitude but extend beyond these latitudes in eastern Africa.

The mean annual temperature for both BWh and BSh climates is greater than 18°C (64.4° F), according to the Köppen system. The climate's warm, dry conditions result from persistent subsidence motion of the subtropical anticyclone. Most of the year skies are clear, and relative humidity is very low. Summer temperatures often surpass 38° C (100° F). The annual range in temperature usually exceeds 17° C (30° F), with daily temperature ranges sometimes spanning 28° C (50° F).

Precipitation is not only low, but variable as well as seasonal. According to Köppen, if the mean annual temperature is 21.3° C (70° F) in a Steppe climate, precipitation cannot exceed 56 centimeters (22 inches), while 28 centimeters (11 inches) is the upper limit in a Hot Desert climate. (See Appendix I for criteria limits of BSh and BWh.) If the rainfall comes during the winter season, lower limits for rainfall are set; conversely, higher limits are set if rainfall comes primarily in the summer because evaporation takes its toll on available moisture.

Figure 9.5

The Tropical Desert (BWh) is exemplified by Alice Springs, Northern Territory, Australia, where the Todd River occasionally runs bank to bank from sudden thunderstorms.

Although rainfall is sparse and unreliable in the dry lands, large amounts may fall in a single cloudburst. On April 8, 1990, Alice Springs, in central Australia, experienced flash floods and high winds bringing 5 centimeters (2 inches) of rain in one hour. The normally dry Todd River flowed with water, covering the channel bank to bank (fig. 9.5). The total annual rainfall for this location rarely exceeds 20 centimeters (8 inches). As you can see, rainfall averages have little meaning in the dry climates (table 9.3).

The season when precipitation occurs differs widely in the dry lands. Winter patterns result from cyclone storms coming in from the oceans on the western sides of the continent. Interior continental locations and east coast dry lands as well as the dry lands bordering the wet tropical climates have summer maximums. In the southern and eastern Mexican deserts, rains come during the summer as convectional thunderstorms from moist air originating over the warm Gulf of Mexico and the Gulf of California.

Table 9.3

Highest Rainfall Intensities in Australia

Station	Years of Complete Records	Period in Hours				
		1	*3*	*6*	*12*	*24*
		mm	mm	mm	mm	mm
Adelaide	79	69	133	141	141	141
Alice Springs	36	75	87	108	133	150
Brisbane	77	88	142	182	266	327
Broome	36	112	157	185	313	353
Canberra	37	40	57	67	76	120
Carnavon	27	44	63	83	95	108
Charleville	35	42	66	75	111	142
Cloncurry	23	59	118	164	173	204
Darwin (Airport)	35	89	138	214	260	291
Esperance	15	23	45	62	68	79
Hobart	75	28	56	87	117	168
Meekatharra	30	33	67	81	99	112
Melbourne	100	76	83	86	97	130
Mildura	34	49	60	65	66	91
Perth	37	31	37	48	64	80
Sydney	71	118	194	200	244	340
Townsville	34	88	158	235	296	319

Source: Pluviograph records in Bureau of Meteorology archives. Bureau of Meteorology, Climate of Australia, October 1989, p. 9

Figure 9.6

Inland Cool Foggy Desert (BWkn) at Windhoek, Namibia, is warm
and dry in contrast to fog-soaked Swakopmund along the coast.

Photo by Dr. Charles Hogue, Curator of Entomology, Los Angeles County
Museum of Natural History.

When we compare Yuma, Arizona, which has winter
maximum precipitation, with Guaymas, Mexico, which receives
7.6 centimeters (3 inches) in August, Guaymas receives nearly
three times as much annual precipitation spread over all seasons.
However, both locations have the same effective water budgets. At
Guaymas, the evaporation takes its toll on the summer rain.

The BWh and BSh climates are the principal source regions of tropical continental air. The Sahara is the only source
region year around for continental tropical air.

The Coastal Dry Lands (BWkn, BSkn)

Along the coasts of North and South America and Africa, where
the interior deserts reach the west coast and meet cold oceans,
the desert is cool, foggy, and very low in precipitation. Cool ocean
currents reduce the temperature and create foggy, stable atmospheric conditions. A strong inversion persists with daily surface
temperatures often remaining 11° C (20° F) to 17° C (30° F)
below air temperature at 1,500 meters (5,000 feet) altitude. As
one moves inland away from the coast, temperatures increase rapidly. Windhoek, at 1,500 meters (5000 feet), in Namibia, is a good
example. Its location is 64 kilometers (40 miles) inland from cool,
coastal Swakopmund where fog persists 150 days per year, giving
high relative humidities and low maximum temperatures (fig. 9.6).
Fog drip is the primary source of moisture along these coastal cool
deserts. Rainfall is infrequent and extremely low.

Coastal Swakopmund, Namibia, is located in the Namib
Desert at 23° S latitude (fig. 9.7). It can be compared to the western
coastal deserts of Baja California, Mexico, and the Atacama Desert
of coastal Chile. All three deserts occur in the same latitude, on
the western coasts, where cool ocean currents refrigerate the coast.

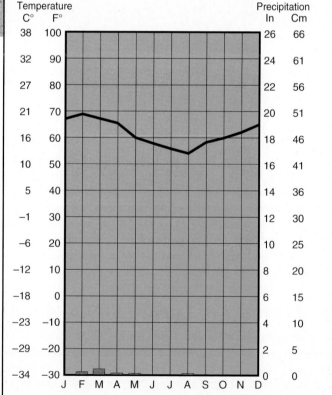

City: Swakopmund, Namibia	Climate name: Cool Desert (BWk)
Latitude: 22° 56'S	Other cities with similar
Altitude: 8m (24 ft.)	climates: Lima, Peru
Yearly precipitation: 2.3 cm (0.9 in.)	Yearly temperature: 62.5° F

Figure 9.7

Mean annual temperatures and precipitation for Swakopmund, Namibia, located along the southwestern African coast. Note the moderating influence of the cool ocean currents. Temperatures are low with little seasonal variation.

Photo by Dr. Charles Hogue, Curator of Entomology, Los Angeles County Museum of Natural History.

Swakopmund throughout the year is under the influence of the semipermanent high-pressure cell occupying the center of the South Atlantic. Subsidence and adiabatic heating warms the air aloft, while the air is cooled from below by the Benguela Current which flows north, bringing cool water from higher polar latitudes. This cool water is cooled even more by upwelling of cold deep water resulting in water temperatures near 12° C (55° F) in all seasons.

Rainfall increases with distance from the chilly coast. Swakopmund has an average annual rainfall of 1.65 centimeters (0.65 inch); Goanckontes, 32 kilometers (20 miles) inland, averages 3.43 centimeters (1.35 inches); and Donkerhuk, 145 kilometers (90 miles) from the coast, receives 17.27 centimeters (6.8 inches). A typical day begins with gray skies of stratus clouds, or

Figure 9.8

The Galápagos Islands (climate BWkn) are refrigerated by the
Humboldt Current. Days are dominated by a roof of stratus clouds
and fog drip. This island group has some of the planet's most
unusual plant and animal life resulting from environmental
conditions and isolation.

Photo by Dr. Charles Hogue, Curator of Entomology, Los Angeles County
Museum of Natural History.

fog, which gradually burn off in the afternoon. The daily range in
temperatures runs from around 11° C (52° F) to around 20° C
(68° F), which is rather small for a desert, due to the moderating
influences of the ocean. The mean summer January temperature
is 17.1° C (62.6° F), and the winter July temperature is 13.8° C
(56.5° F).

The Galápagos Islands, located in the eastern Pacific and
extending about 160 kilometers (100 miles) on each side of the
equator, are another example of a moderate dry climate. Their
equatorial location is not characteristic for this climatic condition.
The same cool Humboldt Current that helps to create the deserts
of northern Chile and southern Peru converges on the islands to
lower temperatures to around 15° C (60° F). The days are dominated by a roof of stratus clouds banked against the higher slopes,
similar to Swakopmund.

As the southeast trade winds blow across the cool ocean,
the air is chilled and stabilized with an inversion. When it does
rain, it comes mostly in January through April as convectional and
orographic showers. Some years no rain has been recorded on the
leeward side of islands in the rain shadow.

Because of the unusual environmental conditions in such
an isolated oceanic region, plant and animal life are distinct and
in some cases unique. You will find 26 types of turtles, for which
the islands are named (galápagos is Spanish for turtle). These
islands host the planet's only seagoing iguana, *Amblyrhynchus*
(fig. 9.8).

Figure 9.9

The Cool Dry Interior climate (BSk) (BWk). Cool dry zones are primarily a Northern Hemisphere phenomenon. Near Carlsbad Caverns National Park, New Mexico, is a classic example of a BSk climate, where winters are cool and snowy and summers warm up to the 33° C (90° F) range with late afternoon convectional showers.

The Cool Dry Interior (BSk, BWk)

The **cool deserts** BWk and **steppe** BSk of the continental interior are strongly influenced by their position far from moisture sources and storm tracks. Continental interiors can be considered the back roads of storm paths, where storms have lost their way and in the process most of their moisture. The size of the continent is important in drying of the air originating over maritime sources. Because the Southern Hemisphere has no large land masses in the higher middle latitudes, this climate is primarily a Northern Hemisphere phenomenon (fig. 9.9).

The major exception is Argentina's Patagonian Desert, where the average temperatures of even the warmest month at Santa Cruz (latitude 50° S) is only 14.9° C (58.6° F). Here we can see the rain-shadow effect of the Andes. Storms lose their moisture on the windward Andean slopes, and the dehydrated air descends the leeward slopes bringing windy, cool, dry conditions to Patagonia.

The rain-shadow effect is a strong influence on the cool, dry interior climates of North America and Asia as well. Storm tracks that reach these interior regions are usually low in moisture but very windy.

The winter season is dominated by high-pressure anticyclonic circulation which contributes to clear skies, low humidities, and wide daily ranges in temperatures. Havre, Montana, and Fort Aleksandrovsk, USSR, both have annual ranges in temperature exceeding 28° C (50° F). In January, temperatures often fall below −18° C (0° F) but rebound to 32° C (90° F) in July.

When precipitation arrives, it usually comes in downpours from brief thundershowers lasting from a few minutes to several hours during the warm summer. Winter precipitation is usually in the snow form in the higher latitudes and elevations and gentle rain in the lower elevations and southern limits of this climate. Prescott, Arizona, has 62 days per year of precipitation coming during both the summer and winter. Winters are cold and snowy with a mean January temperature averaging below 10° C (50° F).

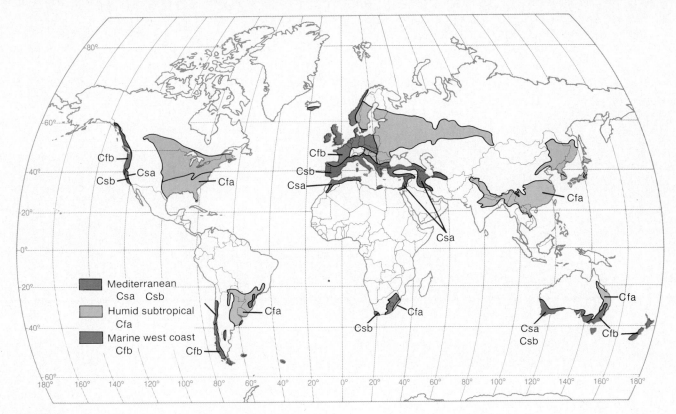

Figure 9.10

In Mediterranean (Csa, Csb) climates, summers are warm and dry, followed by cool, mild, rainy winters. Humid subtropical (Cfa) climates have hot, rainy summers and cool, wet winters. Marine West Coast (Cfb) climates are in the storm track of the middle-latitude cyclone.

Latitude also plays an important role in precipitation patterns of the cool interior deserts. At high latitudes most of the precipitation comes in the summer in convection showers, while lower middle-latitude western continental locations experience a stronger precipitation pattern during the winter when cyclonic polar front weather converges on these latitudes.

Although the winter precipitation is usually light, it has a lasting effect because evaporation is also at a minimum during this time of the year. Snow often remains on the ground for long periods in the low winter temperatures. Therefore, this region in the winter becomes part of the polar continental source region. As summer approaches, the region takes on the characteristics of the tropical continental source region.

WARM TEMPERATE RAINY CLIMATES (C)

The C climates are distinct from the tropical climates in having at least one month of mean temperature below 18° C (64° F). However, no month has a mean temperature below −2.8° C (27° F). There is also enough annual precipitation to distinguish this type from the dry B types.

Consider the warm temperate rainy climates to be the humid climates of the subtropics on both the east and west coasts of the continents, land masses, and islands of the world. These include the **Mediterranean** (Cs), **Subtropical Monsoon** (Cw), **Humid Subtropical** (Cfa), and **Marine West Coast** (Cfb) climates (fig. 9.10).

The C climates are subdivided on the basis of both the seasonal precipitation and temperature patterns of the summer season. See Appendix I for the boundary differentiation.

Mediterranean (Csa, Csb)

The Mediterranean climates (Cs) occur in the subtropical latitudes on the western sides of continents bordering the coasts, with cold ocean currents. These include the lands bordering the Mediterranean Sea, the coast and inland valleys of California, central Chile, the southwestern coast of the Union of South Africa, and portions of the southwestern and southern coast of Australia.

The aridity in the summer reflects the influence of subtropical high pressure that dominates this region from May through September. As the autumn season develops, it shifts equatorward and cyclonic frontal activity increases in frequency.

On the coast of California, precipitation declines from north to south. This results from a seasonal shift in both the subtropical anticyclonic high-pressure cells and the middle-latitude

Figure 9.11

Fog and coastal low clouds dominate the Coastal Mediterranean (Csb) climate. Condensation in the form of fog drip provides up to 50 percent of the moisture supply in the soil.

wave cyclone and jet stream. Precipitation declines southward as the intensity and frequency of cyclones decrease. Southern California remains under the umbrella of high pressure most of the year while the northern part of the state, closer to the jet stream storm tracks, receives a higher percentage of storms. Mean annual precipitation declines from 358.9 centimeters (141.3 inches) at Cordova, Alaska (latitude 61° N) to 26.16 centimeters (10.3 inches) at San Diego, California (latitude 33° N).

The Mediterranean climate varies not only with latitude but also with distance from the coast. Maritime influences moderate the coast during both summers and winters. Coastal areas are mild and moist during the winter and cool and foggy during the summer. In contrast, the neighboring inland locations are in the rain shadow and are hot and drier during the summer and cooler during the winter. Compare precipitation in our *study area* with Ukiah, California, 50 kilometers (30 miles) inland. Our *study area* averages 150 centimeters (60 inches) annually, while Ukiah receives 90 centimeters (36 inches).

The wider annual range in temperature of Ukiah is the result of continential influences or distance from the ocean. The coastal zone reflects the temperature of the ocean water, which runs between 10° C (50° F) and 13° C (55° F) annually. Wider interior ranges in temperature result from the poor heat storage capacity of land and the dominance of high pressure and clear skies, which allow the land to heat up and cool off rapidly on both a daily and annual basis.

We can see this same temperature contrast between coastal and interior locations along the entire coast of North America. A counterpart in the Southern Hemisphere can be found in the Mediterranean climate zone of central Chile.

Fog and low clouds are very significant in the Mediterranean climates. When moist Maritime Polar (mP) air moves into the inland valleys and stagnates, radiation fog forms as a result of cooling by the ground surface. This can cause extended periods of dense fog in the Sacramento and San Joaquin Valleys of California, for instance. Summer coastal fog and stratus clouds are very common along the coast as maritime air is chilled when it moves over cool ocean currents and onshore during the evening hours (fig. 9.11).

In our *study area* of chapter 1, 16 kilometers (10 miles) from the shoreline, fog ensures the coastal redwoods' survival during the dry summer. Condensation on the leaves creates fog drip, which supplies up to 50 percent of the forest's moisture needs between May and October.

Figure 9.12

The Subtropical Monsoon (Cw). The Dry Winter Subtropical Monsoon climate forms a transition between the two great forest belts of the world on the eastern coasts of China (shown here), Brazil, Mexico, Australia, and Africa as well as the interior of northern India. Threshing of rice in the Pearl River Valley near the coast is a labor-intensive activity. Here the monsoon rainy season provides ideal growing conditions between April and October: two crops are harvested during this period.

Photo by Dan Hamburg.

Because the *study area* described in chapter 1 is located on the eastern boundary of the coastal Mediterranean (Csb) marine climate, moisture-loving redwoods decline rapidly east of this boundary.

The Mediterranean climate divides into two subtypes designated Csa and Csb. The Csa is commonly found inland from the coast, away from the marine influences. The coastal Mediterranean Csb occurs where cool ocean currents, coastal fog, and stratus moderate the climate during both the summer and winter, producing a small annual and daily range in temperatures.

Dry Winter Subtropical Monsoon (Cw)

The Cw climate is in a transition zone on the poleward side of the Aw and Am climates of the major tropical belts and equatorward of the Snow Forest (D) climates. Large areas with this climate are located in northern India, southeastern Brazil, northern Mexico, southern Australia, the interior plateau of subtropical South Africa and southern China (fig. 9.12). Where the Cw climate occurs in the tropics, it is usually located on higher plateaus. Elevation

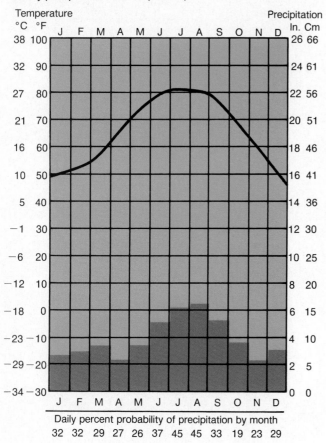

City: Charleston, S.C.
Latitude: 32°46′N
Altitude: (5m) 16 ft.
Yearly precipitation: 120 cm (47.5 in.)

Climate name: Humid subtropical Cfa
Other cities with similar climates:
Sydney, New Orleans, Guangzhou

Daily percent probability of precipitation by month
32 32 29 27 26 37 45 45 33 19 23 29

A

B

Figure 9.13

(A) The Humid Subtropical climate (Cfa) is characterized by hot humid summers and cool rainy winters. Precipitation is convectional during the summer, resembling the tropics, while winter rains are the result of the wave cyclone. (B) Hurricane Hugo flattened a forest near Charleston, South Carolina. These tropical storms are important sources of moisture for this climatic region.

Photo by W. Salizar, U.S. Department of Agriculture, U.S. Forest Service, Southern Region.

in the tropics is the key factor in producing mean monthly temperatures that fall below 18° C (64.4° F). The tropical nature of these plateau stations is reflected in the small annual range in temperature. The precipitation patterns in the tropical highlands, where Cw climates occur, reflect the seasonal shift of the doldrums poleward in the summer. This brings moisture-laden trade winds to these locations. Pueblo, Mexico, elevation 2,176 meters (7,120 feet), and Juliasdale, Zimbabwe, elevation 1,850 meters (6,070 feet), are good examples of high-elevation Cw climates.

Humid Subtropical (Cfa)

The Humid Subtropical (Cfa) climate is located in portions of southeastern Brazil, northern Argentina and Uruguay, New South Wales in Australia, southern Japan, South Korea, along the southeastern coast of China, and the southeastern United States. Charleston, South Carolina, reveals this pattern (fig. 9.13).

All of these locations have several conditions in common. Each is located on the southeastern portion of the major conti-

nents, between 20° and 30° latitude, in both hemispheres. Warm ocean currents flow poleward adjacent to the coast. The mid-oceanic high pumps moist maritime tropical (mT) air into these locations, which becomes unstable in the summer over the hot land surface. The result is heavy thunderstorm activity, usually in the afternoon.

Because of these controls, temperatures are high in the summer and mild in the winter. Occasionally, polar air masses break out of the higher latitudes, causing a rapid fall in temperatures. These cold air outbreaks bring freezing temperatures to areas as far south as Houston, Texas, and even Tampa, Florida. The results can be disastrous to the frost-sensitive citrus industry. In the Southern Hemisphere, cool air can move as far north as the Amazon Rainforest, again triggering thunderstorms.

In summary, winter precipitation results from frontal cyclonic activity as polar air mixes with warm, moist tropical air. Summer rains are more the result of warm maritime tropical (mT) air that is heated by the land's surface to produce convectional instability and thundershowers.

City: Vancouver, Canada
Latitude: 49°17′N
Altitude: (14m) 45 ft.
Yearly precipitation: 146 cm (57.7 in.)

Climate name: Marine West Coast
Other cities with similar climates:
Seattle, London, Paris

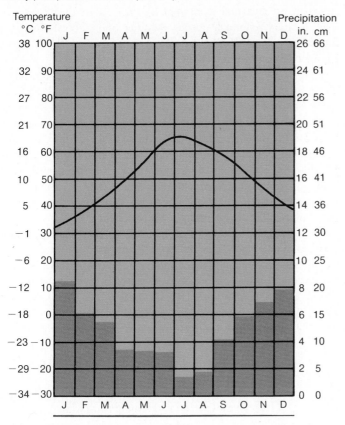

Figure 9.14

Vancouver, British Columbia, Canada, is an example of the Marine West Coast (Cfb) climate. Located on the west coast of the continent in the pathway of the wave cyclone, the Cfb climate is cool and rainy. Summers are mild and short but never dry, while winters are wet and cool with some occasional snow.

Marine West Coast (Cfb, Cfc)

The Marine West Coast (Cfb) climate, located in the upper middle latitudes, has mild humid summers. It is in the pathway of the cooling moist westerly winds that blow onshore. This is the climate of the Pacific Northwest, British Columbia, the British Isles, northwestern Europe, southern Chile, southeastern Australia, Tasmania, and New Zealand. A Cfc subtype with very cool summers, under strong cool marine influences, is found in Iceland, the Gulf of Alaska, the Aleutian Islands, and along the west coast of northern Norway (fig. 9.14).

Precipitation falls as snow and cold rain in the winter and warm rain showers in the summer. Cyclonic systems repeatedly invade this climatic region with accompanying warm and cold fronts. The polar front and jet stream are centered on this climate.

This climatic belt is under the strong influence of the Maritime Polar (mP) and Tropical Pacific (mT) air masses that generate the cyclonic systems originating along the polar front

boundary of these air masses. Because the polar front migrates poleward in the summer, precipitation declines in the lower latitudes of this climate during the summer period, producing smaller annual rainfall totals. Summer cyclonic precipitation reaches only areas poleward of 40° latitude.

Northwestern Europe is the largest region of this climate. Migratory cyclones originating in the Icelandic low-pressure circulation repeatedly fan the continent with summer and winter storm systems. Cyclonic precipitation declines in the Mediterranean European region, especially south of the Alps and toward the interior, where cyclonic systems carry less moisture and occur less frequently. France is a transition location with the Cfb climate dominating the northern half and Mediterranean (Csa) covering the southern half.

Northwestern Europe is one of the Earth's major overcast regions resulting from both cyclonic systems and stratus clouds. Fog and stratus are the norm between storms. For example, Paris averages eight days with fog during January. The most remarkable feature of this climate is its mild winter for its high latitudinal location. This is the result of warm ocean currents and strong maritime westerly winds. Summers are mild for the same reasons. As one moves toward the interior, continental influences are stronger. Winters are much longer and cooler and summers are hotter. Comparing warm humid continental Paris with cool marine Fort Williams, Scotland, on the west coast, Fort Williams has a narrower annual range in temperature and much cooler summer.

Precipitation patterns also shift from a winter maximum cyclonic pattern on the coast to a summer convectional type inland. Summer rains often result from convectional heating in the interior, but weak cold fronts continue to contribute to the summer rains.

SNOW FOREST CLIMATES (D)

By moving either poleward or toward the interior, the C climates make a transition into **Snow Forest climates** (D). In the Southern Hemisphere, the D climate is totally absent. Here cool oceans occupy the same latitudes where this climate is typically found in the Northern Hemisphere. Typically, D climates have cold winters and warm summers. The isotherm of $-2.8°$ C (27° F) for the coldest month separates this climate from the C types along its equatorial side while the poleward limit is defined by the 10° C (50° F) isotherm for the warmest month. If the mean monthly temperature fails to reach 10° C (50° F), the climate is considered to be a polar E type.

Looking at the global climate map in Appendix I, we can see that the D climates dominate most of Alaska, Canada, Eastern Europe, and the Soviet Union. These climates dip as far south as 40° N in the United States, USSR, and east Asia. Poleward, D climates can be found inside the Arctic Circle in both North America and Asia. On the eastern sides of these continents they reach the coast, where the prevailing winds are westerly from the cold continental interior.

Both temperatures and precipitation vary widely in the D climates. Moscow (latitude 55.46° N) has an average July temperature of 18.7° C (65.3° F) in contrast to $-10°$ C (14° F) for January. The annual precipitation, only 55.4 centimeters (21.8 inches), is capable of supporting a forest because low temperatures reduce water loss from the soil and plants. For seven to eight months the land is covered by snow (fig. 9.15).

In middle latitudes such as Chicago (41.52° N), the July average is 24.3° C (75.4° F) and in January drops to $-5°$ C (23° F), with an annual precipitation averaging 84.6 centimeters (33.3 inches). In both Chicago and Moscow, the annual range in temperature is over 28° C (50° F). Both stations reflect a strong continental influence.

Precipitation in D climates comes mainly in the summer. As the land heats up, convectional processes produce heavy downpours from cumulonimbus clouds. Precipitation is greatest on the eastern coasts of the continents near marine influences, especially in southern portions of this zone where more moisture is available. Winter precipitation comes as snow from frontal activity. Because

eastern Siberia and northern China have only one-tenth of the precipitation of the summer season, that area is designated Dw or a dry winter climate. Most of the D regions are noted as Df because sufficient precipitation to support a forest occurs during all seasons. A small zone of Ds or dry-summer Snow Forest climate is found in eastern Washington and Oregon, where the Cascade Range blocks moist air from reaching the interior.

If we examine the distribution patterns of the subdivisions of the D climates, it becomes apparent that latitude and distance from the ocean are important controls. In North America, the warm summer Dfa climate forms the transition from Cfa, then poleward the milder summer Dfb climate of the Great Lakes and southeastern Canada forms another transition to the Dfc climate of cold winter and mild summer of central Canada and Alaska. Killing frosts in Canada can strike as late as early June and again at the end of summer in early September. This can cause complete crop failure either at planting or harvest time.

In Asia the pattern shifts not only latitudinally but also along the lines of longitude. As distance increases eastward toward the interior from the mild maritime Europe, both temperature and precipitation patterns show marked seasonal variation. The Dfb climate dominates European Russia south of 60° N and west of 90° E. The Dfc climate blankets the poleward zone above 60° N until the eastern Siberian dry winter and mild summer climates, Dwc and Dwd, are reached. The dry winters and wide ranges in temperature reflect the monsoon and continental influences. This region has the Northern Hemisphere's greatest annual ranges in temperature. For example Minusinsk, USSR (54°N, 92°E), has a 40.1° C (71.1° F) range in annual temperature while Ust-Maya in Siberia (60° N, 134° E) has a 61.3° C (109.5° F) annual range in temperature. The Dw climate does not extend all the way to the coastal regions in Asia and does not exist at all in North America, where large continental size and the monsoonal influences are absent.

If we look at the chief air masses controlling this Dw continental dry winter climate, we can better understand the reason for such a wide annual variation in temperature and precipitation. During the summer, eastern Asia is dominated by strong onshore maritime subtropical air (mT) that moves over the warm continental surface. This air mass brings torrential rains to southern Asia that decline as distance from the sea increases. The summer maximum pattern persists throughout all of eastern Asia's interior due to monsoonal influence.

Precipitation ceases in October as the interior continent begins to cool off. Maritime air is replaced by cold continental polar air (cP). This region becomes a major source region of this air mass. During January, temperatures plunge to $-45.5°$ C ($-50°$ F). Siberia experiences some of the coldest and longest winters on Earth outside of Antarctica and Greenland. The zone of lowest temperature is found in the region of Verkhoyansk (68° N, 133° E), where the mean January temperature is $-50°$ C ($-58°$ F). Here clear, calm winter skies with low humidity favor the loss of energy from the surface. When this continental polar (cP) air mass moves out of the Siberian high-pressure anticyclone,

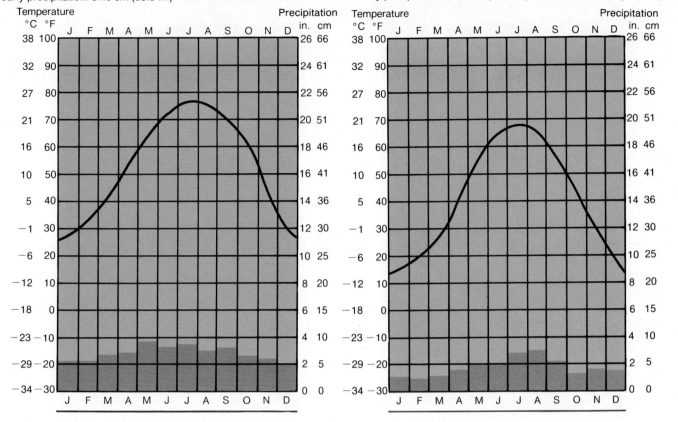

City: Chicago, Illinois
Latitude: 41°52′N
Altitude: (181m) 595 ft.
Yearly precipitation: 84.6 cm (33.3 in.)

Climate name: Humid Continental
Dfa (warm summer)
Other cities with similar climates:
New York, Berlin, Warsaw

City: Moscow, USSR
Latitude: 55°46′N
Altitude: (154m) 505 ft.
Yearly precipitation: 55.4 cm (21.8 in.)

Climate name: Humid Continental
Dfb (cool summer)
Other cities with similar climates:
Montreal, Winnipeg, Leningrad

Figure 9.15

The Snow Forest climates (Dfa, Dfb, Dwa, Dwb) are located in the
Northern Hemisphere only. Winters are cold, with mean
temperature of the coldest month falling below −2.8° C (27° F),
and the warmest monthly mean exceeds 10° C (50° F).

the rest of eastern and southern Asia experiences dry winter monsoon conditions. Only in western Asia and eastern Europe do the winter cyclones bring abundant moisture, usually falling as snow.

THE POLAR CLIMATES (EF)(ET)

Polar climates have two important identifying characteristics: coldness and dryness. The southerly boundary of the polar climates is the 10° C (50° F) isotherm for the warmest month. This line approximates the poleward limit of the coniferous forests. Where mean monthly temperatures rise above freezing (but stay below 10° C or 50° F), tundra vegetation and animal life are abundant. Therefore a subdivision **Tundra climate (ET),** is distinguished from those climatic regions where the mean monthly temperature never rises above freezing. Therefore, EF describes all of the cooler polar lands, and an H is added if it is a polar climate

created by high elevation. These climates dominate not only the high latitudes, but higher elevations as well (fig. 9.16).

In the Southern Hemisphere, the E climates extend to about 50° S over the cold oceans, while in the Northern Hemisphere, the Arctic Circle, 66½° N, is the mean southerly boundary. Here we can see the important influence of land masses in warming up the land in the summer to push this climate boundary further north in the Northern Hemisphere.

A special aspect of this part of the world is not only low temperatures and dry conditions but also the wide range in daylight hours between June and December. At the poles the Sun shines without setting for six months, from equinox to equinox, and then sets for six months. These solar extremes rapidly decline equatorward. Because the atmosphere and snow-covered surface reflect energy, the heating effects of the long duration of sunlight are greatly reduced. In fact, the ground below the top layer is permanently frozen even where the snow melts. This soil condition is called **permafrost.**

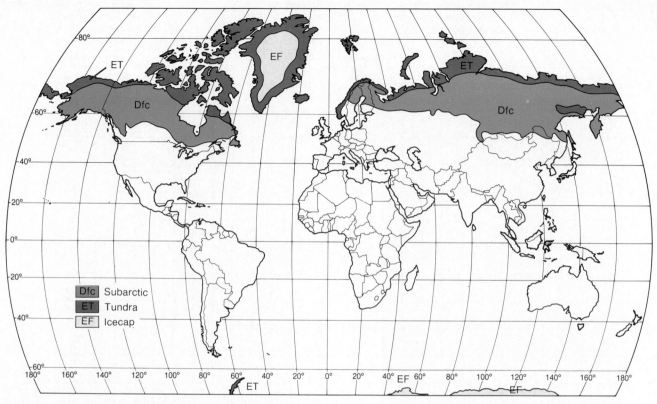

Figure 9.16

Polar Ice Cap (EF), Tundra (ET), and Snow Forest Subarctic (Dfc) climates. Where the forests gradually change to tundra, the 10° C (50° F) high mean monthly temperature isotherm marks the timberline on a mountain slope as well as the poleward limit of the coniferous forest.

Except over the oceans and along the shorelines, this climate experiences a wide annual range in temperature. The Falkland Islands (latitude 52° S) have a 7° C (12.6° F) range in temperature, while Greenland, at the center of the icecap, has an annual range exceeding 36° C (65° F).

Because of very low temperatures and evaporation most of the year, water vapor content is low and therefore precipitation is minimal. Most of this climatic zone experiences less than 25 centimeters (10 inches) of precipitation annually.

However, because temperatures most of the year remain below freezing, heavier snowfall accumulates in the form of large glaciers, especially on the windward slopes where orographic precipitation is enhanced. Both Greenland and Antarctica contain nearly all of the world's ice in the glacial form.

Tundra (ET)

As the name suggests, **Tundra climate** (ET) supports vegetation and animal life because summer temperatures are above freezing. Even dwarf trees in shrub form can be found in sheltered fringes of this frigid land.

In the Northern Hemisphere, Tundra climates extend along the northern shores of Asia, Europe, and North America. Many islands of the Arctic and Antarctic seas also have this climate.

The annual range in temperature can vary widely, ranging from 33° C (60° F) at Lake Harbour, Baffin Island (63° N), to 7.2° C (13° F) at Cape Pembroke, Falkland Islands (52° S), where the maritime influence is very strong. The summer temperatures of tundra regions must fall between the mean monthly temperatures of 10° C (50° F) and 0° C (32° F) according to the definition of this climatic type, but the winter temperatures can reach extreme lows in some arctic locations under strong continental influences.

The Ice Cap (EF)

The **Ice Cap** (EF) climate is located in the polar regions which include Antarctica, Greenland, and the high mountains of the earth above 4,000 to 5,000 meters (13,000–16,000 feet).

The mean temperature of the warmest month, according to the definition of this climate, never reaches freezing (fig. 9.17).

Figure 9.17

Polar Ice Cap (EF) climates are located in Antarctica, Greenland, and many highlands. The mean monthly temperature never reaches 0° C (32° F). Precipitation is low, making this area an ice-covered or "white desert," as seen here over Greenland's coast.

Photo by H. H. Wallen Jr.

This is a region of arid conditions with low precipitation and evaporation, matching some of the driest tropical and subtropical deserts.

Here data are very scanty because these polar deserts represent the last dry frontiers to be investigated. It is a common misconception that arid lands are always hot, yet a review of global patterns of precipitation shows the Earth to have two broad belts of dry lands, located in the middle latitudes and the polar regions. Aridity is the common thread of all desert classification.

Polar lands also exhibit a variety of other arid signs such as wind-formed landforms, xerophytic plant forms, large radiation influxes and effluxes, low humidities, wide daily and annual ranges in temperatures, and irregular and very low precipitation.

Really cold deserts of the EF variety are also typified by harsh environments for organisms. Their adaptive mechanisms reflect the cold, dry conditions. In terms of diversity of species this region is at the bottom of the biological variety scale, and yet plants and animals are here surviving, growing, and reproducing.

Antarctica, the fifth largest continent has three geographic surfaces. It almost doubles in surface area during the winter if we include the expanding ice sheet resulting from the freezing of the seas beyond the shores. The geologic surface below the ice sheet is a continental size similar to Australia. It has the highest average elevation of all continents, about 2 kilometers, and it contains 70 percent of the planet's ice. If all of this ice should melt, global sea level would rise 75 meters (234 feet).

Winter temperatures at Antarctic coastal stations are seldom below −50° C (−58° F). Coastal McMurdo Station experiences a mean summer January temperature of 3.5° C. In July it falls to −29° C (−20.2° F) during the Antarctic winter. Temperatures at the South Pole show a much greater range, in excess of 50° C (90° F). On the other hand, Arctic temperatures are not as low because of two important controls. First, land masses are absent at the North Pole. Second, the mean ocean temperature is at least 3° C (5.4° F) higher than in corresponding southern latitudes.

Precipitation in Antarctica varies with temperature and elevation. The coldest and highest inland stations, such as Amundsen-Scott at the South Pole, elevation 2,800 meters (9,240 feet), show annual precipitation falling as snow, ranging between 2.5 cm (1 inch) and 7 cm (2.8 inches), which is typical for most of the Antarctic continent. These are truly desert precipitation levels, which explains why these lands are called ice or white deserts.

Highland Polar (EH)

The high elevations of the world provide conditions that resemble in many respects arctic conditions. Because temperatures fall at the rate of 1° C per 100 meters (1° F per 180 feet), climates go through rapid transition along a mountain slope. One may experience the entire spectrum of temperature and precipitation range between sea level and the tops of mountains where E climates may be located (fig. 9.18). However, there are some important differences between polar climates and high-elevation climates.

First, E climates outside the polar lands show a much smaller annual range in temperature due to a lower range in insolation. This is especially true in the highland tropics. The length of daylight and solar angle are also quite different with respect to lower-latitude E climates. Slope exposures also vary considerably, giving a wide range of energy per unit surface area. Sunny slopes are much warmer and drier during both the summer and winter while shady poleward-facing slopes with less insolation tend to remain more ice covered and cooler throughout the year. Because of these insolation differences, the E climates of the mountains are quite different and really represent a variety of microclimatic niches that fall within the broader context of the E classification of Köppen's system (fig. 9.19).

Figure 9.18

Highland Polar (EH) climates are actually a variety of microclimates that fall within the broader context of the E climatic classification of the Köppen system. This photograph was taken in Rock Creek, California.

Figure 9.19

World climatic regions.

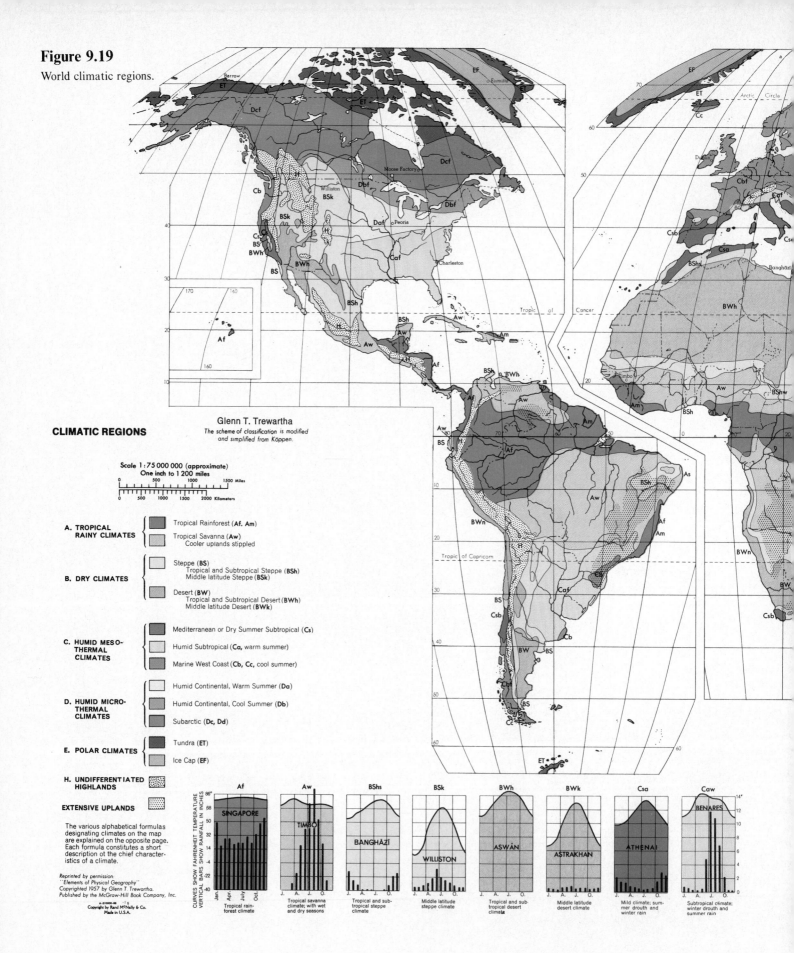

Glenn T. Trewartha

The scheme of classification is modified and simplified from Köppen.

CLIMATIC REGIONS

Scale 1:75 000 000 (approximate)
One inch to 1 200 miles

A. TROPICAL RAINY CLIMATES
- Tropical Rainforest (**Af, Am**)
- Tropical Savanna (**Aw**)
  Cooler uplands stippled

B. DRY CLIMATES
- Steppe (**BS**)
  Tropical and Subtropical Steppe (**BSh**)
  Middle latitude Steppe (**BSk**)
- Desert (**BW**)
  Tropical and Subtropical Desert (**BWh**)
  Middle latitude Desert (**BWk**)

C. HUMID MESO-THERMAL CLIMATES
- Mediterranean or Dry Summer Subtropical (**Cs**)
- Humid Subtropical (**Ca**, warm summer)
- Marine West Coast (**Cb, Cc**, cool summer)

D. HUMID MICRO-THERMAL CLIMATES
- Humid Continental, Warm Summer (**Da**)
- Humid Continental, Cool Summer (**Db**)
- Subarctic (**Dc, Dd**)

E. POLAR CLIMATES
- Tundra (**ET**)
- Ice Cap (**EF**)

H. UNDIFFERENTIATED HIGHLANDS

EXTENSIVE UPLANDS

The various alphabetical formulas designating climates on the map are explained on the opposite page. Each formula constitutes a short description ot the chief character-istics of a climate.

Reprinted by permission:
"Elements of Physical Geography"
Copyrighted 1957 by Glenn T. Trewartha
Published by the McGraw-Hill Book Company, Inc.

Copyright by Rand McNally & Co.
Made in U.S.A.

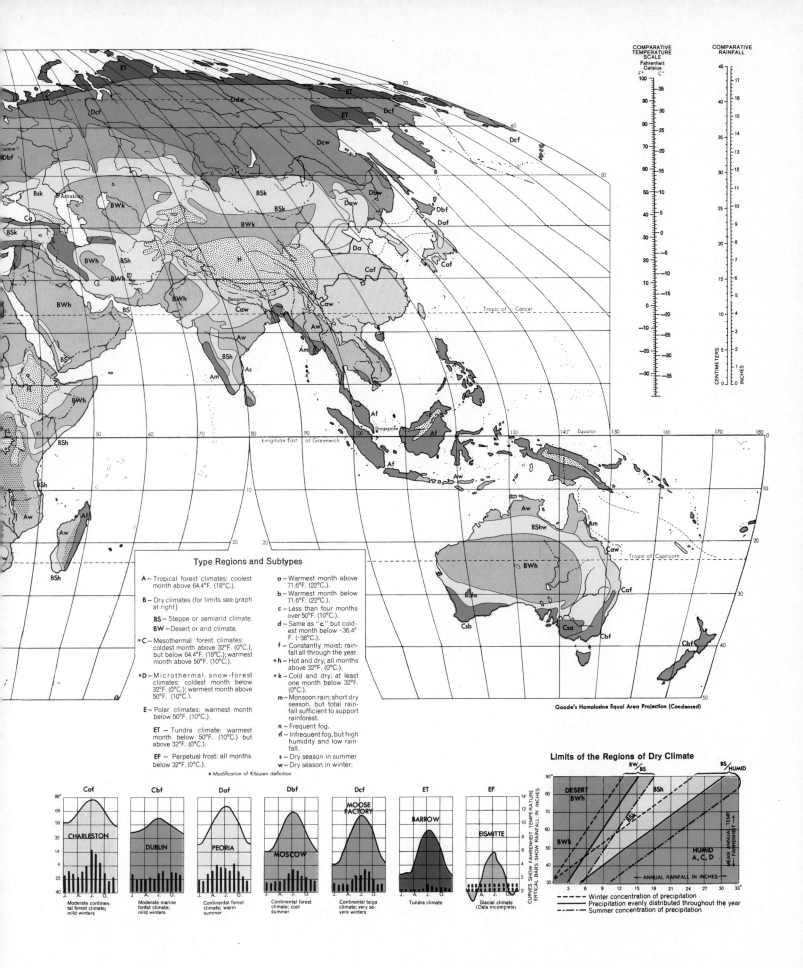

COMPARATIVE TEMPERATURE SCALE
Fahrenheit Celsius
F° C°

COMPARATIVE RAINFALL

Goode's Homolosine Equal Area Projection (Condensed)

Type Regions and Subtypes

A – Tropical forest climates: coolest month above 64.4°F. (18°C.).

B – Dry climates (for limits see graph at right)

BS – Steppe or semiarid climate.

BW – Desert or arid climate.

*****C** – Mesothermal forest climates: coldest month above 32°F. (0°C.), but below 64.4°F. (18°C.); warmest month above 50°F. (10°C.).

*****D** – Microthermal, snow-forest climates: coldest month below 32°F. (0°C.); warmest month above 50°F. (10°C.).

E – Polar climates: warmest month below 50°F. (10°C.).

ET – Tundra climate: warmest month below 50°F. (10°C.) but above 32°F. (0°C.).

EF – Perpetual frost: all months below 32°F. (0°C.).

a – Warmest month above 71.6°F. (22°C.).

b – Warmest month below 71.6°F. (22°C.).

c – Less than four months over 50°F. (10°C.).

d – Same as "c," but coldest month below −36.4° F. (−38°C.).

f – Constantly moist; rainfall all through the year.

*****h** – Hot and dry; all months above 32°F. (0°C.).

*****k** – Cold and dry; at least one month below 32°F. (0°C.).

m – Monsoon rain; short dry season, but total rainfall sufficient to support rainforest.

n – Frequent fog.

n̄ – Infrequent fog, but high humidity and low rainfall.

s – Dry season in summer.

w – Dry season in winter.

*Modification of Köppen definition

Limits of the Regions of Dry Climate

- - - - Winter concentration of precipitation
———— Precipitation evenly distributed throughout the year
-·-·- Summer concentration of precipitation

Caf — CHARLESTON — Moderate continental forest climate; mild winters

Cbf — DUBLIN — Moderate marine forest climate; mild winters

Daf — PEORIA — Continental forest climate; warm summer

Dbf — MOSCOW — Continental forest climate; cool summer

Dcf — MOOSE FACTORY — Continental taiga climate; very severe winters

ET — BARROW — Tundra climate

EF — EISMITTE — Glacial climate (Data incomplete)

CURVES SHOW FAHRENHEIT TEMPERATURE
VERTICAL BARS SHOW RAINFALL IN INCHES

The A climates are located in the tropical world where seasons are lacking. Patterns of winter, spring, summer, and fall are absent. Only summer exists. Temperatures vary more on a daily basis than annually. The Aw and Am climates have distinct dry seasons and are located poleward of the Af Rainforest climate.

Climates of the warm arid world (B) have more potential evaporation than precipitation. They are located over a broad range of latitudes but concentrate in the subtropics along 30° latitude.

The warm temperate C climates are distinct from the A climates in having at least one monthly average temperature below 18° C (64° F). This temperate rainy zone is subdivided into the mild **Mediterranean** (Csa, Csb), characterized by dry warm summers and wet winters, the **Subtropical Monsoon** (Cw) with wet summers and dry winters, **Humid Subtropical** (Cfa) where precipitation falls in all months and summers are hot and humid, and the **Marine West Coast** (Cfb) climate which is a heavily overcast region and has the highest annual precipitation outside the tropics.

Inland from the Marine West Coast climate, in the Northern Hemisphere, the maritime influences decline, allowing moderate temperatures to give way to continental extremes or **Snow Forest** (D) climates. At least one monthly mean temperature falls below −2.8° C (27° F). Typically, the Snow Forest climates have very wide annual ranges in temperature on the order of 28° C (50° F). This is the primary climate of the world's two largest countries, the USSR and Canada, and can only be found in the Northern Hemisphere.

Polar climates (E) are very arid; precipitation totals and even vegetation structure resemble the warmer desert world. Many of the plants of the **Tundra** (ET) climate have the same xerophytic characteristics of the lower- and middle-latitude deserts.

Polar climates in the high elevations, designated by EH, are located on the slopes of the highest mountains, especially on the shaded slopes. The E climates, especially nearer the equator, show less seasonal variation and smaller annual range in temperature.

ILLUSTRATED STUDY QUESTIONS

1. Describe the global precipitation pattern for the Tropical (A) climates (fig. 9.1, p. 172).
2. Describe the general climatic conditions for the Af, Aw, and Am climates (figs. 9.2 and 9.3, pp. 173–74).
3. Describe the general distribution of the B climates (fig. 9.4, p. 175).
4. Explain a major climatic control of the Cool Coastal Desert climate (fig. 9.7, p. 178).
5. Contrast cool, semiarid BSk and BWk interior locations with coastal climates of the same climatic symbol (figs. 9.6, 9.7, 9.8, and 9.9, pp. 177–80).
6. Describe the general distribution of the Mediterranean climate. When does rainfall reach a maximum (fig. 9.10, p. 181)?
7. Contrast the Mediterranean rainfall pattern with the Humid Subtropical climate (fig. 9.13, p. 184).
8. Describe the primary climatic controls of the Marine West Coast climate (fig. 9.14, p. 185).
9. Site a unique location fact about the Snow Forest continental climates (figs. 9.15 and 9.16, pp. 187–88).
10. Contrast Ice Cap (E) climates with High Polar Climates (EH) in terms of temperature, insolation, and variation (figs. 9.17 and 9.18, pp. 189, 190, 192).

The Biosphere

UNIT

Life, energized by the Sun, is unique to planet Earth. It has sprung forth to fill every niche from the driest deserts to the wettest rainforests.

Chirripo National Park, Costa Rica.
© Chip and Jill Isenhart/Tom Stack & Assoc.

Soil-Forming Processes

10

Objectives

After completing this chapter, you will be able to:

1. Identify the four major soil constituents—**organic matter, air, water,** and **inorganic minerals**—and explain their primary role.
2. Describe a typical **soil profile,** and explain the characteristics of each horizon.
3. Define soil **texture** and **structure,** and explain the characteristics of each.
4. Classify soils on the basis of physical properties.
5. Explain the significance of a soil's **pH.**
6. Identify chemical and physical factors influencing the pH of a soil.
7. Define the **clay-humus complex** and its role in the soil's chemical activity.
8. Explain the five soil formation factors: **climate, organisms, parent material, relief,** and **time.**

This soil profile exposed below Chimney Rock, Nebraska, is the product of climate, geology, biological influences, and time.

© David Muench Photography

Energy flows from the Sun to the soil, warming it, activating seeds, inducing moisture to the surface, and transforming lifeless rock into elements that give life to the planet.

Soil is a life support system energized by the Sun. This soil base supports green plants of the **biosphere** that convert solar energy to chemical forms, making great diversity of life possible.

It has been said that humankind is just three feet—the average depth of the soil—away from starvation. This thin veneer is related to the Earth in much the same way the skin of an apple or orange rind is to the fruit, but there are some major differences. The rind of the earth is far less uniform in depth. In some places, soil is entirely lacking, while in others it may be 6 to 9 meters (20 to 30 feet) deep. The colors of the soil vary widely, from red in Hawaii to black in Oklahoma to gray in the arid yet fertile Imperial Valley of California.

Soils also vary from place to place in fertility, but all soils have some things in common. They serve as the interface between the rock interior and the living biosphere. Soils provide the foundation for human activity and the resources for plants to grow.

SOIL COMPOSITION

All soils have four primary constituents. These are **organic matter, air, water,** and **weathered rock materials.** The percentages of the ingredients vary from place to place, yet these four components are always present in true soil (fig. 10.1).

Organic Matter

Organic matter is simply the decomposed remains of plants and animals—**humus**—as well as living organisms. Soils of the western United States generally have less than 2 percent organic matter, while eastern soils developing in forest and grassland regions can have more than 10 percent.

Organic matter includes plant material, roots, bacteria, fungi, molds, microscopic plants (algae) and animals (protozoa), and of course, insects and worms. In a spoonful of garden soil, there can be more individual living things than in the entire human population. Soils are ecosystems, rich and alive.

Air

Air fills the spaces between soil particles. Plants and animals living in the soil need this air for life and exchange oxygen and carbon dioxide with it. Carbon dioxide in the soil is present at over 100 times the concentration in the open atmosphere. Carbon dioxide results from the decomposition of plants and animals in the soil, while oxygen is released from the green tissues of plants in the **photosynthesis** process, discussed in chapter 12.

Water

Water competes with air for the empty spaces. When the spaces are totally filled with air, plants suffer from drought. If water occupies the spaces too long, most plants drown from a lack of oxygen. Moisture is also the key to nutrient uptake. Water is the life blood to plants, providing them with soluble nutrients for their life and growth.

A

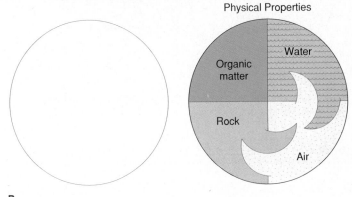

B

Figure 10.1

A true soil consists of various proportions of organic matter, air, water, and weathered rock materials. This yucca plant is growing in recently weathered sandstone.

Inorganic Minerals

Inorganic minerals are weathered rock particles. The development of soil begins with the slow, continuous weathering of rocks, from which many nutrients come. For example, when feldspar weathers, it liberates calcium and potassium.

THE SOIL PROFILE

When the astronauts landed on the Moon, they took samples of the lunar surface. Their probes indicated that the lunar surface was covered with loose rock particles several meters deep, and they described it as "soil." But this **regolith** lacked three of the four principal ingredients. Only inorganic minerals from meteoric impact and volcanic activity were present. Unlike earthly soils, there were no layers that soil scientists, or **pedologists,** call **soil horizons.** The lunar "soil" was simply an assortment of different sized particles resting on a bedrock interior, similar to sand and gravel on a rock slab.

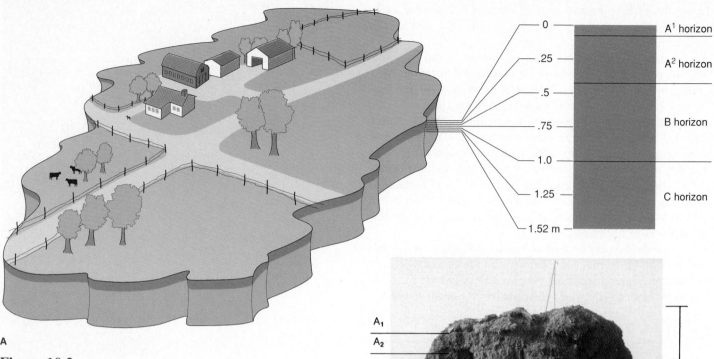

A

Figure 10.2

Given enough time, a true soil will develop into layers known as horizons. The depth and general properties of the horizons are determined by the climate and the mixture of the four components. Each soil horizon is a history of environmental influences over long periods. (A) A single area of a soil type as it occurs in nature with an enlarged sketch of the profile showing its major horizons. (B) A soil profile.

B

In contrast, the soils of the Earth have a profile, or series of layers. The characteristics of the profile are determined by the climate and the mix of air, time, topography, moisture, organic materials, and parent rock minerals. The profile is essentially a testimony or history of the soil-forming process (fig. 10.2).

A soil profile consists of at least three layers, or **horizons.** Each horizon has unique color, particle size, and structure characteristics. Soil profiles range in thickness from centimeters to meters. However, most profiles are 60 to 120 centimeters (2 to 4 feet) thick.

Commonly, soil profiles have three horizons that merge together with one another without sharp, distinct boundaries. The letters, A, B, and C are used to identify each layer (fig. 10.3).

The A and B horizons are called the **solum,** the true soil. Together, they represent the major portion of the soil. The C horizon is the weathered bedrock layer which grades downward into rock less affected by the weathering processes.

In the A horizon, the uppermost layer in the profile, life is most abundant. Plant roots and small animals, ranging from bacteria to rodents, make it their home. Organic matter is most abundant. Because of the surface position, it is more highly leached than the deeper horizons. **Leaching** is the process where soluble minerals are removed as water moves downward into the lower horizons. The A horizon is subdivided into A0, A1, and A2 subhorizons. The top A0 consists of decaying leaves and other plant material, and the A1 and A2 layers have more decomposed organic matter, much like the bottom of a compost pile.

The B horizon, or subsoil, is often lighter in color, having fewer living organisms and less humus than the A horizon. The B horizon may also be harder when dry and stickier when wet, because many clay minerals accumulate in this zone.

The C horizon consists of loose and partly weathered rock beneath the A and B horizons. This zone represents the raw material, or parent rock particles, for the layers above. Given enough time, this layer evolves into the B horizon, as weathering does its work by breaking down bedrock. Each horizon consists of subdivisions that form transition zones.

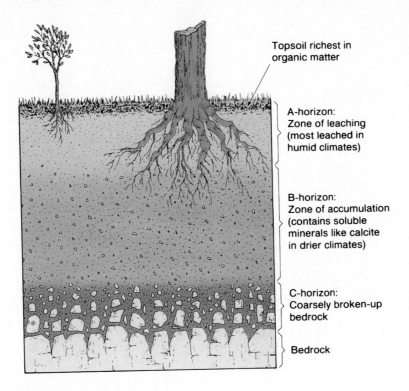

Topsoil richest in
organic matter

A-horizon:
Zone of leaching
(most leached in
humid climates)

B-horizon:
Zone of accumulation
(contains soluble
minerals like calcite
in drier climates)

C-horizon:
Coarsely broken-up
bedrock

Bedrock

Figure 10.3

A soil profile consists of at least two horizons and usually three.
Horizon D is the bedrock zone. The A and B horizons are referred to
as solum, or true soil, while the C horizon is a transition zone of
weathered bedrock. Not all of these horizons are present in any
profile, but every profile has some of them.

PHYSICAL PROPERTIES OF SOILS

When a farmer rubs some soil between the fingers and says it is
heavy or light, **soil texture** properties are being described. **Soil
structure** is another important physical property which characterizes the soil in terms of physical arrangement and groupings of
particles. Together these two physical properties, texture and
structure, play a very important role in determining the nutrient-
supplying characteristics, water-holding capacity, and volume of
air.

Soil Texture

Soil texture is determined by the percentages of various sized particles. A soil can be classified as sandy, loamy, silt, silty loam, clay,
or clay loam. Figure 10.4 gives the U.S. Department of Agriculture and International schemes of particle size distribution.

The soil texture diagram in figure 10.4 has the percentages of sand, silt, and clay on each side of the triangle. By plotting
the percentages of a sample, you can determine the correct soil
class. For example, a soil sample with 40 percent sand, 20 percent
clay, and 40 percent silt is in the **loam** soil texture class, at point
A on the diagram. Loam soils are favored by farmers because they
are more easily managed.

Clay, silt, and sand are terms that refer to particle size
(table 10.1). **Clay** particles have diameters less than 0.002 millimeters. **Silt** particles are between 0.002 mm and 0.05 mm, while
sand particles are greater than 0.05 mm, up to 2 mm. Particles
greater than 2 mm are classified as gravel. Soils contain mixtures
of more than one size class; the percentages of sand, silt, and clay
determine the soil classification.

It is important not to oversimplify or make unqualified
statements about a soil's fertility based on physical properties alone.
Yet, by examining the physical properties, we can find some important relationships. Sandy soils in rainy, tropical climates are
highly permeable and usually are heavily leached of soluble minerals. Their high **permeability** means that water rapidly moves
through them downward into the groundwater table, leaving the
soil to dry out quickly. Plants continually suffer drought, because
the water-holding capacity is very low. Water-holding capacity or
porosity is a function of surface area of the particles, and sandy
soils have the smallest surface area. Clay soils have the greatest
surface area and, therefore, much higher water-holding capacity.

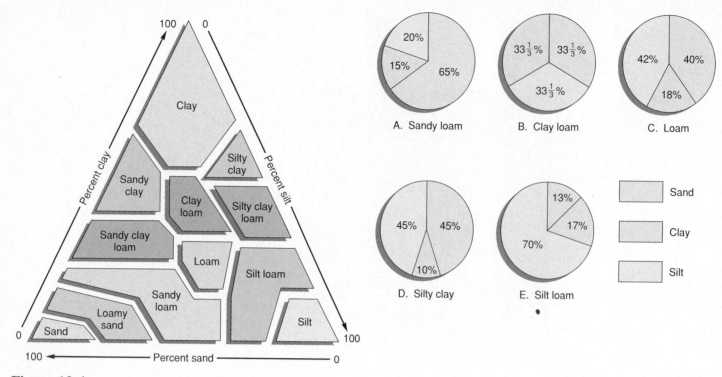

Figure 10.4

Soil texture is determined by the percentages of various sized particles.

Table 10.1		
Size Classes of Soil Particles		
Grade	**Diameter**	
	Inches	*Millimeters*
Coarse gravel	Greater than 0.08	Greater than 2
Fine gravel	0.04–0.08	1–2
Coarse sand	0.02–0.04	0.5–1
Medium sand	0.01–0.02	0.25–0.5
Fine sand	0.004–0.01	0.1–0.25
Very fine sand	0.002–0.004	0.05–0.1
Silt	0.0008–0.0002	0.002–0.05
Clay	Less than 0.0008	Less than 0.002

Source: U.S. Soil Conservation Service, U.S. Department of Agriculture.

They can be described as being very porous, having much more surface area. **Porosity** is a characteristic that describes the water-holding capacity in the soil. Because they hold moisture so well, they pose the opposite problem of sandy soils. Infiltration is often very slow, and they can become waterlogged. When this happens, plants can drown from lack of oxygen. Loams are preferred because they represent a balance in percentages of sand, silt, and clay.

Soil amendments (components added to the soil to improve its productivity) can correct clay or sandy soil problems. Both clay and sandy soil benefit from addition of organic materials. When organic mulches such as manures, wood products, or peat are added to sandy soils, moisture retention is greatly improved, and overly high permeability is reduced. Adding the same amendments to clay soils raises the permeability by opening larger spaces for water to move downward. Gypsum ($CaSO_4$) can also improve a heavy clay soil because calcium in the gypsum removes sodium buildup. Sodium causes clay particles to pack together in such a manner that water cannot get through. This is a serious problem in irrigated regions of the West where the water is loaded with mineral salts. Therefore, gypsum applications, in conjunction with flooding and leaching of fields, provide a way of opening up the pores of the soil and improving its capacity to transmit water (fig. 10.5).

Soil Structure

Soil structure is a physical property that characterizes soils in terms of how grains or particles join together into larger units. These larger aggregates are called **peds**. The shape and size of the peds can vary considerably, but soil scientists have classified all peds into four primary soil structures. The structure can be **blocky, platy, granular,** or **prismatic** (fig. 10.6).

A desirable structure is a stable one, or simply a structure where the soil particles are bound together into stable granules or aggregates. In an aggregate we can observe both large and small

Figure 10.5

Flooding an irrigated field that has developed a hardpan can leach the sodium minerals and improve the permeability of the soil. Irrigation water floods these fields to remove buildup of salts.

Photo by Julie Sowma-Bawsom.

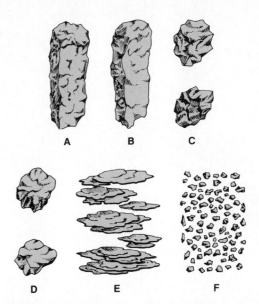

Figure 10.6

Soil structure determines how the particles or grains join together in larger units. Four basic structures can develop—blocky, platy, granular, and prismatic—depending on environmental factors and geologic conditions. (A) *prismatic*, (B) *columnar*, (C) *angular blocky*, (D) *subangular blocky*, (E) *platy, including lenslike*, (F) *granular*.

pores: large pores between aggregates, and small pores in the aggregate's texture. A number of factors affect soil structure, or aggregation. These include roots, organic matter, mineral content of the particles, microorganisms, alternate wetting and drying, freezing and thawing, and human activity.

A plant's root system causes peds to form granular units. When this happens, the soil is better aerated, with higher porosity and more permeability. Soil structure is improved through the action of soil microbes on organic materials.

Microorganisms are responsible for producing strong binding substances through their decomposition of organic residues in the soil. Elements required by plants are released in soluble forms in the process. These organic soil binding substances are not indestructible. Some rapidly decompose, while others break down more slowly. To maintain soil structure, the soil organisms must have a continuous source of organic food.

Controlled studies have shown that earthworms in fine-textured soils with poor aggregating cause plant growth to be stimulated. Earthworm activity in the soil produces large pores, increasing aeration and infiltration of water. Their digestive actions produce soluble nutrients and humus which pass through their bodies as castings. Their numbers and activity in soil depend primarily on the availability of the food supply. Adding organic residue to the soil raises their population and that of microorganisms.

SOIL CHEMISTRY

A soil may have all of the correct physical properties of texture and structure, and even the proper moisture supply, but plants will not thrive if the chemistry is out of balance.

Soil Acidity

When plants do not grow well, one of the first questions the soil scientist asks is whether the soil is *acid* or *alkaline;* that is, whether the soil's pH is too high or too low. The pH can often reveal a possible chemistry imbalance in the soil.

A measurement of **pH,** or **soil acidity,** can be compared to a doctor's physical examination, in which the patient's pulse, blood pressure, and temperature are taken. These measurements reveal if something is wrong, but they may not tell specifically the nature of the illness. Soil pH is determined by measuring the concentration of hydrogen ions and hydroxyl ions in soil solution. Ions in solution, like those in the upper atmosphere are atoms or molecules carrying an electrical charge. **Cations** are positive, **anions** are negative. When there are more positive hydrogen cations than negative hydroxyl anions, the solution is acidic. If there are more hydroxyl anions than hydrogen cations, the soil solution is basic or alkaline. Pure water has equal numbers, so the solution is neutral.

The pH scale, developed by a Danish biochemist, S. P. L. Sorensen, has a range from 0 to 14 based on the number of hydrogen ions per liter of solution. Pure distilled water, pH 7, being neutral, is located at the midpoint of the scale. A pH of 7 means there is 10^{-7} gram (more precisely, 10^{-7} mole) of hydrogen ions per liter of soil solution. A pH of 10^{-6} represents a larger number of hydrogen ions. In the pH scale the negative sign is eliminated as well as the base 10.

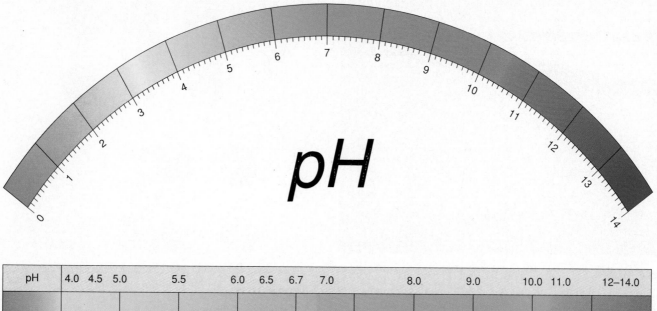

pH	4.0 4.5 5.0		5.5	6.0 6.5 6.7 7.0		8.0	9.0	10.0 11.0	12–14.0
Acidity	Very strongly acid	Strongly acid	Moderately acid	Slightly acid	Neutral	Weakly alkaline	Alkaline	Strongly alkaline	Excessively alkaline	Extremely alkaline

Figure 10.7

The pH scale is logarithmic, increasing and decreasing by powers of 10. The scale has a range from 0 to 14. Seven is neutral; all values greater than 7 are basic or alkaline, and all values less than 7 are acidic. The numbers are actually hydrogen ion concentrations expressed in exponential form: pH 7 is 10^{-7} moles per liter of solution.

If a soil solution has a value below 7, it is acidic because there are more hydrogen cations than hydroxyl ions (fig. 10.7). Put another way, a smaller pH indicates larger concentrations of hydrogen ions. Soil pH values above 7 reveal alkaline conditions, where more hydroxyl ions exist than hydrogen. Hydroxyl ions increase as the pH number increases by powers of 10; thus a solution with a pH of 7 has 10 times the hydroxyl ions as a solution with a pH of 6. You can also say that a sample with a pH of 6 has 10 times the hydrogen ions as a sample with a pH of 7.

Note that when a soil's pH is measured, the pH reading only reflects the *active* acidity, or hydrogen ions in solution. There is a greater *potential* acidity that does not register, because hydrogen ions are held in various chemical combinations and attached to the surfaces of small particles of clay or humus. In this way, hydrogen ions are either neutralized or removed from the soil solution and, therefore, do not register on the pH scale.

Soils become acidic through the processes of leaching. Slowly, over time, hydrogen ions are carried downward in solution, gradually replacing calcium and magnesium ions, which are removed through the groundwater. The more water moving through the soil, the faster is the removal process. Therefore, humid climates tend to have more acid soils, resulting from the leaching process.

Soil texture also influences pH. Clay soils prevent rapid movement of water, and they possess more basic minerals, held by microscopic particles known as **colloids.** The colloids have a negative charge on surface areas less than 0.1 micron in diameter that attract and hold basic cations such as magnesium and calcium, thus resisting leaching.

The pH of the soil gives clues to the nutrient content, or general health, of the soil. Soils that are highly leached have a low pH because basic nutrients have been replaced by hydrogen ions. If a soil's pH is too high, it is an indication that a toxic buildup of basic or alkaline minerals, such as sodium, has occurred. Too much sodium is not only toxic, but it can reduce water movement in the soil by bonding soil particles together.

The pH of soil has a very important influence on the availability of nutrients. In high-pH soils, iron, manganese, copper, and zinc become less soluble. Therefore, in this case, lowering the pH may be more important than adding more nutrients. Without solubility, the nutrient cannot enter the planet. Water is the solvent that transports the nutrient into the root system.

Bacteria and other microorganisms living in the soil are important converters of nutrients from organic compounds to simpler forms that plants can use in the ionic form. By maintaining a pH near 7, the neutral point, the soil environment is more favorable to the development of microorganisms and thus helps improve the availability of nutrients.

The pH of soil, then, is the first question a soil scientist asks in determining the general health of the soil. It is important to remember that additional tests must follow to determine the specific characteristics of the soil's chemistry. Knowing the soil's pH is a starting point in determining the productivity of the soil.

Soil Colloids

Soil colloids, as indicated, are the seat of chemical activity. At this microscopic level of 0.1 micron, clay and humus particles attract and hold, on their negatively charged surfaces, cations present in the soil solution. Because of their small size and large surface area per unit weight, they can hold large quantities of basic nutrients for plant consumption. This colloidal **clay-humus complex** binds calcium (Ca), magnesium (Mg), potassium (K), and sodium (Na) nutrients until the plant draws upon this rich colloidal nutrient bank by exchanging them for hydrogen ions (fig. 10.8). These ions are called exchangeable, or replaceable, cations. *Liming* the soil (calcium carbonate amendment) is a good example of cation exchange. Note that one calcium ion, with two positive charges, will replace two hydrogen ions. The clay-humus complex holds different cations with different strengths: calcium is the most strongly bound, sodium is loosely held, and both potassium and magnesium are moderately held.

Most of the chemical activity in soils is limited to a small proportion of the total soil mass, the clay-humus complex.

Clay soils, rich in humus, have the greatest capacity to hold cations to be exchanged. This value can be expressed as **cation exchange capacity.** Sandy soils lacking a clay-humus complex are easily leached of their basic nutrients, especially in humid climates, because there is nothing to prevent them from being removed.

Nutrient uptake can only occur when nutrients are available in the soil solution. Therefore the clay-humus complex is the soil's temporary nutrient bank. A proper pH is a barometer of the level of the account. If the pH is low, it is a good indication that hydrogen ions have replaced important nutrients in the leaching process. A high pH is an indicator that the bank is full of basic nutrients, held by the soil's colloids.

PLANT NUTRIENTS

Of the 92 natural elements, only 18 are known to be required by plants, and yet the quality of human life is highly dependent on the availability of these plant nutrients in the right proportions during the life cycle of the plant. Plant nutrients are classified as either major nutrients or micronutrients. The word *micronutrient,* referring to elements needed in very small amounts, should not imply that they are less essential for plant vigor (table 10.2).

Soil nutrients occur in solid, liquid, and gaseous forms. In a typical loam soil, 50 percent of the volume is in the solid form, and the remaining 50 percent is represented by pore spaces which are alternately filled with air and water. The organic solid portion rarely exceeds 10 percent.

Plants obtain nutrients from a variety of sources. The atmosphere supplies plants with carbon and oxygen. The carbon comes from carbon dioxide via the **photosynthesis** process described in chapter 12. Hydrogen is supplied by the water molecule

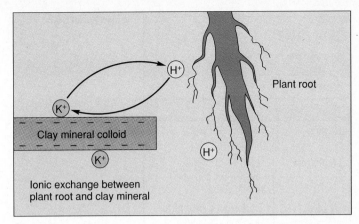

Plant root

Clay mineral colloid

Ionic exchange between plant root and clay mineral

A

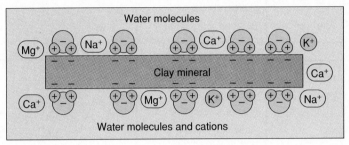

Water molecules

Clay mineral

Water molecules and cations

B

Figure 10.8

Soil colloids are important in attracting and holding water and calcium, magnesium, potassium, and sodium nutrients.

(H_2O). Oxygen is supplied by the soil atmosphere through the root system, which transports all other nutrients from the soil. Think of plants as miners of a low-grade ore, the soil. See Appendix III for a description of the 18 nutrients needed by plants.

SOIL FORMATION FACTORS

Just after the turn of the century, a Russian soil scientist, V. V. Dokuchaiev, proposed five factors that govern the development of thousands of soil types throughout the world. These are **climate, organisms, parent material, relief,** and **time.** In more recent years, an American pedologist, Hans Jenny, considered these same factors, but went on to show how they were functionally interrelated, each factor affecting the other in an overlapping or dovetailing manner.

Jenny expresses these interrelationships in the form of an equation,

$$S = f' (C1,O,P,R,T)$$

in which S, a soil type, is dependent upon or is a function of (f') of **climate** (C1), **organisms** (O), **parent material** (P), **relief** (R), and **time** (T). The importance of this equation is that it is possible to take any soil-forming factor and examine its influence in view of the other factors. None act alone or in isolation.

As we have observed, soil is incredibly complex in mineral and organic content, alive with microbial populations, worms, gophers, and plants ranging from lichen to redwoods. No two soils

Table 10.2
Soil Nutrients

Nutrients	Characteristics
Major Soil Nutrients	
Carbon	From CO_2 via photosynthesis
Hydrogen	From H_2O; main component of sugars and starches
Oxygen	Oxidation; essential for metabolism; plants release it in photosynthesis
Nitrogen	Building material of proteins, promotes rapid growth, gives plants healthy green, increases protein yield
Phosphorus	Stimulates growth, promotes seeds and fruit
	Active ingredient in protoplasm energy system
	Necessary in photosynthesis
	Organisms make it available
Potassium	Helps plants to resist disease and cold
	Needed in making starches and sugars
Secondary Plant Food Elements	
Calcium	Involved in development of roots, cell walls, mitosis
	Corrects acidity (in $CaCO_2$ form) applied on alkaline and saline soils
Magnesium	Essential part of chlorophyll molecule
Sulfur	Part of certain protein components
	Useful in acidifying soil
Micronutrients	
Boron	Necessary for calcium uptake
	High pH inhibits boron uptake
Copper	Catalyst of biochemical reactions
	Promotes vitamin A
Iron	Involved in chlorophyll formation
	pH affects uptake
Manganese	Catalyst for plant growth
Molybdenum	Essential for nitrogen utilization
Zinc	Needed for chlorophyll formation
	Decreases with pH increase
Chlorine	Takes part in metabolic processes; no lack except in humid climates

Note: See Appendix III for complete description of characteristics.

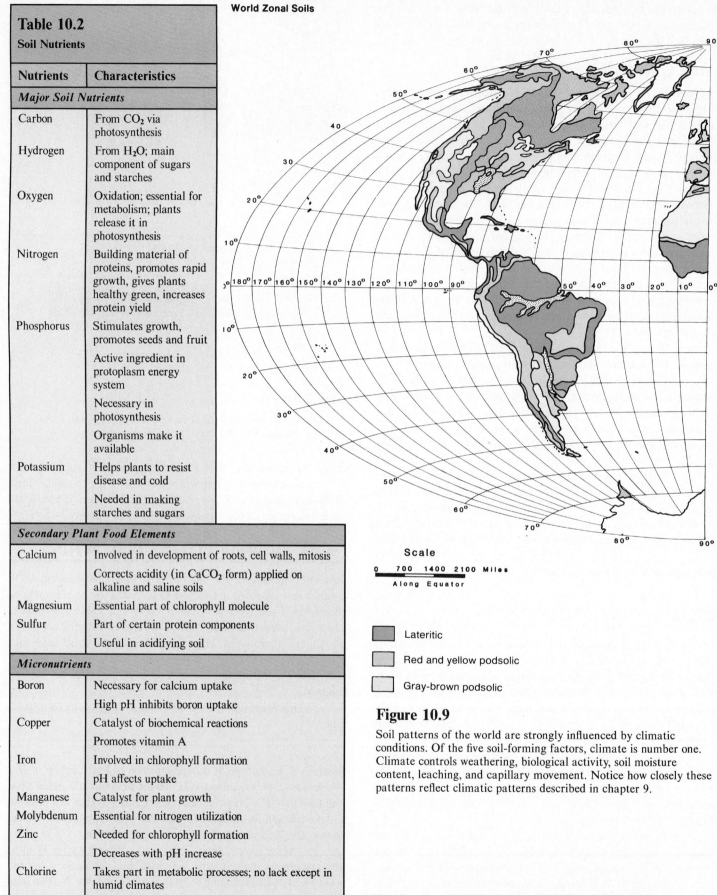

Scale

0   700   1400   2100 Miles

Along Equator

- Lateritic
- Red and yellow podsolic
- Gray-brown podsolic

Figure 10.9

Soil patterns of the world are strongly influenced by climatic conditions. Of the five soil-forming factors, climate is number one. Climate controls weathering, biological activity, soil moisture content, leaching, and capillary movement. Notice how closely these patterns reflect climatic patterns described in chapter 9.

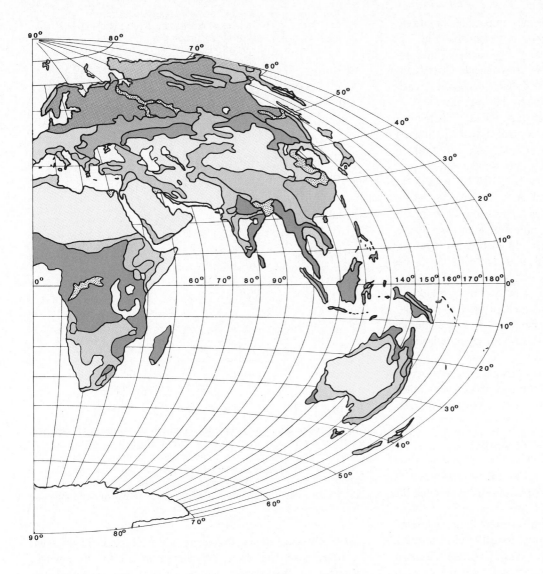

Aitoff's Equal Area Projection

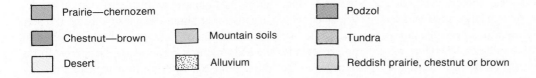

	Prairie—chernozem					Podzol
	Chestnut—brown		Mountain soils			Tundra
	Desert		Alluvium			Reddish prairie, chestnut or brown

are alike. The mix of these five soil-forming factors gives us almost infinite variation in soil patterns. Let us consider these five factors and their interrelationships.

Climate

The major controlling soil-forming factor is **climate.** There is a saying among pedologists: "The thicker the clouds, the thicker the soil." The term "climate" is rather abstract, so let us redefine it in its role of soil formation. To be more specific, *insolation, temperature, wind, humidity, cloudiness, precipitation,* and *evaporation* are the key factors of a climate. Each of these has a marked influence on the soil-forming processes. Note in figure 10.9 how closely the soil patterns of the world resemble climatic zones. Climate strongly influences weathering of parent material, biological activity (plants and animals), soil moisture content, leaching, and capillary movement in the soil.

Temperature and Precipitation

Temperature and precipitation are the most important climate elements governing the rate of weathering of parent material. The equatorial tropical soils, characterized by high temperatures and precipitation, are active chemically and biologically. Heavy rainfall increases leaching and eluviation. Profiles deepen as weathering extends to deeper levels in this moist environment.

With every rise in soil temperature, the rates of chemical and biological reactions accelerate. Animal activity in the soil thrives in a warm, humid environment, and plant decay rapidly increases. Humus development occurs until a mean annual temperature of 25° C (78° F) is reached (fig. 10.10). Beyond this temperature, increased bacterial and insect activity begins destroying potential humus faster than it can develop. Therefore, in humid tropical soils where high temperatures exist, humus is absent. A fallen leaf or twig will be processed by insects and bacterial activity and rapidly converted to soluble materials before it reaches the humus stage (fig. 10.11). In the cooler middle latitudes, humus accumulates because decomposition is much slower, especially during the winter. Bacterial and insect activity is very seasonal. Thus temperature and soil moisture govern the rates of humus development. These same climatic factors also regulate leaching and capillary action.

Soil temperature has a strong influence on seed germination. Table 10.3 lists common American vegetable crops and their minimum, optimum, and maximum temperatures for seed germination.

Wind, Evapotranspiration, and Insolation

Wind not only erodes and deposits soils, but it can quickly dehydrate a soil. Capillary action increases with the wind as moisture is lost by evapotranspiration back into the atmosphere, leaving the soil cracked and hard.

Evapotranspiration is moisture loss from both the ground and plant surfaces, and insolation draws moisture back into the atmosphere. Precipitation figures alone cannot indicate the moisture potential of an area, because evaporation by forces of wind and solar energy removes moisture concurrently as it is being deposited. The moisture in the soil can be compared to a bank account that is actively receiving deposits and withdrawals. Each climate has its own moisture equation. Where large moisture surpluses develop, soils are often heavily leached, while soils in the deficit zones accumulate excessive salts and bases. Therefore soil development is greatly influenced by both evapotranspiration and leaching. In arid zones where evapotranspiration is high, soils are shallow and basic. Tropical soils and soils of the cool, moist middle latitudes are deep and acidic where leaching exceeds evapotranspiration.

Organisms

Climate not only exerts a primary influence on chemical and physical properties of a soil, but it also exerts the major control on biological activity. Note the similarities between global patterns of climate, soils, and vegetation (fig. 10.12). There are some exceptions, but we can observe that humid regions are usually forest

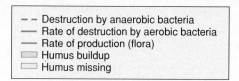

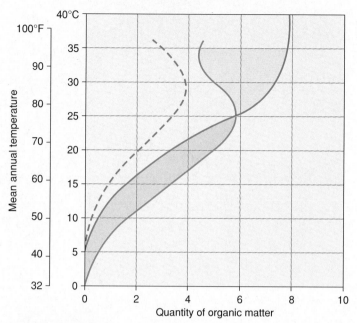

Figure 10.10

High precipitation combined with warm temperatures produce deep soils that are heavily leached and rich in biological activity. The amount of humus begins to decline beyond the mean temperature of 25° C (78° F) because increased biological activity destroys the potential humus faster than it can develop. Humus development is best where a cool season allows decaying vegetation to build up.

environments, developing in contrast with arid regions of grassland and desert vegetation. Therefore, climate governs the development of organisms, the second of five major factors in our equation of soil development.

The role of living **organisms**—plants, animals, insects, bacteria, fungi—including human activity, is important in the development of organic matter or humus. Living organisms also play a key role in profile development, nutrient cycling, and physical properties of texture and structure. For instance, the nutrient forms of nitrogen are added to the soil by nitrogen-fixing bacteria.

Plants and animals mix the soils of different horizons and retard the differentiation of horizons. Deciduous trees reverse the leaching process by nutrient uptake in the B and C horizons, and then deposit these nutrients in the form of falling leaves in the organically rich A horizon.

By comparing soils formed under forest, grassland, and desert vegetation, striking differences can be observed. For example, the pH and organic matter content in grassland soils are much higher than in forest soils, especially in the B horizon. Organic matter is darker in color and higher in cation holding capacity. Minerals such as calcium, magnesium, and potassium are

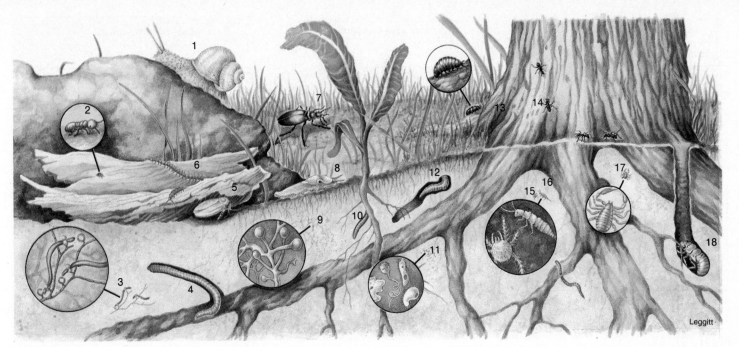

Figure 10.11

Tropical soils are low in humus because insects and bacteria efficiently process forest litter.

Table 10.3						
Soil Temperatures for Vegetable Seed Germination						
Minimum	*32°F*	*40°F*		*50°F*	*60°F*	
	Endive	Beet	Parsley	Asparagus	Bean, Lima	Okra
	Lettuce	Broccoli	Pea	Sweet Corn	Bean, Snap	Pepper
	Onion	Cabbage	Radish	Tomato	Cucumber	Pumpkin
	Parsnip	Carrot	Turnip		Eggplant	Squash
	Spinach	Cauliflower	Celery		Muskmelon	Watermelon
		Swiss Chard				
Optimum	*70°F*	*75°F*	*80°F*	*85°F*		*95°F*
	Celery	Asparagus	Bean,	Bean, Snap	Pepper	Cucumber
	Parsnip	Endive	Lima	Beet	Radish	Muskmelon
	Spinach	Lettuce	Carrot	Broccoli	Sweet Corn	Okra
		Pea	Cauliflower	Cabbage	Swiss Chard	Pumpkin
			Onion	Eggplant	Tomato	Squash
			Parsley		Turnip	Watermelon
Maximum	*75°F*	*85°F*	*95°F*		*105°F*	
	Celery	Beans, Lima	Asparagus	Eggplant	Cucumber	Squash
	Endive	Parsnip	Bean, Snap	Onion	Muskmelon	Sweet Corn
	Lettuce	Pea	Beet	Parsley	Okra	Turnip
	Spinach		Broccoli	Pepper	Pumpkin	Watermelon
			Cabbage	Radish		
			Carrot	Swiss Chard		
			Cauliflower	Tomato		

Source: Bruce E. Bearden, Farm Advisor, Mendocino County Cooperative
　Extension, University of California.

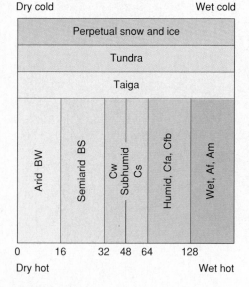

Climatic types

Dry cold   Wet cold

| Perpetual snow and ice |
| Tundra |
| Taiga |

Arid BW | Semiarid BS | Cw Subhumid Cs | Humid, Cfa, Cfb | Wet, Af, Am

0   16   32   48   64   128

Dry hot   Wet hot

Vegetative formations

| Perpetual snow and ice |
| Tundra |
| Taiga |

Desert grasses and shrubs | Steppe | Grassland | Forests | Rain forest

Major zonal soil groups

| Perpetual snow and ice |
| Tundra |
| Podzols |

Sierozems and desert soils | Chestnut and brown soils | Chernozems | Prairie soils— degraded chernozems | Podzols / Gray-brown podzolic soils / Red and yellow podzolic soils / Lateritic soils

32
48
64
128

Figure 10.12

Global patterns of climate, soil, and vegetation are closely correlated. Climate is the controlling factor. For example, a semiarid (BS) climate has steppe vegetation with chestnut and brown soils.

more abundant in grassland and desert soils than in the forest environment where leaching is greater. Arid regions also have greater capillary water movement, which reduces downward movement of these minerals.

The activity of soil microbes on organic materials produces the elements needed by plants in a soluble, usable form. Many toxic substances from plants and other sources are destroyed by organisms. Soil aggregation, or structure, is improved because microbes produce humus and humus colloids, which make the soil less compact, lower in density, and with greater moisture holding capacity.

It is estimated that in one gram of dry soil, bacteria can number from 3 million to 500 million. Table 10.4 gives estimates of other organisms commonly found in a middle-latitude western soil. Their main function is the decomposition of organic materials, all of which are composed of plant nutrients transformed into soluble forms for root uptake.

The soil organisms also make available inorganic plant nutrients. Carbon dioxide from the atmosphere and decomposing humus reacts with water to form carbonic acid in the soil solution. This acidic water lowers the pH and has a much greater solvent, or dissolving, action than pure rainwater on inorganic minerals (although rainwater itself is slightly acid). These microscopic bacteria are not consciously working for the welfare of plants, but are attempting to survive by competing for food obtained in the organic materials they decompose. The major mass of the soil hosting the organic component is weathered bedrock or parent material. Let's examine its key role.

Table 10.4
Microorganisms in Soil

Organism	Number in 1 gram Dry Soil
Bacteria	3,000,000 to 500,000,000
Streptomycetes	1,000,000 to 20,000,000
Fungi	5,000 to 900,000
Yeasts	1,000 to 100,000
Protozoa	1,000 to 500,000
Nematodes	less than 1 to 300
Algae	1,000 to 500,000
Bacteriophages	Unknown numbers
Viruses	Unknown numbers

Source: Dr. James P. Martin, Associate Chemist and Jarel O. Ervin, Laboratory Technician, Division of Soils and Plant Nutrition, University of California Citrus Experiment Station, Riverside, California.

Parent Material

Of the 18 essential elements for plants, only carbon, hydrogen, oxygen, and nitrogen come originally from the atmosphere; the remaining 14 come from rocks and minerals into the soil.

Soils develop from rocks of the Earth's crust which are called **parent material.** In a loam soil sample, 45 percent of the volume comes from rocks that have been weathered into smaller and smaller particles (fig. 10.13). They blend into a mixture of air, water, and organic and mineral material we call the **solum** or true soil.

A

B

Figure 10.13

Rock is the raw material for soil development. (A) In a soil sample, 45 percent of the volume comes from the rock that has been weathered into smaller and smaller particles forming solum. (B) The limestone blocks of Inca walls at Sacsayhuaman fortress, near Cuzco, Peru, have recessed joints from contact edges weathering.

The nature of the rock source can have a marked effect on the soil's productivity, as well as physical properties of structure and texture. For example, soils rich in calcium are formed from parent material such as limestone ($CaCO_3$), which is well supplied with this element. However, where precipitation is high, leaching can rapidly deplete a mineral commonly found in the parent material. Certain soils in the humid tropics and middle latitudes are so severely weathered and leached that the soil may have little weatherable material remaining in the A horizon. Therefore, climate can weather a variety of parent materials until there is little resemblance to the original parent rock.

Parent materials from which soils are weathered are developed from a variety of sources. These include residual, transported, and cumulose deposits (table 10.5). Residual materials originate from rocks and minerals that have remained in a given place long enough to be weathered into soil-size particles. Transported materials are mineral and rock fragments that have moved into a region by wind, water, ice, or gravitational forces. Cumulose deposits are primarily organic materials such as peat that have developed from decaying plants under high water table conditions, or in moist environments such as estuaries, bogs, and meadows.

It is estimated that 3 percent of all soils in the United States are residual soils, having been developed in place from underlying igneous, sedimentary, or metamorphic rocks.

Relief

The lay of the land, or **relief,** plays an indirect role in the soil-forming process. It influences weathering, erosion, soil depth, climate, vegetation, and drainage (fig. 10.14).

Slope Aspect

South-facing slopes (in the Northern Hemisphere) receive more insolation during the year, creating a warmer, drier soil. On these slopes, solar warming increases evapotranspiration. Leaching is less, thus raising the pH values. Because of increased aridity, more drought-resistant, lower density plant cover is established. Thus in turn the organic factor is influenced (fig. 10.15).

Elevation

Climate, soil, and vegetation vary not only with slope aspect, but with changes in **elevation.** As we observed in chapters 6 and 7, changing elevation produces rapid changes in climate and thus vegetation. Generally speaking, increasing elevation produces soils typically found in the higher latitudes where cooler, more humid conditions exist. It is important not to overgeneralize on these points, because there are many factors governing patterns of climate. However, mountain soils are often more highly leached, acidic, and thin, resembling the soils of the higher latitudes.

Table 10.5
Sources of Parent Material

Residual Material	Igneous		Sedimentary		Metamorphic
	Granite Basalt Lava		Limestone Sandstone Shale		Marble Quartzite Gneiss

Transported Material and Erosional Agents	Water	Wind	Ice	Gravity
	Alluvial (running water) Lacustrine (lakes) Marine (ocean)	Eolian Loess	Moraine Till plain Outwash plains	Colluvial

Cumulose material (organic)	Fibric	Folic	Hemic	Sapric
	Peat	Leaf mold	Peat or muck	Muck

Source: U.S. Department of Agriculture.

Figure 10.14

The shape of the land, or relief, influences a variety of environmental factors. These include weathering, erosion rates, soil development and depth, climate, vegetation, and drainage. Deep soils have developed from floodplain deposits of the Urabamba River in the Sacred Valley of the Incas, Peru. Thin soils have formed on the mountain slopes.

Figure 10.15

Slope aspect, or orientation, controls temperatures, evapotranspiration rates, vegetation density and types, soil moisture, pH, and organic and mineral control of the soil. Saguaro cactus are growing on the south-facing slope in southern Arizona near Phoenix. Here soil moisture is less than on a cooler north-facing slope on the opposite side of this valley.

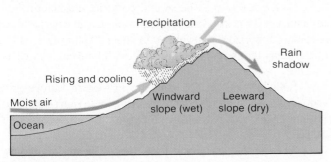

Figure 10.16

Soils on windward slopes usually are higher in moisture content because of the orographic effect. On the downwind leeward slopes, arid conditions prevail.

Soils on windward slopes of mountains tend to be wetter because of the orographic effect described in chapter 6. In contrast, soils located in the rain shadow on the leeward slopes are drier, with less humus and vegetation (fig. 10.16).

Gradient

Slope **gradient** is also a major factor affecting the depth of the profile. Drainage and erosion are rapid on steeper slopes, while deposition of sediments is common on the poorly drained lowlands. The erosional processes constantly remove weathered plant materials, thinning the profile on steeper slopes and producing indistinct, shallow horizons. Poorly drained topography often becomes waterlogged and high in organic content. This is especially true in cool, humid climates. In the desert regions of the interior, saline and alkaline soils develop due to high evaporation causing a buildup of bases in the A horizon and the surface. All other factors being equal, moderately sloping terrain is most favorable to good soil development. Gentle terrain is not easily eroded and yet has satisfactory drainage.

Therefore, relief exerts a significant influence on the nature of the soil developing at a given location, because it influences the local climate and drainage. Each elevation and slope gradient has its own unique environment that governs the soil-forming processes (fig. 10.17). The amount of time needed is one of the greatest variables in the equation.

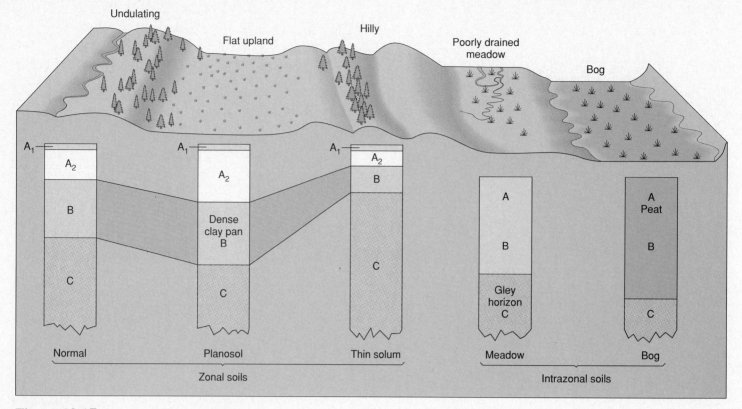

Figure 10.17

Slope gradient controls soil depths and insolation concentrations. Where gradients are steep, soil erosion prevents deep soil development.

Time

Soil can take thousands of years to form, and yet humankind can destroy it in a moment. **Time** is a key ingredient required for soil formation. The amount of time is a function of the other four soil-forming factors. It also depends on where we begin timing the process. A great interval of time is needed if we start the clock of soil development on freshly exposed limestone or granite rock. Millions of years may be required before the parent materials are weathered and horizons formed. The clock speeds up in the more humid tropics and, of course, proceeds very slowly in the cooler, drier, higher latitudes. The rock type, or parent material, is also an influence on the time factor. Granite weathers more slowly than basalt. Larger amounts of time are needed to weather parent material than to produce soil horizons.

Where weathering and erosion have produced fresh **regolith,** or weathered parent material, a soil profile may be formed within a few years. Two soil scientists, R. L. Croker and J. Major, from the University of California, found that a glacial moraine in southeastern Alaska, deposited 30 years earlier, had developed a soil profile with a topsoil horizon darker and higher in organic content than the subsoil. Where the drift has been exposed for 100 years, the soil's organic content in the A horizon rivaled some soils of the eastern United States. This Alaskan example illustrates that, under proper conditions of parent material and climate, soils can form rapidly.

Soils usually must be several centuries old before they have distinct B horizons. Soils formed from loess in eastern Iowa have distinct B horizons that formed in about 20,000 years, as indicated by clay that has accumulated in the B horizons, as well as iron oxides.

Soils formed over Roman ruins in western Europe give us another piece of data. The sands covering these ruins cannot be more than 2,000 years old, and yet the A and B horizons are well developed.

The five factors of soil formation—**climate, organisms, parent material, relief,** and **time**—are not uniformly active over the surface of the Earth. Remember, the soil produced is the end product of the interactions of these soil-forming factors, with each location having its own mix. There are many climates, combinations of living systems, many kinds of parent materials, and a global terrain of multiple slopes.

The end product, or soil type, evolves with time. Time gradually allows climate to do its work to reshape the soil pattern to resemble the climate and vegetation patterns. Youthful, immature soils, still being formed, carry strong influences of the geologic base of their origin. As time passes, weathering, organic activity, and climatic factors dominate to produce soils reflecting the climate of the location more than any other factor. (See box 10.1.)

Sour Soils of the Past

Five hundred thousand years ago the ocean waves cut a beach and deposited sand and gravel. Then the shoreline retreated, and a new beach was cut. This process, repeated, left a series of elevated terraces located along the cool, foggy California coast near the *study area* of chapter 1.

Located on the fourth and fifth terrace from the sea, 120 to 150 meters (400 to 500 feet) above sea level, are some of the world's most acidic soils. They have a pH of 3.5, the same acidity as orange juice.

Here we can see all five soil-forming factors in operation at different stages of time. Terrace number one represents 100,000 years

of soil development. Terraces two, three, four, and five each represent 100,000-year increments back in time with terrace number five being the oldest at 500,000 years (fig. 1).

Table 1 summarizes the five soil formation factors and their influence on the soil formation process on the fifth terrace.

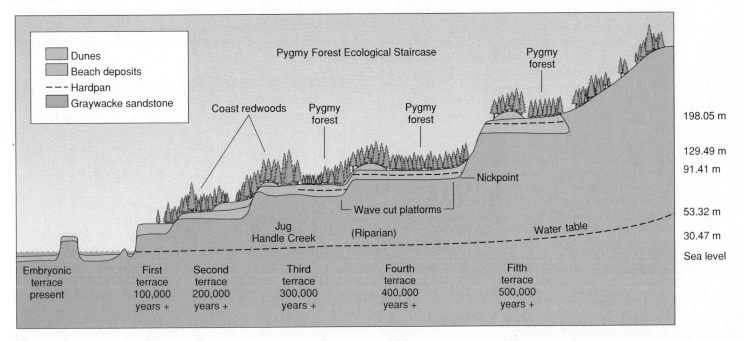

Figure 1

Terraces were cut by wave action over a span of 500,000 years. The first terrace emerged 100,000 years ago, replacing the second terrace for the number one spot on the shoreline. Each terrace is 100,000 years older than the lower one. This 500,000-year time span has produced the world's lowest pH soils and dwarf pygmy forest on terraces four and five. These terraces have been leached for up to 500,000 years, hence removing the bases.

Table 1

Five Soil Formation Factors on a Marine Terrace

Soil Formation Factor	Influence
1. **Climate** (Csb) Cool Mediterranean	Cool, wet, rainy conditions cause weathering and leaching with little capillary action.
2. **Organisms**	Plant and animal decay increases hydrogen ion and lowers pH to 3.5.
3. **Relief**	Gentle slope. Poor drainage causes bog conditions and more leaching.
4. **Parent material**	Mostly sandstone and shale. Highly weathered quartz left as primary mineral in soil with a pH of 3.5.
5. **Time**	500,000 years has been enough time to leach and remove most nutrients.

Just as we can create a multitude of words out of 26 letters in the alphabet, there are hundreds of thousands of different local soil types or combinations. The next chapter examines on a global scale these major soil patterns that resemble so closely the world patterns of climate and vegetation.

SUMMARY

Soils form a thin veneer covering the crust of the Earth with a mix of weathered **parent materials, organisms, air,** and **water.** This thin layer is the food resource base for the planet. A typical **soil profile** consists of subdivisions called **horizons,** which develop as the weathered parent materials undergo change due to climatic and biological influences.

All soils exhibit both textural and structural properties that influence their productivity. **Soil texture** is determined by the percentages of various sizes of particles ranging from fine clay to coarse sand. A loam texture is generally preferred by agriculturists because it represents a balanced blend of sand, silt, and clay, making it more easily tilled. Soil texture can be modified by adding amendments such as manure, gypsum, and wood products.

Soil structure is a term that characterizes how grains or particles join together in larger units. These larger units are called **peds** and can be classified as **blocky, platy, granular,** or **prismatic.**

A desirable structure is a stable one in which soil particles are bound together into granular aggregates, allowing air and water to penetrate the A, B, and C horizons and make available nutrients for plants.

The chemistry of the soil gives us a picture of the potential nutrients available. Of the 92 natural elements, only 18 are known to be required by plants.

Plant nutrients are classified as either major nutrients or micronutrients. Although micronutrients are needed in small amounts, they are essential for plant health.

A **soil's pH,** or **acidity,** gives us clues about the chemical health of the soil. A low pH may indicate low fertility, while a high value implies excesses of basic nutrients. Values in the middle range are more favorable to nutrient uptake and indicate a well-balanced nutrient base.

Soil colloids of organic and mineral origin are at the seat of chemical activity. These microscopic particles hold and attract nutrients until plants need them. We call this nutrient bank the **clay-humus complex.**

The soil's physical and chemical **properties** result from five soil formation factors: **climate, organisms, parent material, relief,** and **time.** These factors are all interrelated. For example, relief influences insolation because solar exposure varies with the angle and orientation of the slope.

Climate is by far the most important soil-forming factor. It controls or influences weathering of parent materials, biological activity, moisture content, **leaching,** and **capillary** movement in the soil. Youthful soils, still being formed, carry strong influences of their geologic parent material. As time passes, weathering processes, organisms, and climatic factors gradually produce a soil reflecting the climate more than any other influence.

ILLUSTRATED STUDY QUESTIONS

1. Locate on a world map regions of deepest soil profiles and absence of humus (figs. 10.9 and 10.10, pp. 205, 206).
2. Describe the soil-forming process that produces the greatest humus (figs. 10.10 and 10.11, pp. 206, 207).
3. Describe the conditions where soils have a high pH. Explain why this condition exists (fig. 10.7, p. 203).
4. Which mountain slopes in the Northern Hemisphere will tend to have more basic soils? Explain your answer (fig. 10.16, p. 211).
5. Describe the four primary constituents of soil (fig. 10.1, p. 197).
6. Compare each soil horizon in a hypothetical soil profile (fig. 10.3, p. 199).
7. Give the percentages of sand, silt, and clay in each soil texture listed: sandy loam, loam, and silt loam (fig. 10.4, p. 200).
8. Why is climate the most important control in the development of soil (fig. 10.9, p. 204)?
9. Why are tropical soils so low in humus (fig. 10.11, p. 207)?

Soil Patterns of the World

11

Objectives

After completing this chapter you will be able to:

1. Explain **laterization, calcification, podzolization, salinization,** and **gleization** soil formation processes.
2. Relate the soil-forming process to the **nutrient cycle.**
3. Compare and contrast the Comprehensive Soil Classification System or 7th Approximation with the 1938 United States Soil Classification System.
4. Describe the chief soil characteristics of the tundra, cool middle latitudes, warm temperate forests and grasslands, desert and steppe, and soils of the humid low latitudes.
5. Contrast **halomorphic, calicmorphic,** and **hydromorphic** soils.

Climate is the chief environmental factor determining soil patterns of the Earth.
© Doug Sherman/Geofile

Soil patterns of the world are primarily the product of climatic and biotic processes acting over long periods of time, gradually transforming the thin veneer of exposed rock into a variety of soils around the world.

There are literally thousands of soil types currently known. In your own backyard you might discover more than one type because of variations in the five principal soil-forming factors described in chapter 10.

Global soil patterns can be broadly grouped into five soil-forming *processes* closely related to the Earth's environmental zones. These soil-forming processes include: (1) **laterization** occurring in warm rainy tropical latitudes, (2) **calcification** associated with dry warm belts of the world, (3) **podzolization** occurring in cool precipitation regions, (4) **salinization** limited to small poorly drained regions of excessive salt accumulation, and (5) **gleization** occurring in areas of high precipitation and poor drainage.

This chapter will examine the soil-forming processes in each of the global environmental zones and then relate these processes to the major soil types.

LATERIZATION

Let us begin our survey in the tropics, where **laterization** processes dominate the humid equatorial belts of Latin America, Africa, the southeastern parts of Asia and North America, northeastern Australia, and the islands of the western Pacific Ocean. It is closely associated with warm rainy climates and tropical rainforest and savanna vegetation (fig. 11.1).

The laterization process can be characterized by high rates of weathering, rapid destruction of organic matter, decomposition of clay minerals, and removal of nutrients by leaching. The soils produced by this process are usually low in basic nutrients, humus, and mineral colloids.

These soils represent the most chemically weathered soils in the world. This weathering is reflected in profiles which are some of the world's deepest, yet lacking distinct horizon development except for a darker A horizon near the surface.

Quartz, one of the most abundant mineral and highly resistant to all forms of weathering, is successfully attacked and removed through the groundwater outflow and surface runoff. In severe cases of weathering, the soil can be depleted of most cations needed by plants as nutrients. All that remains is quartz, iron and aluminum oxides, and hydroxides, producing a reddish-colored soil. Some of the world's richest aluminum ore, or bauxite, is produced by laterization, especially in tropical savanna lands of Africa and South America.

Vegetation and animal organisms play a significant role in maintaining a nutrient balance, which is of paramount importance. Plants and animals literally keep these highly weathered soils alive and productive. Here is how it works. Consider the role of plants as miners or extractors of a low-grade ore, the soil. Insects, bacteria, fungi, earthworms, and many other organisms in the tropical soils are food processors. Because of the warm humid conditions, these organisms thrive and actively convert a plant's

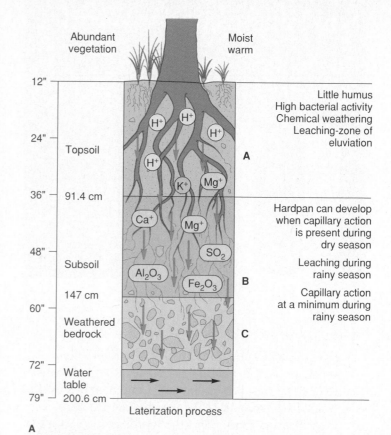

A

B

Figure 11.1

Laterization processes are closely associated with warm rainy climates and tropical rainforest and savanna vegetation.

Figure 11.3

Where virgin forests near the *study area* have been removed, the nutrient cycle is broken and soils can become sterile.

Figure 11.2

Plants and animals are the key to the **nutrient cycle.** As nutrients are removed by leaching, organisms and plant roots extract nutrients and recycle them. Insects, earthworms, bacteria, and root system extract these soluble materials before they can be leached into the groundwater system.

falling leaves and debris back into nutrients ready for uptake by the plant. This is a classic example of a **nutrient cycle.** The principle is universal; both plants and animals tie up released nutrients and keep them recycling (fig. 11.2).

Where forest vegetation has been cleared for agriculture or other uses, the cycle is broken and nutrients are rapidly lost as soluble minerals in the leaching process. What remains is a hard, sterile bricklike crust known as **laterite** consisting primarily of aluminum and iron compounds. Not only are the nutrients lost but soil organisms leave because their food base has been depleted. Even if the forest is allowed to return, second growth does not have the same plant diversity or density as the virgin forest because weathering and plant removal has taken its toll by modifying the nutrient bank (fig. 11.3).

Exceptions do occur where flooding or volcanic activity introduces unweathered soils rich in nutrients from freshly deposited minerals. Therefore, alluvial and volcanic soils of the tropics can and do support agriculture long after the virgin forest vegetation cover has been removed.

CALCIFICATION

Tropical lateritic processes gradually grade into arid and semiarid **calcification** processes that dominate the Mediterranean steppe and desert lands of the middle latitudes (fig. 11.4).

The calcification process is characterized by seasonally high soil temperatures, low precipitation, minimal leaching, and moderate to strong capillary water movement.

Because soil moisture is often limited, profile development is usually shallow and basic minerals tend to accumulate in the upper horizons. The B horizon often becomes rich in cations such as calcium, and to a lesser extent potassium and magnesium (fig. 11.5).

A

A_1
A_2
B
C

B

Figure 11.4

The calcification process dominates the arid and semiarid grasslands of the middle latitudes. Calcium, magnesium, potassium, and nitrogen are nutrients that accumulate in the B horizon.

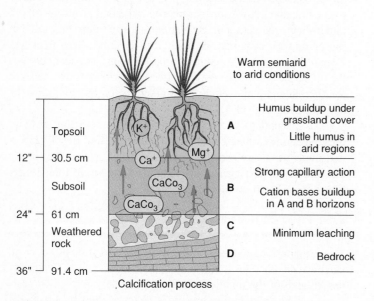

Warm semiarid to arid conditions

Topsoil	A	Humus buildup under grassland cover
		Little humus in arid regions
Subsoil	B	Strong capillary action
		Cation bases buildup in A and B horizons
Weathered rock	C	Minimum leaching
	D	Bedrock

12" — 30.5 cm
24" — 61 cm
36" — 91.4 cm

K^+ Ca^+ Mg^+ $CaCo_3$ $CaCo_3$

Calcification process

Figure 11.5

Calcic soil profiles vary in depth from shallow to deep in the grasslands and are rich in cations such as calcium, magnesium, and potassium. Soil depth varies with climatic and slope conditions. The south-facing slope in the *study area* is where calcification is at work.

In the desert belts where precipitation is low and evaporation is extreme, soil profiles are very thin. They consist of coarse particles of weathered parent material rich in calcium carbonate or lime and very small amounts of organic matter. However, in the more humid mid-latitude grasslands, the A horizons are dark and rich with organic matter while calcium is present but in lower concentrations.

In arid regions where calcium carbonate has accumulated in the B horizons, a **hardpan** or **caliche** often develops. This impervious layer can produce poor drainage and reduce leaching even more. Under extreme desert conditions, caliche forms on the surface. Calcium carbonate bonds together with sand and gravel to produce a crust known as **desert pavement.** This is a very important protective coating on the desert's thin soil, preventing the wind from carrying finer particles away by deflation (chapter 16).

Calcification in the Mediterranean and steppe climates is responsible for producing some of the world's richest soils, located in the mid-latitude grassland regions. The major grain belts of North America, Australia, and Asia are all located where calcification processes are active. Oasis agriculture also develops on soils formed by this process. We can partially attribute high fertility to the fact that leaching is minimal and biological activity recycles

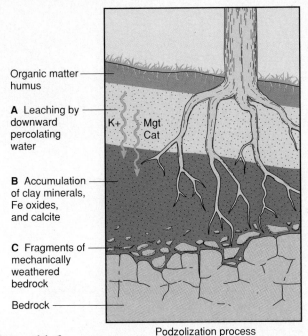

Figure 11.6

Podzolization processes dominate the temperate middle and high latitude forest belts. These soils develop under cool and humid climatic conditions. The *study area's* north-facing slope is undergoing podzolization.

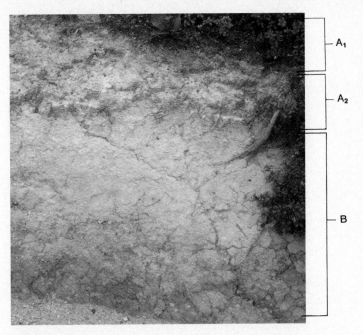

Figure 11.7

This podzol soil profile reveals a distinct ash-gray A2 horizon that gives the soil its name. Podzol means "ash soil" in Russian. The A2 horizon is coarse and gray from weathering and leaching.

nutrients at moderate rates, producing a rich humus layer. Thus, the clay-humus complex is well developed, as an important factor in maintaining nutrients at high levels. Therefore, high yields are possible on a long-term basis providing good soil practices such as crop rotation, fallowing of fields, and erosion control are employed.

PODZOLIZATION

Where grasslands are gradually replaced by forests of the middle and higher latitudes, the **podzolization** process begins to dominate. In the *study area* described in chapter 1, podzolization occurs on the shady north-facing slopes while calcification develops on drier south-facing slopes. The climatological variables producing different environments are also the chief elements influencing these two soil-forming processes on opposite sides of the valley. The shady slopes receive less solar energy, consequently temperatures are lower and evapotranspiration rates are greatly reduced. This produces a cool, moist environment and a more highly leached soil (fig. 11.6).

On a global scale the podzolization process occurs wherever these same cool, moist conditions prevail during most of the year. Therefore, we can find this soil associated with the coniferous and mixed forests of the middle and high latitudes and in the higher elevations of the lower latitudes.

In the podzol soil profile in figure 11.7, we can see a distinct A2 horizon that is coarse grained and bleached a gray ash color. This horizon is leached of bases, oxides, and clays by a soil solution which has accumulated acidic compounds from decomposing humus in the uppermost A0 and A1 horizons. Because temperatures are cool, humus accumulates in the A0 horizon.

Rainwater percolates through this organic layer and becomes strongly acidic and dissolves water-soluble organic compounds which combine to form iron complexes. In this state, iron is easily removed or is said to be mobilized. Aluminum is also removed in the same manner, leading to a complete breakdown of clay minerals which are removed and deposited in the B horizon.

The humus, clay, iron, and aluminum oxides that are removed from the A horizon often form a dense, dark red hardpan in the B horizon. This podzol process produces soils more strongly weathered and leached than those formed from calcification but generally less than lateritic soils. Levels of fertility can vary considerably from moderate to very low. The key to fertility is the leaching factor. Excessive leaching leaves primarily quartz behind in the A horizon, producing a bleached gray soil low in basic minerals and very acidic.

A classic case of extreme podzolization can be observed along the marine terraces of northern California. Wave action coupled with changes in sea level have created a series of five to eight elevated marine terraces along a 30-kilometer (20 mile) section of coast in northern California extending from the Noyo River southward to the Navarro River. On these benches, calcification and podzolization processes can be observed in successive stages, grading from moderate calcification at the youngest and lowest terrace to extreme podzolization on the oldest and highest terrace (fig. 11.8; box 10.1, p. 213).

The higher terraces emerged from the sea first as tectonic processes uplifted the Coast Range. Chapter 15 explains these tectonic processes. Time, one of the five major factors in soil formation is quietly doing its work over a span of 500,000 years.

A

Figure 11.8

(A) Wave action coupled with changes in sea level have created a series of marine terraces on the northern California coast. Each terrace represents a successive stage in soil evolution. The lowest terrace is the most recent to emerge from the sea and represents the youngest soils. (B) The highest terrace is an example of highly leached podzolic soil stripped of its nutrients. The effect of nutrient deficiency is dramatic. Plants are dwarfed, diseased, and decadent.

Eventually soils on the lowest terrace will evolve into soils like those on the next terrace, then the next terrace, and so on. Although the young and old terraces display the same mineralogical parent material, the soil's chemical and physical properties vary on each terrace, producing a progressive evolutionary history of not only soil but an entire ecosystem, described in chapter 12. While soils are changing from step to step in 100,000-year increments, the vegetation is also evolving from coastal grasslands to redwood–Douglas-fir forests to pine, and then to dwarf "pygmy forests" of bishop pine and cypress on the higher terraces.

As podzolization advances, first grasses, then redwoods and Douglas firs, are replaced by bishop pines requiring fewer nutrients. Finally, extreme pygmy forests of these pines result. As much as 25 percent of the ground area is bare or covered with

B

Figure 11.9

Acid-tolerant plants help mobilize iron and aluminum oxides in a soluble form, later to be transported to the B horizon in the leaching process. A hardpan then forms, resulting in little root penetration below and poor surface drainage.

only lichen. Stunted cypresses (*Cupressus pygaea*), gnarled bishop pines (*Pinus muricata*) and Bolander pines (*Pinus bolanderi*) several hundred years old and only 1½ to 3 meters (5–10 feet) tall, dominate these leached sterile soils. Bases are almost absent, and pH is 2.8 to 3.9 in the A2 horizon. Although the parent material sandstone has feldspars in the ratio of 1:18 with quartz, laboratory analysis shows that the base-rich feldspar has been reduced to 0.03 percent, meaning that the A2 horizon is a nonweatherable, inert quartz flour of no nutrient value to the plants.

According to the latest theories, the acid-tolerant plants mobilize iron and aluminum oxides in a soluble form in the A horizon. Once removed by the plant, these oxides are then deposited in the B horizon as an iron hydroxide to form a hardpan cement in combination with clay and quartz minerals (fig. 11.9). This process seems more active where the soil experiences wetting and drying on a seasonal basis. The effect of this hardpan is poor

drainage and lack of root penetration. Thus plant roots are soaked for weeks during the rainy season, and soils are leached of all soluble minerals. During the summer dry season, soils become dry and hard and plants suffer the other extreme of drought.

SALINIZATION

Salinization is limited to small regions of the arid world where excessive salts accumulate in soils of high temperature and poor drainage. Where these two factors occur, a high water table and temporary lakes develop during the unpredictable rainy season. Then evaporation quickly dehydrates the land, lowering the water table. The net effect of this repeating seasonal cycle produces a saline or salty soil. As water evaporates, a portion of its bases and salts in solution are left behind to form a white precipitate in the upper horizon and on the surface. The remaining soil solution

Figure 11.10

Salinization can result from poor drainage, a changing water table, and evaporation. As water evaporates, salts and bases are deposited in the soil, producing a saline soil with a high pH. Only halophytes, or salt-resistant plants, can survive in this environment. This road on an alluvial fan deposit points toward a dry lake bed in Death Valley, California, where salinization processes are at work.

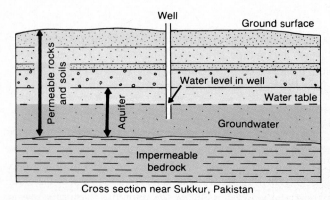

Cross section near Sukkur, Pakistan

Figure 11.11

Pakistan is battling the salt problem in the Indus River valley. A high groundwater table compounds the problem by preventing leaching of salts.

is usually very saline. The nature of the precipitate deposits reflects the parent material. If the parent material is weathered limestone, calcium carbonate coats the surface in high concentrations (fig. 11.10).

One of the greatest problems today in the irrigated agricultural lands of the desert is salt buildup resulting from extreme salinization. The great ancient civilization of Ur, the Chaldean culture, declined as soil salt content increased. Today much of the Tigres and Euphrates River basin is low in productivity because of saline conditions resulting from poor soil management. Pakistan currently faces the same problem of salinization in the Indus River valley (fig. 11.11). The problem there is compounded by a high water table, thus preventing leaching the salts that build up on the surface and top soil horizon.

Both the United States and Mexico continually struggle with saline soils in the Colorado River drainage basin. The problem becomes more acute farther down the Colorado River on Mexico's soil because salinity increases from highly saline irrigation waters returning back into the river from the United States side of the border. Increased sodium chloride content in Mexico's water is a key factor in reducing cotton yields in recent years in northern Mexico in the Colorado River delta.

GLEIZATION

Gleization is at the other end of the rainfall spectrum. High precipitation and poor drainage in low-lying areas produces a soil high in organic matter, low on the pH scale, and very waterlogged. The soil color ranges from coffee brown to blue-gray. If the soil can be drained, agriculture is possible providing lime ($CaCO_3$) is added to replace hydrogen ions and raise the pH. These steps cause microorganisms to multiply in a more favorable environment and process poorly decomposed vegetation into humus and nutrients for plants.

Gleization, like salinization, processes are limited to specific conditions occurring in limited regions. Much of the poorly drained lowlands of northern Europe and the Great Lakes region of North America experience this soil-forming process in low-lying freshwater lake basins and river valleys. These boggy soils formed as a result of the last Ice Age. After melting of the ice cap, poor drainage and numerous lakes dotted the landscape. Here reclamation by drainage management can produce excellent agricultural soils.

SOIL CLASSIFICATION

These five soil-forming processes just described have produced global soil patterns which can be classified in a variety of ways. Every classification system is limited in its ability to accurately describe and define a natural system because nature is far too complex. Simple global classification systems fail to accurately show the complexities occurring at the local level, while complex local systems are inadequate for the broad regional perspective. The importance of all classifications lies in the fact that they serve as a starting point from which further inquiry can progress.

Soil classification, like climate and vegetation classification systems, is still developing. Each classification system reflects the point of view of the author and the author's discipline. As more knowledge of the subject is collected, refinement in the system usually occurs.

Present-day soil science began with V. V. Dokuchaiev of Russia and E. W. Hilgard in the United States. Dokuchaiev developed the first natural soils classification based on observable characteristics such as soil profile development and the origin of soils and their relationship to climate and vegetation. He noted that many soil types could be predicted based upon the geographical factors of climate, vegetation, morphology, and geology.

As a result of this discovery, Dokuchaiev published a world classification system in 1900 which placed an emphasis on zones or environmental belts.

In 1927, C. F. Marbut, chief of the U.S. Soil Survey, developed a revised system by incorporating the ideas of Dokuchaiev with his own to produce a soil system used extensively until the 1950s, particularly the 1938 version.

Figure 11.12 shows the global patterns of the soil orders according to the 1960 Comprehensive Soil Classification, also known as the 7th Approximation. It gives an explanation of the characteristics of each order and suborder. Both systems are in use today. European pedologists lean toward variations of Dokuchaiev's classification system. Americans tend to favor the Comprehensive Soil Classification System.

To maintain our focus on processes, soil patterns, and interrelationships between various elements of the environment, we will survey the soil characteristics of the major climatic and vegetation belts described in chapter 9. Our emphasis will be on soil characteristics produced by laterization, podzolization, calcification, salinization, and gleization processes, and the name of the soil order or suborder will be given in both the 7th Approximation and the 1938 U.S. Soil Classification System. This system is also called the **great soil groups.** Because, on a global scale, the great soil groups correlate so closely with soil-forming processes discussed earlier, we will use the 1938 system as our basis for soil description on a world scale. Not every suborder of each system will be described, but emphasis will be placed on representative examples of each environmental region. See appendix B for a complete description of each system.

HIGH-LATITUDE AND ALPINE SOILS

Tundra Soils

Soils of the high latitudes form the northern border of the North American, European, and Asian continents (fig. 11.12). The importance of climate in the soil-forming processes of the tundra cannot be overstated. Climatic conditions are characterized by long, bitterly cold winters and short, cool summers with light annual precipitation. Below the soil layer the ground remains permanently frozen and is commonly referred to as **permafrost.** This permafrost layer prevents any downward moisture percolation. Therefore, the surface soil often remains saturated with water during the brief summer. If there is a gentle slope, the thawed surface may even creep or flow by a process called **solifluction.** Mounds known as **pingoes** sometimes erupt on the surface as pressures build from surface refreezing, trapping a soggy saturated soil layer between the frozen surface and permafrost below. Water from a slope drains under hydrostatic pressure creating a wedging effect, gradually lifting the frozen surface soil as much as 7 meters, or nearly 25 feet (fig. 11.13).

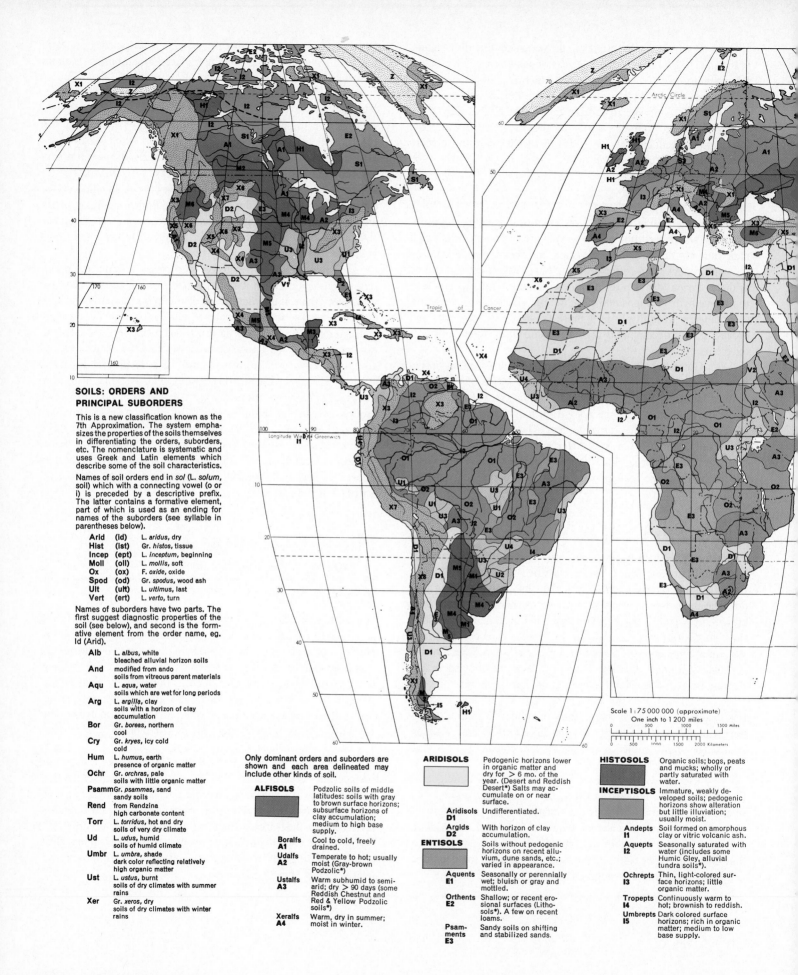

SOILS: ORDERS AND PRINCIPAL SUBORDERS

This is a new classification known as the 7th Approximation. The system emphasizes the properties of the soils themselves in differentiating the orders, suborders, etc. The nomenclature is systematic and uses Greek and Latin elements which describe some of the soil characteristics.

Names of soil orders end in *sol* (L. *solum*, soil) which with a connecting vowel (o or i) is preceded by a descriptive prefix. The latter contains a formative element, part of which is used as an ending for names of the suborders (see syllable in parentheses below).

Arid	**(id)**	L. *aridus*, dry
Hist	**(ist)**	Gr. *histos*, tissue
Incep	**(ept)**	L. *inceptum*, beginning
Moll	**(oll)**	L. *mollis*, soft
Ox	**(ox)**	F. *oxide*, oxide
Spod	**(od)**	Gr. *spodus*, wood ash
Ult	**(ult)**	L. *ultimus*, last
Vert	**(ert)**	L. *verto*, turn

Names of suborders have two parts. The first suggest diagnostic properties of the soil (see below), and second is the formative element from the order name, eg. Id (Arid).

Alb	L. *albus*, white bleached alluvial horizon soils
And	modified from ando soils from vitreous parent materials
Aqu	L. *aqua*, water soils which are wet for long periods
Arg	L. *argilla*, clay soils with a horizon of clay accumulation
Bor	Gr. *boreas*, northern cool
Cry	Gr. *kryes*, icy cold cold
Hum	L. *humus*, earth presence of organic matter
Ochr	Gr. *orchras*, pale soils with little organic matter
Psamm	Gr. *psammas*, sand sandy soils
Rend	from Rendzina high carbonate content
Torr	L. *torridus*, hot and dry soils of very dry climate
Ud	L. *udus*, humid soils of humid climate
Umbr	L. *umbra*, shade dark color reflecting relatively high organic matter
Ust	L. *ustus*, burnt soils of dry climates with summer rains
Xer	Gr. *xeros*, dry soils of dry climates with winter rains

Only dominant orders and suborders are shown and each area delineated may include other kinds of soil.

ALFISOLS

Podzolic soils of middle latitudes: soils with gray to brown surface horizons; subsurface horizons of clay accumulation; medium to high base supply.

Boralfs A1	Cool to cold, freely drained.
Udalfs A2	Temperate to hot; usually moist (Gray-brown Podzolic*)
Ustalfs A3	Warm subhumid to semi-arid; dry > 90 days (some Reddish Chestnut and Red & Yellow Podzolic soils*)
Xeralfs A4	Warm, dry in summer; moist in winter.

ARIDISOLS

Pedogenic horizons lower in organic matter and dry for > 6 mo. of the year. (Desert and Reddish Desert*) Salts may accumulate on or near surface.

Aridisols D1	Undifferentiated.
Argids D2	With horizon of clay accumulation.

ENTISOLS

Soils without pedogenic horizons on recent alluvium, dune sands, etc.; varied in appearance.

Aquents E1	Seasonally or perennially wet; bluish or gray and mottled.
Orthents E2	Shallow; or recent erosional surfaces (Lithosols*). A few on recent loams.
Psamments E3	Sandy soils on shifting and stabilized sands.

HISTOSOLS

Organic soils; bogs, peats and mucks; wholly or partly saturated with water.

INCEPTISOLS

Immature, weakly developed soils; pedogenic horizons show alteration but little illuviation; usually moist.

Andepts I1	Soil formed on amorphous clay or vitric volcanic ash.
Aquepts I2	Seasonally saturated with water (includes some Humic Gley, alluvial tundra soils*).
Ochrepts I3	Thin, light-colored surface horizons; little organic matter.
Tropepts I4	Continuously warm to hot; brownish to reddish.
Umbrepts I5	Dark colored surface horizons; rich in organic matter; medium to low base supply.

Scale 1 : 75 000 000 (approximate)
One inch to 1 200 miles

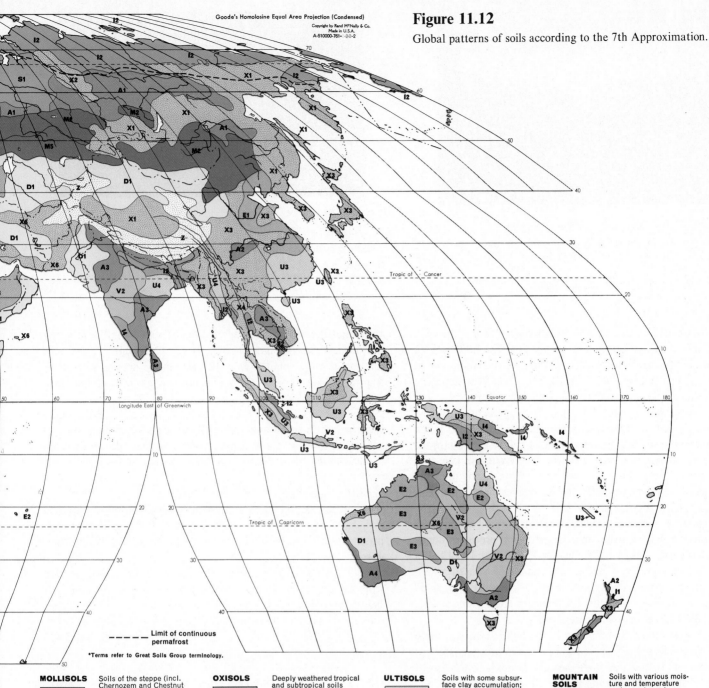

Figure 11.12

Global patterns of soils according to the 7th Approximation.

Tropic of Cancer

Longitude East of Greenwich

Equator

Tropic of Capricorn

- - - Limit of continuous permafrost

*Terms refer to Great Soils Group terminology.

MOLLISOLS
Soils of the steppe (incl. Chernozem and Chestnut soils*). Thick, black organic rich surface horizons and high base supply.

Albolls M1	Seasonally saturated with water; light gray subsurface horizon.
Borolls M2	Cool or cold (incl. some Chernozem, Chestnut and Brown soils*).
Rendolls M3	Formed on highly calcareous parent materials (Rendzina*).
Udolls M4	Temperate to warm; usually moist (Prairie soils*).
Ustolls M5	Temperate to hot; dry for > 90 days (incl. some Chestnut and Brown soils*).
Xerolls M6	Cool to warm; dry in summer; moist in winter.

OXISOLS
Deeply weathered tropical and subtropical soils (Laterites*); rich in sesquioxides of iron and aluminum; low in nutrients; limited productivity without fertilizer.

| Orthox O1 | Hot and nearly always moist. |
| Ustox O2 | Warm or hot; dry for long periods but moist > 90 consecutive days. |

SPODOSOLS
Soils with a subsurface accumulation of amorphous materials overlaid by a light colored, leached sandy horizon.

Spodosols S1	Undifferentiated (mostly high latitudes).
Aquods S2	Seasonally saturated with water; sandy parent materials.
Humods S3	Considerable accumulations of organic matter in subsurface horizon.
Orthods S4	With subsurface accumulations of iron, aluminum and organic matter (Podzols*).

ULTISOLS
Soils with some subsurface clay accumulation; low base supply; usually moist and low inorganic matter; usually moist and low in organic matter; can be productive with fertilization.

Aquults U1	Seasonally saturated with water; subsurface gray or mottled horizon.
Humults U2	High in organic matter; dark colored; moist, warm to temperate all year.
Udults U3	Low in organic matter; moist, temperate to hot (Red-Yellow Podzolic; some Reddish-Brown Lateritic soils*).
Ustults U4	Warm to hot; dry > 90 days.

VERTISOLS
Soils with high content of swelling clays; deep, wide cracks in dry periods dark colored.

| Uderts V1 | Usually moist; cracks open < 90 days. |
| Usterts V2 | Cracks open > 90 days; difficult to till (Black tropical soils*). |

MOUNTAIN SOILS
Soils with various moisture and temperature regimes; steep slopes and variable relief and elevation; soils vary greatly within short distance.

X1	Cryic great groups of Entisols, Inceptisols and Spodosols.
X2	Boralfs and Cryic groups of Entisols and Inceptisols.
X3	Udic great groups of Alfisols, Entisols and Ultisols; Inceptisols.
X4	Ustic great groups of Alfisols, Entisols, Inceptisols, Mollisols and Ultisols.
X5	Xeric great groups of Alfisols, Entisols, Inceptisols, Mollisols and Ultisols.
X6	Torric great groups of Entisols; Aridisols.
X7	Ustic and cryic great groups of Alfisols, Entisols; Inceptisols and Mollisols; ustic great groups of Ultisols; cryic great groups of Spodosols.
X8	Aridisols; torric and cryic great groups of Entisols, and cryic great groups of Spodosols and Inceptisols.

| z | Areas with little or no soil; icefields, and rugged mountain. |

Soil Patterns of the World   225

Figure 11.13

Mounds known as pingoes result from freezing and thawing of the soil layer. This railroad track in the Northwest Territories of Canada shows permafrost damage.

Photograph by L. A. Yehle, U.S. Geological Survey.

Figure 11.14

An arctic gray soil profile (inceptisol) in the arctic tundra of Alaska reveals poor horizon development due to cryoturbation.

Reproduced from the Marbut Memorial slide set, 1968, SSSA, by permission of Soil Sciences of America, Inc.

Because temperatures are below freezing most of the year, vegetation is sparse and soil organisms are few. The little organic matter present is very slow in decomposing, so the soil is low in humus with very little chemical weathering. Nitrogen supply to plants is often low because nutrients are so slow in being recycled.

Horizons are poorly developed, being constantly disrupted by a process of churning or **cryoturbation.** Because freezing occurs first on the surface, a saturated layer will be thrust upward as the expanding freezing surface ruptures. This churning process, with solifluction, keeps the tundra in a state of flux during the short thawing period. The soils are literally in a freezer most of the year, remaining inactive and nonproductive.

Since the areas of good drainage are limited in the high-latitude tundra, **podzolization** and the **gleization** processes dominate much of the Canadian and Russian tundra. Figure 11.14 shows a profile of an arctic gley soil in the tundra of Alaska where drainage is poor. Tundra soil then varies considerably with slope and drainage. According to the 7th Approximation, tundra soils are classified as **Entisol** and **Spodosol** orders.

As one moves equatorward in the high elevations, these soils can be found above the timber line where arctic conditions prevail. The main difference along the equator is a lack of seasonal variation. Freezing and thawing is on a diurnal basis.

SOILS OF THE COOL MIDDLE LATITUDES

Soils formed by the **podzolization** process dominate the higher middle latitudes of the Northern and Southern Hemispheres. The great coniferous and deciduous forest belts thrive on these soils. These forest soils have distinct horizons easily distinguished by vertical changes in physical and chemical properties. Podzolic soils have a rich humus layer in the top A0 and A1 horizons resulting from the accumulation of loose leaves and organic debris. Because

the climate has distinct seasons, biological activity and weathering are cyclic. Like the tundra soils, these soils go into deep freeze during the winter and then come alive in the spring. Therefore, four to six months out of the year soil-forming processes are active and then quiet down as winter sets in.

Since *eluviation* is active in the A2 horizon, this zone may become light colored and sterile of nutrients as soluble minerals are lost to lower levels in the compact B horizon. Bases are often leached by humus-related acids. The podzolic soils may become very acidic and low in fertility; however, if they are properly managed they can support agricultural activity.

The dominant characteristic of a podzol soil is a leached A horizon resulting in low basic nutrient levels. In the 7th Approximation these podzolic soils are known as **Alfisols** and **Spodosols.**

In areas where drainage is poor, bog soils or gleys develop from the gleization process. Many of the poorly drained soils of Wisconsin and Canada's glacial region exhibit gleization. These soils are referred to as Histosols in the 7th Approximation soil classification system and Hydromorphic soils in the Marbut 1938 system.

SOILS OF THE WARM MIDDLE-LATITUDE CLIMATES

Soils in the mid-latitudes vary considerably because of a wide range of climates. We can divide the middle latitudes into broad environmental zones stretching from east to west across the North American and Eurasian landmass. Vegetation cover changes from warm temperate forest soils bordering the east and west coasts to temperate grassland, steppe, and deserts in the interior of the continents. In this soil pattern both temperature and precipitation variation contribute great variation in soil types. Again, it is climatic variation that controls soil development and type (fig. 11.15).

Figure 11.15

Soils of the warm middle-latitude climates can be subdivided from east to west into the warm temperate forest soils of the east coast and the temperate grassland soils, the steppe and desert soils, and the Mediterranean soils of the west coast. Mollisols such as this chernozem develop under grasslands of the prairie. The rod shows depth in feet.

Reproduced from Marbut Memorial slide set, 1968, SSSA, by permission of Soil Sciences of America, Inc.

Warm Temperate Forest Soils

The soils of the temperate forest belt are formed by podzolization processes, but because these soils are more active biologically, humus development is less than in soils of the cooler forest of the middle latitudes. It is not unusual to find characteristics of laterization occurring in the tropics. Temperate forest soils can be found in the southeastern parts of the United States and China, eastern Australia, southeastern Brazil, and northeastern Argentina.

Hot, humid summers with heavy convectional showers and mild, rainy winters result in extensive chemical weathering. Thus, minerals are easily weathered and leached where a vegetation cover has been disturbed. Surface runoff is often heavy during showers, causing severe erosion and flooding. Because of chemical weathering, the soils are generally red or yellow. The red soils, colored

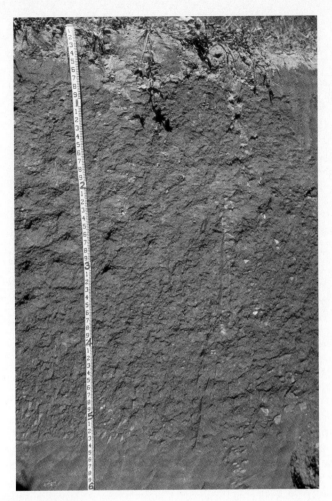

Figure 11.16

The warm temperate forest soils (ultisols) such as this red and yellow podzol are the product of high rainfall and warm temperatures. These soil profiles have many of the same chemical and physical properties of the tropical soils developed by the laterization process.

Reproduced from the Marbut Memorial slide set, 1968, SSSA, by permission of Soil Sciences of America, Inc.

by oxidation of iron minerals, tend to be in drier areas while yellow conditions are more common in more humid zones where iron is removed. Figure 11.16 shows a yellow podzolic soil near Sydney, Australia. According to the 7th Approximation, these red-yellow podzolic soils belong to the **Ultisol** order, suborder **Udults.**

Temperate Grassland Soils—Prairie

Where precipitation falls below 75 centimeters (30 inches) annually, the rich black-earth belt stretches across the North American prairie and tall-grass Russian steppes. These soils can also be found in Argentina and Uruguay, resulting from a blend of soil-forming factors that produce high fertility. Temperature in concert with precipitation and vegetation and animal life in the soil are responsible for the development of the world's most productive agricultural soils.

Figure 11.17

Prairie soils develop where precipitation is less than 75 centimeters (30 inches) under semiarid conditions of grassland vegetation. Prairie soils represent some of the world's most productive agricultural land.

Photo by Juanita Creekmore.

The moderate precipitation promotes spring and summer annual and perennial grasses, which become dormant in the winter. This means that humus develops from the annual decay of grasses. Nutrients collect in the upper A horizon because leaching is moderate in a more neutral pH soil condition. Because leaching is minimal, bases such as calcium, magnesium, and nitrogen remain in the soil along with rich humus development. Figure 11.17 illustrates a typical profile.

The Russian and revised Marbut American system classifies these soils as chernozems and chestnut soils produced by the calcification process. Under the 7th Approximation they belong to the **Mollisols** order.

Steppe Soils—Short Grasslands

As one approaches the more arid steppe in the interior of the continents, a short grass dominates the landscape. The soils there are often referred to as chestnuts and brunizems in the older classification system and belong to the Ustolls suborder of the Mollisols order. These soils are also the product of calcification but lack the rich humus of the tall-grass prairie. They also have greater salt buildup because of lower rainfall and higher evaporation. Profiles are also shallower because weathering fails to reach lower depths (fig. 11.18).

Chestnut and brunizem soils extend from the south Ukraine eastward across the southern Russian plains along latitude 50°N. They are also widely distributed in eastern Mongolia and northern Manchuria. The North American chestnut soils form a long arc from northern Saskatchewan southeastward to the pla-

Figure 11.18

Soils of the short grasslands, or steppe, are lacking in the abundant humus and fertility of the prairie lands. Due to higher evaporation and lower rainfall, greater concentrations of salt result in the top horizons. These soils are also shallow and easily eroded if the vegetation cover is altered.

teaus of northern Mexico. These soils can also be found in the Mediterranean lands where grasses form the plant cover. The soils of the steppe are easily disturbed if cultivation is practiced without sufficient supplemental irrigation. The Dust Bowl of the 1930s on the Great Plains of North America is still a reminder of poor agricultural practices. Large tracts of virgin steppe were plowed under and converted to agricultural crops. All went well until drought produced crop failures, which were followed by high winds. The soil was literally blown off the land eastward, with fallout reaching as far as Washington, D.C.

The Soviet Union experienced the same devastation 30 years later, in the 1960s, when vast tracts of virgin Asian steppe land were severely eroded by spring winds.

Desert Soils

Soils of the desert cover over 17 percent of the Earth's surface, according to Russian pedologists. If we look at the percentage covering each continent, Australia takes first place and Africa is second followed by Eurasia, North America, and South America. Africa has the greatest total desert acreage. Most of the major desert

Figure 11.19

Soils of the desert (aridisols) are very thin and low in organic content, averaging less than 2 percent. Soil development is very slow because of the low amounts of moisture and biological activity. The ruler is marked in inches.

Reproduced from the Marbut Memorial slide set, 1968, SSSA, by permission of Soil Sciences of America, Inc.

belts are located along the 30th parallels of latitude and on the western sides of continents. Both plants and animals are well adapted to the climatic extremes of high temperature and low irregular precipitation. Long periods of drought are common, broken by brief torrential thundershowers. Temperatures, while being high, can fall below freezing and the daily range may exceed 40°C. Since moisture is in short supply, weathering is primarily mechanical, producing a coarse regolith by processes described in chapter 17.

Wind is a very important environmental factor. It redistributes weathered material in a sorting process, with sand, silt, and clay-sized particles settling out in that order downwind.

Because the land is covered by sparse vegetation, wind erosion is a continual problem if the surface is disturbed by human or animal activity. Desert pavement, described in chapter 17, forms a protective seal on the surface as long as it remains intact, preventing wind from deflating and eroding the surface.

If we can examine the profile development of a desert soil, in most cases, it is very thin or almost absent (fig. 11.19). The raw material desert plants are growing in is primarily weathered bedrock rather than true soil. Organic content averages from 0.5 to 2 percent by weight. Desert soils develop under the process of calcification but because of low precipitation, the process works very slowly. Only in an alluvial oasis are desert soils deep and productive for agricultural activity.

Halomorphic and Calcimorphic Soils of the Deserts

Salinization processes dominate the poorly drained desert areas and salt-rich geological substratum, producing **halomorphic** soils. These soils are rich in salts which cause toxic effects to plants.

Salt buildup can occur if the irrigation water used in desert agriculture is loaded with saline minerals. Whatever the source of salt, halomorphic soils are the result. The soils are low in organic content, yet rich in the cations sodium, calcium, and potassium and the anion chlorine.

Commonly a white salty crust, resulting from strong capillary action and evaporation, coats the soil. If the geologic substratum is principally limestone, the salinization process produces calcimorphic soils with high calcium content. These soils can be quite fertile if other nutrients are introduced or are currently present in the profile.

Figure 11.20

Tropical soils of the low latitudes are deceptively low in productivity. Luxuriant vegetation gives a false impression of fertility. Because of high rainfall and warm temperatures, rapid weathering and leaching coupled with biological activity cause these soils to lose their

nutrients to runoff and groundwater. This is especially true if the ecosystem is altered through vegetation removal.

Photo by Dr. Charles Hogue, Curator of Entomology, Los Angeles County Museum of Natural History.

The 7th Approximation classifies these desert soils as **Entisols** and **Aridisols** because they are weakly developed, lacking horizon development, while the Marbut system classifies them as sierozems and desert soils.

SOILS OF THE HUMID LOW LATITUDES

Lateritic soils form where lush luxuriant growth of the humid tropics produces a vegetation paradise. The ecosystem of the humid tropics is a testimony to the stability of a fragile plant community. No other region has comparable plant and animal diversity and density.

At first glance one is likely to rate the tropics as some of the planet's most productive real estate. In terms of biomass pro-

duction it truly is in first place. Yet the soils tend to be low in nutrients and unproductive when cleared of native vegetation. Soils are easily leached of cations, and humus development is restricted by rapid breakdown of organic matter. High temperatures and abundant moisture promote rapid chemical weathering, organic decomposition, and bacterial and insect decomposition. These two environmental factors are the keys to conversion of both inorganic minerals and organic matter into soluble nutrient forms.

The native plants and animals of the forest and grasslands are able to thrive in these environments because they rapidly absorb the nutrients being released before leaching removes them. If the soil and vegetation are left undisturbed, nutrients are recycled and fertility remains. However, this equilibrium is fragile. When middle-latitude agricultural practices are introduced, success seldom follows because the cycle is interrupted (fig. 11.20). Plants and animals are nutrient banks above ground. Vegetation

A Soil Myth in the Amazon

Fears about **laterite** emerged in the 1960s when the first programs to clear large tracts of the Amazon discovered soil that turned rapidly into the rock-hard mineral upon exposure. News stories picked up the theme that a denuded Amazon would become a stony desert.

However, Herbert Otto Roger Schubart, director of Brazil's National Institute for Amazon Research, said that at most only 4 percent of the basin has the mixture of clays and water content required to form laterite. It just happened that deforestation started in a laterite area.

This does not mean that clearing the rain forest uncovers good soil, he added.

Without nutrients recycling through an intact forest, and impoverished by the leaching effect of thousands of years of heavy rain, most Amazon soils quickly lose the ability to support fast-growing vegetation and are subject to severe erosion (fig. 1).

Source: Charles Petit, San Francisco Chronicle, September 28, 1989.

Figure 1

The Ucaylali River is a light coffee-with-cream color, resulting from heavy runoff of exposed tropical soils in this agricultural area of the Shipibo Tribe near Pucallpa, Peru.

---

removal takes away the community's primary source of nutrition because each year a part of the biomass dies, decays, and returns to the forest in the form of nutrients.

Where this rapid nutrient recycling process has been going on for thousands of years on the same site, the forest is luxuriant, but as soon as it is deforested, nutrients are quickly leached and agriculture quickly declines.

Profile development gives a strong clue to the environmental characteristics. A profile sample will reveal a lack of humus and a permeable, heavily leached A horizon. Under warm, humid conditions all but oxides of iron and aluminum are removed by weathering processes. In the B horizon are high concentrations of nearly pure iron and aluminum oxides with silica (SiO_2), while clay and humus colloids are missing. A hardpan layer may form at this level; however, horizon differentiation is difficult to distinguish. Profiles may exceed 6 meters (20 feet) in thickness where drainage is moderate and bog conditions are absent.

The **laterization** process dominates the humid tropics, producing a variety of tropical soils (see box 11.1). These include true laterites, red-brown laterites, and yellow-brown laterites. In the humid subtropical zones, the laterites gradually grade into red-yellow podzolic soils typically found in the southeastern United States. According to the 7th Approximation, the red-yellow podzols and red-brown laterites correlate with the **Ultisol** order. **Oxisols** of the 7th Approximation relate to true laterites formed in the wet-dry savannas and tropical rainforest climates.

SUMMARY

Global soil patterns are the result of five soil-forming factors: **climate, relief, parent material, organisms,** and **time.** Each of these five factors blend in various proportions to produce five soil-forming processes closely associated with the Earth's environmental belts.

The **laterization** process is characterized by high rates of weathering, rapid destruction of organic material, decomposition of clay minerals, and removal of nutrients by leaching. The role of plants and animals in the nutrient cycle is most evident in this process, especially where vegetation cover has suddenly been removed.

Calcification occurs in regions of seasonally high temperatures, moderate to low precipitation, and high evapotranspiration. Leaching is minimal while capillary activity is strong. Soils tend to be shallow and basic. Calcification produces some of the richest soils beneath the grasslands of the middle latitudes.

Podzolization is characterized by moderate to high acidity, considerable leaching, and moderate to low fertility. It occurs wherever cool, moist conditions prevail most of the year, producing a forest vegetation cover. Where podzolization is extreme the primary nutrients are leached, leaving behind sterile quartz minerals in the A horizon. A classic case study of podzolization can be observed along the coast of our *study area,* where wave action and changes in sea level have created a series of five elevated marine terraces, each in an evolving stage of podzolization.

Salinization is significant in areas characterized by high temperatures and poor drainage, producing temporary lakes, marshes, and high water table conditions during periods of excessive unpredictable precipitation. The soils are excessively salty. Only salt-tolerant plants known as halophytes can thrive or even survive in most cases. Where poor irrigation practices occur in arid regions, salinization becomes the soil-forming process replacing calcification.

Gleization occurs at the wet end of the rainfall spectrum in poorly drained regions of the cool upper middle latitudes once glaciated. The soils are rich in organic matter, acidic, and very soggy, often supporting a peat bog.

Soil classification systems are still being modified. Currently two global systems dominate the field. Present-day soil science, or pedology, began with V. V. Dokuchaiev of Russia and E. W. Hilgard of the United States. The 1938 U.S. Soil Classification System incorporates the ideas of Dokuchaiev and C. F. Marbut. This system stresses formative factors of climate and vegetation. In 1960 a new system known as the Comprehensive Soil Classification System or 7th Approximation (see appendix III) was developed by the U.S. Soil Conservation Service. It stresses the current physical and chemical properties of the soil.

World soil patterns closely resemble climate and vegetation regions with some exceptions. Only in regions of alluvial deposits and recent geologic activity, where fresh unweathered parent material are exposed, do soil patterns lack conformity to global patterns of climate and vegetation. Therefore, we can find zonal soil characteristics in each latitudinal environment being developed by one of the major soil-forming processes.

The high-latitude soils of the tundra vary considerably with slope and drainage. Gleization processes dominate the poorly drained areas while podzolization is more evident on better drained rolling terrain. Because these soils remain frozen most of the year, there is little humus and biologic activity. They tend to be rocky, poorly weathered, and low in nutrients.

Soils of the middle latitudes vary considerably because of extensive variations of climate and vegetation. Soils formed by podzolization dominate the higher middle latitudes and upper elevations where coniferous and deciduous forests thrive. The dominant characteristic is a highly leached A horizon resulting in low nutrient levels. Where leaching is moderate, a rich humus layer develops and fertility is good. Areas of poor drainage produce bogs from the gleization process and only meadows rather than forests exist.

Soil types of the warm middle latitudes range from podzols of the temperate forest to desert soils, low in humus and rich in bases and salts. Warm temperate forest soils tend to be heavily leached with decreasing humus as one approaches the humid tropics. The temperate grassland soils produced by calcification are rich in humus with only moderate leaching. These soils represent some of the world's most productive regions. As one moves into the more arid short-grass prairie and desert, soils tend to have shallow profile, low in organic content, and less weathered parent material.

Salinization processes produce *halomorphic* and *calcimorphic* soils in arid regions of poor drainage or poor irrigation practices.

In soils of the humid low latitudes, vegetation is lush and yet in most cases agriculture, using middle-latitude practices, has not been successful. **Laterization** processes rapidly leach the soil when the original vegetation cover is removed. The plants and animals of the forest play a vital role in the nutrient cycle. Clearing the forest breaks the cycle.

ILLUSTRATED STUDY QUESTIONS

1. Explain the laterization soil formation process (fig. 11.1, p. 216).
2. Describe the role of plants and animals in the nutrient cycle (fig. 11.2, p. 217).
3. Viewing both virgin tropical forest and cleared land in Peru, explain the effect of clearing on the nutrient cycle (fig. 11.20 and box 11.1, fig. 1, pp. 230, 231).
4. Describe the role of climate on the calcification soil formation process (figs. 11.4 and 11.5, p. 218).
5. Contrast vegetation cover for podzolization and laterization (figs. 11.6, 11.7, 11.8, 11.9, 11.20, and box 11.1, pp. 219, 220, 230, 231).
6. Explain the chief cause of salinization (figs. 11.10 and 11.11, p. 222).
7. Draw a soil profile of podzolization. Label each horizon (figs. 11.6 and 11.7, p. 219).
8. Viewing figure 11.19, p. 229, explain why desert soils are very thin and often low in humus material.
9. Are deep soils necessarily fertile (fig. 11.20, p. 230)?

Environmental Factors of Ecosystems

12

Objectives

After completing this chapter you will be able to:

1. Explain the four major environmental factors of **ecosystems,** which include the **climatic, edaphic, biotic,** and **geomorphic.**
2. Explain the interrelationships between climate, soils, landforms, and biological factors that shape the biosphere.
3. Define **plant succession** and the role environmental factors play in the succession process.
4. Explain the role of energy in moving materials through the biosphere.
5. Define and explain the unique nature of the **biosphere.**
6. Explain how a tree grows and records its environmental history in its structure and internal anatomy.
7. Explain the roles of fire and flood in **plant succession** and **community** development.

Some of the planets' oldest members are millennium survivors. Ancient Bristlecone Pine Forest, White Mountains of Eastern California.
© David Muench Photography

All living systems on the Earth ultimately are powered by the Sun and are strongly affected by variations of incoming solar energy. It is continually transformed into other forms, such as heat, light, fossil fuels, plant starches, and proteins. We have made advances in understanding the environmental factors that influence the great diversity of life scattered in a multitude of environmentally different places. We call this living zone of the Earth the **biosphere.**

Our purpose in this chapter is to understand the major environmental factors of the biosphere that are interwoven to produce a variety of unique living communities on the Earth. John Muir, an American naturalist, once said, "When we try to pick out anything by itself, we find it hitched to everything else in the universe."

These environmental factors occur in different proportions, producing a variety of communities or habitats. The processes we observe can be inferred to exist and operate any place in which life exists. The combinations of living systems differ and the environmental factors such as soil, water, temperature, and insolation vary over the surface of the Earth producing unique communities, which are the topic of the next chapter.

THE BIOSPHERE

What is so special about the biosphere? First, it is found nowhere else in the solar system. Without this special zone, the Earth would be as sterile and barren as the Moon. Second, it is a unique region where water can exist in large quantities as solid ice, liquid water, and gaseous vapor. Third, it is the recipient, user, and converter of an outside source of energy, ultimately the Sun. Fourth, the biosphere represents a zone of interaction between the lithosphere, atmosphere, and hydrosphere.

Both plant and animal organisms actively capture and store energy through very elaborate systems. The environmental adaptability of organisms is remarkable. The only water in the neighborhood of some desert or Antarctic organisms may be the water contained in their tissues. Yet, at some point, the organism will need some outside source of water such as dew on a plant or water metabolized in digestion. As we look at the environmental factors that influence the plant community, the forest, or the entire biosphere, all have three common denominators: *food material, energy,* and *water consumption.* Plants and animals cannot exist without these basic three (fig. 12.1).

ENVIRONMENTAL FACTORS SHAPING PLANT COMMUNITIES

All environmental factors can be divided into four broad categories. **Climatic** factors primarily include *light, temperature* and *heat, atmospheric gases, wind, evaporation, condensation,* and *humidity.* **Edaphic** or soil factors deal with *physical* and *chemical* properties of the soil. **Geomorphic** factors include *elevation and terrain shape,* or *relief.* **Biotic** factors consist of all biological organisms, including *people,* exerting an influence on the community of plants and animals (fig. 12.2).

Figure 12.1

All organisms have three fundamental requirements for life: food materials, energy, and water. These three ingredients are utilized in the biochemical factory of this coconut seedling on Cocos Island, located near the equator southwest of Costa Rica.

Photo by Dr. Charles Hogue, Curator of Entomology, Los Angeles County Museum of Natural History.

The arrangement of species of plants and animals in today's world is a very recent event and is undergoing continual change. Each environment today is quite different from a hundred, a thousand, or a million years ago. Geologic and climatic change, plus plant and animal migration, fire, competition, and human influences all cause continual change. Until we learned to harness matter and energy on a large scale, human-induced change was very slow, and most natural systems were relatively stable. Human activities have become so extensive and far-reaching that we cannot at this point even predict what kind of new changes will occur in the next few hundred years. Let us examine these four major environmental influences shaping the living system of plants and wildlife.

The Climatic Factor

Climate, the average, long-term weather conditions of a place, is the foremost influence in determining the physical environment of a living system or community. We often fail to appreciate the strong influence it exerts on the living systems of the Earth. Every plant and animal requires *food material, water,* and *energy.* Where does this supply originate? The environment is both a resource and manufacturer of the needs of the biosphere. As you read this page,

Figure 12.2

Environmental factors shaping plant communities fall into four broad categories: climatic, edaphic, geomorphic, and biotic. The plant is the organizer of these factors, blending them into various processes. When any single factor is altered, the plant becomes endangered. A landslide is undermining the root system of this Bishop pine.

perhaps hunger and thirst are developing, or even fatigue. This may be a signal to refuel on energy and materials. Because the modern human community is extremely complex with global interaction, the resources you choose may originate in Brazil as a coffee bush or sugar from Hawaiian cane fields. Energy stored in these plants traveled even farther from its ultimate source, the Sun, in the form of light.

Light: Energy for Life

Light is generated throughout the universe, but the particular radiant energy that reaches us from the Sun touches life on the Earth. The biosphere postpones the effects of increasing entropy of this energy.

Think of a living organism as a series of chemical reactions that operate together to utilize energy and materials for activity and reproduction. The quality of energy is just as important as the quantity. The solar energy received at the ground is highly variable in quantity and quality from place to place and season to season. Changes in clarity of the atmosphere and variations in the angle of the sunlight cause the variability. Light intensity is altered by a number of environmental factors.

In the *study area* of chapter 1, summer fogs greatly reduce the amount of light reaching the treetops. Under a blanket of fog, the coastal redwood is one of nature's most efficient users of light, requiring far less than most western conifers. It grows vigorously where the sunlight is less than one percent of the full intensity of a clear summer day.

Light entering the forest is used in several ways. First, it warms the plant and increases the motion of the molecules in the system. Second, it is both reflected and reradiated as long-wave infrared radiation in the wavelengths unique to the plant. Third, it is utilized to accelerate chemical reactions. And, fourth, light is utilized to manufacture chemical energy or food in the photosynthesis process.

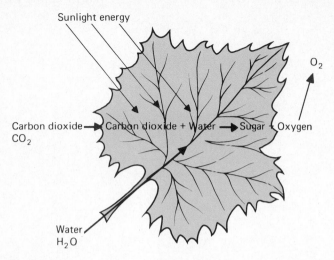

Figure 12.3

Photosynthesis is a combining process. Light energy is the key to the conversion of water and carbon dioxide to chemical energy. When the proper mix of water, carbon dioxide, and energy from the sun react together, the result is sugar and oxygen.

Photosynthesis: Energy Transformation

The word literally implies a combining process, synthesis by light (photo). Plants grow by a unique process which calls for light energy, carbon dioxide, combined with water, to produce sugars and starches stored for use in the plant system. Practically all organic matter in the present biosphere originates in this **photosynthesis** process. A by-product of this reaction is free oxygen, which is also essential for the maintenance of most life (fig. 12.3).

When animals eat other animals or plants, they are able to release the energy stored in the plant. This energy is expressed as work, metabolic processes, and reproduction.

The amount of light actually reaching the forest varies not only in quantity but in quality. Increased quantities of water vapor and carbon dioxide below the treetops filters out energy in the infrared region of the electromagnetic spectrum. Atmospheric dust reduces and scatters shorter wavelength violet light more than the longer wavelength red. Of the total potential solar energy or insolation reaching the ground, only one-quarter is capable of stimulating photosynthesis, and only a small fraction of that is actually used by plants in photosynthesis. Most of the energy reaching the surface is either reflected or absorbed by the forest. The portion reflected back to space makes no contribution to the energy of the forest. The absorbed portion not used in photosynthesis is exchanged through evaporation of water, conduction in the soil and air, and reradiated back into the atmosphere and space.

During the day, incoming energy is greater than losses, and the air heats and stirs by convection, mixing the heat received. At night, the surface cools faster than the air above it, and the atmosphere stabilizes. If the sky is overcast, the Earth's infrared radiation is absorbed by the water vapor and droplets of clouds in the atmosphere and reemitted, producing a greenhouse effect. Carbon dioxide also produces this same effect. Thus the window to solar energy is open primarily in the visual portions of the spectrum.

Figure 12.4

Photosynthesis, hence life itself, depends on the quality of the atmospheric window. Air pollution over Phoenix, Arizona, has become a serious problem, where inversions are the norm in this high-pressure desert climate.

Photosynthesis, and life itself, hangs in the balance depending on the quality of the atmospheric window to space. Pollution of the atmosphere dirties the window. Moderate shifts in the amounts of ultraviolet or infrared may cause great changes in the environment of the biosphere. At this point it is unclear what we are doing to the window, but it is clear that it is fogging up a bit (fig. 12.4).

Temperature: A Marker of Heat Energy

Temperature is a conversation piece: "My, it's cold today." The temperature affects our activities and the efficiency of our work. Temperature indicates the rate of growth and general speed of

have a short, one-season life span and generate seeds to sprout in the next growing season. The arctic lands have few annuals because the growing season is so short. Plants merely go dormant and then spring to life when temperatures rise.

The forest vegetation varies in form and size because of a number of environmental factors, but temperature is a key underlying control. Each plant has its ideal range in temperature. The coastal redwoods are an excellent case in point. They are limited to a narrow coastal strip stretching 80 kilometers (500 miles) from the southwestern corner of Oregon to the central coast of California. Rarely are they found growing naturally farther than 50 kilometers (30 miles) from the coast or above 600 meters (2,000 feet). Higher than that, the air is very dry and warm throughout the summer months and too cold during the winter. Farther inland, seasonal high temperatures and evaporation are often too great. The northern limit is marked by frost and competition from other cold-resistant plants. To the south, the higher evaporation and lower precipitation reduce the water supply. Summer fogs are less frequent, and rainfall is not only lower but more variable.

Temperature differences help explain many plant patterns all over the world. By journeying up the mountain slope of the Andes in eastern Peru, one can easily see the effects of temperature (fig. 12.5). It is quite similar to traveling from the equator to the Arctic Circle. Starting at sea level, we pass lush tropical forests where average monthly temperature never falls below 18° C (64.4° F). We then move into mid-latitude broadleafs and conifers at 1,800 meters (6,000 feet). Then at about 2,700 meters (9,000 feet), vegetation grades into stands of pure conifers. Finally at 4,600 meters (15,000 feet) the environment changes into a treeless tundra, a transition between forests and the arctic snowfields at the mountain summit.

The vegetation changes occurring with elevation are principally caused by changes in temperature. This same vegetation change with elevation occurs around the world. However, timberline is reached at 4,000 to 4,600 meters (13,000 to 15,000 feet) in the tropics, 3,000 to 3,300 meters (10,000 to 11,000 feet) in the Sierra Nevada of California, and 1,500 to 2,100 meters (5,000 to 7,000 feet) in the Cascades and the European Alps. Higher latitudes tend to bring the vegetation boundaries down to lower levels. We could begin at the equator and travel toward the North or South Pole and experience the same influences of temperature on vegetation. Plants of the low latitudes and frost-free zones are known as **megatherms.** At the other extreme, **microtherms** thrive in frosty lands of the high latitudes or high elevations. Between these extremes are the **mesotherms,** lovers of moderation in temperature.

Moisture

Plants are always seeking a drink: Some know how to conserve while others are big spenders.

By backing away from the Earth and looking at it from space, one is impressed by the carpet of grasslands and forests arranged in shades of green. Where water is sparse, the green disappears. Starting at the equator and traveling north or south, rainfall grades into desert conditions.

metabolic processes in all organisms. Again we must look to the Sun, for most of the radiation reaching the surface is either reflected or converted to heat energy, thus maintaining the energy level of our environment. Each environment has a different set of energy requirements, but temperature maintenance is near the top of the list. Life lives in a very narrow temperature range and could not exist unless insolation is consistent. However, within this narrow temperature spectrum, plant and animal environments vary considerably. World patterns of temperature produce a variety of environments for plant and animal species. If a plant is transplanted out of its environment, it may not adapt. For example, avocadoes might grow well in Maine all summer, but as soon as winter comes, they would die quickly in the low winter temperatures.

Plants have adapted in a variety of ways to changes in seasonal and daily temperature. Some lose their leaves and become dormant. Others just cease to grow until it warms up. Annuals

Figure 12.5

Energy in the form of heat is a key control determining the species of a plant community. The impact of heat is evident on a mountain slope in the tropics of Peru. For every 300 meters (1,000 feet) of change in elevation, temperatures change on the average of 2° C (3.5° F). Vegetation reflects this rapid variation. In the tropics, one can experience climates of all latitudes on a mountain slope that extends to heights above 4,600 meters (15,000 feet). The moderating influence of Lake Titicaca permits the cultivation of corn and wheat on the lower terraces. Their normal limit is 3,500 meters (11,500 feet) at this latitude.

Photo by Sheri Wallen.

Abundant water is the one substance making the planet Earth unique in the solar system. Without water, no chemical reactions in the cell or organisms can occur. A seed will lie dormant for years if it is deprived of water. Where water is absent, life is also absent. Contrast the dry deserts with an occasional oasis. The oasis is a beehive of activity of both plants and animals, made possible by water. Although light, temperature, and soils are important, water is the single most important influence on the distribution of plants.

Looking at the Earth from space, we note that green belts are confined to the tropics and higher middle latitudes (fig. 12.6). When we compare global precipitation with vegetation, the pattern correlation becomes clear. The world's largest plant forms, forests, are in the regions of most abundant available water.

Patterns of precipitation tell only part of the story of moisture availability. The coastal redwoods would not grow to great size if they were dependent solely on precipitation. There is a saying, "Where the fog flows, the redwood grows." Mark Twain once said that the coldest winter he ever spent was a summer in San Francisco; this is because summers along the northern California coast are foggy, moist, and cool. Rain gauges remain empty all summer, but moisture from fog drip can equal the precipitation in the forest. **Fog drip** is the key to the primary distribution of the coast redwoods; it gives them a summer drink.

However, if we only examine water sources or supplies, only half of the story has been told. Remember that the Sun's rays heat the Earth and take great quantities of moisture from the ground and plant surfaces by **evaporation** and transpiration. Solar energy is the lifting force. It takes about 540 calories for every gram of water evaporated and returned to a vapor form. **Transpiration** of water from leaf surfaces into the atmosphere is simply another evaporation process. The combined evaporation by soils

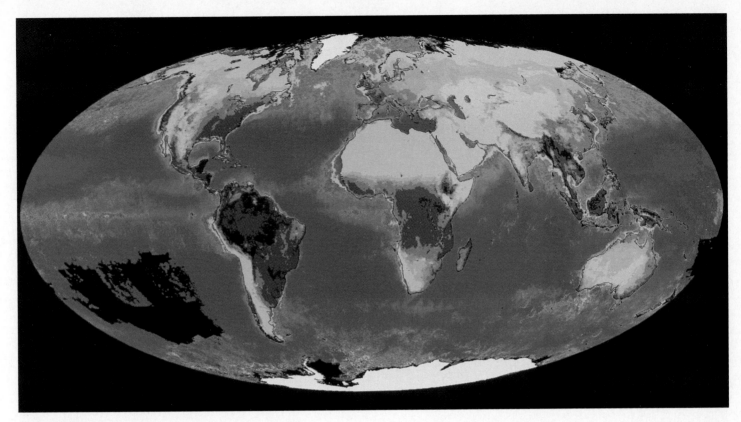

Figure 12.6

This image, produced by combining data from two different satellite sensors, shows the productive potential of the Earth's vegetative biomass. The ocean image is a composite of data collected from November 1987 through June 1980 from the Coastal Zone Color Scanner (CZCS) on NASA's Nimbus-7 satellite. Note the clear delineation of the equator through increased plant abundance. The land-vegetation image is a composite of three years of data from the Advanced Very High Resolution Radiometer (AVHRR) on the NOAA-7 satellite, which measured land-surface radiation in the visible and near-infrared bands to estimate chlorophyll and leaf mass. The dark areas (rainforests) show the highest potential for chlorophyll production. Deserts, high mountains, and arctic regions reflect barren conditions. This view from space reveals that the green belts of the Earth are closely related to zones of abundant moisture. In the false-color infrared, all plants are reflecting red: "If it's not red it's dead."

Photo by NASA.

and plant surfaces is referred to as **evapotranspiration.** Plants and soils lose not only moisture but also energy by these evaporation processes.

The role of evaporation is very significant in determining the water available for plant environments. First, consider the climatic factors influencing evaporation rates. The highest evaporation rates are found in regions where temperatures are high and moisture content or humidity in the atmosphere is low. Warm air has a greater moisture-holding capacity than cool air, therefore evaporation is potentially highest in the lower latitudes where insolation is high. The equatorial tropics experience high evaporation but it is offset by high precipitation, while tropical deserts are real losers of moisture because evaporation far exceeds precipitation.

Plants can be classified according to their water needs. Plants that have adapted to high water loss potential and limited water availability are known as **xerophytes** (Greek *xeros,* dry, and *phyton,* plant). Most desert plants and some arctic tundra plants are considered xerophytes (fig. 12.7). These plants are real misers when it comes to conservation of water.

Figure 12.7

Xerophytes are plants well adapted to water loss because of their miserly water conservation design. A prickly pear cactus stores moisture in its tissue and taps this biological tank during periods of low soil moisture.

Figure 12.8

Hygrophytes are big spenders of water. They can only thrive when abundant water supplies exist. The roots of these cypresses in Florida must have their feet in water.

Plants with very high water availability and consumption and loss through transpiration are commonly referred to as **hygrophytes** (Greek *hygros,* wet). These are the big spenders of water (fig. 12.8).

Plants found in areas of moderate moisture sources are known as **mesophytes** (Greek *mesos,* middle). All vegetation has upper and lower limits of moisture demand. If these limits are exceeded the plant shuts down, by dying or going dormant. Most plants of the temperate middle latitudes fall into this category (fig. 12.9).

Wind

Solar energy heats the Earth unevenly, stirring the air and waters constantly. **Winds** mix the air, redistributing moisture, energy, and gases. A region may have ample precipitation but the wind factor may be very high, giving the area a more arid climate by accelerating evaporation. Winds may also deform or damage a plant.

Winds perform beneficial pruning services to trees by removing dead and diseased wood. The wind of a forest transports seeds and pollen as well as moisture in the vapor and cloud form. Along the coast on most summer afternoons the sea breeze transports cooling moist air onshore and into the interior (fig. 12.10).

The effect of wind is especially noticeable on an exposed mountain slope as timberline is reached. High winds twist and deform the trunks of trees. Only in very protected areas do the trees grow large. The windy places on the higher slopes are often covered by short dwarfed trees and shrubs. Because the ground surface acts as a frictional brake, wind speeds decrease with lower elevation.

Wind also cools a surface when evaporation accelerates moisture loss. Stirring of the atmosphere can also warm the ground by mixing cool ground air layers with warmer air aloft. Wind machines do this artificially to raise orchard temperatures above the frost danger levels.

Figure 12.9

Mesophytes are plants located where moisture availability is moderate. A homestead cabin near Celina, Tennessee, nestles in second-growth deciduous hickory, oaks, hemlocks, and evergreen white pine.

Photo by Dr. H. H. Wallen, Jr.

Figure 12.10

As the atmosphere stirs, moisture and energy are redistributed. These date palms are receiving a cool Mediterranean sea breeze through a wind gap in the Jordan Valley of Israel.

The Geomorphic Factor— Shape and Relief

As we have seen in the case of wind, the shape of the land exerts a very strong control on the distribution of plants. **Elevation, slope orientation** or **aspect,** and **gradient** influence the fundamental climatic elements and controls of insolation, winds, precipitation, and air-mass movement. Major regions of the Earth are mountainous or hilly. These rugged zones represent complex plant patterns, because the elements of climate are highly variable over the hilly region. Each ridge and valley is a unique microclimate. Let us examine some of these geomorphic factors in the mountains of North America.

Elevation

The Sierra Nevada has many examples of geomorphic influences on vegetation. When you change elevations, temperature and precipitation patterns are also altered. In general, temperature decreases 1.8° C (3.3° F) for each rise in elevation of 300 meters (1,000 feet) up a mountain slope. However, if the air is in motion and rises up a mountain slope, it can cool as much as 3° C (5.5° F) per 300 meters (fig. 12.11). On a typical summer day in the Sierra Nevada, the temperature may reach 38° C (100° F) at Mariposa, elevation 600 meters (2,000 feet). If we continue up the western slope to Tioga Pass in Yosemite National Park, the temperature

may be 18° C (65° F) the same day as a result of air cooling as it rises up the western slope. Summer average maximums at 3,000 meters (10,000 feet) range from 22.8° C (73° F) to 29.4° C (85° F) while at Sonora, in the foothills, average maximums run from 31.1° C (88° F) to 38.3° C (101° F).

Elevation also influences patterns of precipitation. Both rain and snow increase with elevation up to subarctic levels, then taper off. In the lower elevations, precipitation averages 15 to 50 centimeters (6–20 inches), while at 1,500 meters (5,000 feet) in the central Sierra precipitation averages 150 to 200 centimeters (60–80 inches); however, at 3,000 to 3,300 meters (10,000–11,000 feet) precipitation tends to be less, averaging 75 to 125 centimeters (30–50 inches). This is the orographic influence discussed in chapter 6.

Precipitation not only varies with elevation, but also with exposure to the prevailing winds. Windward slopes receive greater amounts of precipitation due to the orographic influence. If we look at the 1,675-meter (5,500 feet) elevation in the central Sierra, we find that the windward side receives about 190 centimeters (75 inches) while the leeward side has only 50 centimeters (20 inches). The mountain barrier forces the air to rise, cool, and condense, giving abundant precipitation as rain and snow on the west-facing slopes. As the air rises above 2,700 meters (9,000 feet), the major amounts of moisture in the air have already condensed, explaining the lower precipitation on the high slopes. Once the air reaches the

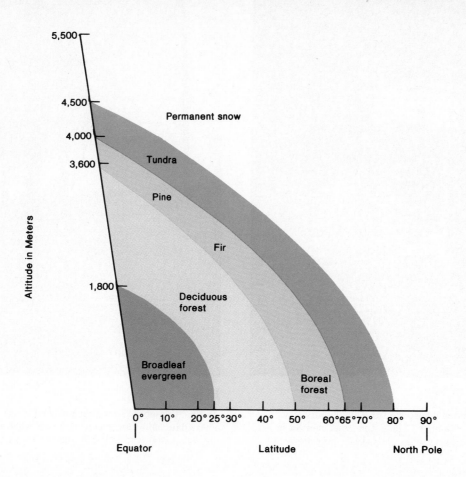

Figure 12.11

Elevation changes influence temperature at the rate of 0.6° C (1° F) per 100 meters (300 feet). However, if the air is rising, expansion results in additional cooling, causing temperatures to drop at the rate of 3° C (5.5° F) per 300 meters (1,000 feet). Thus, it is possible to experience the same climatic and vegetation changes by traveling up a slope or by moving from the lower latitudes of the Arctic Circle.

summit and begins to descend down the leeward slopes, it heats at the same rate it cooled before, and precipitation rapidly declines. This leeward region is referred to as the rain shadow.

If we follow a parcel of air from the Pacific coast across the Coast Range, the Sierra Nevada, and the Basin and Range country to the east, we can see increased precipitation on the windward west-facing slopes on each mountain barrier, followed by more arid conditions on the leeward side.

The conifer forests of the western United States grow on the mountain slopes only where rainfall exceeds 50 centimeters (20 inches). If we examine plant patterns on a larger scale in the Sierra Nevada, each conifer species flourishes at a specific elevation. Digger pine is found in the foothills below 600 meters (2,000 feet); then we encounter yellow pine from 1,200 to 2,400 meters (4,000 to 8,000 feet), followed by firs, spruce, and lodgepole pine above 2,400 meters (8,000 feet) on the Sierra slopes (fig. 12.12). Entire plant communities change with elevation resulting from change in precipitation and temperature and slope orientation or aspect.

Slope Orientation

Slope orientation or the aspect of the slope refers to the direction a slope faces with reference to the Sun. On or near the equator, slopes are equally exposed to solar rays because the Sun is near the zenith at noon most of the year. The farther north one travels, the lower the angle at which solar rays reach the ground. In a hilly region a south-facing slope receives more insolation than one facing in the opposite direction or poleward. Figure 12.13 shows contrasting vegetation in the *study area* of chapter 1. Here on the west coast of North America, the north-facing slopes are wooded, often with evergreens, while the hotter south-facing slopes support grasses, chaparral, and other plants such as poison oak, chamiso, toyon, and sages. All are drought-resistant **xerophytic** plants.

Slope orientation influences disappear not only near the equator but along the foggy Pacific coast. Fog drip restores water losses due to evaporation, and cloudy conditions reduce insolation and evaporation to a minimum on all slopes. Therefore plants in the fog belt do not suffer from excessive water loss.

Figure 12.12

Each conifer species flourishes at its own favorite height and temperature. The red fir grows at the timberline.

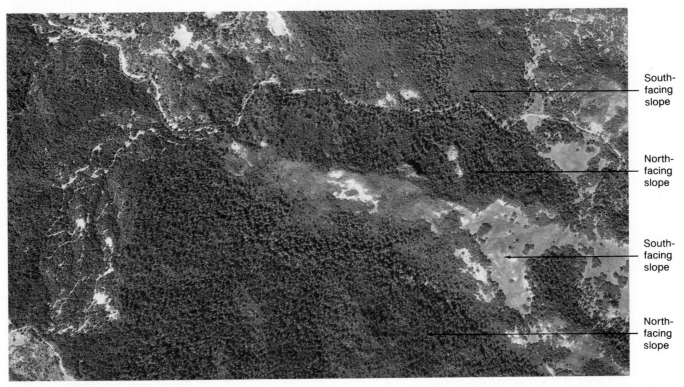

South-facing slope

North-facing slope

South-facing slope

North-facing slope

Figure 12.13

The *study area* gives a classic picture of variation in slope orientation and its influence on insolation at the surface. North-facing slopes are cool, moist, and heavily forested while south-facing slopes are warm and support drought-resistant oaks, chaparral, and grasses.

Photo by NASA.

Figure 12.14

Edaphic (soil) factors. Plants are miners of the soil. They mine water, oxygen, and minerals of inorganic and organic sources.

Edaphic (Soil) Factors

Plants are miners of the soil's water-soluble elements (fig. 12.14). The abundance or scarcity of these elements, as shown in chapter 10, is a determinant factor in the development and distribution of plant communities. The natural vegetation of any habitat is adapted to the nutrient, water, and climatic conditions.

Although no two soils are alike, they all have a number of qualities in common. They can hold large quantities of water, up to 40 percent of their weight, and they store plant nutrients. Soils are also suitable environments for a whole host of organisms ranging in size from bacteria to gophers and other burrowing animals. Recall, too, that soils tend to reflect climatic patterns more than the underlying geologic conditions because climate is the most important determinant factor in soil formation.

However, rocks that have an abundance of certain minerals tend to exert a very strong influence on plant species. In regions of limestone or serpentine, or soil altered by high concentrations of minerals such as salt from evaporation, plant patterns reflect mineral patterns. In the Coast Range near Santa Cruz, California, coastal redwoods are conspicuously missing and large ponderosa pines, typically found in high colder mountain climates, are growing near sea level. The ponderosas are growing in porous, acidic quartz-rich sandy soils. Redwoods prefer richer clay soils with greater water-holding capacity and moderate acidity (fig. 12.15).

On the global scale, forests are found growing on acidic soils that are fairly well drained. Grasses prefer alkaline soils, which can be heavy and poorly drained, while desert plants prefer soils that are generally porous, alkaline, and low in organic content.

Soils support the primary producers or plants in a community association. Each soil hosts a unique plant community.

Biologic Factors

All the plants and animals at any one location are interrelated members of a biological **community.** The location may be a pond, fallen log, estuary, sand dune, or meadow. Each member has a unique function or "job" that contributes to the total community (fig. 12.16). Plants are *primary producers.* They transform solar energy and nutrients from the soil into foods for the animals in the community. The plant feeders are *primary consumers,* while some animals and even plants prey on other animals. Thus a coyote is classified as a *secondary consumer* because it feeds on other animals. A number of organisms feed on dead plants or animals, serving as *scavengers* or *decomposers.* It has been said that the forest would literally choke in dead wood if it were not for the scavenger insects that reduce the wood to humus and soluble nutrients. Consider the effect of accumulation of dead animals if vultures and other scavengers were not actively cleaning up the decaying matter. In the cleanup process organic matter is broken down to humus. Plants recycle this humus matter by soaking up nutrients being released in the decomposition process. Therefore,

Figure 12.15

Plant patterns or species distribution are strongly influenced by mineral content. Ponderosa pines have filled a niche in the Coast Range of California where the coastal redwood is climatically better suited. The key is soil type. The Ponderosa pines are growing in very porous acidic sandy soils that hold very little soil moisture. Because redwoods can't tolerate drought or low moisture conditions, they cannot compete with the ponderosas here.

Figure 12.16

We are all members of a family or community. This is true of all plant and animal organisms. Interdependence, spatial interaction, and internal cohesion are the norm. These saguaro cactus are growing in a small niche of soil moisture and temperature range. The saguaro is limited by low water availability at lower elevations and by frost at higher elevations in southern Arizona. Survival of seedlings is aided by shade of smaller shrubs and rocks.

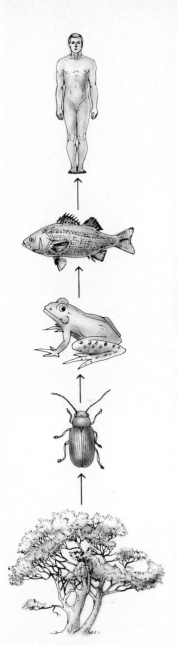

Figure 12.17

The food chain is nature's linkage of survival. If the linkage is broken, the flow of food and energy stops at that point. Plants are primary producers, mining the soil and transforming solar energy into chemical forms. The plant feeders, such as ourselves, are primary consumers. Carnivores are secondary consumers in the food chain. The garbage collectors of the planet (insects, bacteria, vultures) decompose and reduce dead plants and animals to soluble forms. Thus, energy and materials are continually being recycled.

in any biotic community the flow of energy and materials is cyclic in a **food chain** (fig. 12.17). If we look at the biological community in terms of total numbers, the members form a **pyramid of numbers.** Those at the bottom are abundant but small, and those at the top are large but few (fig. 12.18). In a North American grassland community the pyramid starts with the grasses, then the animals that feed on them such as insects, rodents, and seed-eating birds. Next come the predators such as insect-eating birds and rodent-eating owls. Hawks occupy the top of the pyramid and prey upon the birds and rodents below them.

From this brief description of a food chain we can observe that life is extremely interdependent. The environment is not segregated units of atmosphere, rocks, soil, water, and living systems, but a complex of interacting components, living and nonliving, organized into a completely integrated community.

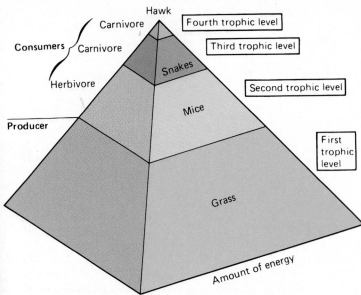

Figure 12.18

Every ecosystem has a pyramid of numbers and energy, reflecting the relationship between energy flow and loss of the organisms in the food chain. The smallest organisms are the most numerous and consume the greatest energy, while large life forms, such as secondary consumers, are lowest in population and represent the least energy.

If any of the energy and material users of the community change in numbers, the populations of other members are also influenced because the food chain is altered. Drought may reduce the browsing food for deer, causing a higher mortality rate. This in turn reduces the predators such as the mountain lion and coyote, because their food supply is also reduced. However, vultures and other scavengers may increase because of a more abundant food supply. Nature is never in perfect balance. Feast and famine occur in irregular cycles because the environment is in constant change. These changes may take centuries or days or minutes, but change is the norm. The plant community we observe now is the product of all past events. These include **climatic, geomorphic, edaphic,** and **biotic** changes.

ENERGY AND MATERIAL MOVE THROUGH THE BIOSPHERE

The entire community is involved in the flow of energy and materials through different levels of the system. Energy comes from the Sun and flows with the materials, giving plants and animals the ability to carry on activities and maintain the organization. Unless energy is put into the community for maintenance and reproduction, all systems tend toward **entropy** or disorder and eventual loss of organization. This is true of a home, a factory, or a road, and energy is no less valuable for organisms and biological communities.

PLANT COMMUNITY SUCCESSION AND CLIMAX

The majestic forests of Maine or New Hampshire, the luxuriant Peruvian rainforests of the Amazon Basin, or the xerophytic plants stabilizing a drifting sand dune in New Mexico each represent a history of successive stages of plant and animal invaders making their impact on the biological community. An important aspect of this change is that it is not just a modification of a single community, but a succession of communities in one area. Human activity is a significant force too, producing changes in the environment of plant and animal communities.

Let's take a walk up a series of wave-cut terraces along the northern California coast, described in chapter 11, where we

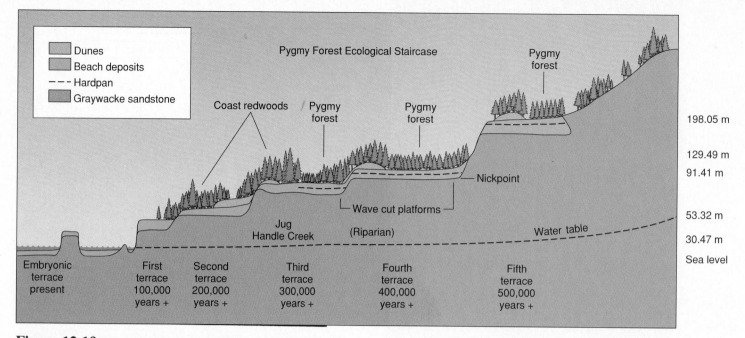

Figure 12.19

The pygmy forest represents plant succession in all stages. The highest terrace is in a climax stage or state of stability.

will observe successive stages of development of a plant community. Here trees on each terrace represent successive ages of the community. The oldest terraces are covered with stable dwarf trees known as the **pygmy forest,** resulting from poor soils, a hardpan, and excessive leaching. The forest in this condition is in a climax stage (fig. 12.19).

A series of five elevated wave-cut marine terraces is the stage for our study of soils and plant communities experiencing change or succession. On each terrace we can observe firsthand the integrated evolution of soils, landforms, and ecosystems. The cycle starts on the wave-cut bench, exposed only at low tide. Here salt spray and wave action are etching and sculpturing a sixth terrace yet to be uplifted. Just above the splash zone, graywacke and sandstone cliffs mark the first riser to the wave-cut terrace number one. Each of the five steps represents a history of changing sea level, caused by expansion and recession of glacial ice, and tectonic uplift of the land.

Sea level fell when precipitation was locked up on the land as ice during the Ice Age, and rose again when the ice melted. The terraces formed when sea level rose and waves cut into the sandstone rock, making benches several hundred meters wide. In addition to the fluctuation in sea level by about 60 meters (200 feet) or more over periods of thousands of years, this part of North America is tectonically active, rising several centimeters per century. This movement can be related to faulting along the San Andreas fault, which runs parallel to the coast just offshore. It appears that this part of the coast has been rising for 500,000 years with each step separated by about 100,000 years.

As uplift occurred and the Ice Age advanced, the oceans receded, dropping a variety of clay, sand, and gravel on each step.

(If the land rises or sea level drops, step number one will become step number two, going up like an escalator.) It appears that rising sea level cut the terraces, while a receding sea left beach material up to 9 meters (30 feet) thick on each of the five terraces.

Soil evolution and **plant succession** occur as communities, represented by a variety of living systems, respond to similar environmental conditions. This change involves every member and occurs because of changes in climatic, biotic, edaphic, or topographic factors. On the first stairstep, the grasses represent the **pioneer** community growing in prairie soils. As we progress up the staircase conditions change, and the results represent intermediate communities and podzolic soils. The forest is a biological adaptation to new soil conditions and drainage. The higher terraces are called **climax communities** because they represent a stable condition until major environmental change occurs.

LIFE AND DEATH OF A TREE
Growth of a Tree

Trees represent the oldest and largest living systems on the Earth. Their lives can span many centuries. Civilizations come and go but the forest remains, grows, reproduces and changes, living longer than any one organism in its plant community.

The coastal redwoods, *sequoia sempervirens,* are not quite the oldest nor quite the largest living trees, but they have been around a long time. Some of the living trees of the coastal redwood community were here when the Roman Empire was at its peak of development. Many of the giants of today were mature trees when the first English, Spanish, and Russian explorers found this part of the Pacific coast (box 12.1).

Growth Rings of a Tree Tell the Life History

BOX 12.1

Rings give us clues about the tree's rate of growth, climatic environment, fire, insect attacks, and other historical events in the life of the forest. This tree shows an adjustment to lean.

(A,B,C) Buttress is formed.

(A) 112 rings are crowded into 8 inches (see buttress enlargement).

(B) On a vertical line through B, 100 rings occupy 36 inches indicating an accelerated growth rate to hasten formation of the buttress.

(C) Sixty rings are crowded into 4 1/2 inches. On the vertical line through B these 60 rings occupy 20 inches.

(D) Scar of fire of 1595. Subsequent new growth coming together from two sides of the wound has imprisoned some bark in a depression.

(E) Scar of fire in 1789.

(F) Scar of fire of 1806. The fires of 1789, 1806, and 1820 delayed formation of the buttress. This indicates that the tree started to lean more than 100 years before its fall.

(G) Scar of fire of 1820. This fire must have been severe. It killed the tree from G and C to N, burning off bark, sapwood, and some heartwood. Very likely, however, earlier fire contributed to the great scar and may have begun it; the evidence has been burned off.

(H) Completely healed injury made by fire in 1147 left a weakness which later caused radial crack.

(J) Healed fire wound. Preceding healing, "brown heart rot" attacked old heartwood and subsequently a crack developed along the ring.

(K) Similar to J. Fire of 1595, manifested at four points.

(L) Brown heart rot ("dry" rot) following fires of 1820 and earlier.

(M) "Ring shake" development along fire scar of 1147 and "stringy white rot" infected surrounding wood. Ring shakes and radial cracks are normal in large old redwoods and develop as result of internal stresses. Had this tree not developed ring shakes, the central rift crack would have been longer and the separation wider.

(N) New growth working its way over the surface burned in 1820; interrupted by fires in 1848, 1866, 1883, and 1895. This point, being behind the axis of the lean, was not stimulated to form a buttress.

Facts about the Fallen Redwood:

Average diameter of stump, inside bark: 12 feet

Height of tree: 310–320 feet

Age at stump section: 1,204 years (total age of tree at least 25 years greater)

Gross volume of tree: 95,000 board feet (not deducting for rough top, longs, cracks and falling breakage)

Probable weight of tree: 500 tons

Source: E. Fritz, "Story Told by a Fallen Redwood," Save the Redwoods League.

Figure 1

Growth rings of a tree tell the life history of the tree: its rate of growth, climatic environment, fire, insect attacks, and other historical events in the life of the forest. This tree shows an adjustment to lean.

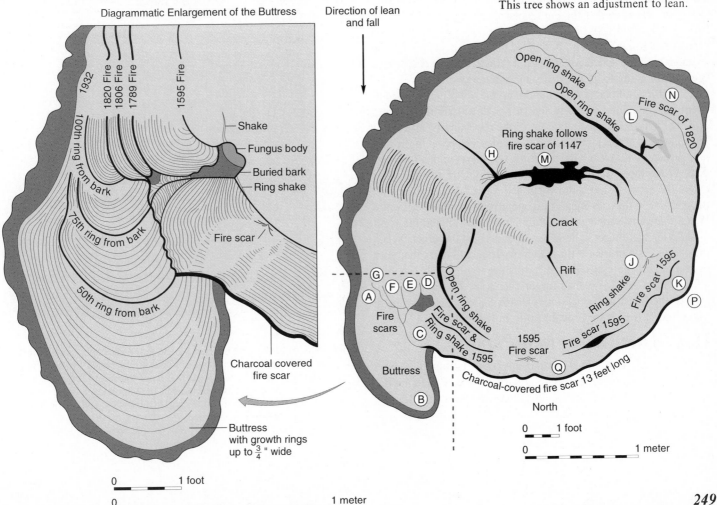

Diagrammatic Enlargement of the Buttress

1932 Fire / 1820 Fire / 1806 Fire / 1789 Fire / 1595 Fire

100th ring from bark
75th ring from bark
50th ring from bark

Shake
Fungus body
Buried bark
Ring shake
Fire scar
Charcoal covered fire scar
Buttress with growth rings up to 3/4" wide

0 — 1 foot
0 — 1 meter

Direction of lean and fall

Open ring shake
Open ring shake
Ring shake follows fire scar of 1147
Fire scar of 1820
Crack
Rift
Open ring shake
Fire scar & Ring shake 1595
1595 Fire scar
Fire scar 1595
Ring shake
Fire scar 1595
Charcoal-covered fire scar 13 feet long
Fire scars
Buttress
North

0 — 1 foot
0 — 1 meter

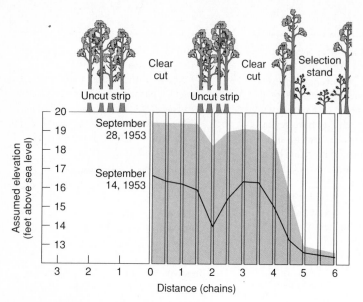

Figure 12.20

Transpiration from a forest works on the water table like water pumps. Note the draw down in this series of temporary wells extending from a clearcut area into a selectively cut stand of loblolly pine in eastern North Carolina. The depth variations are due to the trees.

The tree also has a long fossil history which indicates much more extensive distribution reaching back to the Cretaceous period. Let us examine some of the physiological elements of this tree to better understand how redwood trees have survived in time. The processes of growth can also be applied to other trees of the middle latitudes with the understanding that no two trees, even of the same species, are going to grow exactly the same way.

First, all plants grow by mining the soil of water and nutrients and soaking up energy and carbon dioxide in their leaves. Nutrient ions or charged atoms enter the root hairs by **osmosis**, a process where atoms pass through molecule-sized pores in the root hair surface. These root hairs penetrate the soil searching for moisture. When moisture is present and their cells are thirsty, water enters the pores. Osmosis draws the soil water into the plant until the pressure in the cells equals the counterpressure of the cell walls. When the water supply is plentiful all cells are full and the plant is perky or has turgor. The leaves and stems stand erect. When the plant loses more water through **transpiration** than it can absorb by osmosis in the root zone, cell walls begin looking like underinflated tires, and the plant's turgor decreases. This is what happens to the forest in late summer. Death can come if not enough water and nutrients are brought up from the roots (fig. 12.20).

How Water Rises in a Plant

The upward movement of water in plants occurs because the forces driving water up are greater than gravity. First, root pressure develops through osmosis. Soil water diffuses into root cells diluting the dissolved sugars, salts, tannin, and other sap substances until

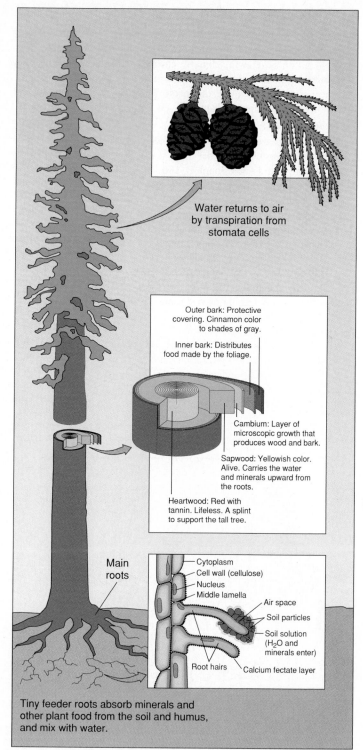

Figure 12.21

Water rises to the tops of trees by the pull of transpiration. Each molecule of water exerts electrical bonding forces on the next in a long string of molecules that stretch from the soil to the stomata in the leaf. As molecules are released in the vapor form, others come up and the chain remains unbroken, lifting water and nutrients hundreds of feet in the sapwood.

the cells are stretched tight or become turgid. This force is not enough to lift water the 90 meters (300 feet) to the top of a tall redwood. Ninety meters up, the plant is losing hundreds of liters per day resulting from transpiration through the leaf pores or **stomata.** A single corn plant may transpire 2 liters per day.

Replenishment starts in the ground through osmosis, then the water moves up the sapwood layer by leaf suction. The leaf creates a suction force when the threads of water in the leaves are transpired through the pores. Water has cohesive properties (its molecules stick together) and adhesive properties (molecular forces attract water molecules to the cell walls). This tendency to adhere and be cohesive is the basis for **capillary action** or upward movement (fig. 12.21).

Therefore, the leaf can effectively exert a pull of hundreds of atmospheres on the plant's circulatory system. Solar energy is driving the suction pump of the leaves. When night falls transpiration nearly ceases, and the plants reduce water uptake as well.

Energy and Oxygen

The energy used to evaporate water keeps the water moving up the tree to the leaves. Here **photosynthesis** produces carbohydrates by the chemical bonding of water and carbon dioxide during the light absorption process as described earlier.

Stating this as a simple chemical formula,

$$6H_2O \text{ (water)} + 6CO_2 \text{ (carbon dioxide)} + \text{light} \rightarrow$$
$$C_6H_{12}O_6 \text{ (sugar)} + 6O_2 \text{ (oxygen)}$$

Oxygen (O_2 molecules) is a by-product of the energy conversion process. When plants use energy, they convert carbohydrates back into water and carbon dioxide, tapping the food stored from the photosynthesis process. Plants are unique in that they not only consume chemical energy but produce an excess, which becomes the fuel for all other animal organisms. They also give us a second bonus in oxygen. The process of converting carbohydrates into energy, or **respiration,** cannot take place without oxygen molecules present. Oxygen and the world's food supply are both tied to the green plants of the world. The next time you walk across a lawn, remember that you are stepping on your oxygen life support system. A typical tree- and grass-covered college campus supports dozens of students by the presence of its green plants.

The results of energy processing adds height and diameter as layers of wood fibers are added on the outside of the old lifeless heartwood. Most growth occurs when water, light, and temperature reach optimum. Photosynthesis increases as the days grow longer and warmer. Winter is a time of slowest growth because of lower temperature and luminosity intensity and shorter days. In the spring and early summer, photosynthesis increases and wood is added to the tree. The long dry summer begins to take its toll on growth during late summer and fall. This is when a dark, hard, narrow ring of wood is produced compared to the lighter, wider bands produced during spring and early summer. This ring pattern is characteristic of forests beyond the tropics. The rings reflect seasonal variations in growth and can be used to count the age of the tree and infer its growing environment.

Figure 12.22

Death of a redwood results either from natural causes or from its few enemies. These giants can be toppled by wind, flooding, drought, and fire. Compared to its competitors, this tree is especially resistent to fires, floods, and insect attacks. Silt deposits from flooding have stained these trees several meters up the tree trunk.

Death of a Tree and Preservation

A tree, like all living things, eventually dies of natural causes or is killed by an enemy.

Death comes to the redwood mainly by losing its balance with the help of wind or movement of the soil. However, some just lose their balance, and then gravity does the rest without the slightest wind. These giants do not have big feet, so the root systems can't hold them up if they begin to lean. Perhaps the reason they are still with us is because they have so few causes for death. The greatest enemy today may be those trying to preserve the redwood forest. Some of the largest and oldest trees located in the state and national parks are faced with new problems of extinction brought about by flood control.

FLOODING IN THE FOREST

Redwoods growing on old river bars and terraces represent the tallest and usually oldest trees in the redwood community. These stands are often uniform in age. Seldom a year passes without high water jumping the river bank and depositing sand, silt, and gravel, rich in mineral nutrients, as much as a meter (3 feet) thick (fig. 12.22). Although flooding topples some of these giants, most survive and thrive following these fresh applications of fertilizer. They

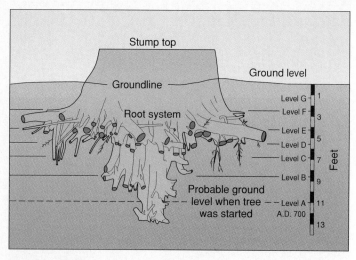

Figure 12.23

Flood insurance built into the tree's design gives the redwood the ability to not only survive flooding but thrive and generate new root systems while its competitors are being toppled. At point B a fresh deposit formed a layer 1 meter (3 feet) deep and stimulated a series of lateral roots. Other floods brought the level to D, E, F, and G and finally to its present ground level.

have a flood survival kit built into their physiology. Other trees in the community do not fare so well. The floods are like a tonic, and new growth rings are added along with new sprouts.

Let us examine a fallen redwood's root system to gain some insights into the environment that favored its survival for so many years. When the tree in figure 12.23 started its life in the year A.D. 700, the ground level was at least 3.4 meters (11 feet) below the current level. The trees growing at that location are standing on a river terrace or flat built up by deposits from past floods.

Sediment studies indicate that at least seven great floods and a number of minor ones built the current base 3.4 meters (11 feet) above the original site. Redwoods, when partially buried, send out a new and higher root system from the main trunk; there is no taproot. Redwoods have a very shallow root system because water is usually close by in the shallow water table. At level B, a thick layer of sediment was deposited a thousand years ago. The older root system continued to function until a new system was eventually formed in the new deposits. Another hundred years passed until another large flood occurred, bringing the level to C, and again a new root system developed above the old one.

Other floods brought the level to D, E, F, and G and finally to ground level in 1933, where the tree fell. Each new layer of sediment triggered lateral root development; at the same time the trunk portion below the sediment ceased its growth in diameter. The flood history of the fallen giant is a good example of how a redwood tree survives and thrives in an environment we tend to consider negative to most vegetation. Yet here the process of flooding is intricately interwoven into the survival ecology of the redwood growing on a stream bank.

Figure 12.24

Fire clears the land of plant invaders and forest debris. Fires have swept over most of the land repeatedly. It is an integral part of the planet's natural system.

Photo by Gary Bowman.

Many of the plants associated with the redwood community are absent from the stands on the alluvial flat. For example, a young tan oak or Douglas fir will die if a meter of silt wets its ankles. The redwood can survive where oxygen levels fall very low in the saturated soil conditions. Where organic matter is buried with the sediment, it can be an inhibiting growth factor. Redwoods thrive on inorganic mineral soils or soils without much humus. Because redwoods prefer soils low in organic content, seedling regeneration occurs best on flood deposits and recently disturbed soils. Redwood seedlings, growing on the steep slopes above the stream floor, get their start on soils that are unstable. Old landslide scars that have stabilized become nurseries for young seedlings. Erosion, mass wasting, and flooding are all important in the life cycle of the redwood plant community. It is interesting to note that the rivers of the coastal redwood community rank among the most sediment-laden streams on the North America continent.

Flood-control dams planned for many of the redwood watersheds will reduce the beneficial silt deposits that act as a tonic

By reducing fires we have introduced a whole new set of problems for grasslands, chaparral, and forests. The next chapter will examine fire, and other environmental factors, in these plant communities.

SUMMARY

The **biosphere** is the living zone of the Earth. Solar energy is the driving force. It gives life and mobility to all living systems through energy transformation in the **photosynthesis** process. **Climate,** which includes all the meteorologic elements, is the most important environmental force.

All living systems find their special niche where each environmental factor (**climatic, edaphic, biotic,** and **geomorphic**) provides the right ingredients for a complex or community to carry on activities, as energy and materials flow through **food chains.** The chain is an energy system of producers, consumers, and decomposers.

In time all communities change because environmental factors change. Eventually a stable state may be reached for a period, and the community is said to be in a **climax** state. However, this state is rarely reached because change in nature is the norm. The **pygmy forest** represents such a state.

Next we examined the influence of **fire** and **floods** on the coastal redwood community and how a tree survives these forces and records in the growth process its environmental history in the plant tissue. Where humans interfere by flood and fire control, the forest is placed in jeopardy because these natural forces are interrupted.

ILLUSTRATED STUDY QUESTIONS

1. Describe the four major environmental factors shaping plant communities illustrated in figure 12.2, p. 235.
2. Explain the role of light energy in the forest (fig. 12.3, p. 236).
3. Explain the photosynthesis process and the resulting by-products (fig. 12.3, p. 236).
4. How is the quality of life affected by the clarity of the atmospheric window (fig. 12.4, p. 236)?
5. Explain the role of elevation on vegetation patterns (figs. 12.5 and 12.17, pp. 238, 246).
6. What does a view of Earth from space reveal regarding vegetation patterns (fig. 12.6, p. 239)?
7. Describe xerophytic, mesophytic, and hygrophytic plant structures and their adaptation to water needs (figs. 12.7, 12.8, and 12.9, pp. 239, 240, 241).
8. Figure 12.13, p. 243, shows a windward and leeward slope. Where is precipitation likely to be lower? Why?
9. Explain why temperatures are likely to be cooler on north-facing slopes in the Northern Hemisphere and south-facing slopes in the Southern Hemisphere (fig. 12.13, p. 243).
10. Why can it be correctly stated that plants are miners of the soil (fig. 12.14, p. 244)?
11. Describe the flow of energy and nutrients or materials in the food chain (fig. 12.17, p. 246).
12. Describe the relationship between organism numbers and organism size (fig. 12.18, p. 247).
13. Describe the changes in plant communities on each terrace in figures 12.19a–c, p. 248.
14. Explain some of the clues a tree ring pattern reveals regarding its history (box 12.1, fig. 1, p. 249).
15. Describe the effect of fire on an ecosystem. How does it play an integral role in the system (fig. 12.24, p. 252)?

yet drown the competitors. In summary, if floods are reduced, other plants will quickly claim the site where a redwood has fallen. The flood is the friend of the redwood. It reduces competition and fertilizes the forest.

FIRE

Fire is nature's scalpel. It clears the land of plant invaders and forest debris. Our discussion of environmental factors determining the green patterns of the Earth would be incomplete without considering the impact of fire on plant communities.

The evidence of beneficial fire sweeping most of the land periodically is abundant. Before the establishment of European colonies on the American, Australian, and African continents, fires swept over most of the land (fig. 12.24).

Fire was deliberately used as a tool by early populations to clear the land, drive wild animals into traps, and improve hunting habitats. Our perspective of fire has changed from time to time. Only recently have we begun to reconsider the beneficial effects of fire in light of research indicating that fire prevention may actually increase the destructive characteristics of a major fire. Where fuel from dead matter accumulates from overprotection, fires are extremely hot and destructive.

The Green Patterns of the Earth

13

Objectives

After completing this chapter you will be able to:

1. Explain the special conditions for life and describe the environmental adaptations of plant and animal communities.
2. Explain the role of solar energy, temperature, and moisture availability in each major plant community in terms of shaping plant patterns and structures.
3. Explain the difference between a *floristic* and *structural* approach to vegetation classification.
4. Divide the Earth into the major environmental zones based on the world's climatic belts.
5. Explain the environmental factors involved in shaping each of the world's major plant communities, ranging from the tropical rainforest to the arctic tundra.

Hall of Moses, Hoh Rain Forest, Olympic National Park, Washington.
© David Muench Photography

Have you ever received a birthday package and unwrapped it to find another smaller package inside? Then after you unwrapped it, there was another, then still another inside. The largest package is the entire biosphere of the Earth. The next smaller size represents the land communities on one side and the aquatic communities on the other. By continually changing scales we can go down to the local community of a meadow or pond, and finally the microscopic partnership of algae and fungus within a lichen on a rock that is exposed to sun, wind, and rain.

This chapter examines communities on a global scale as well as on the scale of the *study area* of chapter 1. Three questions should come to mind: (1) How does energy from the Sun influence each plant and animal community in terms of patterns and plant forms? (2) What are the special conditions of life here? (3) How are they met by environmental adaptations within the community?

The book you are holding is a storehouse of knowledge, but did you ever consider it to be a storehouse of energy from the sun? Consider the paper that it is made from is transformed solar energy through the photosynthesis process described in chapter 12. Solar energy is being captured by green plants, converted to chemical energy, and used by living systems for their life processes. The Earth's plant communities represent unique responses to variations in solar energy. From the sun-baked tropical low latitudes of forests and deserts to the seasonal middle-latitude zones to the cool high latitudes, one can observe the effects of light on the plant and animal world. Both plants and animals adapt structurally, growing and reproducing in response to the flow of energy through the biosphere. This response is revealed in global vegetation patterns.

VEGETATION CLASSIFICATION

When we evaluate the Earth's vegetation patterns and communities, it is important to approach the study in a systematic way. Plants and their communities vary widely in terms of dominant plant species, numbers, densities, and associations. For example, a tropical rainforest in the Amazon Basin may have hundreds of different tree species per square kilometer. At the other end of the forest spectrum, the great Canadian forests may have as few as five (fig. 13.1). A species census, or *floristic* study, will reveal the complexity of plants in a given plant community, but it leaves us with questions related to structure, pattern, density, and form of the plant community. Geographers tend to focus on the entire community. A *structural* description includes the physical forms of plants and inner workings of the community. Our approach will concentrate on plant structure, the forest, grassland or desert vegetation, and the interrelationships between climate and soils. We will examine key individual plant members in the community and the roles each play in shaping the environment and adapting to it.

FLORISTIC HISTORY PATTERNS

Similar environments do not always have the same plant types. However, one can see many similarities in plant structure and form if similar environments are studied. The deserts of Australia, South Africa, and Mexico are different floristically, but they share very similar structural characteristics, despite different evolutionary histories.

The evolutionary development that produces these similarities in structure is known as "convergence." Cacti, for example, are common on the North American deserts, but unknown in Africa. African deserts are populated with euphorbias, which look like cacti but are genetically unrelated. Both cacti and euphorbias are succulents, spiny plants well adapted to arid conditions. For another example, the jackrabbit is a common animal of the North American grassland and desert. Argentina has no jackrabbits; instead the viscacha has filled a similar set of environmental conditions, or **niche.** The viscacha looks and even behaves like a rabbit but actually is a cousin to the guinea pig. Convergent adaptation has been strongest where the environment has placed special demands upon its inhabitants. The worldwide examples of convergent evolution are numerous, and they can be used to teach us about environmental influences and relationships. Because convergence is so common, plant classification by structure is possible. Plants and animals of similar environments often have common structures (fig. 13.2).

The floristic or species pattern may reflect earlier continental connections or positions of the continents. As discussed in the next unit, the continents are giant slabs of rock drifting on the conveyer belt of the Earth's mantle. Scientists have reconstructed earlier positions of the continents, which will shed more light on the current patterns of plant species. Perhaps plants of today will help unlock the secrets of the earlier locations of the continents.

FLORISTIC REALMS VERSUS STRUCTURAL PATTERNS

Today botanists have developed six floristic realms. Eurasia, Greenland, and North America have small floristic differences so they are grouped as one floristic realm, known as the Holarctic. Plate tectonic theory supports the concept of one land mass consisting of the above-mentioned land masses.

Much larger differences exist between the tropical floras of the Old and New Worlds, so two floristic realms are named, the *Paleotropic* of the Old World and the *Neotropic* of the New. Floristically, the southern parts of Africa and South America and the continent of *Australia* have less in common with the rest of the continents, so they have been grouped into three separate realms.

A

Figure 13.1

Amazonian (A) and Canadian (B) plant communities. The
complexity of the plant community increases as one approaches the
tropical rainy zone.

B: Photo by Bruce Carroll.

B

Mountain viscacha

Figure 13.2

Convergence occurs when plants and animals have similar appearance or structure but very different geographies and evolutionary histories. The North American jackrabbit and the South American mountain viscacha fill similar environmental niches but are not ancestrally related.

acteristics. New Zealand contains both Paleotropic and Antarctic elements. There are no sharp boundaries and many overlaps.

Australia is our best example of uniquely developed species, because isolation has been a major factor. The organisms of Australia came at different times from different places and evolved uniquely in that environment once they arrived. Historical biogeography is concerned with the historical development of life.

Our focus in this chapter is on the current plant structural patterns and the elements shaping the plant communities. We will use structural features as clues to the environment and boundaries of the community. Remember the concept of change, presented in chapter 1: past history is just as important as present events in understanding vegetation patterns. By combining our knowledge of plant history and the current environmental forces, today's makeup of plant patterns will become clearer.

VEGETATION PATTERNS OF THE EARTH FROM SPACE

A space view of the Earth's plant communities reveals the major climatic environmental zones described in chapter 9. We will first examine the equatorial belt, where winter is unknown, to examine the oldest and most complex plant community, the tropical rainforest (fig. 13.3). In the **tropics,** where the latitudinal zone is located between the Tropics of Cancer (23½° N) and Capricorn (23½° S), the Sun is directly overhead at noon at least once a year. Here solar energy drives the hydrologic cycle, producing large annual convectional precipitation.

The Antarctic realm comprises the southern tip of South America and the subantarctic islands. The *Australian,* which encompasses the entire continent, and the *Capensic,* the smallest realm but very rich in species, in the southwest corner of Africa, make up the rest of the Southern Hemisphere. These realms often merge in char-

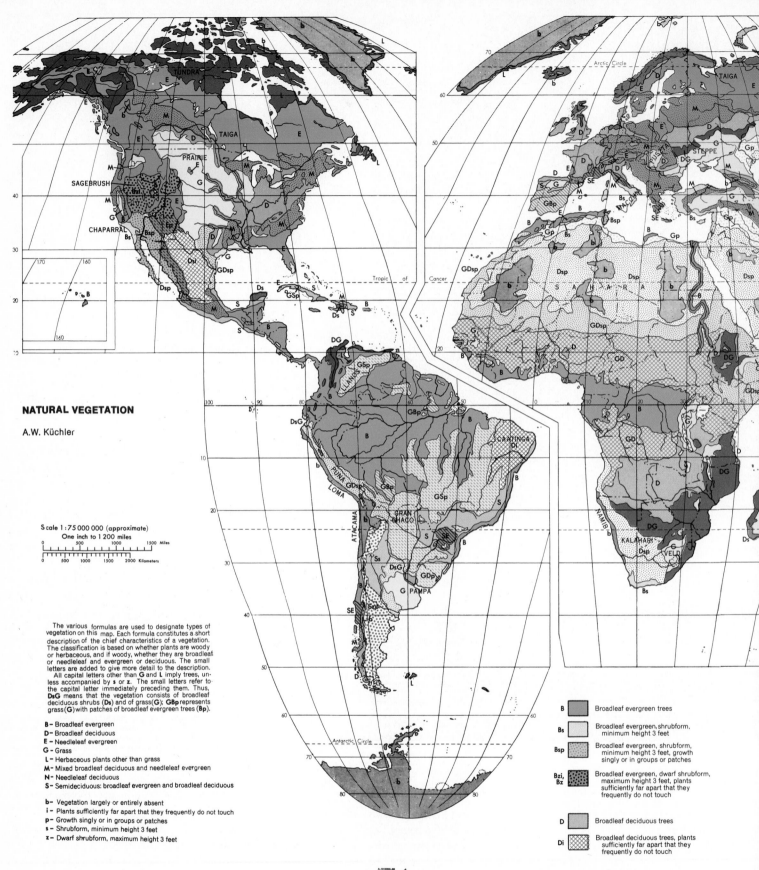

NATURAL VEGETATION

A.W. Küchler

Scale 1:75 000 000 (approximate)
One inch to 1 200 miles

0   500   1000   1500 Miles

0   500   1000   1500   2000 Kilometers

The various formulas are used to designate types of
vegetation on this map. Each formula constitutes a short
description of the chief characteristics of a vegetation.
The classification is based on whether plants are woody
or herbaceous, and if woody, whether they are broadleaf
or needleleaf and evergreen or deciduous. The small
letters are added to give more detail to the description.

All capital letters other than **G** and **L** imply trees, unless accompanied by **s** or **z**. The small letters refer to
the capital letter immediately preceding them. Thus,
DsG means that the vegetation consists of broadleaf
deciduous shrubs (**Ds**) and of grass(**G**); **GBp** represents
grass(**G**) with patches of broadleaf evergreen trees (**Bp**).

B – Broadleaf evergreen
D – Broadleaf deciduous
E – Needleleaf evergreen
G – Grass
L – Herbaceous plants other than grass
M – Mixed broadleaf deciduous and needleleaf evergreen
N – Needleleaf deciduous
S – Semideciduous: broadleaf evergreen and broadleaf deciduous

b – Vegetation largely or entirely absent
i – Plants sufficiently far apart that they frequently do not touch
p – Growth singly or in groups or patches
s – Shrubform, minimum height 3 feet
z – Dwarf shrubform, maximum height 3 feet

B	Broadleaf evergreen trees
Bs	Broadleaf evergreen, shrubform, minimum height 3 feet
Bsp	Broadleaf evergreen, shrubform, minimum height 3 feet, growth singly or in groups or patches
Bzi, Bz	Broadleaf evergreen, dwarf shrubform, maximum height 3 feet, plants sufficiently far apart that they frequently do not touch
D	Broadleaf deciduous trees
Di	Broadleaf deciduous trees, plants sufficiently far apart that they frequently do not touch

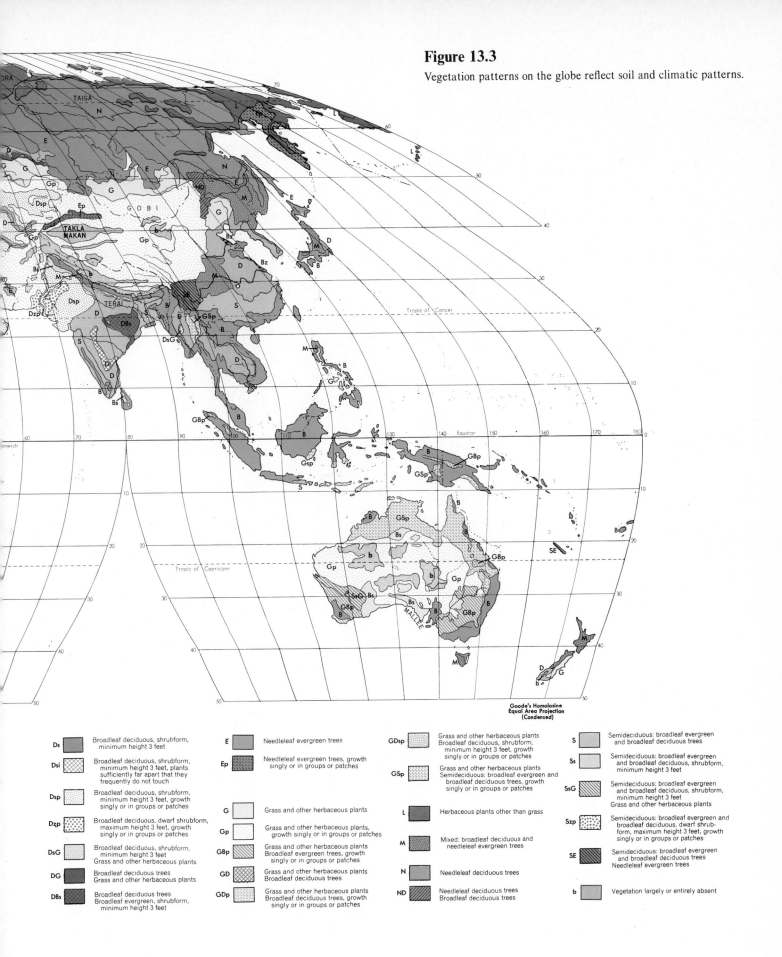

Figure 13.3

Vegetation patterns on the globe reflect soil and climatic patterns.

Ds	Broadleaf deciduous, shrubform, minimum height 3 feet	
Dsi	Broadleaf deciduous, shrubform, minimum height 3 feet, plants sufficiently far apart that they frequently do not touch	
Dsp	Broadleaf deciduous, shrubform, minimum height 3 feet, growth singly or in groups or patches	
Dzp	Broadleaf deciduous, dwarf shrubform, maximum height 3 feet, growth singly or in groups or patches	
DsG	Broadleaf deciduous, shrubform, minimum height 3 feet Grass and other herbaceous plants	
DG	Broadleaf deciduous trees Grass and other herbaceous plants	
DBs	Broadleaf deciduous trees Broadleaf evergreen, shrubform, minimum height 3 feet	

E	Needleleaf evergreen trees	
Ep	Needleleaf evergreen trees, growth singly or in groups or patches	
G	Grass and other herbaceous plants	
Gp	Grass and other herbaceous plants, growth singly or in groups or patches	
GBp	Grass and other herbaceous plants Broadleaf evergreen trees, growth singly or in groups or patches	
GD	Grass and other herbaceous plants Broadleaf deciduous trees	
GDp	Grass and other herbaceous plants Broadleaf deciduous trees, growth singly or in groups or patches	

GDsp	Grass and other herbaceous plants Broadleaf deciduous, shrubform, minimum height 3 feet, growth singly or in groups or patches	
GSp	Grass and other herbaceous plants Semideciduous: broadleaf evergreen and broadleaf deciduous trees, growth singly or in groups or patches	
L	Herbaceous plants other than grass	
M	Mixed: broadleaf deciduous and needleleaf evergreen trees	
N	Needleleaf deciduous trees	
ND	Needleleaf deciduous trees Broadleaf deciduous trees	

S	Semideciduous: broadleaf evergreen and broadleaf deciduous trees	
Ss	Semideciduous: broadleaf evergreen and broadleaf deciduous, shrubform, minimum height 3 feet	
SsG	Semideciduous: broadleaf evergreen and broadleaf deciduous, shrubform, minimum height 3 feet Grass and other herbaceous plants	
Szp	Semideciduous: broadleaf evergreen and broadleaf deciduous, dwarf shrubform, maximum height 3 feet, growth singly or in groups or patches	
SE	Semideciduous: broadleaf evergreen and broadleaf deciduous trees Needleleaf evergreen trees	
b	Vegetation largely or entirely absent	

Goode's Homolosine
Equal Area Projection
(Condensed)

Figure 13.4

The tropical rainforest as viewed from the top shows the Ucaylali River meandering through a sea of green.

Figure 13.5

Tropical agriculture exposes the soil to the pounding rain, resulting in nutrient loss through weathering and leaching of the soil.

Tropical Rainforest Plant Community

High rainfall and high temperatures are the main environmental forces that define what we mean by the word *tropical*. Abundant moisture and warm temperatures are the norm. The tropical rainforests occur where the mean monthly temperature is about 18° C (64.4° F) and frost-free. Abundant precipitation occurs in most months of the year, producing large water surpluses and continual stream runoff. In some places annual rainfall exceeds 7.6 meters (300 inches). The tropical rainforest covers most of Central America, equatorial South America and Africa, southeastern Asia, the East Indies from Indonesia through New Guinea and the Philippines, and on the northeast coast of Australia. There is no place on Earth where life is more varied.

A tropical rainforest frequently consists of hundreds of species of plants and animals. As many as 500 different tree species per square mile (2.59 square kilometers or 259 hectacres) have been identified within such a forest. A flight over the Amazon

rainforest gives one the impression of an ocean of green (fig. 13.4). The forest appears quite uniform in structure and density, except for the interruptions of brown silt-laden rivers and clearings for settlement. If we view the forest from the ground, conditions for life are quite different. Even at high noon, the forest floor is poorly illuminated. The tallest trees reach 40 meters, spreading their branches to filter out the sunlight and the rainfall. On the forest floor, only a few plants with minimal light requirements grow. Humidity is high and conditions are dripping wet most of the time. Temperatures are very constant throughout the day and year, averaging between 27° C and 32° C (80° F to 90° F). The forest floor is relatively uncluttered, lacking decaying plant matter. Humus is entirely lacking because bacterial action and chemical decay or weathering quickly reduces organic matter to water-soluble minerals.

The warm, moist environment creates a soil that is always wet and alive with organisms that recycle decaying plant material. The luxuriant forest maintains itself as a balanced system. How-

Figure 13.6

The tropics take first place in plant diversity or variety. No other place on Earth has such abundance and variety of plant life, as illustrated by this coastal swamp in Costa Rica.

Photo by Dr. Charles Hogue, Curator of Entomology, Los Angeles County Museum of Natural History.

ever, when agricultural activities replace the forest cover, nutrients are quickly leached out and washed away because the protective cover of the forest is missing. Agricultural ground cover is no match for the pounding rains, which produce heavy runoff and rapid leaching into the groundwater table. The root systems of agricultural crops are able to recycle only a small fraction of the nutrients formerly utilized by the forest (fig. 13.5).

Plant Structures of the Tropical Rainforest

The tropical rainforest's great variety of plant forms range from the simple lichen, which represents a partnership between fungi and algae to the giant mahogany. Lianas or climbing woody vines are especially characteristic of rainforests. They climb from the dark forest floor high into the canopy of the tallest trees where they soak up the light. Both the trees and the vines are **broadleaf**

evergreens. They lose their leaves but not in one season, therefore the forest has a continuous green cover. The broadleaf evergreen structure is a common plant form in this environment.

Diversity of Plants and Animals

The variety of organisms in the tropics far exceeds the diversity of life in the temperate and higher latitudes (fig. 13.6). The number of species of butterflies in all of North America totals about 600. On the smallest island of Trinidad, just north of Venezuela, it is possible to equal that number. There are over 1,200 species of trees in Costa Rica, while all of North America falls well below that number. The same is true of mosquitoes. North America's mosquito species number around 180, while in a tropical region the size of Costa Rica there are over 200. Why does the tropical rainforest have such diversity? Some believe that time is the chief factor.

Life or Death in the Tropical Forest?

BOX 13.1

Life is in the forest. Over half of this planet's life forms (plants and animals) live in the forest. It is estimated that no fewer than 1.2 million species will likely vanish or become extinct by the year 2000.

In California, the last remaining California condors now reside in cages for their own protection. Currently 254 animals and 173 plants are on the endangered species list in the United States alone. This is a minute number compared to the global picture.

Because forests are ecosystems, their destruction means massive extinction for their inhabitants. Tropical forests are especially sensitive to high loss rates (fig. 1). The tropical forests have a much greater plant and animal diversity than any other region. Also, a very high proportion of the plant and animal species are endemic; that is, they cannot survive outside of their tropical habitats.

Extinctions have been estimated by some to average more than 100 species per day. Most of these, it is thought, will disappear before they have even been discovered. This extinction period has been compared to the high rate of 65 million years ago, when the dinosaurs and many other species vanished.

A microcosm of this doomsday projection can be observed on the island of Madagascar in the Indian Ocean. This island has been isolated from the continent of Africa for over 150 million years. As a result, like Australia and New Zealand, its native plants and animals represent a very high ratio endemic to the island. Seventy-five percent of its 8,500 known species of plants are found nowhere else. Ninety-three percent of its animals and 43 percent of its birds are also found nowhere else.

Today, as a result of expanded cattle ranching and rice growing, the land has been cleared of 90 percent of its virgin forest. The impact on species has been catastrophic. Fourteen types of lemurs have become extinct, leaving 22 species endangered. Plant survival is also uncertain, because the count is still coming in with new discoveries still being made. One particular plant, the rosy periwinkle, has played a major role in combating and arresting Hodgkin's disease, lymphatic cancer, and leukemia. Who knows what valuable plants will be gone forever before they can be found?

Source: James P. Jackson, "The Edge of Extinction," *American Forest,* November/December 1988.

Figure 1

NASA's Coastal Zone Color Scanner (CZCS), operating aboard the Nimbus-7 satellite, completed a computer-generated color image showing, for the first time, the distribution of microscopic plant life, or phytoplankton, in the surface waters of the entire North Atlantic Ocean. The image also shows the Land Vegetation Index for the continents bordering the North Atlantic Basin, making it the first image to display indices for land and ocean plant abundance. The Land Vegetation Index was acquired using the Advanced Very High Resolution Radiometer (AVHRR) on the NOAA-6 satellite. Dark green colors show high-density vegetation; light patches in the Amazon Basin are areas of deforestation.

Photo by NASA.

The older the community, the more species it will contain. Evidence shows that the tropical rainforest has been around a long time, dating back to the *Cretaceous Period*. It is believed that time has created this diversity through biological evolution and has added continuously to it (box 13.1). The temperate and higher latitudes are regions that have greater variations in temperature and precipitation. Harsher, cooler, and more arid conditions have restricted evolution by narrowing the environmental conditions for rapid biological change.

Another theory points to the greater variation in **niches** or habitats in the tropics, which provides for a more complex plant distribution.

One fact is clear: the tropics have been relatively stable environments. They have been here a long time, and the environment's stability tends to maintain rather than eliminate species. Environmental change always brings extinction to some less adaptable plants and animals.

The living elements of the tropics can be divided into three important functioning units of a **nutrient cycle.** These are the **producers, consumers,** and **decomposers.** The producers are plants ranging from trees to herbs. They take solar energy, carbon dioxide, and water and turn them into sugars through photosynthesis. Without this stage of the nutrient cycle there would be no animals. This is the primary stage for all other life. The consumers of the forest consume and process plant food and in turn provide food for other animals or secondary consumers, such as parasites, insects, and predators.

All living things eventually die. This is where the decomposers step in. A leaf falls on the ground or a tree is uprooted by the wind. Next, minute organisms such as termites, wood-boring beetles, bacteria, and fungi go to work decomposing cellulose into soluble molecules. Then the rain carries these nutrients into the soil to be soaked up by the roots of trees, herbs, and shrubs. The nutrient cycle is then continuous. The tropical forest is the world's greatest nutrient recycler. More biomass material is processed here than any other place. **Biomass** is the net dry weight of vegetation of a community. Deserts and tundra have the lowest biomass and the fewest species, while the tropical rainforest plant community represents the greatest biomass production on the planet.

Light and Zonation

Plants of the tropics grow in niches or environments favorable to them. Light is the controlling factor determining both the vertical and horizontal plant patterns. Along the edges of a stream or clearing, light comes in and floods the area, changing the vegetation in the process. Light lovers thrive in these bright spots, including ferns, the balsawood tree, and the cecropia tree with its jointed trunk and large leaves. Here the forest is described as a jungle because of dense vine-covered trees interwoven with herbs and shrubs. This is not typical of the true tropical rainforest. It represents an edge effect, or **ecotone,** created by light access to the forest floor.

The vertical zonation from the forest floor to the tops of the trees or canopy is developed in levels or **tiers** of vegetation. The forest may at first appear densely packed, but on closer inspection it can be divided into five layers that grade together (fig. 13.7).

The lowest zone consists of herbaceous plants with soft leaves, often dark green or reddish in color, making use of all the light they can obtain. Next, the woody shrubs, including large ferns, are found growing in this dimly illuminated zone. Above the woody shrubs and ferns are three levels or tiers of trees designated A, B, and C. Zone C is the lowest, ranging between 4.5 and 7.5 meters (15 and 25 feet) in height. In Zone B, trees range from 11 to 23 meters (36 to 75 feet) in height. These tiers form the canopy. Level A consists of the great giants, reaching up over 45 meters (150 feet). Trees of the lower levels grow slowly because light is in shorter supply. If a lower tier tree is exposed to more light after a taller canopy tree falls, its growth is accelerated. Seedling trees of the A tier are rare except where light can penetrate to the floor. The forest is a realm of competition. Plants must have nutrients and a place to plant their feet. This is why seedlings of the large trees are rare and the floor of the forest is open and dark.

Because the forest is tiered, we can also expect animal life to reside at different levels above the ground. A great variety of birds and monkeys populate the treetops. On the ground are numerous insects: ants, termites, flies, butterflies, beetles, and moths. Frogs and snakes are also abundant, and tree-climbing carnivores, especially cats such as the leopard of Africa and jaguar of South America, are common but often their numbers are exaggerated. Because ground cover is sparse resulting from poor light, ground-dwelling herbivores are less common than in other vegetation regions.

One of the greatest surprises for most first-time visitors to the forest is the lack of animal life evident during the day. However, as nightfall approaches, the forest begins to hum with sounds of birds, monkeys, grasshoppers, frogs, anteaters, and pigs. Most animals are nocturnal (night active) because days are oppressively hot and humid. As the relief of evening comes, the animals begin to go about their business of looking for energy in the **food chain** of their **ecosystem.**

The Tropical Semideciduous Forest

The tropical rainforest gradually gives way first to the **tropical semideciduous forest** and then to the **scrub woodland** and **savanna,** where moisture supply begins to fluctuate seasonally and also decline annually. Temperatures still are high and constant. On the margins of the tropical rainforest, a seasonal rhythm in precipitation occurs because the equatorial convergence zone shifts with the sun's position. A poleward shift brings heavy precipitation to the margins of the tropical rainforest.

The tropical semideciduous forest is dense yet consists of smaller trees than the tropical rainforest (fig. 13.8). More light reaches the lower vegetation because of the sparse canopy of taller

Figure 13.7

Light availability produces zonation of plants in both horizontal and vertical directions.

Photo by Dr. Charles Hogue, Curator of Entomology, Los Angeles County Museum of Natural History.

trees. The dry season occurs during the period of lowest sun. The semideciduous trees partially lose their leaves and grow dormant, and may lose all foliage. During this season, the forest takes on a brownish appearance. At onset of the rainy season, when the Sun traces a higher arc across the sky, life springs back and the forest begins to sprout new leaves, buds, and flowers. Animal life also revives.

Tropical Scrub Woodland and Savanna

Let's move poleward a few more degrees to 10° to 20° latitude, in either Africa or South America. Here we can find an even more pronounced dry season with a shorter, more unpredictable rainy period during the period of high Sun.

Here we find a parkland environment, trees and grassland blended together. Just as we observed in the tropical semideciduous forest, seasonal change in precipitation produces the dense shorter forest cover (fig. 13.9).

Shortly after the rains begin, the trees sprout new leaves, flowers bloom, and dry river beds become muddy channels that often overflow their banks during torrential downpours. Then, as abruptly as the rainy season began, the convectional showers shut down. This results from a shift equatorially of the tropical convergent zone with high pressure moving in behind. The land becomes parched and ground cover dries up, except along perennial streams. Here a **gallery forest** follows the river out in the grasslands. The tropical plants of the rainforest survive in this drier region along the rivers.

Fire in the Tropical Scrub Woodland and Savanna

The location of the tropical scrub woodland and savanna is intermediate between the tropical forest and the sparse dry deserts of the subtropics. Fires spread more easily here than in the rainforest or the desert. It has been suggested that the scrub woodland and savanna's vegetation is the result of fire as well as climate (fig.

Figure 13.8

This tropical semideciduous acacia forest in Nairobi National Park, Kenya, reflects seasonal variation in rainfall in its semideciduous plant life. Plants adapt to drought stress by dropping some of their leaves.

Photo by Dr. Charles Hogue, Curator of Entomology, Los Angeles County Museum of Natural History.

13.10). The scrub forest and grassland is very combustible, especially at the end of the dry season. The young trees are killed, though older ones survive, and grasses gradually take over. Thus, it is believed that fire, started by both lightning and humans, is the basic cause of the existence of these tropical scrub woodlands and savannas. If fires were reduced, woodlands would be able to gain a greater footing in this transition zone between deserts and forest.

Homer L. Shanty and C. F. Marbut, in their studies of tropical vegetation of Africa, categorized three kinds of transition vegetation between the Congo tropical rainforest and the Sahara desert: (1) acacia, short desert grass savanna, (2) acacia, tall-grass savanna, and (3) high-grass low-tree savanna, adjacent to the rainforest. The rainfall increases rapidly from north to south and the vegetation changes from grasses and thorn scrub woodland to dense low forest and less tall grass. Total annual evapotranspiration exceeds the rainfall. It is in this area of Africa, known as the Sahel, where fire and overgrazing are believed to be causing the desert to advance southward 25 to 32 kilometers (15 to 20 miles) per year (fig. 13.11). It is believed that overpopulation, which contributes

Figure 13.9

In tropical scrub woodland and savanna, trees and grasslands blend together to produce a parklike setting.

Photo by Dr. Charles Hogue, Curator of Entomology, Los Angeles County Museum of Natural History.

Figure 13.10

The scrub woodland savanna is shaped by both climate and fire.
Fires prevent woody plants from dominating the grasses.

Photo by Jennifer Locke.

to overuse of the land, has interrupted the normal hydrologic cycle.
Removal of vegetation causes more runoff, less transpiration, and
therefore change in the water vapor content in the atmosphere.
This means less moisture for condensation and precipitation. Increased population has also caused a lowering of the water table
near wells as drought persists. Herders have concentrated around
the wells, tapping them, in many cases, to depletion. This overuse
compounds the drought problem.

Animals of the Tropical Scrub Woodland and Savanna of Africa

The open parkland and thorny woodland forms a habitat favorable
to a unique variety of native wild animals. This is an interesting
paradox, for here is a vegetation environment created by human
action through fires. It is this land where we can observe firsthand
the African elephant uprooting trees to feed on the roots while the
giraffe plucks its lunch out of the tops of trees far above the ground.
The zebra, antelope, and buffalo all feed on their favorite plants.
Preying on these **herbivores** are predator **carnivores** that balance
the food chains by preventing plant eaters from increasing beyond
food supplies (fig. 13.12).

Figure 13.11

Overgrazing is believed to be a major factor in causing the desert to
advance by 25 to 32 kilometers (15–20 miles) per year. Wildlife is
being pressured by livestock for pasture.

Photo by Jennifer Locke.

Figure 13.12

Wildlife of the tropical scrub woodland rely on a vegetation environment created by human action through fires. However, today many animals face extinction due to human action.

Photo by Jennifer Locke.

Figure 13.13

The dry lands are the heat islands of the continents. Clear skies, low humidity, and high temperatures characterize this region.

Photo by Dr. Charles Hogue, Curator of Entomology, Los Angeles County Museum of Natural History.

This is "big game country," where many animals today face extinction as we hunt them or alter their habitat. It has been suggested by many ecologists that wild game could be managed to improve the protein diets of this part of the world. Cattle tend to destroy the grassland by overgrazing, while wild native animals feed on a greater variety of plants without destroying their habitat and food supply.

The Dry Lands

Low-Latitude Steppe and Deserts

The subtropical and tropical deserts are the middle-latitude (20°–40°) heat islands of the continents. Clear skies, low humidity, high temperatures, and wide daily ranges in temperature are typical climatic characteristics (fig. 13.13). The wet-and-dry rhythm of the tropical scrub woodland and savanna gradually shifts to a longer dry period with lower annual rainfall. These warm, sunny regions are under the constant influence of high pressure with subsiding dry air from the global circulation of subtropical high-pressure cells. The vast Sahara of Africa, deserts of the Middle East, the Outback of Australia, and deserts in southwestern North America and the west coast of South America all fall into this global high-pressure zone. Here high pressure, mountain barriers, and cold ocean currents often combine to reduce moisture on the land. These regions are typified by high potential evaporation as well as low, unpredictable rainfall. When rain does come, it may fall as torrential cloudbursts causing severe flooding. The rapid runoff problem is compounded by sparse vegetation and a lack of well-defined drainage systems.

The vegetation of the desert is **xerophytic,** or drought resistant (fig. 13.14). Plants survive by a variety of unique adaptations. Annuals scatter seeds until rainfall causes them to germinate. They complete their life cycle in a matter of weeks. The **perennials** survive by lying dormant until rains come. Then they send out new

Figure 13.14

Xerophytes—plants designed to survive within zones of low precipitation and high evaporation—can be classified as annuals or perennials. Annuals have a short life cycle, growing, reproducing seed, and dying in one season. Perennials survive over longer periods of time by conserving water until the rains come. Water is stored in the structure of these Joshua trees on the Mojave Desert of California.

Photo by Dr. Charles Hogue, Curator of Entomology, Los Angeles County Museum of Natural History.

Figure 13.15
Drought resisters of the desert include these Joshua trees of southern California.
Photo by Wilshire, U.S. Geological Survey.

root systems, which spread out over a large area near the surface to pick up moisture. The creosote bush of the Mojave Desert is a plant of this type.

Plants and animals of the desert must adapt to extremes of temperature and "boom and bust" water availability. Living systems basically fall into two adaptive types. There are those we describe as "drought evaders": plants that persist as seed until it rains, flower quickly, and then produce seed and die.

Animals evade drought by remaining underground. Reptiles such as the lizard, sidewinder rattlesnake, and chuckwalla are nocturnal on the desert. They metabolize moisture from the foods they eat.

The "drought resisters" are water storers. When the rains come, succulent plants such as the saguaro cactus swell up and continue to grow during the dry season, using the water stored in their stems. Leaves on desert plants offer clues to their adaptive abilities. It is through the leaves that most moisture is lost by transpiration. When the daily temperature begins to climb, the leaves on many desert plants turn and curl to reduce the surface area facing the Sun. As the day cools, they open again.

Many desert plants have hairy stems and leaves which serve as condensation surfaces for moisture in the air; a good example is the sand verbena. The hairs also act as shields from direct sunlight. Desert plants often have leathery surfaces with a waxy coating to reduce water loss. The creosote bush has a shiny wax coating which reflects heat and seals the surface.

A large number of desert plants are spine covered. Many cacti, acacias, mesquite, ocotillos, and yuccas have them. Only on the Australian desert are spine-covered plants sparse or missing. It is conjectured that spines evolve as a defense against browsers, since Australia is not inhabited by many native browsers.

Plants such as the long-branched ocotillo adapt by dropping their leaves after a brief period of growth. The stem is drought resistant, being covered by a resin coat under the bark.

Every desert has a few common plants distributed broadly over the region. In Africa the acacias, saltbushes, and sages are found on many deserts. The Mojave Desert of California is famous for creosote bush; it occupies more territory than any other single species. This same plant is also well established in Argentina in a similar environment. If you look at the spacing of the creosote bush, it appears as if someone planted them at approximately the same distance apart. Some believe that the roots release toxins into the soil to kill off competitors. The roots virtually fill up the entire space between plants, lying very near the surface to collect any moisture released from a shower.

Animal drought resisters have the adaptive advantage of mobility over plants. Plants must stay in one place and adapt to the environment. On the other hand, animals move about, go underground, and migrate to more favorable sites for food and water. Animals such as the lizard and snake produce no liquid waste by recycling it internally or by using the liquids of their food sources.

Modern Impact on the Desert

The weekend traveler has discovered the desert as a place for refuge, relaxation, and recreation. Ever-increasing numbers are now purchasing land, building houses, and retiring around oases such as Palm Springs, California, Tucson, Arizona, and Las Vegas, Nevada.

Desert vegetation has also been discovered as a landscape requiring little maintenance in the southwest. Arizona and California have serious problems with illegal plant and animal removal; many natives are becoming endangered species (fig. 13.15).

Recreation vehicles are also seriously damaging the ecosystem of the desert. Offroad vehicles are forming permanent scars on the land. Vegetation altered in this climate returns very slowly to its previous state.

Cattle grazing for the last hundred years has left its mark on the desert. The short grasses in many areas have disappeared to be taken over by mesquite. Cattle left to browse on mesquite greatly multiply germination because digestion prepares the seed for the soil. This has caused many desert areas to change from marginal short-grass rangeland to mesquite thickets which support less cattle or wildlife.

These changes caused by human occupation have altered the existing plant types and, in some cases, expanded the desert into areas once occupied by grasslands.

The Mediterranean Woodland and Scrub Community

Located on the fringe of the steppe and desert, the Mediterranean woodland and scrub community occupies only 1 percent of the world's land surface, yet the unique community in this small area supports a surprising 6 percent of the world's population (fig. 13.16).

The climate is truly unique. When most of the world's population is experiencing the dry season, the Mediterranean lands are preparing for winter rain. The Mediterranean climate is transitional between the hot, dry deserts and the moist, cool higher middle latitudes.

This intermediate position produces broadleaf, evergreen **sclerophyll** (waxy) plant forms. Whether you are in Australia, Europe, or North or South America, the vegetation looks similar in form and structure. In all of these regions, frost is infrequent and winter days are mild and sunny between storms. Plants, therefore, continue to grow and flower through the rainy winter months. The winter is much like spring. Hillsides turn green and wildflowers bloom while the interior parts of the continents remain brown or white in the dead of winter. The forty-niners, searching for gold in California, were surprised to find green hills in January.

When summer drought begins in this Mediterranean zone, plants adapt in the same manner as the xerophytes in the arid lands. Some Mediterranean plants store water in their tissues under

Figure 13.16

The Mediterranean woodland and scrub community. Its transitional position between the deserts and moist, cool marine environments has been a key factor in determining its unique animal and plant adaptations.

heavy layers of protective woody fiber. Others lose their leaves to conserve moisture, and annuals adapt by completing a rapid life cycle, leaving seeds behind to spring to life in the next rainy season.

Because of heavy and increasing populations in this mild climate of moderate rainfall, many changes are occurring in the vegetation pattern. The scrub woodlands are nearly gone in many regions of this zone and so are many bordering stands of forests such as the cedars of Lebanon. In California, the coastal sage is almost gone, and most of the live oaks and black walnut have been removed to clear the land for urban settlement and agriculture.

In the Western United States the scrub woodland is referred to as **chaparral** (a Spanish word meaning brush), while in Europe the French describe it as **maquis.** This growth consists of a variety of drought resisters such as chamiso, toyon, ceanothus, scrub oak, live oak, and manzanita. These plants tolerate extreme conditions: shallow soils, steep slopes, and long, hot, dry summers.

They are evergreen sclerophylls whose small leathery waxy leaves conserve water. Root systems penetrate deep into the rocky hills.

Fire in the Chaparral

Most of the vegetation in the Mediterranean woodland and scrub community depends on fire for regeneration. It is natural for the chaparral to burn as often as every 15 to 20 years and at least every 30 to 40 years. Some plants have seeds that will not germinate without the heat of fire to open the pods and remove the dense ground cover. Other plants resprout at the root crown with new vitality. Following a fire, the brush grows with vigor for about 15 to 20 years, then, unless a fire comes along, it begins to decline. Animal populations find less to browse on. In another few years the community is dominated by large woody plants that are decadent and producing little new growth. Litter covers the ground, and conditions are set for a new burn to start another cycle.

Figure 13.17

Fire in the chaparral. How beneficial is it?

Some then argue that since fire is a natural part of this plant community it should be permitted as a part of the balance of nature. Moreover, more frequent, smaller fires mean fewer big and disastrous ones (fig. 13.17). Let's examine some of the pros and cons of fire on this plant community.

Improved Habitat for Plants and Wildlife

This heading may appear contradictory, especially after you see an area recently burned. It looks bad, black and bleak. A mature, 40-year-old chaparral community is a biological desert, a decadent and dying system. Ecologists have taken censuses of wildlife numbers before and after burning to discover exploding young populations of deer, quail, rabbits, rodents, and birds. Burning produces open space and soil water for young, tender shoots of higher protein content. Deer population densities have been observed to increase by nearly four times when brush was cleared with fire.

Very few species of birds and other animals, such as rabbits, ground squirrels, and mice, seem to adapt to extensive stands of mature chaparral. After a fire these populations thrive, especially if there is a mixture of ages of brush or if there are openings with grass and other herbaceous plants. Many species of birds and mammals use the brush for protection by escaping and resting during the heat of the day but feed in nearby clearings. The dense mature stands provide little food in the form of grass or herbaceous plants because of the shade and competition from the sages and wood scrub.

If the chaparral is burned too frequently and extensively, it will convert to a grassland that is not as favorable to birds and mammals that hide in the brush cover and feed in the clearings. Too much fire erases the *ecotones* or transition zones between the brush and grass. Ecotones appear to be very important to the animal life of the chaparral.

Figure 13.18

Mudflows follow fire in the chaparral.

Watershed Protection

A large fire can destroy the ability of this plant community to hold back rapid runoff and soil erosion. Smaller fires with less fuel cause only minor erosion. Smaller fires burn smaller areas and generate less heat, leaving some brush intact. Large fires leave the ground bare and unprotected.

Water Availability

Plants recycle soil moisture back into the atmosphere by transpiration. Studies have shown that areas converted from chaparral to grass can increase groundwater and surface runoff up to 50 percent when properly managed with fire. Springs and streams that normally dry up during the late summer now flow year-round. Grass cover removes less water from the soil than the native chaparral.

These are certainly strong reasons for controlled burning of this brush, but there are prices to be paid when we take over the work of nature.

Soil Erosion and Mudflows

After a fire the bare soil is exposed to the forces of erosion. Brush serves as a canopy by intercepting the falling rain, filtering the wind and sunlight. Each of these forces has a strong influence on the environment, and the fire is only the first wave of destruction. When the rains come, the canyon bottoms become rivers of mud from runoff and landslides occur on the steep, bare slopes. Many foothill residential communities of southern California have experienced rivers of mud that literally leave destruction in their paths (fig. 13.18). A problem that seems to compound runoff is severe baking of the soil by the fire. Moisture is repelled by the ash-covered, rock-hard soil. The heat causes the soil to become water-repellent. Rain, instead of slowly soaking in, runs off all at once, carrying the soil and ashes with it.

Air Pollution

In smoggy regions of the Mediterranean zones, fire is not only a threat to property but a contributor to watery eyes and low visi-

Figure 13.19

Browsing animals take their toll on young sprouts. Springbuck in Gemsbok Park, Namibia.

Photo by Dr. Charles Hogue, Curator of Entomology, Los Angeles County Museum of Natural History.

bility. Air pollution, natural or artificial, is harmful to your health and can leave a layer of ash and soot on your property.

Scenery

Fire leaves black scars and a sense of destruction even though it is only temporary. The natural beauty of an oak woodland is lost to this generation, although it will recover in time. It takes hundreds of years to reproduce the largest coastal live oaks often found completely charred in a severe chaparral fire. There is no assurance that a controlled burn will remain controlled. Unpredictable winds can cause the most carefully planned controlled burn to leap the fire lines and burn thousands of acres downwind.

The Temperate Grasslands

The temperate grasslands form a large transition zone between both the Mediterranean Woodland and temperate forests and the dry interiors on each of the major continents.

As in the case of the tropical savannas, these grasslands could probably support more forests, but they seem to have been encouraged by the fires of earlier cultures and lightning strikes of nature. So, to understand the patterns of today, it is important not only to evaluate current climatic data but also to consider historical settlement and land uses of each grassland region. Here we see an excellent application of the concept of change described in chapter 1, "The Geographic Perspective."

Grassland Formation Processes

The forest does not stop suddenly like the border of an alpine meadow; there is a gradual transition through a forest grassland zone. Trees gradually decline in numbers or density and size as conditions become drier. Precipitation ranges from over 75 centimeters (30 inches) per year to less than 25 centimeters (10 inches) on the desert margins. The temperate grasslands form a fringe around the forests produced by the drier climate, plant competition, grazing, and fire.

Competition

At one edge of the transition zone, the forests dominate with tall grasses or prairie forming small islands or clearings. The climate itself does not favor either forest or grasses in this transition zone, but soils, drainage, slope, and competition with other plants shape the local community. Forests dominate the well-drained habitats on higher ground and shaded slopes or slopes facing poleward, while grasses do best on poorly drained soils and slopes more exposed to solar rays.

Forest also tends to predominate along the banks of streams and rivers where drought is not a problem and soils are sandy, porous, and enriched by alluvial materials. Grasses and tree seedlings compete with one another for soil moisture during the first few years, and the grasses often win out unless they are removed.

Grazing

Livestock, deer, and fires are the major enemies of young trees. Sheep, cattle, and deer browse the succulent young sprouts of oaks or conifers. Where animals have been fenced out, survival of young trees is markedly increased (fig. 13.19).

Fire

Fire is also very important as a forest inhibiting factor in this transition zone.

When a fire burns an open grassland, it tends to hold back the spread of woody plants, favoring the grasses. A grass fire removes the dry annual residue of grasses, leaving ashes to fertilize the soil. The shrubs, which have taken several years to grow, are severely set back even though many shrubs sprout rapidly after burning. Grasses reseed rapidly after the next season's rain. Therefore, frequent fire tends to favor grassland. Many ecologists believe that significant grassland belts of the continents owe their existence to fires started by both lightning and human populations. However, it is important not to overgeneralize because soils, climate, and competition with other plants are all significant environment factors determining a grassland environment. The concepts of *spatial interaction* and *internal coherence,* described in chapter 1, can be applied to better understand a plant community where many factors are shaping it.

In the arid Southwest and prairies of the Midwest, occasional fires have served to maintain the grasslands over other vegetation forms. Fire control and increased pressure from livestock grazing appear to be reducing the grasslands in many of these regions. Mesquite, a shrub in the grasslands of western Texas and eastern New Mexico, is spreading rapidly until it now covers over 28 million hectares (70 million acres) of former grassland. The question of major cause is yet unresolved, but it appears that fire control and overgrazing have favored the decline of grasses and increase in mesquite.

Drought

Drought is another inhibiting factor for trees on the prairie. The effects of the drought of 1934–1941 and again in 1988 on the North American grasslands are still visible today. These recurrent periods of drought, every century or so, are undoubtedly responsible for the absence of trees on the prairie. Drought kills off the very young and aging trees first. Young mature trees that are still growing rapidly may be set back by drought, but usually survive.

Vegetation Forms

The vegetation consists of perennial grasses and herbs reaching three meters (10 feet) tall in the more humid regions. A thick sod and some of the world's richest soils, described in chapter 11, have developed in this grassland environment. As one moves deeper into the continent's dry interior, tall **prairie** grasses are gradually replaced by the shorter varieties known as **steppe** grasses. The thick sod also gradually disappears, and more and more bare soil and rocky, sandy surfaces are exposed. Xerophytic plants such as cacti, mesquite, and short-lived flowering plants then dominate the dry landscape.

Animal Communities

The temperate grasslands of the world host a wide variety not only of plants but also of animal life. This life zone typifies the food chain concept. Here one can see the browsing herbivores such as deer, antelope, and bison gain energy from the rich grasses and herbs. Smaller animals such as ground squirrels, rabbits, mice, and gophers also feed on these nutritious plants. The plains also support a rich variety of birds and insects, some of which spend only the summers while others adapt to year-round living.

The bison, probably more than any other mammal, symbolizes the plant and animal community of the tall-grass prairie. This vegetation community, more than any other on the American or Eurasian continents, has been almost totally destroyed along with the animal community.

Until the Europeans began invading this region, local human populations had little impact on it. However, when the horse and gun came into use, the pace of disturbance accelerated. The horse and gun improved mobility and hunting ability. At one time it is estimated that at least 60 million bison roamed the prairie and steppe grassland of North America. It is estimated that by 1889, there were only 150 bison alive in the wild.

The food base was lost by fencing, fires, and agriculture, and the animal was quickly slaughtered for hides and trophies and to clear the land. Today the bison is making a comeback on preserves and wildlife refuges, where representative habitats can still be seen, in Montana and North Dakota.

Temperate Forests

The temperate forest belts of the middle latitudes form a transition from the drier prairie. The rhythm of the seasons is reflected in the fall and spring colors. As spring comes, new life rises in dormant plants and hibernating animals. The summers are noted for rapid periods of growth charged by long, sunny days and abundant precipitation. Then, as the days grow shorter and cooler, leaves begin producing a color display of golds, orange, and reds. Finally the forest rests under a coat of intermittent frost or snow (fig. 13.20).

It is unwise to overgeneralize about the temperate forests, because there are several varieties growing in a wide range of conditions. Temperate forests occur in the eastern half of the United States, much of Europe, eastern Australia, eastern China, New Zealand, Japan, Chile, Argentina, and Brazil. All of the above locations have deciduous trees, which lose their leaves in the winter (fig. 13.21). Another type of temperate forest is found on the west coasts of continents. Here the trees are needle-leaf and broadleaf evergreens. In North America this belt follows the west coast from coastal California into Alaska.

If we examine the locations more closely, we find that west coast locations, regardless of hemisphere, share common latitudinal zones starting at 35° and extending poleward to 60°. The east coast temperate forests also have common latitudinal locations between 25° and 45°. If mountain barriers are absent, the temperate forest extends inland until the climate becomes too cold and dry. The influence of moist marine air masses typifies this vegetation belt.

The temperate forest can also be found growing in the tropics. The highlands of equatorial Peru, southern Mexico, Guatemala, and Costa Rica provide similar cool, moist conditions with one difference. Seasons are almost nonexistent in terms of temperature and energy variations. However, tropical highlands do experience seasonal variations in precipitation similar to the tropical lowland climates of the same latitude.

Eastern Temperate Forests of North America

The eastern forests of the middle latitudes reflect the strong seasonal rhythm of insolation. Days are noticeably longer in the summer and shorter in the winter. The summer days are not only longer but very warm and humid, as if the tropics have moved north. In the southeastern part of the United States, moist maritime tropical air moves north around the Bermuda high-pressure center. Summer then brings abundant convectional showers mixed with very high humidity. Nights often do not cool off very much. The land is engulfed in tropical air.

Winters do not come suddenly, but gradually the days grow shorter and temperatures become cooler. Leaves reduce photosynthesis and begin displaying fall colors. Nutrient uptake also slows down. Finally in about late October or early November, the first winter cyclonic storms come, bringing perhaps some snow and cold clear nights. Now it's as if the subarctic has paid a visit to this land. These climatic conditions of cool temperatures, shorter daylight periods, and snow and frost continue through March. Then life erupts in blossoms and leaves on the deciduous trees, and wildflowers spring up and dormant grasses turn green in the clearings and meadows.

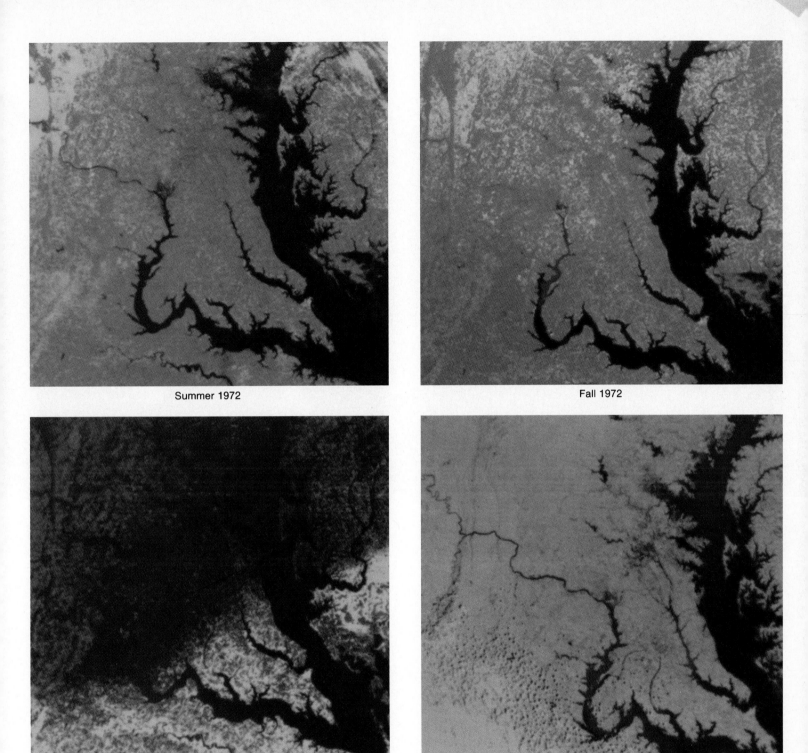

Summer 1972

Fall 1972

Winter 1972

Spring 1973

Figure 13.20

Temperate forests reflect the rhythm of the seasons in the middle latitudes. The deciduous tree is this forest's emblem of seasonal change in the Washington, D.C. region, shown in these false-color images from the ERTS-I satellite.

Photo by NASA.

Figure 13.21
This temperate deciduous forest in Rothenberg, Germany, responds
to seasonal change in autumn.
Photo by Elizabeth Wallen.

The plant associations are numerous in the deciduous temperate forest. Oak and hickory associate together in the Central Plains and Midwest extending to the Middle Atlantic states. In the South, oak and chestnut is the natural association, although the American chestnut tree was nearly wiped out by a fungus blight accidentally imported from Japan in 1904 on a few Japanese chestnut trees. The blight first appeared in New York, then swept across the eastern temperate forest like a forest fire. By 1950 most of the 9 million acres remaining in America had serious disease or were dead. Research long underway will undoubtedly bring back the chestnut, but it will be many years before the oak-chestnut association can be restored.

A flight over jack pine forests of Michigan, Minnesota, and Wisconsin reveal a widespread uniform structure resulting from the days of early logging and fire. The original forest, before logging, consisted of both jack pine and broadleaf hardwoods. Loggers selectively cut the pines and burned the slash and remaining hardwoods. All that was left was ashes and the hardy, fire-resistant seed of jack pine. When the cone of jack pine is heated

by fire, it opens and releases its seed. Fire favors this species and tips the balance of vegetation toward a single forest species. Jack pine thus has filled the vacuum created by fire and logging and no longer competes with other plants.

Pure stands of Douglas fir forest also require fire to compete with other plants. The young seedlings need light and will not grow in a mature, dimly illuminated stand. If a mature tree falls, more shade-tolerant hemlock and cedars will compete and become the climax vegetation of the community. A fast-moving small fire will burn out cedars and hemlock and mature firs, opening up the forest to fir seedlings and a pure stand of Douglas fir. Where fires have been excluded, the forest is dominated eventually by cedars and hemlocks. Fire not only opens the forest, encouraging the light lovers, but it also cleans the forest of debris and removes competing saplings. Thus, the remaining larger trees grow faster and have a greater survival potential when the next fire sweeps the area because there is far less fuel on the ground. One of the main reasons forest fires today are often catastrophic is that great quantities of fuel have accumulated. When the fire moves through the

A

Figure 13.22

(A) The great fires of Yellowstone, visible wavelengths. The great Yellowstone fires are an example of overmanagement and fire control for 90 years. As a result, fuel buildup coupled with a dry (humidities down to 6%) summer of 1988 created the longest burning and most extensive forest fire in the region's history. In the previous 116 years of park history, only 146,000 acres had burned. By August 20, 1988, over half of the park's area had been scarred by eight huge fires. (B) Color infrared of the same region in 1988.

Photos by NASA.

B

unburned forest of today, the entire stand of trees may be destroyed. The Yellowstone National Park fire of 1988 is a reminder of the principle (fig. 13.22).

Walking into a deciduous forest is an unforgettable experience summer or winter, spring or fall. The tallest oaks, maples, and hickory form a shady canopy. There may be lower tier consisting of flowering dogwood, blue beach, and hophornbeam. If the upper canopy and tiers are dense, the underlying shrub layer will be sparse. Gooseberry, spice bush, and a variety of herbs near the ground need light. The herb layer begins dying back in June when light is reaching its maximum at the canopy level but, paradoxically, it is at a medium on the ground.

Unlike the tropical rainforest, the ground is full of debris, fallen leaves, branches, and downed trees. Decomposition slows down during the winter when insects, bacteria, and chemical activity are greatly reduced. The soils in the forest are formed by **podzolization.** Leaching is common, removing water-soluble basic minerals into the root zone and water table, and leaving behind soluble iron and aluminum oxides and acidic conditions which favor forest development.

Many characteristic mammals of the forest such as the raccoon, eastern chipmunk, southern flying squirrel, gray fox, and bobcat are rapidly disappearing at the same time as the original forest. Today there is less than 25 percent of the eastern temperate forest of North America left and only one-tenth of one percent in virgin condition. The Great Smoky Mountains and Ozark Plateau are the two best examples of nearly virgin conditions.

The forest belt is rich in bird life. Some of the more common birds include the great horned owl, arcadian flycatcher, several varieties of woodpecker, and thrushes. Reptiles are also abundant. Some of the more common ones include the deadly copperhead and timber rattlesnake. Numerous salamanders and frogs live along the streams. Each of these members of the forest community are important consumers and processors. They take food

Figure 13.23

The Douglas fir–redwood community grows best where the fog belt dominates the coast of California. During the dry summer, coastal fogs provide moisture until the winter rains arrive.

Photo by David Swanland, courtesy of Save-the-Redwoods League. Used by permission.

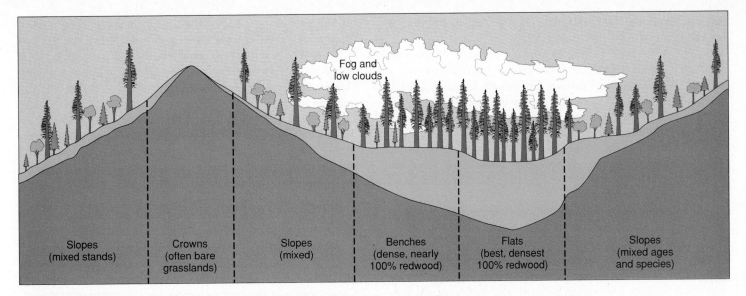

Fog and
low clouds

| Slopes (mixed stands) | Crowns (often bare grasslands) | Slopes (mixed) | Benches (dense, nearly 100% redwood) | Flats (best, densest 100% redwood) | Slopes (mixed ages and species) |

Figure 13.24

A cross section of a coastal valley reveals where the redwoods grow. The tallest trees are located along the valley bottoms where flooding is common. Here fresh nutrients are deposited and the trees' chief competitors are removed or severely damaged by the flood waters and silt deposits. Redwoods are smaller and find more competition on the higher slopes.

and energy, process it, and return it to the soil. The forest represents a community of interdependence between all of its members. Plants supply food, yet they need other organisms such as birds, bees, and bacteria to recycle nutrients back into their soluble forms.

West Coast Temperate Forests

The temperate forests of the west coast of North America represent one of the major remaining timber regions of North America. Along the Pacific coast from Monterey, California, to Anchorage, Alaska, grow some of the tallest and fastest-growing trees on the continent. Coastal redwood, Douglas fir, and Sitka spruce dominate the forest. Sitka spruce, grand fir, and Douglas fir associate from Oregon northward while coastal redwood and Douglas fir dominate the California coast and southwestern corner of Oregon. The term rainforest has been used to describe this region because of the abundant precipitation and dense forest conditions. Rainfall on the Olympic Peninsula in Washington generally exceeds 250 centimeters (100 inches) annually. Where rainfall is less in the southern regions, fog drip almost daily in the dry summer season accounts for 50 percent of the moisture.

Let us examine more closely the Douglas fir–redwood community as a prime example of a plant community in the west coast temperate forest. The coastal redwood favors a cool, moist environment. "Where the fog flows, the redwood grows" is a local saying in redwood country. Summers along the California coast are cool and moist with frequent drizzle, fog, and mist. When winter approaches, the fogs disperse, and the next months pass with alternate days of sunshine and rain as the cyclonic systems move across the land dumping large amounts of precipitation (fig. 13.23).

Location plays a key role in determining the environment and distribution of this plant community. Above 600 meters (2,000 feet), the air is too dry and warm throughout the summer months and too cold in the winter. Cold ocean currents moving south along the California coast refrigerate the air, causing fog and low clouds and a temperature inversion, but above it the summer sunlight dries the higher slopes and bakes the soil. Fog drip returns great quantities of moisture to the soil during the long rainless summer in the lower marine layer. To the east and above the marine layer, fog and cool coastal breezes turn to dry, hot winds. Winter precipitation falls short of enough moisture to tide the tree over the long hot periods of the inland valleys. So the coastal redwood is rarely found growing more than 50 kilometers (30 miles) from the coast and only in the fog belt. The small valley of our *study area* in chapter 1 is representative of a small, isolated inland grove located on a cooler north-facing slope and along a perennial stream. To the north, groves in the Chetco River area of Oregon's southwest corner mark the northern limit, where frost and competition from other plants began driving out the redwood seedlings. Figure 13.24 shows a cross-sectional distribution of the coastal redwood in a typical river valley along the coast of northern California.

Redwoods begin to disappear in the south, just north of San Simeon in the Santa Lucia Mountains of central California, because of dwindling precipitation. The redwoods stop just short of the coast because of all the salt spray, high winds, and lower fog drip and rainfall along the rocky shoreline. Fog and rain, however, are not the only environmental factors favoring the redwoods. Soils must be well drained and reasonably fertile. The best groves are located along the river flood plains and stream terraces where flooding is common. Here nature renews the soil fertility by bringing in fresh deposits of rich alluvial material.

A return visit to the *study area* of chapter 1 reveals that the coastal redwood–Douglas fir community is associated with a number of plant species. However, Douglas fir, a conifer, the tan oak, and madrone are its most frequent tree companions, especially on the upland hilly slopes in the central and northern Coast Range of California.

The stream banks and river flats have their own unique species associated with the redwoods. Bigleaf maple, red alder, and California laurel (bay pepperwood) are the most frequently found competing with the redwood.

Common shrubs include California huckleberry, creek dogwood, and poison oak and are often seen in the redwood belt, both on the flats and slopes.

For us to understand why the redwood is growing with such a variety of plants, we need an understanding of its physiological capabilities in combination with the environmental controls, such factors as seed production, fire resistance, growth response to soil nutrients, temperature and moisture stress, tolerance to flooding, and resistance to insects and other organisms of the forest.

The coastal redwood community is a good example of a habitat surviving today partially because of recurrent fires. The evidence of fire scars is present on most old trees and stumps. Redwood continually competes with other tree species for nutrients. Fire kills seedlings and youthful trees of redwood and its competitors while more mature redwoods survive. Mature redwood is resistant to fire. The bark is thick and burns poorly. If the crown is killed, a new one often emerges from buds along the stem and branches. A very hot fire can penetrate the bark protection and leave scars and points of entry for heartrot and more fires, which can burn out the heart of the tree and leave only a shell susceptible to wind toppling. Again we can see that small fires in a stand of redwood open the forest, reduce competitors, and clean excessive debris from the floor of the forest.

Fire prevention paradoxically works against the preservation of these giants. Mature trees that continue to survive without fire often do not become susceptible to heartrot. However, eventually they fall but when they do, almost without exception they become completely uprooted. In contrast, old fire-scarred, infected trees break off above the ground leaving a root system in place, and sprouts rapidly begin to grow around the stump. These sprouts become the next generation of mature redwoods. In 60 to 100 years sprouts can grow 60 meters (200 feet) into the air.

A mature tree that topples and is completely uprooted leaves no sprouts; on the other hand, sprouts are stimulated when a tree has been injured. Fire then actually stimulates regeneration by sprouting.

The Temperate Forest of Asia

Traveling north from the tropical rainforests of Indonesia into northern China and southern Japan, the climate shifts to a seasonal temperate pattern. The name "laurel forest" is applied to this transition zone because of similarities to the temperate evergreen oak and magnolia forest associations of the humid subtropical Gulf region of the southern United States. In Asia most of this original forest has given way to the ax for fuel, timber, and cropland. The lands are now under intensive cultivation. The same can be said of European temperate evergreen and deciduous forest. However, in small isolated places survivors of the ax are still standing.

One such place can still be observed in central China where laurel forest grades into more deciduous cold-resistant types. In Sichuan province, a variety of common deciduous oak, maple, and chestnut trees grow along with a deciduous redwood named the dawn redwood, described in box 13.2.

Boreal Forest or Taiga

Along the poleward side of the temperate mixed forest of evergreen and deciduous species lies a worldwide circumpolar belt of conifers, or cone-bearing trees. These evergreen trees carpet the North American continent where temperatures are harsh and cold a good portion of the year. In almost every environmental respect, the northern coniferous forests are in stark contrast with the tropical rainforest.

Vegetation Communities

Because the environment is harsh, only a few plant species can survive, in contrast with the great variety in the tropics. Spruce, fir, larch, and pine dominate the boreal or **taiga** northern lands.

Some of the common associations are spruce-fir forests in North America; larch, cedar, and maple occur with spruce and fir in the lowlands and poorly drained areas of western Siberia ("taiga" is a Russian word). This belt also extends south along the western coasts of Europe and North America and the eastern coast of Asia, stopping only where moisture declines or competition with plants or people forces the taiga to give way to other species. In North America, the boreal belt unfurls like a green ribbon down the west coast from Alaska to Central America. In the Pacific Northwest, the boreal forest merges with the temperate forests at or near sea level. As you trace this green ribbon south into Mexico, its elevation increases. Near Mexico City one must climb to over 2,400 meters (8,000 feet) to leave the temperate forest and grasslands and enter taiga conditions. The boreal forest in the lower latitudes takes to higher ground where temperatures are cooler and moisture is abundant. It is in this niche, regardless of latitude, that the boreal forest competes successfully.

The poleward limits of the boreal forest are controlled by temperature and available moisture. These same climatic limits restrict the growth of taiga forest at the upper limit on the mountain slope. We refer to this limit, where trees become sparse and only shrubs remain as the **timberline.**

In the arctic regions of Alaska, Canada, and Eurasia precipitation is less than 25 centimeters (10 inches) annually. Winters are long and cold. The 10° C (50° F) isotherm for the warmest month is approximately the upper limit where trees become dwarfed and the shrubs and grasses of the tundra begin to dominate. The southern limit of the taiga corresponds very closely to places having at least three months averaging above 10° C (50° F) and at least a fourth month not too far below it. Oaks need at least four months above 10° C to compete with conifers.

The Lost Redwood Forest—Alive and Well

Until 1944, a small valley in central China, enclosed by steep canyon walls, hid from the world a third species of redwood. Before this discovery, the dawn redwood, *Metasequoia*, was known only in fossil form. Its fossils occur in northern Siberia, Spitzbergen, Russia, Greenland, Alaska, Japan, Canada, and the Rocky Mountains of North America (fig. 1).

This discovery revealed a remarkable life span dating back 100 million years. The dawn redwood was thought to have become extinct 20 million years ago. Its actual fossil history shows it was a thriving tree 110 million years ago during the Early Cretaceous Period when dinosaurs roamed the continents. As the arctic climates cooled, the tree's distribution moved south. On the North American continent, it reached Nevada, but it was finally forced out of all locations except in China.

One of the clues to climatic conditions of a place is plant form or structure, and a study of the dawn redwood was no exception. Dr. Ralph Chaney observed deciduous, needle-leaf foliage, similar to the redwood's broadleaf hardwood companions of oak and chestnut. They also grow to about 50 meters (160 feet)

in height with branches that bow upward (fig. 2).

A visit in 1981 by Dr. Bruce Bartholomew in a joint Sino-American botanical expedition to western Hubei province revealed that these deciduous redwoods are barely surviving. The small valley is 200 kilometers (12.5 miles) long, running north and south, and is completely enclosed on all sides. A small river drains the valley southward through porous limestone formations. The valley bottom is occupied mostly by rice paddies, with farmhouses on both the east and west sides. The majority of the dawn redwood trees are located in the more protected side canyons along small tributaries. It is here that they reach 50 meters (160 feet) in height, with trunks up to 1.5 meters (5 feet) in diameter. A recent Forest Bureau survey lists 5,426 trees with diameters over 20 centimeters (8 inches). Only recently has the government protected these trees, while many of the companion hardwoods are still being cut for firewood and furniture.

As mentioned, clues to climate can be found in the tree itself. Bartholomew reports that the tree's rapid growth follows the monsoonal rhythm of warm summer rain and rapid

green leaf growth, followed by autumn color change to golden leaves, concluding in winter with deciduous leaf loss as dry north winds bring snow to the bare forest. March and April mark a change with the influx of southerly winds and the weather proceeds to warm. The dawn redwood then begins to bud. New seeds, from winter cone drop, spring up in the moist and warming soil of weathered limestone. The forest comes alive again to reflect the rhythm of the strong seasonal climate in this protected niche in China. Thus a deciduous forest reflects strong seasonal change.

There are some unanswered questions. Why is this the only place where the dawn redwood survives in a natural setting? Is it a unique combination of isolation, climate, soils, and geologic events? What role has human activity played in its current limited distribution? According to *Natural History* magazine in 1948, Ralph Chaney stated, "I hope to go back to Central China myself in the near future to help unravel the mystery of the nearly worldwide extinction of the dawn redwood and its companion trees." Dr. Chaney never did go back, so the questions are still unanswered today.

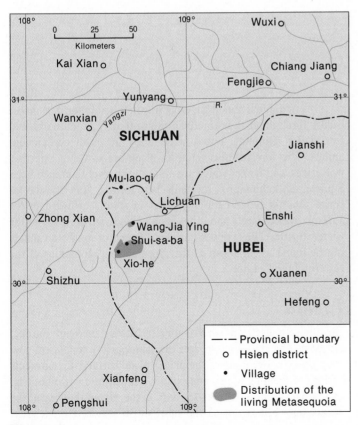

Figure 1

Sketch map showing the limited geographic area of *Metasequoia*.

Figure 2

The dawn redwood, survivor of the age of agriculture.

Figure 13.25

Tundra or grasses and shrubs ring the shorelines of the Arctic Ocean and cap the high mountains of the world. This shrub is sheltered from wind.

Forest Environment

If we examine the taiga forest floor, we find it is thickly carpeted with needles, many dead branches, and downed trees in contrast with the absence of debris in the tropical rainforest. When the land remains frozen for six to nine months, biological and chemical activity is limited to a short period in the summer. Therefore, the forest has a very cluttered look beneath the low canopy trees. Fires, started by summer lightning, can burn extensive areas where little natural control exists.

Because the taiga region was extensively glaciated in North America and Eurasia, the forest appears in patches where the soil has not been scraped away by ice. This is especially true in Scandinavia and eastern Canada.

The taiga, like the higher latitude tundra, is almost lacking in the Southern Hemisphere because the growing season ranges between three and six months, long enough for a magnificent hardy forest. Coniferous trees, especially spruce, dominate the landscape, along with alder, birch, aspen, and juniper thickets. Stands of aspen and birch fill niches, cleared by fires, which later are overcome by invading conifers.

Wildlife is abundant, especially in the summer. The moose occurs throughout the American taiga. It may be outnumbered by smaller mammals but it is certainly the most conspicuous animal. This forest is also noted for black bears, wolves, martins, wolverines, and lynxes which are almost exclusively confined here. Squirrels and chipmunks thrive in these forests. This is also a favorite summer habitat for birds that migrate southward in the autumn. Most insects and other invertebrates are dormant in the winter months. The hemlock-hardwood forests of southern Canada can be found as far south as the Appalachian Mountains of the eastern United States.

Tundra

The tundra spans the northern shores of North American, Europe, and Asia. The land is harsh, treeless, and sparsely populated, covering nearly a twentieth of the planet's land surface. It represents a fringe of teeming life thriving under temperature extremes (fig. 13.25). The ground freezes rock-hard in the winter, thawing at the surface only briefly during the short warm season. Here a great variety of sedges, grasses, mosses, and lichens grow on a thin, rocky, windswept soil base. Tundra plants grow where trees cannot. The Arctic Ocean is ringed by this plant community. It can also be found above timberline on alpine slopes scattered throughout the world.

Environment

Temperatures in January often approach −40° C (−40° F) and sometimes −57° C (−70° F). Yet summer temperatures can reach 21° C to 28° C (70° F to 80° F) when the Sun never sets for several weeks. The yearly range can be as much as 56° C (100° F). The growing season lasts no more than 60 days, but sunlight is nearly constant during this period, so plants grow rapidly.

Not only are temperatures harsh but precipitation is very sparse. Rainfall comes during the short growing season with amounts ranging from 13 to 25 centimeters (5 to 10 inches) annually. Yet, because of low temperatures, three-quarters of the year, little moisture is lost due to evaporation. In fact, great quantities are locked up in the frozen state most of the time.

Plant and Animal Communities

Summer literally erupts in the tundra. The Sun's warming rays cause plants and animal life to come alive, flourish and reproduce before the chilly death of winter advances over the land coating it with an icy blanket for nine to ten months of the year.

Plants and animals adjust to this cool land in a variety of ways. First, many tundra plants during the brief growing season bloom and go to seed before winter comes, similar to desert annuals. Some have highly specialized roots which spread horizontally in the shallow thawed zone of soil, or active layer. The plants of the tundra are often colorful and dwarfed, such as the dwarf willow, dogwood or bunchberry, and mountain cranberry. Lichens, lacking roots, form a plant community within a single organism composed of algae and fungi. This plant community is one of the most adaptive forms of life, clinging to wood, stone, or soil and growing in the tropics, desert, and polar regions. It represents a symbiotic relationship of interdependence.

The only trees to be found in this region are like bonsai: alders, birch, and willow grow hugging the ground. They may appear more like young shrubs, but some may be very old, on the order of hundreds of years. Growth is imperceptibly slow. Most of the plants are perennials, passing the winter under the protection of snow. Their roots are thick storage compartments allowing them to survive on their stored supply during the long winters.

When spring comes, birds arrive from all over the world to the tundra, their breeding grounds. Only a few species such as the snowy owl and raven remain the year around. Most come for a few weeks when insects are abundant, to lay eggs in soft pillows of lichen, raise their young, and then return on their long journey back to the United States, Mexico, and southward, even to the tip of South America.

Mammals of the tundra show similar adjustments to the long, cold winter. Because there are few places of refuge, little hibernation takes place. Small animals burrow beneath the snow, and others, like the caribou, migrate south into the taiga. The Arctic fox grows a thick white coat, the Arctic ground squirrel hibernates, and the lemming burrows under the snow for food. Some animals have heavily furred paws and small ears and tails to reduce heat loss, while others develop thick layers of fat to insulate against the cold and store energy for the winter.

Summers stimulate many biological changes such as shorter coats, darker colors, and increased appetites for both food and mates.

Although insects are few in species, they become numerous as the weather warms. Mosquitoes, black flies, and deer flies reach unbelievable numbers when the permafrost thaws on the surface.

SUMMARY

Looking at the world's major plant communities, several important themes or concepts can serve as unifying concepts. The chief control or influence on vegetation is climate. Each community is the product of environmental and floristic influences that are in a dynamic state of change.

The tropical communities are energized with abundant solar energy. Precipitation can vary tremendously from the wet extremes of the tropical rainforest to the arid deserts of the Sahara. Water variation is the key variable in defining the structure of plants in the tropics. Where temperatures and precipitation are high, the most luxuriant and diverse plant species are found. Moving poleward toward the tropical deserts, plant and animal species decline in number and variety as water availability declines. Plant forms reflect this water shortage by developing structures designed to survive drought.

Plant communities of the middle latitudes reflect not only variability in water availability but variation in seasonal temperatures. Plants grow in rhythmic seasonal cycles. The deciduous forest of the eastern and western coasts of continents are synchronized to the Sun's path. The deserts and steppe of the middle latitudes differ markedly from the tropical dry lands in seasonal temperature patterns. Summers may be hot but winters can be exceedingly cold. Plant forms reflect these environmental characteristics.

The taiga or boreal forests are located where the climate is cool and moist a major portion of the year. Conifers provide an evergreen carpet across the northern continents that is structurally uniform with few plant species. Here, low temperatures produce conditions that exclude most broadleaf sun-lovers. Near the northern coasts, the trees of the taiga give way to the harsh cold tundra of the Arctic. Plants of the tundra thrive for only a few weeks during the summer's long daylight hours. Both moisture and heat are in short supply, so as we might expect, plant and animal species are few in numbers.

Consider plants (and animals to a lesser extent), because of their lack of mobility, to be indicators of environmental factors of climate and soils. They are products of their environment and reflect the mix of the major environmental influences discussed in chapter 12.

Water is the key to life. It is the blood of plants. Where it is abundant, so is life. As we might expect, the green belts of the Earth are in the major precipitation zones of the middle latitudes and in the wet equatorial tropics. However, if one simply travels up a mountain slope in any latitude, one can rapidly move from one environment to another.

1. Contrast the complexity of the Canadian and Amazonian plant communities (fig. 13.1, p. 256).
2. Give one example of convergence (fig. 13.2, p. 257).
3. Make an overlay of vegetation patterns for Australia on a climate map and describe the patterns of each in terms of high or low correlation (fig. 13.3, pp. 258–59, and Appendix I).
4. Locate the major vegetation regions and explain the influence of climate on each plant community (fig. 13.3, pp. 258–59).
5. Locate global high and low pressure patterns and describe the vegetation associated with each barometric condition (figs. 13.3, 13.15, and 13.16, pp. 270, 272).
6. Which vegetation patterns are associated with cold ocean currents of the middle latitudes (figs. 8.4 and 13.3, pp. 158, 258–59)?
7. Describe a location factor all temperate forests have in common. How does this factor affect seasonal temperature patterns (fig. 13.3, pp. 258–59)?
8. Plot the 10° C (50° F) mean annual isotherm on a vegetation map and describe the correlation with the equatorward boundary of the tundra plant community (figs. 13.3 and 13.25, pp. 258–59, 284).
9. What effect does light availability have on plant zonation and plant tiers (fig. 13.7, p. 264)?
10. Contract tropical scrub woodland and savanna with tropical semideciduous forest (figs. 9.10 and 13.8, pp. 181, 265).
11. Give three environmental factors that characterize the dry lands (fig. 13.13, p. 268).
12. What are two benefits of fire in the chaparral and scrub woodland savanna plant communities (figs. 13.10 and 13.17, pp. 266, 273)?
13. Explain why the great Yellowstone fires of 1988 are an example of overmanagement (fig. 13.22, p. 279).
14. How close does the coastal fog pattern resemble the coastal redwood plant community (figs. 13.23 and 13.24, pp. 280–281)?
15. Describe a vegetation cross section of a coastal valley where redwoods grow. Where are the grasslands located (fig. 13.24, p. 281)?

The Lithosphere

UNIT IV

The lithosphere forms the crust of the Earth. This crust is always in a state of change, experiencing dynamic internal forces and weathering and erosion on its surface.

Park Avenue, Arches National Park, Utah.
© Kerrick James

Rocks and Minerals of the Earth's Crust

14

Objectives

After completing this chapter you will be able to:

1. Explain how **minerals** are part of the rock formation process.
2. Explain the general characteristics of the three major rock families.
3. Distinguish between **igneous, metamorphic,** and **sedimentary rocks.**
4. Explain how rocks are formed according to the **"rock cycle."**
5. Describe the origin, distribution, and mining methods for energy fuels, metal ores, and nonmetallic minerals.

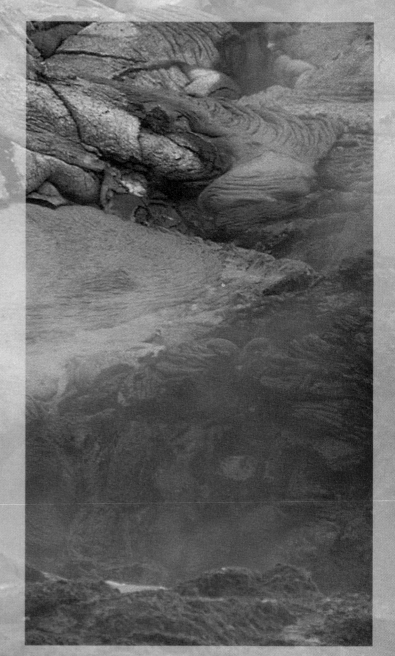

The birth place of rocks and minerals. New lava reaches the sea, Kalapana Coast, Hawaii Volcanoes National Park.
© Stephen Trimble

Consider that there are only 92 naturally occurring elements in the Earth's crust, atmosphere, and oceans, and yet an enormous diversity of solids, liquids, and gases. When you pick up a rock from a cold mountain stream or wave-swept beach, the beauty and variety of rocks are amazing. What gives the unique qualities to the rock you selected are the variety of minerals that combine to make it up. Let's examine what a mineral is.

MINERALS, THE BASIC COMPONENTS OF ROCKS

A **mineral** can be simply defined as a naturally occurring, inorganic solid with specific physical properties and a consistent internal crystal structure. "Naturally occurring" means nature produced it. "Inorganic" tells us it is from nonbiological processes. It must be in the solid state: water can be a mineral, but only if it is ice. The chemical properties can range from a single element to complex compounds consisting of several elements. *Diamond* and *graphite* are examples of a single element, carbon. The crystalline structures are determined by arrangement of its atoms. Although graphite and diamond have the same carbon chemistry, the arrangement of the element carbon gives these two unique minerals their physical properties.

Think of minerals, then, as the fundamental building blocks of the Earth's crust and rocks. Nearly all minerals have a crystalline form (fig. 14.1). They have a **crystal form,** unique **chemical composition, specific hardness** and **density, luster,** and **cleavage** or **fracture** characteristics. These unique physical and chemical properties help us to identify them.

When rocks and minerals form, heat energy is either released or absorbed in the process, so again we see that energy is very much at the heart of all mineral and rock formation processes, which we will examine in this chapter. We will also examine the distribution and formation processes of the ore bodies and energy resources upon which we depend.

Let's examine two key physical properties that can be observed in minerals. These include crystal form and hardness. Once we have identified the mineral or minerals, our rock's identity is no longer a mystery.

Crystal Form

Crystal form reflects the internal atomic structure of the mineral. As we have mentioned, diamond and graphite are chemically identical, consisting of pure carbon, but their internal atomic structures are vastly different. This difference gives graphite its black, opaque color, its softness, and its greasy feel. Diamond has an atomic structure where each carbon atom bonds to every adjacent atom. This bonding makes diamonds transparent, colorless, and extremely hard. Crystal form refers to the assemblage of flat faces that make up the surface (fig. 14.2). There are seven crystal systems. The same mineral will always show the same geometric design and fall into one of the seven crystal systems (table 14.1).

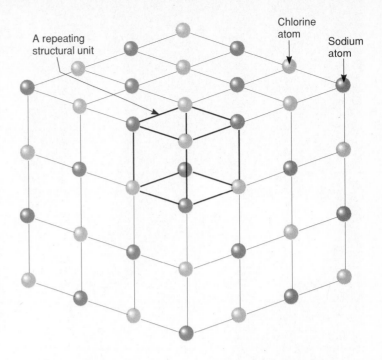

Figure 14.1

General physical properties of minerals are determined by the internal arrangement of the atoms. Crystal form, hardness, specific gravity, cleavage, fracture, tenacity, color, luster, and even magnetism are the result of atomic structure. Here the basic structure of halite (sodium chloride) determines the cubic crystal form. *The cleavage directions of halite are parallel to the cube faces of its crystals.*

Photo © Wm. C. Brown Publishers/Photography by Bob Coyle.

Hardness

Hardness is resistance to scratching or abrasion. Friedrich Mohs (1773–1839), a German mineralogist, introduced a relative hardness scale now called the Mohs Scale of Hardness. It consists of 10 common minerals arranged in order of increasing hardness

Figure 14.2

Quartz crystals are always shaped with the same angle between two adjacent faces. The angle 120° forms a six-sided pillar. The faces of the "pyramid" meet the sides at an angle of exactly 141°45′.

Photo by U.S. Geological Survey.

Table 14.1
Characteristics of the Seven Crystal Systems and Some Examples

CRYSTAL SYSTEM	CHARACTERISTICS	EXAMPLES*
CUBIC (ISOMETRIC)	Three mutually perpendicular axes, all of the same length ($a_1 = a_2 = a_3$). Four-fold axis of symmetry around a_1, a_2, and a_3.	Halite (cube) Pyrite Fluorite Galena — Magnetite (octahedron) — Pyrite — Fluorite (twinned)
TETRAGONAL	Three mutually perpendicular axes, two of the same length ($a_1 = a_2$) and a third (c) of a length not equal to the other two. Four-fold axis of symmetry around c.	Zircon — Zircon
HEXAGONAL	Three horizontal axes of the same length ($a_1 = a_2 = a_3$) and intersecting at 120°. The fourth axis (c) is perpendicular to the other three. Six-fold axis of symmetry around c.	Apatite — Apatite
TRIGONAL	Three horizontal axes of the same length ($a_1 = a_2 = a_3$) and intersecting at 120°. The fourth axis (c) is perpendicular to the other three. Three-fold axis of symmetry around c.	Quartz — Corundum — Calcite (flat rhomb) — Calcite (scalenohedron) — Calcite (steep rhomb) — Calcite (twinned)
ORTHORHOMBIC	Three mutually perpendicular axes of different length. ($a \neq b \neq c$). Two-fold axis of symmetry around a, b, and c.	Topaz — Staurolite** (twinned)
MONOCLINIC	Two mutually perpendicular axes (b and c) of any length. A third axis (a) at an oblique angle (β) to the plane of the other two. Two-fold axis of symmetry around b.	Orthoclase — Orthoclase (carlsbad twin) — Gypsum — Gypsum (twinned)
TRICLINIC	Three axes at oblique angles (α, β and α), all of unequal length. No rotational symmetry.	Plagioclase

Source: K. G. Cox, Price, and Harte. *An Introduction of Practical Study of Crystals, Minerals, and Rocks,* revised edition. Copyright © 1974 McGraw-Hill Book Co.

Table 14.2
Mohs' Scale of Hardness

Hardness	Standard minerals	Common objects
1	Talc	
2	Gypsum	
2½		Fingernail
3	Calcite	
3½		Copper penny
4	Fluorite	
5	Apatite	
5–5½		Knife blade
5½		Glass plate
6	Orthoclase (Feldspar)	
6½		Steel file
7	Quartz	
8	Topaz	
9	Corundum	
10	Diamond	

Source: C. C. Plummer & D. McGeary, *Physical Geology,* 5th ed., p. 37. Dubuque, Ia.: Wm. C. Brown, 1991.

(table 14.2). Graphite, for example, is harder than talc, and calcite is harder than gypsum but not as hard as fluorite. If an unknown mineral will scratch calcite but not fluorite its hardness is said to be between 3 and 4 on the hardness scale. A knife blade will test 5½, a steel file 6½, and your fingernail is 2½. Nothing can "touch" or scratch diamond, with a Mohs hardness of 10.

Additional properties include **specific gravity, cleavage, fracture, tenacity, luster, color optical properties, taste, feel,** and **chemical reactions** with other chemicals. For example, bubbles will form when calcite is treated with a weak solution of hydrochloric acid. This bubbling action is a test of calcium, not only in calcite but also limestone.

MINERAL CLASSIFICATION
Silicates

All minerals can be grouped into either **silicates** or nonsilicates. As we see in table 14.3, the two most common elements are silicon and oxygen. The silicate group represents the most abundant minerals found. The silica (SiO_2) structure gives each mineral in this group unique physical properties.

The simplest silicate structure is that of *quartz,* containing only silicon and oxygen. First place in abundance goes to **feldspar,** a family of closely related minerals that are rich not only in the elements silicon and oxygen but also in aluminum and either sodium and potassium or calcium. These are all abundant elements in the Earth's crust. Ferromagnesian silicates are rich in magnesium or iron (table 14.4) and tend to weather or decompose more easily.

Table 14.3
Chemical Grouping and Composition of Some Common Minerals

Chemical Group	Mineral Name	Chemical Formula*
Elements	Native Copper	Cu
	Graphite	C
	Diamond	C
Oxides	Quartz	SiO_2
	Hematite	Fe_2O_3
	Magnetite	$FeO \cdot Fe_2O_3$
	Limonite	$Fe_2O_3 \cdot nH_2O$
	Corundum	Al_2O_3
Sulfides	Pyrite	FeS_2
	Chalcopyrite	$CuFeS_2$
	Galena	PbS
	Sphalerite	ZnS
Sulfates	Anhydrite	$CaSO_4$
	Gypsum	$CaSO_4 \cdot 2H_2O$
Carbonates	Calcite	$CaSO_4$
	Dolomite	$Ca,Mg(CO_3)_2$
Phosphates	Apatite	$Ca_5(PO_4)_3F$
Halides	Halite	$NaCl$
	Fluorite	CaF_2
Silicates Olivine Group	Olivine	$(Mg,Fe)_2SiO_4$
Amphibole Group	Hornblende	Ca,Na,Mg,Fe, Al Silicate
	Asbestos (fibrous serpentine)	Mg,Al Silicate
Pyroxene Group	Augite	Ca,Mg,Fe,Al Silicate
Mica Group	Muscovite	K,Al Silicate
	Biotite	K,Mg,Fe,Al Silicate
	Chlorite	Mg,Fe,Al Silicate
	Talc	Mg Silicate
	Kaolinite	Al Silicate
Feldspar Group	Orthoclase (K-Feldspar)	$K(AlSi_3O_8)$
	Plagioclase (Ab,An)	Mixture of Ab and An
	Albite (Ab)	$Na(AlSi_3O_8)$
	Anorthite (An)	$Ca(Al_2Si_2O_8)$

*Some common elements and their symbols:

Al — Aluminum	Fe — Iron	O — Oxygen
C — Carbon	H — Hydrogen	P — Phosphorus
Ca — Calcium	K — Potassium	Pb — Lead
Cl — Chlorine	Mg — Magnesium	S — Sulfur
Cu — Copper	Mn — Manganese	Si — Silicon
F — Fluorine	Na — Sodium	Zn — Zinc

Source: J. H. Zumberge and R. H. Rutford, *Laboratory Manual for Physical Geology,* 8th ed., p. 16. Dubuque, Ia.: Wm. C. Brown, 1991.

Table 14.4
Crystal Abundance of Elements

Element	Symbol	Percentage by Weight	Percentage by Volume	Percentage of Atoms
Oxygen	O	46.6	93.8	60.5
Silicon	Si	27.7	0.9	20.5
Aluminum	Al	8.1	0.8	6.2
Iron	Fe	5.0	0.5	1.9
Calcium	Ca	3.6	1.0	1.9
Sodium	Na	2.8	1.2	2.5
Potassium	K	2.6	1.5	1.8
Magnesium	Mg	2.1	0.3	1.4
All other elements		1.5	—	3.3

Source: C. C. Plummer & D. McGeary, *Physical Geology,* 5th ed., p. 28. Dubuque, Ia.: Wm. C. Brown, 1991.

Clay minerals have sheetlike structures, with stacks of silica sheets that are weakly bonded together. Because of this atomic structure, clay minerals can absorb large quantities of water between the sheets. This property may cause the mineral to expand when wet and shrink when dry. Soil that is rich in this type of clay may be very hazardous for a building site.

Nonsilicates

Nonsilicates include minerals that can be grouped by common chemical characteristics. These include oxides, sulfides, sulfates, carbonates, halides, and native elements (table 14.4).

ROCKS—KEYS TO THE PAST

To a geologist, rocks are the documents that record the Earth's history, signposts of past geologic events. In the eighteenth century science began to answer some of the mysteries held secret in the rocks. In 1788 James Hutton, of Scotland, first published his "Theory of the Earth," based on the principle expressed in the maxim, "the present is the key to the past." Geologists refer to this concept as the principle of **uniformitarianism.** It means that in reconstructing the geologic past, we assume that the forces now operating to change the face of the Earth have worked in the same way, and at roughly the same rate, over all of geologic time. His views strongly shaped all later studies and discoveries in geology because this principle, though not the only principle in geology, has guided a vast amount of very fruitful research.

THE EARTH'S CRUST

When we look at the Earth's crust, several striking facts stand out. First, most of it is drowned by oceans. Nearly three-fourths of the surface is covered by water. Take a globe and hold it so that you are focusing on France. In this position, 81 percent of all the land area is within the hemisphere you are viewing, with France representing the geographic center of the continental masses.

Now turn the globe over to the opposite side, and you will be focusing on New Zealand in the watery world of the other hemisphere. The continents of the world stand out as floating islands when viewed from space. Actually, this perspective is correct in another sense. The crust consists of two major kinds of rocks, making up the continents and the ocean floor respectively. The volcanic rock of the ocean floor is rich in silicon and magnesium hence is referred to as **sima** or **mafic rock.** The continents are composed of **sial** or **felsic rocks,** such as granite, rich in silicon and aluminum (table 14.4). The crust, both sima and sial, is lighter and literally floats on the hot, plastic rock below, within the upper mantle, known as the **asthenosphere,** 100 kilometers (60 miles) or more below the surface (fig. 14.3). The continents, being made of lighter rock, ride higher than the ocean floor.

THE ROCK FAMILY— IGNEOUS, METAMORPHIC, AND SEDIMENTARY ROCK

To the untrained eye, when you've seen one rock, you've seen them all. They tend to merge into a great assortment of sizes, shapes, and colors; but when viewed through the eyes of a geologist, order and consistency replace confusion. The crust of the Earth is really a thin veneer overlying a massive mantle. The crustal rocks can be classified into just three general groups according to origin. These major groups are **igneous, sedimentary,** and **metamorphic rocks** (fig. 14.4).

Igneous Rocks

Mineral grain size is used to distinguish between two classes of igneous rocks; **plutonic** or **intrusive** rocks formed deep in the Earth, and **volcanic** or **extrusive** rocks formed from cooling magmas on the surface. Grain size reflects a rock's history and gives a simple, rational basis for classification.

The molten material or **magma** from which igneous rock forms is thick and viscous, resembling melted glass in both properties and composition. In some instances, molten lava cools very rapidly to a natural glass, **obsidian.** Usually cooling is slow enough to allow some crystals to form. Rapid cooling produces fine-grained crystals or part crystal and glass mixtures. If cooling is very slow, the minerals crystallize and grow large to form a coarse-grained rock (fig. 14.5). Examples of fast-cooling volcanic extrusive rocks include **rhyolite, andesite,** and **basalt,** all of which form on the Earth's surface.

The intrusive counterpart to rhyolite is **granite,** which has a coarse-grained texture. Granites have cooled very slowly deep in the Earth and are now exposed through uplift and erosional processes.

Figure 14.6 shows two of the most commonly found igneous rocks, fine-grained **basalt** and coarse-grained **granite.** These, like all materials of the Earth, are made of one or more minerals,

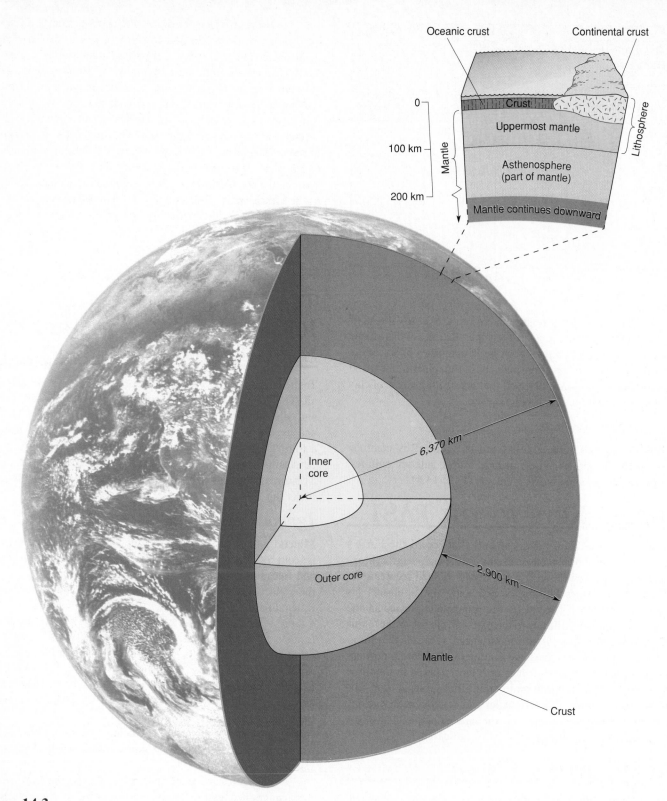

Figure 14.3

The continents rest on the plastic layer of the mantle or
asthenosphere 100 kilometers below the surface. The crust ranges
from 8 to over 50 kilometers in thickness.

A

B

C

Figure 14.4

The crust of the Earth can be classified into three major rock groups or families: **igneous, sedimentary,** and **metamorphic** rocks. (A) The Incas of Peru shaped this volcanic rhyolite into perfect fits. (B) Sandstone blocks support a bank in Bern, Switzerland. (C) Marble-granite pillars at Cesarea, Israel, were once imported by the Romans. The granite came from Egypt at Aswan and the marble was imported from Greece to this fortress site.

A

B

Figure 14.5

Rapidly cooled lava forms natural glass or obsidian (A) with no crystal structure. These two volcanic rocks cooled at very different rates. In porphyritic andesite (B), crystallization had already begun before eruption; hence, large phenocrysts (light colored) were formed, giving this rock a porphyritic texture.

Photos by C. C. Plummer.

each with specific chemical and physical properties. The mineral *texture* and *chemical composition* of the magma are indications of the rate of cooling and the physical environment in which these rocks formed.

Figure 14.7 shows a more detailed classification of the major igneous rocks. The change in mineral composition from the top to the bottom of the graphs parallels a change in chemical composition. Note that granite and rhyolite are rich in quartz, making up 70 percent by weight, while basalt and gabbro are lacking quartz. Rocks rich in quartz are generally low in ferro-magnesian minerals and are referred to as **felsic** rocks or siliceous, meaning they have high silica (SiO_2) and aluminum (Al) content. Gabbro and basalt, which contain an abundance of the basic oxides of iron, calcium, and magnesium, are called basic or **mafic** rocks. Basalt represents 97 percent of all volcanic rock. It has high iron and magnesium content and densities around 3 grams per cubic centimeter.

Figure 14.6

In A, fresh pillow basalt, an extrusive rock, forms on the ocean floor.
In B, El Capitan in Yosemite National Park is intrusive granite,
exposed by erosion.

A: Photo by U.S. Geological Survey.

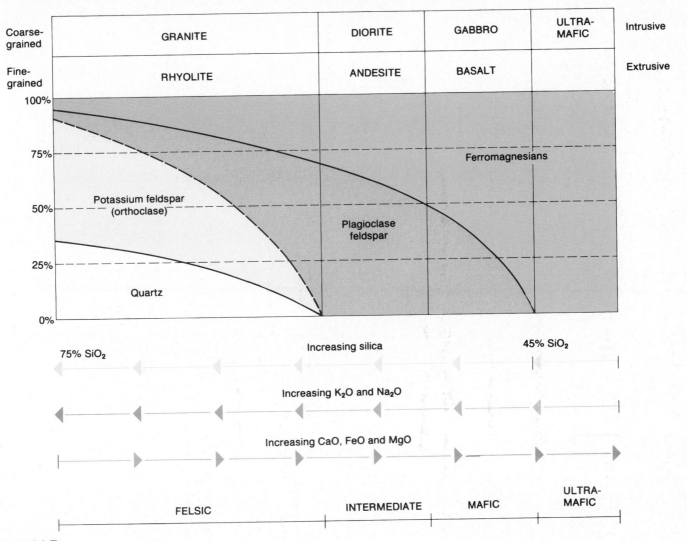

Figure 14.7

Commonly found igneous rocks arranged by texture and mineral composition.

GRAVITATIONAL ENERGY MAKES ITS DEPOSITS

Sedimentary Rocks

Sedimentary rocks are the most commonly found rocks on the continents. Covering over 70 percent of the North American conti-

nent, they are striking because of their layered appearance (fig. 14.8). Leonardo da Vinci, artist, inventor, and scientist of the sixteenth century, observed and recorded in his notebook the close similarity between sedimentary rocks high in the mountains of northern Italy and the sand and mud of the seashore. He noted fossils of seashells in the mountains similar to the living shellfish along the beach. He came to the correct conclusion that layers of rock seen in the mountains of northern Italy are seafloor deposits lifted high above sea level.

Sedimentary rocks, as the name indicates, are formed of layers of material such as coarse sand and gravel, silt, and clay washed into lakes, streams, and the ocean floor. Particles are then cemented together by substances such as calcium carbonate, silica, and iron oxides. Along with **cementation,** compaction is going on as particles are continually buried under newer deposits.

When igneous rocks are weathered and eroded, new rocks are eventually formed. These new rocks—sedimentary rocks—are formed by several important energy processes. The Sun's energy, gravity, and heat from the Earth's interior all play a role in their making. Let us examine where the work of gravity has aided in the removal, transportation, and deposition of sediments to form sedimentary rocks.

Figure 14.8
The Grand Canyon of Arizona gives us a cross section of sedimentary rocks.

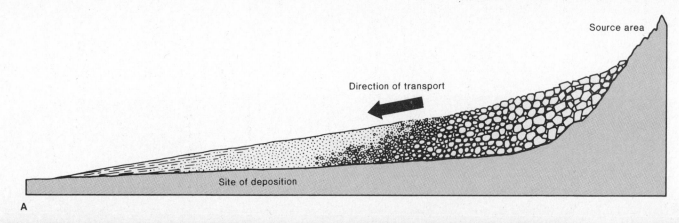

A

B

Figure 14.9

Graded bedding results from the sudden loss of energy of a stream, causing larger particles to deposit first. A mudflow carried these large boulders downslope until the gradient change caused the heavier rocks to be deposited on a park lawn in Azusa, California.

Stratification and Graded Beds

One of the most striking features of sedimentary rocks is their **stratification** or bedded appearance. How do rocks become bedded? Why do some beds consist of mainly gravel while others are clay?

To answer this question for yourself, take some garden soil and put it into a jar of water. Shake the mixture well and then let it stand. The heaviest particles will settle out first and form a deposit on the bottom of the jar. This is stratification. The beds are *graded* according to particle size from coarse to fine. This is also a good way to determine the percent of sand, silt, and clay particles in the garden. Graded bedding is rather uncommon on land but does occur when streams suddenly lose their energy following flood stage (fig. 14.9). Graded bedding is very common offshore where sediments laid down by rivers move down the continental shelf in occasional large flows. It is in large, shallow nearshore basins that most sedimentary rocks form (fig. 14.10).

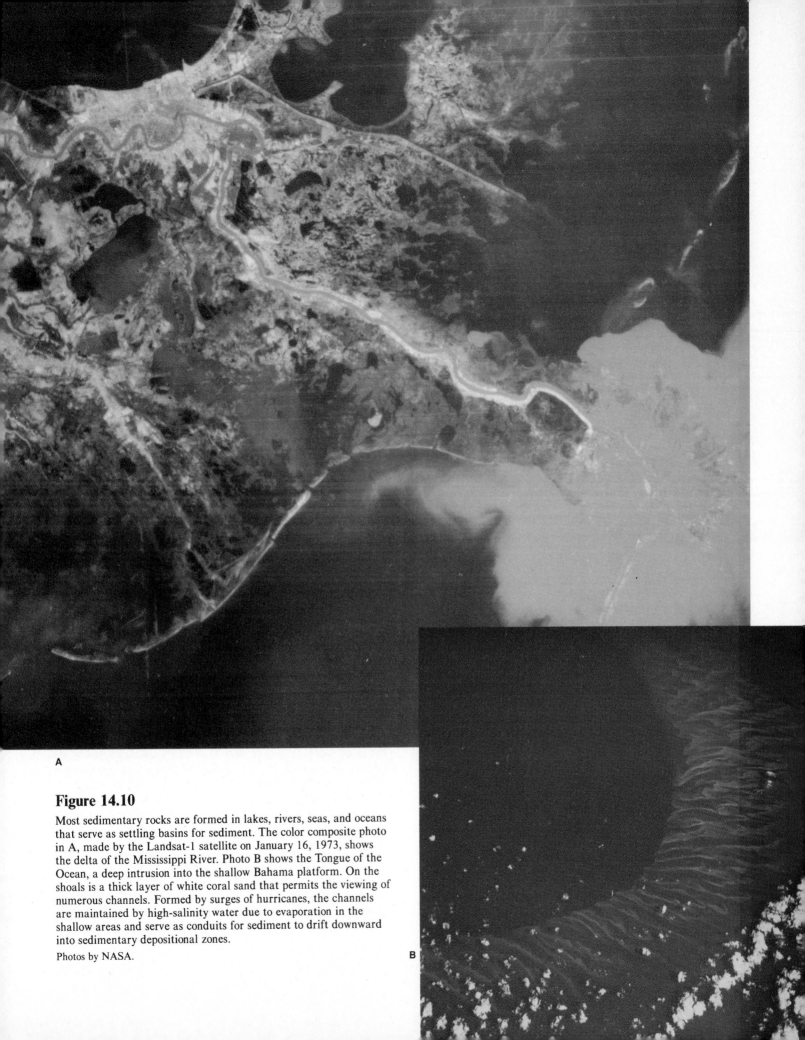

A

Figure 14.10

Most sedimentary rocks are formed in lakes, rivers, seas, and oceans
that serve as settling basins for sediment. The color composite photo
in A, made by the Landsat-1 satellite on January 16, 1973, shows
the delta of the Mississippi River. Photo B shows the Tongue of the
Ocean, a deep intrusion into the shallow Bahama platform. On the
shoals is a thick layer of white coral sand that permits the viewing of
numerous channels. Formed by surges of hurricanes, the channels
are maintained by high-salinity water due to evaporation in the
shallow areas and serve as conduits for sediment to drift downward
into sedimentary depositional zones.

Photos by NASA.

B

Figure 14.11

An unconformity results when deposits, such as these tilted beds, have been eroded away and younger beds are laid down on top of the erosional surface.

Much more common than graded bedding is laminated bedding of many different layers, or laminae. One bed may consist of sand, another of gravel, and another of clay-sized particles. In laminated beds, changes in grain sizes result from fluctuations of energy in a stream. Some beds change when time gaps occur between deposition. Erosion of a bed is followed by more deposition. Thus a portion of the older bed is missing, and an erosional surface forms the boundary between the beds in contact. We call this contact an unconformity (fig. 14.11).

Origin of Sedimentary Rocks

By revisiting the *study area* described in chapter 1, we can observe the processes involved in the making of sedimentary rock.

The whole process began several million years ago when the Pacific plate's leading edge collided with the North American plate at a rate slow enough to buckle the crust, raising a young mountain range between the two plates. The plate concept is covered in chapter 15.

The coastal mountains consist of crustal material that folds under compression between the two plates. This process is still going on while the destructive forces of weathering and erosion are removing the mountains and returning them to the sea in the form of sediment carried by streams and rivers draining the mountains. Sand, silt, and clay become continental shelf sediment along the coast. Some of it is carried by currents and deposited on the beaches by waves. In deeper, quieter waters, compaction and cementation over long periods of time form new sedimentary rock.

In this entire sequence, the forces of the Earth's interior, coupled with energy from the Sun and the pull of gravity, pulverize, transport, and compact the Earth's material to form new rocks.

Not all sedimentary rocks are the by-products of transported sediments. Some consist of mineral matter deposited out of solution, and others consist of organic substances like plants and animal remains. The limestones of the Rockies and of the Midwest and Colorado Plateau are of this type.

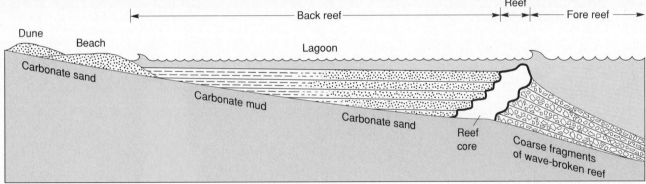

Figure 14.12

The Carlsbad Caverns of New Mexico formed from an ancient coral reef, which was later uplifted and exposed to weathering and erosion. The caverns are a series of honeycombed cavities in the reef core.

Photo by Thomas Clabe Campbell.

The Carlsbad Caverns in New Mexico were formed in weathered coral reefs which grew in a shallow, warm ocean (fig. 14.12). As the oceans withdrew around Carlsbad and water tables lowered, limestone caverns became visible above the water table. Slowly dripping groundwater from the ceiling gradually deposited limestone ornaments or *stalactites* decorating the ceilings of the caverns (fig. 14.13). In chapter 17 we will learn more about the weathering processes involved in the formation of these features.

GRAVITATIONAL ENERGY GRADES THE ROCKS

Classes of Sedimentary Rock

Sedimentary rocks are classified according to mineral, fragment size and shape, texture, and chemical composition. Table 14.5 lists the principal kinds of sedimentary rocks. Many form from fragments of other rocks; these are referred to as **clastic** sedimentary rocks.

Figure 14.13

At Endless Cave, New Mexico, slow dripping groundwater with
calcium carbonate in solution deposited limestone layer upon layer in
much the same way icicles form from a dripping roof of melting
snow.

Photo by Jerry Trout, U.S. Department of Agriculture, U.S. Forest Service,
Lincoln National Forest.

Table 14.5

Classification and Identification Chart for Hand Specimens of Common Sedimentary Rocks

Dominant Constituents	Textural Features	Composition and/or Diagnostic Features	Rock Name
Inorganic Detrital Materials (Clastic Sedimentary Rocks)	Pebbles and granules embedded in a matrix of cemented sand grains	Angular rock or mineral fragments.	Breccia
		Rounded rock or mineral fragments.	Conglomerate
	Course sand and granules	Angular fragments of feldspar mixed with quartz and other mineral grains. Pink feldspar common.	Arkose
	Sand-size particles	Rounded to subrounded quartz grains. Color: white, buff, pink, brown, tan.	Quartz Sandstone
		Calcite and/or dolomite grains. Light colored.	Calcarenite
	Sand-size particles mixed with clay-size particles	Quartz and other mineral grains mixed with clay. Color: dark gray to gray green.	Graywacke
	Fine-grained, silt and clay-size particles	Mineral constituents not identifiable. Soft enough to be scratched with fingernail. Usually well stratified. Fissile (tendency to separate in thin layers). Color: variable.	Shale / Mudstone
		Mineral constituents not identifiable. Soft enough to be scratched with fingernail. Massive (earthy). Color: variable.	
Inorganic Chemical Precipitates (Chemical Sedimentary Rocks)	Dense, crystalline, or oolitic	$CaCO_3$; effervesces freely with dilute HCl. May contain fossils (fossiliferous). Some varieties are **crystalline.** Some varieties are **oolitic.** Color: white, gray, black. Generally lacks stratification.	Limestone
	Dense or crystalline	$Ca,Mg(CO_3)_2$; powder effervesces weakly with dilute HCl. May contain fossils. Color: variable; commonly similar to limestones. Stratification generally absent in hand specimens.	Dolomite
	Dense, porous	$CaCO_3$; effervesces freely with dilute HCl. Color: variable. Contains irregular dark bands.	Travertine
	Dense (amorphous)	Scratches glass, conchoidal fracture. Color: black, white, gray.	Chert
	Crystalline	$CaSO_4 \cdot 2H_2O$; commonly can be scratched with fingernail. Color: variable; commonly pink, buff, white.	Rock Gypsum / Rock Salt
		$NaCl$; salty taste. White to gray. Crystalline. May contain fine-grained impurities in bands or thin layers.	
Organic Detrital Materials (Organic Sedimentary Rocks)	Earthy (bioclastic)	$CaCO_3$; effervesces freely with dilute HCl; easily scratched with fingernail. Microscopic remains of calcareous organisms. White color.	Chalk
		Soft, crumbles, but individual grains are harder than glass. Resembles chalk but does not react with dilute HCl. Commonly stratified. Gray to white. Microscopic siliceous plant remains.	Diatomite
	Bioclastic	Calcareous shell fragments cemented together.	Coquina
	Fibrous	Brown plant fibers. Soft, porous, low specific gravity.	Peat
	Dense	Brownish to brown black. Harder than peat.	Lignite
	Dense	Black, dull luster. Smudges fingers when handled.	Bituminous Coal

Source: J. H. Zumberge and R. H. Rutford, *Laboratory Manual for Physical Geology,* 8th ed, p. 32. Dubuque, Ia.: Wm. C. Brown, 1991.

Figure 14.14

Conglomerate is a mixture of sand, gravel, and even boulders or pebbles resembling concrete.

Conglomerates

Rounded boulders, gravel, sand, silt, and clay-sized particles occasionally become cemented together to form a mixture called **conglomerate.** It is easy to identify because of its strong resemblance to concrete (fig. 14.14). The pebbles and boulders are rounded during transport by streams or through the pounding of the surf along the shoreline. Conglomerates consist of any kind of rock, but they most frequently have pebbles rich in resistant and durable quartz.

Breccia—Product of Gravitation

Breccia, at first glance, looks like conglomerate except that most of the particles are angular instead of rounded. Sometimes the two blend together and it becomes difficult to distinguish between breccia and conglomerate. Breccias are commonly found where rocks have been fragmented in a landslide, transported by gravity to a lower slope, and buried. Cementation with silica, clay, iron oxide, or calcium carbonate occurs as water solutions release the salt minerals (fig. 14.15).

Sandstone

Sandstone is made of cemented sand grains, often consisting of quartz. At one end of the size spectrum, coarse sandstone grades into conglomerate. Very fine sandstone grades into shale. In many rocks, sandstones have mixed particles of clay and are classified as sandy shale. The cements are again silica, calcium carbonate, or iron oxides (fig. 14.16).

Shale

Shale is made of compact clay- and silt-sized particles so fine they appear homogeneous to the unaided eye. Shale is one of the most common sedimentary rocks. It can feel smooth and soft or greasy with colors ranging from shades of gray, green, red, and brown to black. The bedding planes can be very thin and flaky to massive blocks. Most shale forms in deep, quiet waters of lakes and seas. Particles are so fine that they remain in suspension for long times and need very quiet waters to settle out (fig. 14.17).

Figure 14.15

Breccia is formed of rock fragments that have been cemented together following a land movement. The scale is in centimeters.
Photo by David McGeary.

Figure 14.16

Sandstone is rock consisting of cemented sand-sized particles of resistant quartz and other minerals. A poorly sorted sediment of sand, silt, and clay grains produces a "dirty sandstone." The scale is in centimeters.
Photo by David McGeary.

Chemical and Organic Rocks

Limestone

Limestone is a very common rock and one of the few rocks that consist primarily of a single mineral, calcite. However, a number of impurities can give it a variety of colors and textures. Some are very fine textured or **aphanitic,** probably forming from a chemical precipitate. Other deposits are coarse textured, resulting from fossils and crystallization of calcite.

Salt

East of the Guadalupe Mountains on the plains of New Mexico are large deposits of rock salt found where the area was once covered by a shallow sea. As evaporation increased, leaving behind

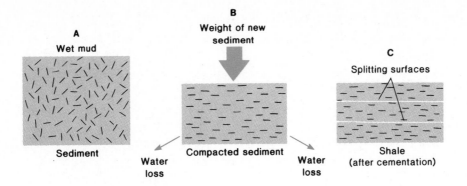

Figure 14.17

Shale particles are so fine they feel like a powder. This rock is generally formed in deep quiet waters where particles slowly come out of suspension.

precipitates, great beds of mineral salts accumulated on the continent. Each mineral precipitated out of the seawater solution at its particular level of saturation. Calcite, halite, and sylvite ("potash") each precipitated out of a solution at a given concentration. Traveling down a 260-meter (850-foot) shaft into a potash mine near Carlsbad, New Mexico, one passes through 15 meters (50 feet) of thick limestone, then halite, then "potash," then more limestone, and the pattern continues. Each mineral bed represents a period of chemical precipitation of that mineral. Fortunately for mining companies, each mineral has its own fallout schedule producing rather pure beds of each mineral.

The Carlsbad site is being prepared as a site for nuclear waste disposal because these salt beds are thought to represent a stable environment (see box 14.1).

Where the Earth's crust is energized by heat and pressure, a third class of rocks form.

THE EARTH'S HEAT AND PRESSURE FORM A NEW ROCK CLASS

Metamorphic Rocks

The Earth's crust is in a constant state of flux driven by energy deep in the interior, producing new patterns of rocks and minerals. When rocks are altered by this energy, new rocks are formed. We call these **metamorphic rocks.** The Alps of Europe are perhaps the best-studied mountains in the world. Here we find layers of sedimentary rocks that have been folded, buckled, and crumpled by the African plate's leading edge. Here the African plate is closing the Mediterranean Sea and uplifting the Alps. By following a bed of sedimentary rock such as shale along the ground, we find it grading into a slate.

In another part of the world along a Pacific beach cliff, a volcanic intrusion or dike has changed the surrounding rock's color, hardness, and mineral structure. On closer examination it becomes clear the rocks on each side of the dike are altered to a

new rock, fired in a natural kiln. A tap with a pick produces a ring like the clink of a bottle, while farther away from the dike it is more of a dull thud or thump (fig. 14.18).

These two examples from the Alps and the Pacific coast tell us that **metamorphism** is brought about by two different processes: (1) deformation at high pressures and temperatures and (2) high temperature, similar to the firing of clay in a furnace. The first is known as dynamic or **regional metamorphism,** the second as thermal or **contact metamorphism** resulting from volcanic intrusion.

Metamorphic rocks are also classified as either foliated or nonfoliated, according to the arrangement of minerals.

Foliated Rocks

Many of the metamorphic rocks consist of minerals set in parallel bands or streaks resulting from pressure on the rock. The rock has a flaky appearance and splits into thin sheets. This is **foliation** (Latin *folium,* leaf).

Slate, phyllite, schist, and **gneiss** are examples of foliated forms. Returning to the Alps for a moment, we can infer from observations that metamorphism of shale proceeds through stages. Shale grades into slate, to phyllite, to mica schist. Different stages are reached because the intensity of stress varies from place to place. Chemical analyses of shale, slate, phyllite, and mica schist from this alpine belt are very similar: each has approximately the same percentages of silica, aluminum oxide, iron, calcium, potassium, magnesium, and sodium. In metamorphism, these chemical elements stay in the rock but shift into new combinations.

Nonfoliated Rocks

Nonfoliated metamorphic rocks, marble and quartzite, are classic examples of metamorphism where no layering is present. **Marble** is produced when limestone is heated or under stress. Marble, then, is recrystallized limestone or dolomite (fig. 14.19). Quartzite is a metamorphic form of sandstone and conglomerate. In metamorphism, silica has filled in the spaces between the grains of the sandstone.

A Salty Solution for Nuclear Waste Disposal

BOX 14.1 Salt deposits have been recommended as one of the leading candidates for the permanent disposal of radioactive nuclear wastes generated from defense programs of the United States. The principal advantages of salt include the fact that: (1) most deposits of salt are found in stable geological areas with very little earthquake activity; (2) salt deposits demonstrate the absence of flowing fresh water, because water would have dissolved the salt beds had it been present; (3) salt is relatively easy to mine; and (4) it has the ability to heal fractures because of its plastic quality and the presence of minute amounts of saturated brine. That is, salt formations will slowly and progressively move in to fill a void or seal a waste repository. As salt takes on small amounts of water, crystals grow, expanding the salt mass.

The New Mexico salt formations were deposited in 900-meter (3,000-foot) thick beds by the evaporation of an ancient ocean. It is mostly sodium chloride. The deposits of salt in this region were formed approximately 225 million years ago and are an excellent repository rock. The large expanse of uninterrupted salt beds provides a stable environment free from the disturbances of large earthquakes.

Salt, while easy to mine, is stable and provides good shielding from radioactivity. At the 650-meter (2,150-foot) depth of the repository, the salt will very slowly "flow" under pressure from overlying rock and eventually encapsulate the buried waste in stable rock.

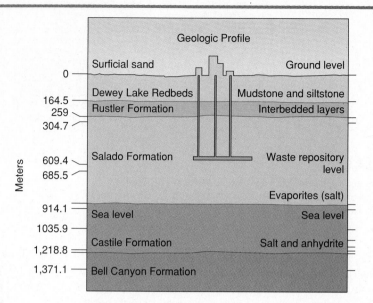

Figure 1

Geologic profile of a nuclear waste disposal site.

In 1962, the U.S. Geological Survey reported that the Permian Basin, which includes salt beds in southeastern New Mexico, parts of west Texas, Oklahoma, Colorado, and Kansas, was one of the most likely locations for such a repository.

Searches for petroleum and potash in the southeastern portion of New Mexico has led to extensive data concerning the geology of this area (fig. 1). The remote location favors its choice for a nuclear waste repository. Salt beds in the vicinity are accessible without disturbing established communities or residences, and the area is removed from any potentially disruptive geological features.

Source: U.S. Department of Energy, Office of Public Affairs.

THE ROCK CYCLE: DRIVEN BY GRAVITY AND THE EARTH'S HOT INTERIOR

We have had an opportunity to examine three rock families. It is also important to understand how these three rock families are related in a cyclic process. By examining the **rock cycle,** we will be able to see that all rocks come from other rocks (fig. 14.20).

Rocks are in a constant state of change because energy is flowing through them, producing new environments. Chemical combinations, temperature, pressure, and weathering are in a state of flux. The Earth's crust may appear static, but when movements are measured over long periods of time we find it to be very dynamic.

All rocks can be traced back ultimately to **magma.** As magmas cool and solidify, igneous rocks are formed. The type of igneous rock is determined by the chemistry of the magma and the environment where it cooled and crystallized. Because we live on a planet with not only solar energy but also moisture, winds, waves, and currents, igneous rocks are attached by weathering and erosion, which begin reducing the rock to fragments and soluble solutions. The material is then easily transportable by streams, rivers, ocean currents, winds, and glaciers. Gravity and solar energy are the driving forces for these agents of erosion. Deposition is the next step in the cycle. The sediments of erosion are carried to quiet waters including lakes, meandering streams, and seas, slowly settling to the bottom. Next, cementation and compaction results in lithification, or the formation of sedimentary rocks. Then, when these sediments are almost forgotten, the Earth's crust buckles,

Figure 14.18

Metamorphic rocks result from reheating of older rocks. Rocks of metamorphic origin often look like the parent material but are often harder, and mineral structures are rearranged or deformed. This is an example of contact metamorphism where surrounding rock was heated by a volcanic dike to become altered chemically into a metavolcanic tuff.

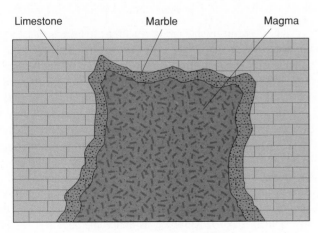

Zone of contact metamorphism (aureole)

Figure 14.19

When a magma intrudes limestone, marble forms along the contact.

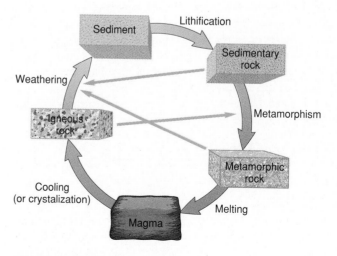

Figure 14.20

The rock cycle.

uplifting these beds above their ocean birthplace to be weathered and eroded again. When they undergo severe stress or come into contact with hot magmas, the rocks are changed into metamorphic rocks.

We have evaluated the rock formation processes. This understanding is basic to knowing how rocks and minerals are influencing human activities both now and in the past.

MINERAL RESOURCES OF THE EARTH

Mineral resources lie buried in rocks in concentrations of ore quality. By examining some of humankind's major minerals in terms of origin and recovery, we can appreciate the significance of their patterns around the world. By examining some key mineral resources, we will gain a better understanding of their formation processes. Second, we will be able to relate these minerals to their major rock families and the "rock cycle."

The Earth's crust is a storehouse of energy and materials in greater and greater demand as the needs of civilization continue to expand. How extensive is this mineral storehouse? What kinds of rock concentrate the major minerals in demand? What are the future prospects for this supply? How have minerals been used in our progress?

The use of metals and fossil fuels dates back to Babylonian times. Archaeologists have discovered that the Babylonians were using petroleum instead of wood for fuel nearly 4,500 years ago. The Chinese were mining coal and drilling wells hundreds of feet deep for natural gas 1,100 years before Christ. Minerals have always been in the marketplace. The use of gold, silver, and bronze predate the Inca empire in Peru by a thousand years as tools and art works. However, most cultures, until just a few hundred years ago, were dependent on primarily wood for fuel and housing and the muscles of humans and animals for work.

Today the picture is quite different. A small minority of Americans and Europeans and a smaller percentage of Asians consume over 50 percent of the world's energy and mineral resources. The frightening thing about this picture is that we are reaching limits on resources. Energy, metals, and nonmetals are becoming shorter in supply because most are nonrenewable. Coal, oil, and natural gas represent solar energy stored in chemical compounds. Once it is used, it cannot be replaced. As for iron, copper, and other essential metals, they are very limited in location and amount when compared to the size of the Earth's surface and the future population, expected to reach 8 billion by the year 2000.

ENERGY FOSSIL FUELS

Coal

Coal occurs in bedrock interbedded with shale and sandstone of the Appalachians and Colorado Plateau. It is the most abundant fossil fuel in the United States. Geologists estimate that the United States has several hundred years of reserves. It can be refined to a liquid for fuel. Most of the coal mined is used to heat boilers, making steam for power plants and space heating. Coal is also used as coke in steelmaking. It is the chief raw material from which nylon, some plastics, and many chemicals are produced. Its location in the Appalachian Mountains played a key role in the development of the steel industry of the northeast.

Coal is a fossil material formed right where the plants grew. The coal beds contain fossil remains of tree stumps, roots, leaves, and humus. It is produced by bacterial and chemical processes that reduce the organic matter to increasing proportions of carbon. The process begins with peat. It is gradually converted to lignite, then through increased pressure of overlying sediments, subbituminous coal is formed, which under continued pressure converts to bituminous coal. A metamorphic form is *anthracite*.

Petroleum

The sedimentary rocks of the Earth are the storehouses for oil and natural gas, collectively called petroleum. More than two-thirds of all energy consumed in the United States comes from these rocks in the form of oil and gas. Until just a few years ago we were an exporting nation, but our demands have outstripped our supplies. Over 90 percent of our consumption is consumed as fuel; the remainder goes into lubricants and petroleum-based chemicals.

Petroleum is a complex mixture of gaseous, liquid, and solid hydrocarbons. A century of mining and research has revealed some important characteristics about its occurrence.

For oil or gas to accumulate, four essential geologic conditions must be met (fig. 14.21):

1. There must be a reservoir rock, usually sandstone or limestone. Since oil is a fluid, it saturates rocks like a sponge. Oil pools are not ponds but more like saturated permeable bricks, which hold water between the grain spaces. The rock must be permeable so that oil can flow through it into wells.
2. The reservoir rock must be overlain by an impermeable cap rock such as shale to prevent upward escape.
3. The oil pool must have a trap that prevents the oil from escaping, not only vertically but laterally as well.
4. Source rocks must be present to provide oil for the reservoir.

These requirements are very similar to an artesian water system, described in chapter 18. The two systems have reservoirs, roofs, and traps. Although these characteristics occur many places, it is no guarantee of oil. Therefore a fourth characteristic, the presence of a source rock to provide oil for the reservoir, is necessary.

Oil is 95–99 percent carbon and hydrogen with special chemical structures that come only from organic compounds. When oil is discovered, it is nearly always found in marine sedimentary rocks on both the plains and plateaus of the continents and beneath the seas along the continental shelf (fig. 14.22). It is believed that oil forms as a result of accumulation of marine organisms, mostly of plant origin. Where the water is stagnant and oxygen is low, the marine organisms are decomposed by anaerobic bacteria, leaving residual carbon and nitrogen. Burial of this material destroys the bacteria and provides heat and pressure to cook the material and force it out into nearby layers of sandstone.

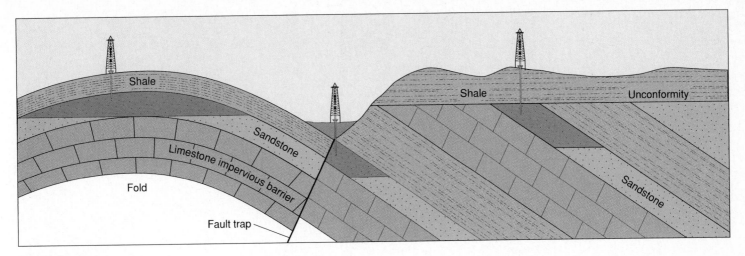

Figure 14.21

Sedimentary rocks trap petroleum. The sandstone is the reservoir rock. Traps occur where sedimentary beds are folded, faulted, or form unconformities that prevent petroleum from moving vertically. Traps are also formed in sandstone lenses.

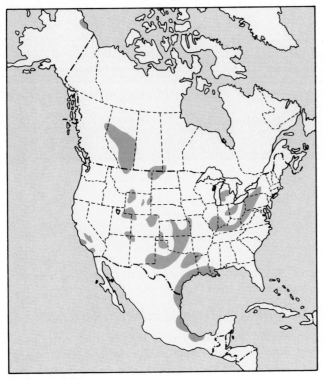

Figure 14.22

Petroleum originates in marine sediments from the accumulation of marine organisms. Major oil fields in North America are located in marine sedimentary rocks from the North Slope of Alaska to the coast of southern Mexico.

METALS AND NONMETAL RESOURCES

One of the fundamental facts about ore minerals is their obvious relation to igneous rocks. Ore deposits are aggregates of minerals from which one or more metals can be extracted profitably. Since technology, market, and costs fluctuate, minerals come and go as ores. Mines are worked and then closed, and towns go from thriving commercial centers to quiet, ghostlike communities (fig. 14.23).

Concentration within Magma with the Aid of Gravity

When molten magmas begin to cool, it is believed that minerals separate out, with heavier ones settling to the bottom concentrating under confining pressure of the surrounding rocks.

The surrounding rocks resemble a classic pressure cooker in which leaks are common. The famous Kiruna iron deposits of Sweden seem to have been concentrated in a magma and then squeezed out under pressure. Smaller iron ore deposits of magnetite in the Adirondack Mountains of New York also appear to have formed by this process.

Nonmetals such as diamonds from the mines of South Africa are thought to have formed through a similar concentration method.

Alteration of Rock in Contact with Magma

In the copper mining district of Cananea, Sonora, Mexico, copper ore is recovered in limestone at the contact of an intrusive body.

Figure 14.23

Ghost towns such as Calico, California, symbolize the economic impact of a community dependent on the supply and value of ore deposits.

The igneous body produced a halo of altered sedimentary rock in which fissures were filled with veins of copper ore introduced from very hot fluids originating in the magma.

This process is credited for producing ores of iron, zinc, and lead. In addition, copper, rubies, and sapphires are frequently discovered in geologic zones of this type.

Geothermal Waters

Hot waters from cooling magmas carry mineral deposits into open spaces in rocks. This is by far the best known process by which ore deposits are concentrated and formed. The major gold fields of the famous Mother Lode in California, Kirkland Lake in Ontario, Canada, as well as metal ores of silver, copper, lead, and zinc are found in deposits of this sort. Hot waters infiltrate and deposit

minerals in joints and cracks and small faults. The ores are referred to as fissure veins (fig. 14.24).

Ore Deposits from Lake Solutions

Clinton iron ore occurs as a lenslike body in sedimentary beds extending from New York state southward into Alabama. Fossil bodies show that the strata were deposited in a shallow sea, as deposits of iron-bearing minerals that weathered out of basic igneous rocks. Streams carried the dissolved minerals, precipitating them in a shallow sea as oxides. The rich iron ore beds of Lorraine in France formed by the same geologic process. The Lake Superior iron ores, which form the backbone of the American steel industry, were also formed in this manner. Weathering has enriched them by concentrating the ore.

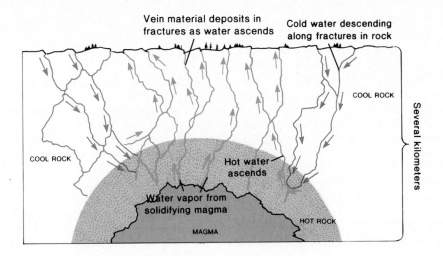

Figure 14.24

In the ore formation process, silica-rich hydrothermal waters ascend to the surface. Microscopic ore minerals, such as gold and silver, are deposited in the fractures as quartz veins.

Concentration by Weathering and Groundwater

When mineral environments change, they become more vulnerable to chemical attack. Minerals that have concentrated in fissure veins remain quite stable until erosion exposes these deep-lying minerals to weathering and decay. When metallic minerals such as copper are attacked and transported by groundwater downward into a zone of saturation out of reach of the air, it adds to the existing ore, enriching and concentrating low-grade ore to an economically workable ore. This process is known as secondary enrichment.

The copper mines at Douglas, Arizona, represent deposits from hot solutions coming from magma beneath the deposits. Through a zone more than 30 meters (100 feet) thick, secondary enrichment has improved the upper level of the ore.

Residual Concentration

The large bauxite deposits of Venezuela and Jamaica are a by-product of chemical weathering. Aluminum has been concentrated by the removal of silica in the original compound of aluminum silicates. The groundwater solution decomposes the aluminum silicates in the rock, carrying away silica in solution and leaving aluminum and iron behind as oxides. Aluminum may reach 20 percent purity.

Mechanical Concentration

When gold was discovered in 1848 in California, it was not found in veins but as small nuggets in a tributary of the Sacramento River. Gold reaches the stream after chemical and mechanical weathering of a vein at its outcropping. The gold is released and eventually washes into a stream. Because it is very heavy, gold settles quickly to the bottom of the stream. This type of deposit is referred to as a placer. The term "gold dust" describes the form of most placer gold. It comes down streams in grains the size of silt particles. Pebble-size fragments are called nuggets.

Gold is recovered in Mexico and Australia from placers concentrated by wind. Quartz and other light minerals have been deflated or removed in these arid regions, leaving behind the heavier gold particles.

As we can see, the Earth's basic rock families, igneous, metamorphic, and sedimentary, supply us with a rich assortment of useful minerals. Perhaps the fundamental question of concern is: how should we allocate our present mineral resources for the benefit of future generations?

Our power needs alone are doubling every decade. Although the United States represents only 6 percent of the world's population, we are currently consuming 35 percent of its energy and mineral resources to maintain our level of living. Between now and the year 2010, about 20 years away, the United States will consume more energy than it has in its entire history, and the worldwide demand will probably triple. These demands are going to tax our ability to discover new sources and recycle waste, extract and refine fuels, and dispose of wastes from these processes without greatly harming the planet.

SUMMARY

Minerals are the fundamental building blocks of the Earth's crust. By their specific physical and chemical characteristics, minerals can be classified and therefore identified.

Rocks may be divided according to their origin into three major classes: **igneous, metamorphic,** and **sedimentary.** Igneous rocks solidify from a molten state. Molten magma that do not reach the surface but slowly cool underground are referred to as **plutonic** or **intrusive** igneous rocks. When **magma** reaches the surface and cools, it is called an **extrusive** volcanic igneous rock. Igneous rocks constitute the bulk of the ocean floor and continents, with volcanic mafic rock occupying primarily the ocean floor and felsic intrusive rocks making up the bulk of the continents.

Sedimentary rocks are composed of particles and chemical precipitates derived from previously existing rock. Through the processes of weathering and erosion, deposits are transported by water, wind, or ice to be deposited and cemented under pressure over periods of time. Sedimentary rocks can be classified as clastic, organic, or chemical.

Metamorphic rocks are rocks that have been changed or altered by tremendous heat and pressure accompanying mountain-building processes. The parent material can be rocks of igneous, sedimentary, or metamorphic origin. Contact metamorphism occurs where magma intrudes older rock heating it up enough to produce change.

The impact of minerals on humankind's history is impressive. Gold, silver, iron, coal, and petroleum are just a few examples of minerals responsible for major changes in the history of nations.

ILLUSTRATED STUDY QUESTIONS

1. Which mineral property gives a diamond its hardness (fig. 14.1, p. 289)?
2. Apply the definition of a mineral to the following minerals by identifying each characteristic in the photograph (figs. 14.1 and 14.2, p. 290).
3. Describe the mineral structure that gives clay minerals water-holding capacity (fig. 14.1 and table 14.4, pp. 289, 293).
4. Describe the location of the Earth's two crustal hemispheres (fig. 14.3, p. 294).
5. Identify the rock family for each of the following photographs (fig. 14.4, p. 295).
6. Describe the key characteristics of each rock type (figs. 14.5 and 14.6, p. 296).
7. Describe the texture of volcanic vs. plutonic rocks (figs. 14.5, 14.6, and 14.7, pp. 295, 296).
8. Describe the birthplace of sedimentary rocks (fig. 14.10, p. 301).
9. Explain the events that produced this unconformity (fig. 14.11, p. 302).
10. Explain the formation process of a stalactite (fig. 14.13, p. 304).
11. Describe the similarities and differences between conglomerates and breccia (figs. 14.14 and 14.15, p. 306).
12. How does sandstone differ from shale in terms of particle size (figs. 14.16 and 14.17, pp. 306, 307)?
13. Describe the environment for contact metamorphism to occur (figs. 14.18 and 14.19, p. 309).
14. Contrast a foliated and nonfoliated metamorphic rock (fig. 14.19, p. 309).
15. Starting with magma and using the rock cycle, explain how each of the rock families is formed (fig. 14.20, p. 309).

Plate Tectonics, Seismic Waves, and the Earth's Interior

15

Objectives

After completing this chapter you will be able to:

1. Explain the major sources of energy shaping the Earth's crust.
2. Explain two driving forces of plate motion.
3. Explain the current evidence in support of **plate tectonic theory.**
4. Describe the anatomy of the Earth's interior from the **crust** to the **core.**
5. Define "P," "S," and "L" waves in terms of their path and motion through the Earth.
6. Explain the **rock cycle** in the light of plate tectonic theory.

The Earth's crust is continually in a state of stress and motion induced by gravity and internal heat energy.
NASA

A

Figure 15.1

An oceanic ridge system is similar to a seam on a baseball winding its way around the globe. Along this seam the oceanic plates are separating and causing new ocean floor to appear. Iceland lies on this ridge. This Icelandic fire plume reaches 200 meters high.

A: Photo by Moore, U.S. Geological Survey.
B: World Ocean Floor by Bruce Herzen and Marie Tharp, 1977, and Copyright by permission of Marie Tharp, Washington Avenue, South Nyack, NY 10960.

PLATE TECTONICS

According to plate tectonic theory, the Earth's outer layer is a series of large slabs that are in motion. The nature of the contact between plates determines the types of features being produced. Where plates are drifting apart, a fissure is produced where new crustal material forms. In contrast, plate collisions cause the heavier plate to sink below the lighter one and return to a molten state in the mantle. In the process, trenches, island arcs, earthquakes, and explosive volcanoes are produced. The supportive evidence for this unifying model of geologic processes has come from evidence discovered in the last half of this century.

A Unifying Model in Patterns of the Earth

Today most geologists are convinced that the arrangement of the continents and ocean basins is of rather recent origin. A German meteorologist, Alfred L. Wegener, suggested that South America and Africa may have once been connected. His ideas were based on circumstantial rather than direct evidence. In his book, *The Origin of Continents and Oceans,* published in 1911, Wegener wrote: "The continents must have shifted. South America must have lain along Africa . . . the two parts must then have become increasingly separated over millions of years." He believed that the continents, consisting of blocks of lighter rock, must have somehow plowed through the heavier rocks of the ocean floor like giant ships, driven by the Earth's spinning action. Wegener based his hypothesis not only on the fit of the continents but also on fossil patterns. He also found climatic evidence in sedimentary rock, which preserves signs of depositional environments. Although most geologists reacted with disbelief, a few took him seriously enough to keep his theory alive. The major weakness of his model was in the lack of evidence of disruption of the ocean floor from continental motion.

After World War II the sonar technique was perfected to map the ocean floor. Through this new mapping technology, a chain of huge mountains and rifts was traced down the middle of the Atlantic Ocean where, we now know, the plates of Africa and Europe were once connected with the Americas (fig. 15.1). Geologists found that a ridge system 64,000 kilometers (40,000 miles) long ran through the Earth's oceans like the seam on a baseball. This oceanic ridge, if seen in a profile, is divided in turn by a deep, narrow **rift zone** along its centerline or axis. This mapping technology has provided supportive evidence in a variety of ways.

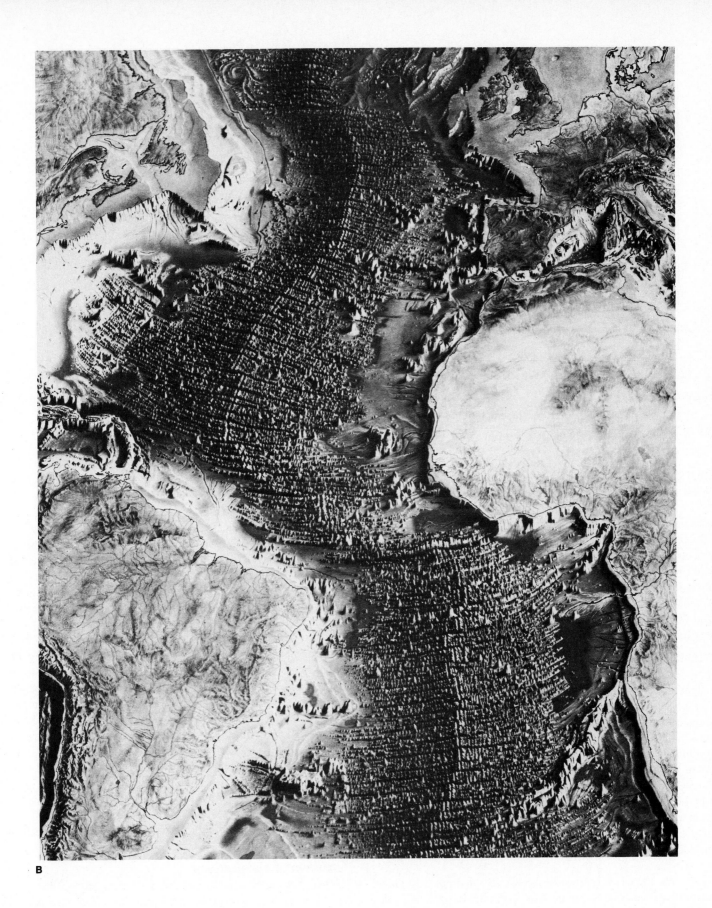

B

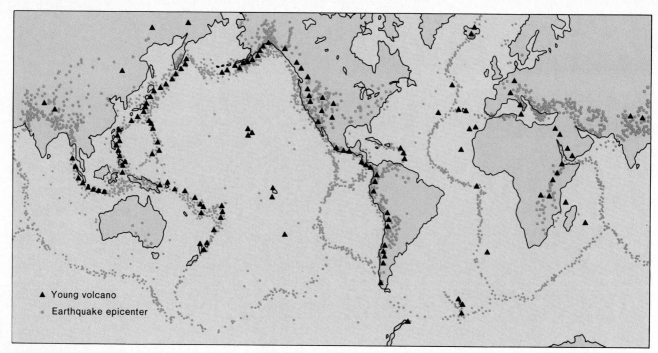

Figure 15.2

Active tectonic regions of the Earth include the Pacific Ring of Fire and the Eurasian-Melanesian belt.

▲ Young volcano

• Earthquake epicenter

Earthquake Patterns

While mapping this mid-oceanic ridge system, it was noted that strong correlations exist between shallow earthquakes and this oceanic ridge pattern (fig. 15.2). Deeper earthquakes, as deep as 700 kilometers (450 miles), were also known to cluster along the rim of the Pacific beneath the volcanic island chains, the "Pacific Ring of Fire." In 1961 it was suggested that the sea floor is spreading out from the oceanic ridges and descending back into the mantle in the deep oceanic trenches. This **subduction** of ocean floor was postulated to be the cause of the deep earthquakes and volcanic eruptions around the Pacific.

Magnetic Field Reversal Patterns

The next clues came from the Earth's magnetic field—lines of magnetic energy or force running between the North and South Magnetic Poles. This field influences the magnetism of rocks being formed from molten lava. Rock layers being formed now will preserve as a permanent record the direction of the current magnetic field unless reheated beyond a certain temperature called the **Curie point.** When scientists in the late 1950s began comparing the past magnetic poles indicated by rocks of both Europe and North America, rocks of the same age on different continents seemed to point to different locations. A British team later demonstrated that the rocks on both sides of the Atlantic could be precisely matched by closing the Atlantic and thus bringing the continents back together.

In the early 1960s, researchers recorded not only magnetic polar wanderings but nine *reversals* in the Earth's magnetic field—or a flip-flop of the magnetic poles—during the last 3.5 mil-

lion years. The reversals were recorded in rocks collected from all over the world. Two British scientists, Frederick J. Vine and Drummond H. Matthews of the University of Cambridge, proposed that if new crustal rocks were cooling and preserving the magnetic field along the mid-oceanic ridge and spreading outward, a permanent record of the magnetic reversals would be preserved in the sea floor. They wrote in *Nature,* "If spreading of the ocean floor occurs, blocks of alternately normal and reversely magnetized material would drift away from the center of the ridge and parallel to the crest of it."

Seafloor spreading became an established fact in 1966. Seafloor magnetic orientations lined up in parallel stripes along the ocean ridges and also appeared at depth in sediment cores from the ocean floor. Detailed surveys found similar magnetic stripes on both sides of the oceanic ridges, like mirror images. For example, if a magnetic reversal was found 400 kilometers east of the Mid-Atlantic Ridge, a matching stripe would be found 400 kilometers to the west of it. By measuring ages of rocks and distances from the ridge, the rate of seafloor spreading can be determined (fig. 15.3).

The Atlantic Ocean was shown to be opening at the rate of 5 centimeters per year. Parts of the Pacific Ocean floor off the coast of South America are spreading at the rate of 13 centimeters per year.

Continental Connections

A British scientific team at Cambridge, led by Sir Edward Bullard, used a computer to make models of the period when Africa and South America fit together. Instead of comparing today's

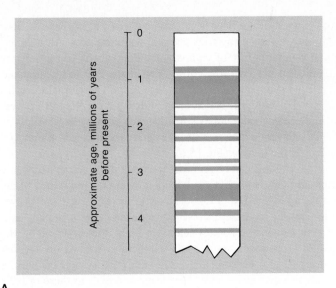

A

Figure 15.3

Magnetic reversal provides a clue to continental drift. Graph shows a portion of the reversal history of the Earth's magnetic field for the relatively recent geologic past. Times of reversed magnetic orientation are shaded in green.

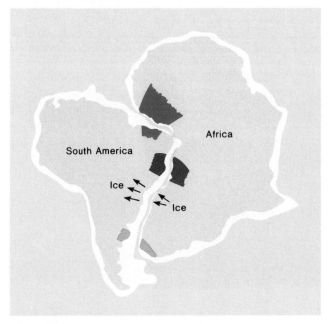

Figure 15.4

The rocks on both sides of the Atlantic Ocean can be matched like a puzzle by closing the Atlantic. Alignment is aided by comparing magnetic polarity in the rocks and the movement in age and type of rocks. Glacial striations, or grooves in the rocks, also match.

coastline, they used the 900-meter (3,000-foot) depth line along the continental slope where the edge of the continent is actually located. Until the actual continent edge below current sea level was used, the match was difficult. Bullard's team brought the mosaic of the continents together in a close fit along this revised boundary.

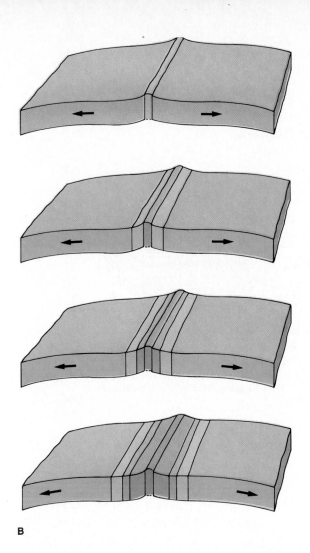

B

Then with this new boundary, researchers from the Massachusetts Institute of Technology matched the ages and rock layers precisely. Bands of iron ore, tin, and gold lined up even though they are currently an ocean apart on two different continents (fig. 15.4).

Fossil Patterns

In Antarctica, in 1967 and 1969, explorers found a fossil of the sheep-sized reptile *Lystrosaurus* and a leaf of *Glossopteris*. Remains were also found in South Africa and India. Because this reptile could not have crossed the sea to reach Antarctica by walking or swimming, the conclusion was clear. There was one landmass, now referred to as Gondwanaland, which later split into several segments yielding our continents (fig. 15.5). The *Glossopteris* leaf was telling evidence because it was found in the same continental sandstones, shale, and coal beds on each continent.

More evidence came in the form of mud and rock cores from drilling sites all over the world's oceans. By dating the fossil skeletons of microscopic sea organisms that collect as residual sediments, paleontologists can determine the time at which the sea floor was formed.

After examining thousands of cores, paleontologists showed that the oldest sea floors formed in relatively recent geologic times, less than 200 million years ago, and that everywhere

A leaf of *Glossopteris*

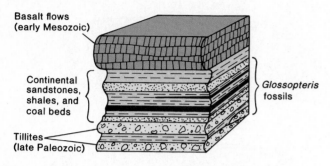

Basalt flows
(early Mesozoic)

Continental
sandstones,
shales, and
coal beds

Glossopteris
fossils

Tillites
(late Paleozoic)

Figure 15.5

A plant, *Glossopteris,* has been discovered in Antarctica, South Africa, and India, all of which were once part of a single southern land mass now referred to as Gondwanaland. Rock sequences have also been found in this same order at these locations.

the youngest ocean floor is the mid-oceanic ridges. The sea floor is literally sliding away from the mid-oceanic ridges. At the maximum estimated rate of seafloor spreading, about 16 centimeters (6.3 inches) per year, the entire floor of the Pacific Ocean could be created in a span of only 100 million years. Therefore the ocean's floor is the youngest section of the Earth's crust. Its birth occurs at mid-oceanic ridge and its death takes place where two plates collide and subduction recycles the plates and scraped-off material back into the mantle (fig. 15.6).

It is easier to document the birth of a rock in a volcanic eruption than its burial or subduction in an offshore trench in the Sea of Japan, the Sea of Okhotsk, or the Java Sea. Little is known of the specific processes involved. No one has ever seen a plate exit. Current evidence indicates that as a plate goes down along the edge of a continent, some of its sediment is scraped off and becomes part of a coastal range. The mountain system behind the trench also displays a series of volcanoes similar to those in New Zealand, Indonesia, the Philippines, and the Cascades of North America.

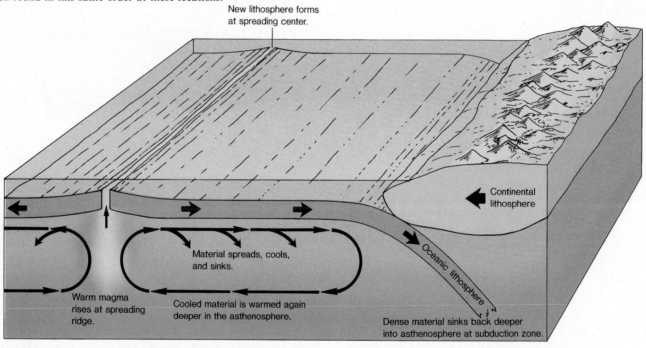

New lithosphere forms
at spreading center.

Continental
lithosphere

Oceanic lithosphere

Material spreads, cools,
and sinks.

Warm magma
rises at spreading
ridge.

Cooled material is warmed again
deeper in the asthenosphere.

Dense material sinks back deeper
into asthenosphere at subduction zone.

Figure 15.6

Birth and death of a crustal plate. The driving forces are gravity in the subduction zone and heat in the spreading zone, hence convection.

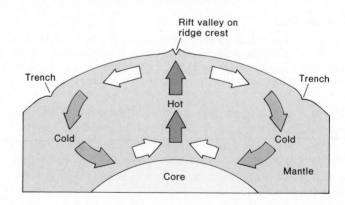

Rift valley on
ridge crest

Trench

Trench

Hot

Cold

Cold

Core

Mantle

The destruction of the crustal plate in the ocean's depths has helped explain a perplexing problem in geology. The oceans contain very little fossil sediment older than 80 million years and no sediment older than 150 million years. Now it is easy to explain because the sediment is either buried with a plate in the mantle or piled up against the continents. Evidence for this can be found in the types of marine sedimentary rocks found along the coasts of these burial sites and in the types of lava released by the chains of volcanoes.

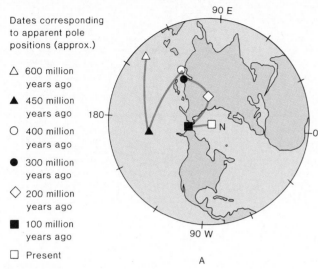

Figure 15.7

Ancient rocks record geomagnetic directions that differ from today's. If the assumption is made that the geographic and magnetic poles have always remained close together, then the continents have moved and rotated as shown here.

The volcanic rocks reflect burial and subduction activity in their mineral content. Andesite is rich in silica and is more typical of continent rocks. It is believed that subduction and melting of sediments enriches the magma with silicates to produce the andesites or sialic magmas.

Now that we have seen the evidence needed to become a theory, let us summarize how the theory works.

Plate Dynamics

According to plate tectonic theory, the Earth's crust is divided into huge moving segments supported by the mantle. One way we know that these plates move is that some rocks indicate the direction of north in their magnetic minerals. Rocks of different ages show that the *apparent* position of the North Pole has changed over time (fig. 15.7). This is in fact the result of plate motion.

When two plates move apart, a fissure or spreading center issues hot plastic material from the mantle that cools to form the rocks of the trailing edge of each plate. The leading edge or oldest portion of the plate, farthest from the spreading center, pushes against another plate. Where plates are in collision, the leading edge may be deflected downward into the soft plastic upper mantle located 100 kilometers (60 miles) below the surface. This process destroys the plate while at its other end it is being created at the mid-oceanic ridge spreading zone (fig. 15.8).

Geologists estimate that the plates encompass the entire surface of the Earth down to a depth of some 50 to 100 kilometers. The surface consists of a horizontal mosaic of perhaps 20 major plates. These plates include the Earth's crust and part of the upper mantle. The plates move and circulate like a pot of boiling soup on a stove with clusters of froth drifting about. The driving forces are still under investigation and scientific debate.

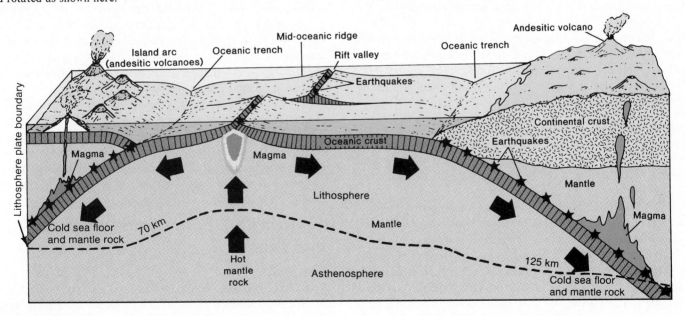

Figure 15.8

Burial of a plate occurs in the asthenosphere. This is how the continents have drifted apart from a single continent, Pangea.

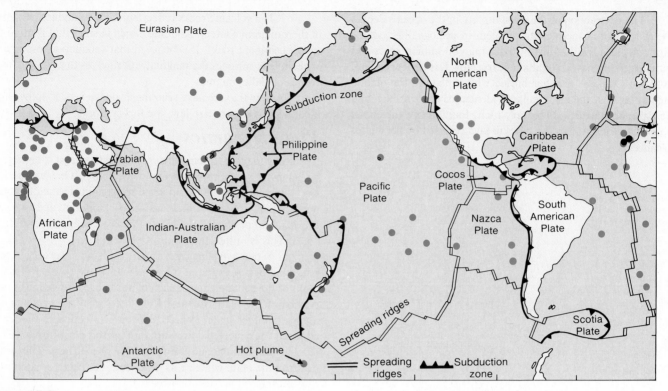

Figure 15.9

Hot plumes in the mantle are the energizers for plate motion. Gravitational energy is the downward pulling force returning the higher density, cooler crust back to the mantle.

The Earth's central core is the source of heat, trapped from its early formation and preserved by the insulation of the outer layers. This energy source is the key factor in shaping the Earth's rugged outer skin. The ocean floor, continental slabs or plates, mountain chains, rift valleys, and volcanic peaks are all testimonials to this energy source that has set the whole crust in convectional motion. There is general agreement that energy from within provides the driving power, but questions remain regarding the details of this motion.

Causes of Plate Motion

First a caution. No one knows with certainty what drives the crustal plates. The assumption that heat-driven convection is the chief cause is a good starting point; however, gravity-driven subduction may be the primary driving force.

When "continental drifters" began popularizing their hypothesis, it was first suggested that broad convectional cells are fueled by heat from the core. This causes magma to rise along major ridge zones where tensional cracks and seafloor spreading had been observed. Others proposed that subduction is caused by gravitational forces pulling heavier oceanic plate sections down under continental plates, allowing lighter material to distill off it at subduction boundaries. This would produce a pulling force where plates are in collision. Rising magma in the spreading zones would be a result rather than a cause of plate motion, and gravity would be the primary driving force.

Hot Plumes or Chimneys

Convection occurs off the ridges as well, in the form of *mantle plumes* or *hot spots*. Energy and matter from the hot interior rises in narrow columns and fans outward in all directions. This type of convection is a key driving force observable in both the oceans and atmosphere when heating occurs from below. A good example in the atmosphere is a thunderstorm, when the cumulus clouds build to great heights as the heated less dense air ascends.

Hot spots do not appear to move significantly; instead, the plates move over them causing the hot spots to leave trails of geothermal activity, lava flows, and uplift. It is thought that the radial energy flow outward from the hot spot plays a part in breaking up the crustal plates and setting them in motion. Gravitational energy is the downward pulling force returning the higher density crust to the mantle (fig. 15.9).

Continental Plates: Past, Present, and Future

The Past

Two hundred million years ago the land areas assembled into a single continent which geologists call **Pangaea** ("all lands"), washed by a universal ocean, **Panthalassa** (fig. 15.10). Pangaea broke into continent-sized fragments that began drifting across the face of the Earth. Back in time, 135 million years ago, the northern landmass, **Laurasia,** split off from the southern segment known as

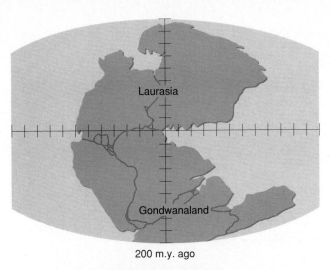

Laurasia

Gondwanaland

200 m.y. ago

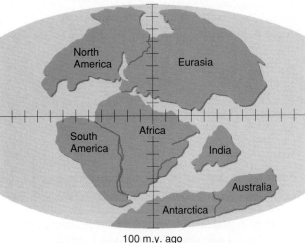

North America

Eurasia

South America

Africa

India

Australia

Antarctica

100 m.y. ago

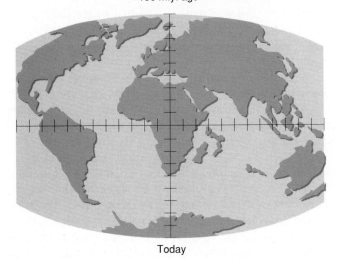

Today

Figure 15.10

Past and present positions of the continents over a 200-million-year period. First Pangaea drifted apart, forming Laurasia and Gondwanaland. Then today's continents drifted apart. This illustration shows all areas of continental crust, some of which are submerged.

Gondwanaland. India also broke away from Gondwanaland and headed for the Eurasian plate, which represents several plates bonded together. Then 65 million years ago the North Atlantic and Indian Oceans began to take shape and the South Atlantic began to widen. Australia was still connected to Antarctica but has since detached and moved northward.

Present

Today we see Europe and North America have separated, forming the Atlantic Ocean. India has pushed into Eurasia, thrusting up the Himalayas. North and South America have parted the waters of the Atlantic, and Antarctica stands alone, isolated in the Southern Hemisphere (fig. 15.10).

Future

What will our continent look like 50 million years from now? First, we will see a section of California, west of the San Andreas fault, detached from the mainland and shifted north to the Gulf of Alaska and Aleutian trench. The Atlantic and Indian oceans will have widened, while the Mediterranean will have shrunk as Africa continues to move into Europe. The Alps are being uplifted and the European and African continents are being sutured in the Mediterranean Sea.

The Rock Cycle Revisited

Figure 15.11 interprets the rock cycle described in chapter 14 in plate tectonic terms. Thus we have a truly unifying theory.

Unanswered Questions

This beautiful and simple view we call plate tectonics is in its embryonic stages. The causes of motion and many details of plate motion are not well understood nor as simple as described. Nor does plate tectonics directly explain the dynamics at the base of the mantle or the core. Dr. Maurice Ewing, founder of Lamont Geological Observatory, said in 1974, "Nothing is going to be as simple as plate tectonics now makes it seem, just as no atom is exactly like the Bohr atom. But the Bohr atom was an idea capable of sufficient refinement and so is plate tectonics." For example, recent (1991) geologic evidence suggests that Antarctica was attached to North America's east coast. If true, this will change the puzzle pieces of Pangaea. Therefore there are still many unanswered questions, but this model certainly unified our view of a dynamic surface of the Earth.

EARTHQUAKE WAVES AND THE EARTH'S INTERIOR

Plate boundaries are stress zones where most earthquakes occur. These events provide most of our information about the Earth's interior. Slow buildup of crustal strain and sudden release of this strain produces the earthquakes and shifting of the crust along fissures or faults.

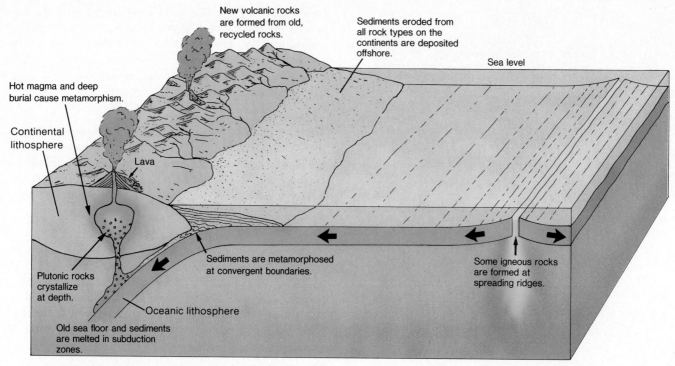

New volcanic rocks
are formed from old,
recycled rocks.

Sediments eroded from
all rock types on the
continents are deposited
offshore.

Sea level

Hot magma and deep
burial cause metamorphism.

Continental
lithosphere

Lava

Plutonic rocks
crystallize
at depth.

Sediments are metamorphosed
at convergent boundaries.

Some igneous rocks
are formed at
spreading ridges.

Oceanic lithosphere

Old sea floor and sediments
are melted in subduction
zones.

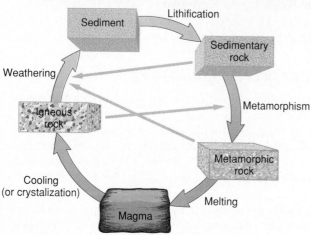

Sediment

Lithification

Sedimentary
rock

Weathering

Metamorphism

Igneous
rock

Metamorphic
rock

Cooling
(or crystalization)

Melting

Magma

Figure 15.11

The *rock cycle* revisited in the light of the unifying theory of plate tectonics. *Metamorphic rocks* form at convergent zones. Spreading and convergent zones are formation environments for *igneous rocks*. *Sedimentary rocks* form from erosion of all rocks where these sediments are cemented in the offshore waters.

Mechanisms of Shaking

The study of the vibrations emitted during an earthquake is conducted by **seismologists.** When the Earth shakes, vibrational waves are sent out in all directions. There are two fundamental types of waves. First, **body waves** are **seismic waves** that travel through the Earth's interior. Second, **surface waves** spread away from the **focus**

or **hypocenter,** the point where the earthquake starts, like ripples made by a pebble falling into a pool. Seismologists have found unique characteristics for each wave type.

After repeated laboratory and field tests, seismologists have found that seismic waves not only have their own velocity but also refract or bend in a manner governed by the nature of the material they pass through. Therefore, both density and, to a degree, rock chemistry can be inferred throughout the Earth's interior.

Body waves consist of "P" and "S" waves. The "P" or primary wave is similar to a sound wave (fig. 15.12). It is the first to reach the surface, traveling at about 5.6 kilometers (3.5 miles) per second. Shaking is a push-pull force on rocks as it passes. Rocks are first compressed (pushed) and then stretched (pulled).

The "S" or shear waves arrive next and create an up-and-down motion at right angles to the direction of motion traveling at about 3.2 kilometers (2 miles) per second near the surface. The motion is best illustrated by snapping a rope like a whip. The motion is up-and-down, but the wave travels horizontally (fig. 15.12).

The surface wave or "L" or long waves travel slowly and are felt only at long distances. Tall buildings 50 kilometers (30 miles) away swayed in downtown Los Angeles from the 1971 San Fernando Valley earthquake under the influence of "L" waves.

Seismographs around the world record these patterns, and calculations based on time lags between arrivals allow the seismologist to plot the distance from where the earthquake occurred. The **epicenter** is the point on a map directly above where the earth-

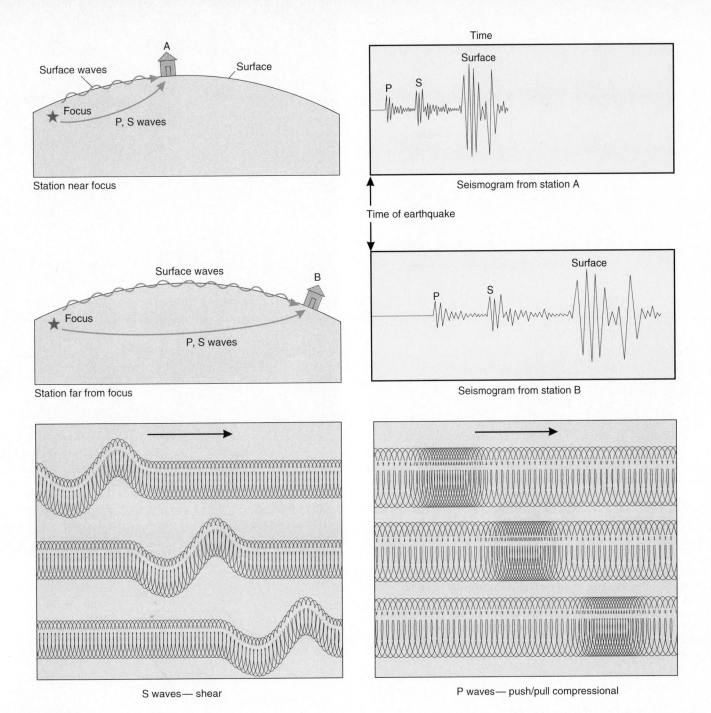

Figure 15.12

P or primary waves and S or shear waves along with L waves all give clues to the secrets of the Earth's interior. Each wave travels a unique velocity and its path is determined by the nature of the wave and environment.

quake started. When three or more stations share data, the epicenter can be pinpointed (fig. 15.13). The procedure can be understood by likening the seismic waves to a foot race. Runners leave the starting blocks together each running at a different rate. The finish times vary in one, two, three order. If we know only their velocity and time intervals between finishes, we can compute the distance to the point of origin or epicenter.

The **focus** or **hypocenter** is the actual point of origin below the surface (fig. 15.14). The hypocenters are very deep for most earthquakes occurring in areas where plates are being subducted. At locations where seafloor spreading is occurring, earthquakes are usually shallow. Therefore, in the Gulf of California where a new sea floor is being formed, most earthquakes observed are only a few miles deep.

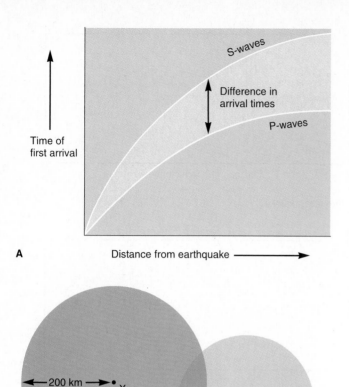

A

Time of first arrival

Distance from earthquake ⟶

S-waves

Difference in arrival times

P-waves

200 km ·X

Y ·150 km

·Earthquake

100 km ·Z

B

Figure 15.13

The focus, or hypocenter, of an earthquake is its point of origin below the surface. It can be located by seismographic data based on lag times between P and S waves. The greater the lag time between P and S waves, the greater the distance. Data from three seismic stations allow one to calculate the location of the earthquake's epicenter, the surface point directly above the hypocenter.

The Richter Magnitude Scale

The Richter magnitude scale is named after its developer, Professor Charles Richter. It measures vertical ground movement of the earthquake on a seismogram (box 15.1). The seismogram is the record of the shaking as registered by a seismograph. The reading is corrected for distance because wave energy naturally weakens with distance from the epicenter. A station at 100 kilometers is the standard for wave adjustment. Table 15.1 gives some of the major earthquakes of the century.

Magnitudes are expressed in whole numbers and decimals on a logarithmic scale, ranging from less than 1 to over 8. Chile and Japan have registered earthquakes of over 8.6. There is

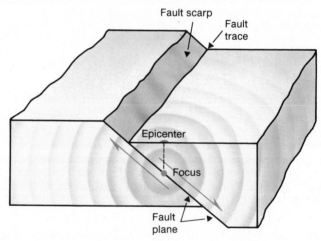

Fault scarp

Fault trace

Epicenter

Focus

Fault plane

Figure 15.14

Most of our information about the Earth's interior comes to us from earthquakes that occur at the focus along a fault plane.

Photo by Dr. Carol Prentice, U.S. Geological Survey.

Earthquake Magnitude Determination

In 1935, Dr. Charles F. Richter of the California Institute of Technology introduced to seismology a scale which made it possible to compare earthquakes throughout the world. The scale is now known as the **Richter magnitude scale.** This standardized measure of earthquakes is based on the logarithm of the maximum amplitude of seismic waves recorded by a Wood-Anderson seismograph; plus a factor which takes into account the weakening of the waves as they spread away from the earthquake focus. Hence, nearly the same value of magnitude can be determined by seismologists throughout California in a matter of minutes. Although Dr. Richter used a Wood-Anderson seismograph to develop the earthquake magnitude scale, it is possible to empirically relate the recordings produced by other types of seismometers to this scale (fig. 1).

Figure 2 illustrates how a seismologist would determine the magnitude of a particular earthquake from a Wood-Anderson seismogram. Two measurements are required: (1) maximum trace amplitude in millimeters

Figure 1

This seismograph at the University of California, Berkeley, is a detector and recorder of ground motion. The larger the ground motion, the larger the oscillation of the pin.

Photo by Juanita Creekmore.

(point "A") and (2) difference between arrival times of the primary "P" and secondary "S" seismic waves (point "B"). The values of these two measurements are then found on each side of the nomograph and connected by a line. That point where the line crosses the middle column is the value of the Richter magnitude of the earthquake, which in this example is magnitude 5.

Because magnitude is based on a logarithmic scale, a unit change in magnitude corresponds to a change in trace amplitude by a factor of ten as shown by the dashed line on the nomograph. Note that a magnitude 3 earthquake is defined on the Richter magnitude scale as an earthquake that will cause a peak amplitude of 1 mm to be recorded on a Wood-Anderson seismograph at an epicentral distance of 100 km (shown by dotted line on nomograph).

Source: R. Sherburne, *California Geology,* July, 1977.

To determine the magnitude of an earthquake, connect on the chart

A the maximum amplitude recorded by a standard seismometer, and

B the distance of that seismometer from the epicenter of the earthquake (or the difference in times of arrival of the P and S waves) by a straight line, which crosses the center scale at the magnitude.

- - - A unit change in magnitude corresponds to a decrease in seismogram amplitude by a factor of ten

.... Definition of a magnitude 3 earthquake

The Richter Scale

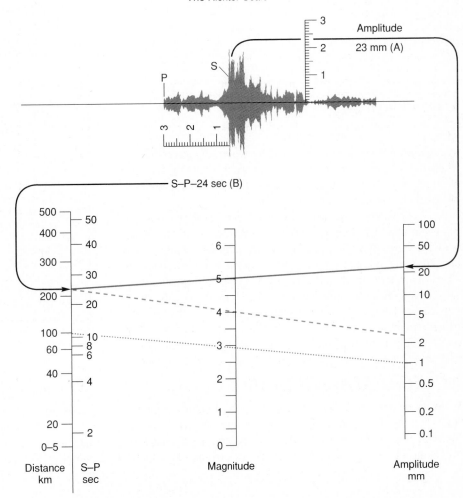

Figure 2

This diagram illustrates how seismologists determine earthquake magnitude using a Wood-Anderson seismograph recording and a magnitude determination chart.

no theoretical upper limit on the Richter scale. An increase of one whole number, let's say from 5 to 6, means a wave energy increase of 30 times and a wave height increase of 10 times. A magnitude of less than 2 is seldom felt by humans.

The energy released at the source of an earthquake is even more variable than the seismic records indicate. Duration or time is also a very important factor in determining the effects. When analyzing magnitudes, the record does not take into account the effect of location, depth of the hypocenter, soil, geology, and structural conditions in the affected area. Therefore, the Richter numbers cannot be used to estimate damage. An earthquake centered in a metropolitan area may be far more damaging, killing dozens and leveling many buildings, while an earthquake in the desert may only frighten a few animals and cause a rift in the ground.

Modified Mercalli Scale

Intensity or damage is best measured by the Modified Mercalli Scale first developed in 1902 by an Italian seismologist, Giuseppe Mercalli, and later modified for American conditions by Harry O. Wood and Frank Neumann (table 15.2). An earthquake can have only one Richter magnitude while it can have several intensities (fig. 15.15). The intensity is highest nearest the epicenter and gradually decreases outward in all directions.

Earthquake waves are as important to the seismologist as light waves are to the astronomer. Analysis of this form of energy has given us three important facts about the Earth's interior. First,

Figure 15.15

Intensity varies with distance from the epicenter and with geologic conditions and construction design. The magnitude 7.5 Guatemalan earthquake of 1976 caused damage of IX on the Mercalli Intensity scale.

Photo by U.S. Geological Survey.

Plate Tectonics, Seismic Waves, and the Earth's Interior   *329*

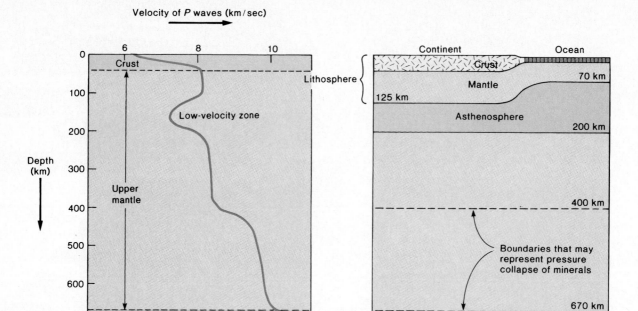

Velocity of P waves (km/sec)

Figure 15.16

The asthenosphere is located 100 kilometers below the surface. It is a low-velocity zone of seismic waves, indicating a plastic zone, and is the conveyer belt of the crustal plates.

the Earth is layered somewhat like an onion. Second, each layer is unique in chemistry and density. Third, there is a low-velocity zone in the mantle that may account for the dynamic movement of the plates. This low-velocity zone is an indication of plastic, semifluid conditions in the high-density rock. The three major layers detected by earthquake waves include the **crust, mantle,** and **core.** Let's consider each of these zones using the path and velocity of "P" and "S" waves as sensors of the Earth's anatomy.

ANATOMY OF THE EARTH

The Earth's Crust

We know the crust consists of continents and ocean basins, each uniquely different. Not only do the continents stand higher but they are made of very different rocks. We refer to the solid zone of crust and upper mantle as the **lithosphere,** a high-velocity seismic wave zone. At 100 kilometers (62 miles) depth, the low-velocity zone is reached indicating a plastic layer, the **asthenosphere** (fig. 15.16).

The ocean crust is a higher density (3.0 g/cm³) rock, rich in compounds of silica and magnesium. It is referred to as **sima** or **mafic** rock, basalt being the most abundant type (fig. 15.17).

Continents consist of lower density rock (2.8 g/cm³) called **sial** or **felsic,** rich in silica and aluminum. Granite is the most common rock making the major mass of the continent.

About three-fourths of our planet's surface is covered by the oceans, but not all of that fraction is ocean basin for the continents consist of a broad shelf often extending many miles under a shallow sea. The ocean basins actually cover only about half the Earth's surface.

With these known contrasts in mind, let us review the current working hypothesis of the origin of the continental crust. First, it is believed that the original surface consisted only of ocean basin or sima rock. It was from the oceans that the continents began to rise. When oceanic plates collided, perhaps buckling produced fissures, ridges, and troughs above sea level. Then weathering and erosion of the raised land produced sediments, rich in felsic minerals, which were deposited as a thin veneer over the oceanic plates. When two oceanic plates continued to collide, the heavier plate was subducted into the hot asthenosphere.

The boundary between these plates formed an oceanic trench. As the lower plate melted in the asthenosphere, magma was thrust to the surface to form the volcanic rock **andesite,** which built an island chain. Andesite is a blend of both mafic and felsic minerals intermediate in density and silica content. The rock is named for the Andes of South America. Most of the island arcs of the Pacific as well as the Cascade volcanic peaks of Oregon, Washington, and California are formed from this rock.

This underthrusting of one plate under another thickens the crust. The andesite and granitic material in the upper plate is much lower in density than the oceanic plate being subducted. The

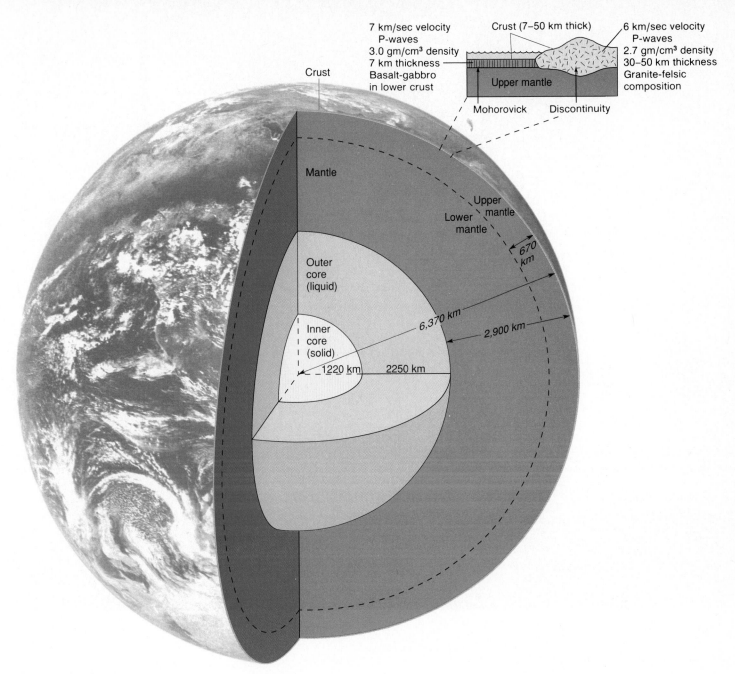

7 km/sec velocity
P-waves
3.0 gm/cm³ density
7 km thickness
Basalt-gabbro
in lower crust

Crust (7–50 km thick)

6 km/sec velocity
P-waves
2.7 gm/cm³ density
30–50 km thickness
Granite-felsic
composition

Upper mantle

Mohorovick   Discontinuity

Crust

Mantle

Upper mantle

Lower mantle

670 km

Outer core (liquid)

6,370 km

2,900 km

Inner core (solid)

1220 km   2250 km

Figure 15.17

Mafic and felsic rocks of the Earth's crust.

lower plate in the process lost its lighter density andesite and was also colder, so it continued sinking deeper into the mantle until it bottomed out at 700 kilometers.

As the continents thickened and rose because of buoyancy, more sediments were deposited into the trenches to form **geosynclines** or large geologic troughs. When the oceanic denser plate moved downward, it pulled these sediments under the lighter continental plate to melt and form **felsic** rocks. Perhaps the Sierra Nevada, a great granite batholith, was formed in this manner. A **batholith** is a large mass of granitic rock which forms from slowly cooling magma at great depths. Compressional forces carried the Sierra up to the surface and exposed it to erosion (fig. 15.18).

The Himalayas resulted from collision and compressional forces between the subcontinent of India and the Asian continent. As erosional processes reduce the mountains of the Earth, more sediments are carried to the sea to form more low-density magma, giving the continents andesite and granitic rocks. Continents are continually growing on their margins where plates are in collision.

Figure 15.18

The Sierra Nevada batholith is a felsic granite exposed by uplift and erosion. Yosemite Falls flows over the batholithic granite.

The Mantle

The boundary separating the outer crustal shell from the mantle, within the lithosphere, is marked by a significant change in density, as noted by earthquake wave refraction and changes in velocities. It was first discovered by a Yugoslavian geophysicist in 1909, Andrija Mohorovičić, who gives us the name for the **Mohorovičić discontinuity,** or *Moho* for short.

The second shell, the **mantle,** contains most of the Earth's mass and extends to a depth of 2,900 kilometers (1,800 miles). Recent seismic studies in the oceanic ridge zones and other continental locations have confirmed that the mantle is much denser than the crust, composed mainly of olivine and pyroxene, iron- and magnesium-rich minerals, and varies widely in plasticity or flow characteristics (fig. 15.19). Temperatures are high enough to melt most rocks, but pressures are so great at these depths that most of the mantle is in a solid state. As mentioned earlier, seismic studies have also discovered the asthenosphere, a zone of low strength or high plasticity at depths ranging from 100 to 350 kilometers. Here seismic waves slow down, meaning the zone is very plastic.

Plate tectonic theory states that the lithospheric plates, consisting of the crust and upper mantle, move on top of the asthenosphere. Many of the earthquakes and volcanic eruptions originate in the asthenosphere along the leading edge of the plates, where one plate is being subducted beneath another.

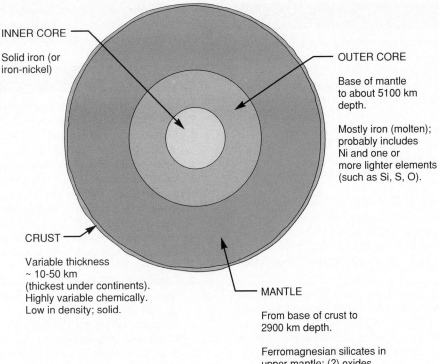

INNER CORE

Solid iron (or
iron-nickel)

OUTER CORE

Base of mantle
to about 5100 km
depth.

Mostly iron (molten);
probably includes
Ni and one or
more lighter elements
(such as Si, S, O).

CRUST

Variable thickness
~ 10-50 km
(thickest under continents).
Highly variable chemically.
Low in density; solid.

MANTLE

From base of crust to
2900 km depth.

Ferromagnesian silicates in
upper mantle; (?) oxides
of Fe, Mg, Si, and minor
elements below.
Localized partial melting within
upper mantle above ~ 250 km;
otherwise solid.

Figure 15.19

A model of the Earth's layered interior and composition of each
layer.

The Earth's Core

As seismic waves move through the mantle, another sudden change
in wave velocities and direction occurs at a depth of 2,900 kilometers from the surface. This is the boundary between the mantle
and liquid outer **core.** This abrupt change in velocity results from
the change in matter from solid to liquid. This high-temperature
layer is the source of the Earth's magnetic field and polarity. Its
fluid motion, perhaps driven by the Earth's rotation, must be responsible for producing magnetic fields that wander and reverse.
It is not well understood how this process works.

The core has a radius of 3,400 kilometers (2,100 miles)
and consists of a liquid iron outer shell, as indicated by "S" waves
(fig. 15.20), and a solid inner core. The inner core is believed to
be composed of extremely hot iron and nickel, perhaps with metallic silicon and oxygen as minor ingredients, that sank downward
and displaced lighter elements upward. Current plate tectonic
theory states that the high-temperature core ultimately provides
the energy for crustal plate motion.

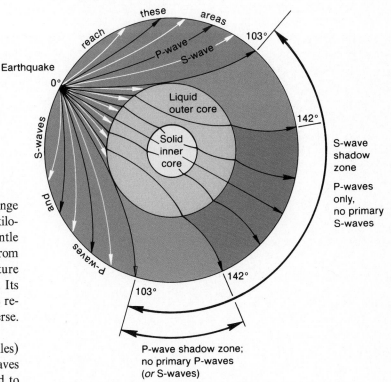

Figure 15.20

The core consists of a liquid outer core and a solid inner core of iron
and nickel. The S waves are stopped by the liquid zone while P
waves continue, reflected and refracted. Note the shadow zones for
P and S waves.

SUMMARY

The evidence in support of plate motion is still being gathered, but we can summarize the following clues in support of plate tectonic theory. These include:

1. Radioactive and **fossil dating** of rocks and sediments indicate the oldest sediments are farthest from the mid-oceanic axis where spreading is occurring.
2. The discovery of the reptile *Lystrosaurus* and the plant *Glossopteris* in Antarctica, South Africa, and India point to a common single land mass, **Pangaea.**
3. Matching of the continents shows a close fit between Africa and South America.
4. Evidence of **sea floor magnetism** and **magnetic field reversals** show that the Atlantic Ocean was once closed, which could only mean the ocean basin is rather young, and the continents here have drifted apart. The causes of these movements is still under investigation, but it appears that the Earth's hot interior provides the energy.

Earthquakes provide the geologist with information about the Earth's interior in the same way light gives the astronomer clues about the universe. **Shaking** produces three types of waves, "P," "S," and "L" waves, each of which reveal different aspects about the interior. From the science of **seismology** we have learned that the Earth is layered like an onion. The outer layer forms the **crust,** which covers the **mantle.** Next in this series of layers is the liquid **outer core** of iron and nickel, which surrounds the solid iron and nickel **inner core.**

ILLUSTRATED STUDY QUESTIONS

1. Explain how the island of Iceland was formed (fig. 15.1, p. 316).
2. Describe the active tectonic regions of the Earth (fig. 15.2, p. 318).
3. Explain the role of magnetic field reversal in revealing seafloor spreading (fig. 15.3, p. 319).
4. Give several clues to continental drift theory (fig. 15.4, p. 319).
5. Name the continents where the plant *Glossopteris* has been discovered in fossil form (fig. 15.5, p. 320).
6. Describe the locations of birth and death of an oceanic crustal plate (figs. 15.6 and 15.8, pp. 320, 321).
7. Explain the role of hot plumes in the mantle (fig. 15.9, p. 322).
8. Explain the *rock cycle* in light of plate tectonic theory (fig. 15.11, p. 324).
9. Contrast the "P" or primary wave and "S" or secondary wave in terms of velocity, amplitude, and pathway (figs. 15.12 and 15.13, pp. 325, 326).
10. Determine the magnitude of an earthquake if the $S - P = 10$ seconds and the amplitude is 15 mm (box 15.1, fig. 2, p. 327).
11. Describe the Earth's layered interior (figs. 15.17 and 15.19, pp. 331, 333).

Faulting, Folding, and Volcanism

16

Objectives

After completing this chapter you will be able to:

1. Explain the major tectonic processes shaping landforms.
2. Define **faulting** and **folding** and give examples of each process.
3. Compare and contrast **reverse** and **normal faults.**
4. Explain the causes of volcanism in terms of plate tectonic theory.
5. Describe various types of volcanic forms and their causes.
6. Explain the current hypothesis for volcanic island chain formation.
7. Describe the global patterns of volcanic activity and relate this pattern to plate tectonic theory.
8. Explain a variety of volcanic erosional forms.
9. Describe the geothermal energy sources and their relationship to volcanic activity.

The Earth's internal heat releases steam and drives the continental size plates, Namarskard, Iceland.
© Kevin Schafer/Tom Stack & Assoc.

Figure 16.1

This collapsed building in the Marina district of San Francisco was built on fill debris from the 1906 San Francisco earthquake. The October 17, 1989, Loma Prieta earthquake registered 7.1 magnitude and was centered 100 kilometers (60 miles) south of here.

Photo by Michael Molligan, Lake County Record Bee. Used by permission.

FAULTING AND ENERGY RELEASE

The Loma Prieta Earthquake

At 5:04 P.M. on October 17, 1989, the earth shook for 15 seconds along the San Andreas fault. The earthquake, named for a local mountain peak near the epicenter, violently shook the San Francisco Greater Bay Area from Monterey to Mendocino. The quake registered 7.1 on the Richter scale, sending the recording pens off the graph at many local seismographic stations.

In the first second, there was a deep rumbling. In the second and third seconds the ground began to shift both horizontally and vertically. Buildings, bridges, freeways, and a stadium packed with people for the 1989 World Series, suddenly seemed

out of control. By the fourth and fifth seconds, rumbling deepened and as its full fury was felt, gas lines in San Francisco, Santa Cruz, Oakland, and Alameda ruptured, triggering fires. The entire city of San Francisco suddenly lost power, and then water pressure failed. By the tenth second, streets buckled, and buildings and freeway overpasses collapsed. The 50-year-old Oakland–San Francisco Bay Bridge partially collapsed, losing a 50-foot section.

In that 15 seconds, 63 people died or were mortally injured, 30 thousand were suddenly homeless, and property damage of over a billion dollars was inflicted on the economy. Seven counties were declared disaster areas by the state of California (fig. 16.1).

Let's examine this event in the context of the geologic setting in order to understand **strike-slip faulting** and its geologic affects.

Figure 16.2
Strike-slip fault motion is horizontal, similar to two blocks rubbing together in opposite directions. This was the primary cause of the Loma Prieta earthquake.

Juanita Creekmore.

The Geologic Setting on the San Andreas Fault

The San Andreas fault is perhaps the best known fault in the world. It certainly was dramatically publicized in 1906 when an estimated 8.2 magnitude earthquake left a path of destruction from Monterey to the south to Fort Bragg to the north. A surface rupture was measured over 430 kilometers (270 miles) long along the fault.

The fault is a large fracture some 1,000 kilometers (600 miles) long and reaches a depth of over 30 kilometers (20 miles). It has many parallel smaller faults that also release energy from time to time. This fracture represents the plate boundary between the Pacific plate and North American plate. When the 1906 earthquake occurred, the Pacific plate shifted northward 5 to 6 meters (18 to 20 feet) relative to the North American plate. This type of horizontal motion is referred to as strike-slip motion, similar to rubbing two blocks of wood together producing shearing forces (fig. 16.2). This **strike-slip fault** is considered a **right-lateral fault** because the land on the opposite side of the fault shifts to the viewer's right, as illustrated in the offset of the Great Wall of China in figure 16.3.

Figure 16.3
In the 1906 San Francisco earthquake, a right-lateral strike-slip motion produced offsets up to 5 to 6 meters (18 to 20 feet). The Great Wall of China has been offset by similar motion in the Ningxia Hui Autonomous Region. The fault lines intersect both the wall and bridge almost at right angles.

Photo by U.S. Geological Survey.

The Loma Prieta earthquake displayed strike-slip motion of 1.8 meters (6 feet), normally expected along this portion of the plate boundary. Unexpected was a vertical movement of 1.3 meters (4 feet) (fig. 16.4). Another unusual feature was its depth, several kilometers deeper than most earthquakes on this fault. Apparently the shape of the fault is not a simple straight line.

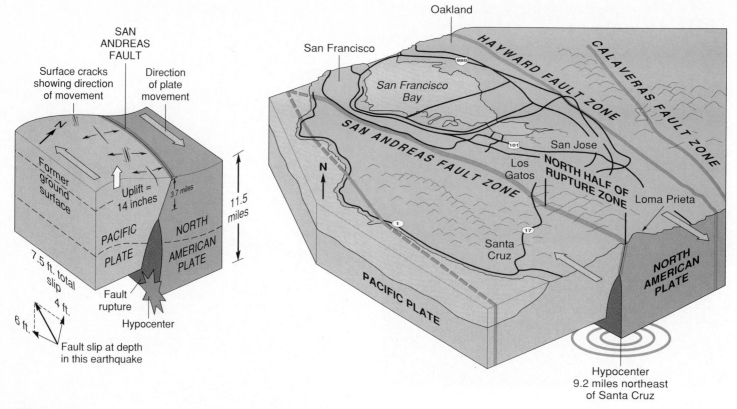

Figure 16.4

Changes in elevation and horizontal strain measured after the Loma Prieta earthquake show that the Pacific plate moved 1.7 meters (6 feet) to the northwest and 1.3 meters (4 feet) upward over the North American plate. This motion was not observed along a single fault break at the surface but occurred in a complex series of cracks and fractures. The upward motion is related to deformation of the plate boundary at a bend in the San Andreas fault.

Geologists noted that this portion of the San Andreas fault had been storing energy, as shown by a scarcity of seismic activity in this section until recently. It is also a section of the fault that has a bend, which causes the plates to stick and build up energy. When this stuck area suddenly broke free, energy was released, much the same way a tree limb vibrates after a branch is snapped off. The Pacific plate rode up onto the North American plate as it shifted northwestward.

The energy released was manifested in both ground shaking and crustal movement. Rupture along the San Andreas fault started at 19 to 24 kilometers (12 to 15 miles) below the surface, sending energy waves upward at 3.2 kilometers (2 miles) per second. The internal rupture was measured to be 35 kilometers (22 miles) long and 16 kilometers (10 miles) toward the surface. No surface rupture was located, but many fissures and landslides in the region resulted from the earthquake. The epicenter was 11.5 kilometers (7.2 miles) northeast of the beach community of Santa Cruz.

Areas most severely damaged were generally on alluvial stream deposits, bay estuary deposits, or bay fill where land had been reclaimed from San Francisco Bay. There was also severe damage from shaking near the epicenter. Here also large fissures and landslides along the ridge tops of the Santa Cruz Mountains were found. The greatest damage and loss of life and property occurred 90 kilometers (55 miles) from the epicenter on bay fill and old estuary deposits. In this type of ground, with a high water table and sandy conditions, seismic waves are highly magnified, especially the "L" waves, creating the process known as **liquefaction,** where the soil loses all strength and creates a weak base for buildings and bridge structures.

Aftershocks

Following the main event, the Earth is still capable of releasing pent-up energy. Thousands of lesser earthquakes, or **aftershocks,** may occur. Almost immediately after the Loma Prieta main shock, the U.S. Geological Survey installed a number of portable seismographs. The zone of high-intensity aftershock locations defined a plane dipping about 70 degrees to the southwest (fig. 16.5). Aftershocks ranged from almost at the surface to a depth of 18 kilometers (11 miles). The hypocenter is located at the bottom edge of this plane of aftershocks.

Between October 17 and October 25, the aftershocks consisted of 21 with a magnitude greater than 4 and two greater than 5. Table 16.1 lists the events of magnitude greater than 4.

Strike-slip faulting is the major motion along the San Andreas fault resulting from shearing forces. Other kinds of fault motion occur where the two sides move toward or away from each other. If the fault plane is steep and most of the displacement is vertical, the fault is referred to as either normal or reverse.

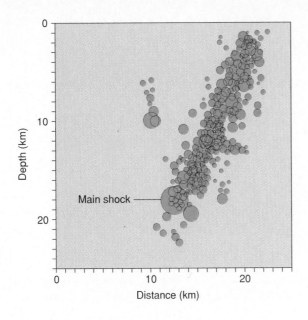

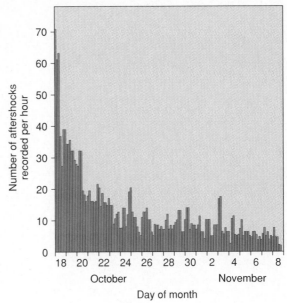

Figure 16.5

The epicenter of the Loma Prieta earthquake was pinpointed at 15 kilometers (9.3 miles) northeast of Santa Cruz. By noon on November 7, 1989, 4,760 aftershocks of the Loma Prieta earthquake had been recorded. The diminishing number of aftershocks with time is typical for large California earthquakes.

Table 16.1

Loma Prieta Earthquake Sequence

Greenwich Mean Time				Pacific Daylight Time			Latitude		Longitude		Magnitude
MO	DA	HR	MN	(DA	HR	MN)	DEG	MIN	DEG	MIN	
10	18	9	4	(17	17	4)	37	1	121	54	6.9
10	18	0	9	(17	17	9)	37	1	121	51	4.3
10	18	0	12	(17	17	12)	37	7	122	1	4.5
10	18	0	25	(17	17	25)	37	2	121	48	4.8
10	18	0	30	(17	17	30)	37	5	122	0	4.2
10	18	0	30	(17	17	38)	37	10	122	1	4.3
10	18	0	41	(17	17	41)	37	10	122	3	5.2
10	18	0	45	(17	17	45)	36	55	121	43	4.0
10	18	2	15	(17	19	15)	37	4	121	44	4.5
10	18	2	26	(17	19	26)	37	1	121	46	4.2
10	18	4	16	(17	21	16)	37	3	121	54	4.1
10	18	4	50	(17	21	50)	37	8	122	3	4.3
10	18	5	18	(17	22	18)	37	1	121	51	4.2
10	18	10	22	(18	3	22)	36	59	121	51	4.5
10	19	8	45	(19	1	45)	36	57	121	51	4.3
10	19	9	53	(19	2	53)	36	55	121	41	4.5
10	19	10	14	(19	3	14)	36	57	121	50	5.0
10	19	12	25	(19	5	25)	36	55	121	41	4.0
10	21	0	49	(20	17	49)	37	1	121	53	4.3

Source: U.S. Geological Survey.

Figure 16.6

At Carson Valley near Carson City, Nevada, normal faulting results
from crustal tension or stretching. This view is looking west toward
the Sierra Nevada, which forms the uplifted block.

Normal Faulting

The **normal fault** is the result of tension or crustal stretching where
one block falls relative to the other. The **footwall** partially supports
the downward-moving **hanging wall.** The force of gravity produces
the downthrow. Figure 16.6 shows the Basin and Range country
near Carson City, Nevada, where normal faulting has occurred.
This is also occurring in the spreading zones of the world where
new ocean floor is being created. The Gulf of California is widening as a result of this type of faulting; it is also producing the
Imperial Valley in California.

Reverse Faulting

Reverse faults occur when compressional forces cause the **hanging
wall** to rise relative to the **footwall.** Note how the hanging wall
block hangs over the footwall block (fig. 16.7). Where plates are

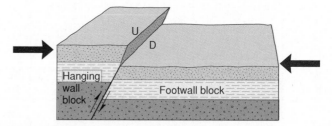

Figure 16.7

Reverse faulting produced by compressional forces has raised the
earth's major mountain ranges. There was a component of this
motion in the Loma Prieta earthquake.

in collision, reverse faulting is common. The Andes of South
America and the Coast Range of North America have many examples of this form of faulting.

Figure 16.8

The San Gabriel Mountains rose 1.2 meters (4 feet) and shifted south the same amount on February 9, 1971. The Van Norman Dam did not completely fail; however, the top service road slid into the lake and 30,000 people had to be evacuated.
Photo by R. Wallace, U.S. Geological Survey.

Thrust Faulting

Thrust faulting also results when one block moves up over another. It results from compression, but with a greater horizontal component of motion. Both the San Francisco Bay Area's 1989 earthquake and the San Fernando earthquake of 1971 had this type of movement.

In the 1971 earthquake, the San Gabriel Mountains of southern California shifted about a meter (3 to 4 feet) upward and also southward toward the Pacific Ocean on a fault plane sloping at about 45 degrees. Such a low-angle reverse fault is called an **overthrust fault.**

Shortly after six o'clock in the morning on February 9, 1971, about 10 million Americans were shaken out of bed by a sudden terrifying shock. It lasted about a minute, doing most of its damage in about 10 seconds to residents in the northern portion of San Fernando Valley. Structures collapsed, causing 64 deaths, and water, gas, and sewer lines ruptured, leaving this suburban region in a state of destruction. Possibly 50,000 homes would have faced complete destruction if the Van Norman Dam in the San Fernando Valley had failed (fig. 16.8).

Fortunately, the San Fernando earthquake was not a great earthquake. It released only one-hundredth as much energy as the Anchorage, Alaska, earthquake of 1964. However, the San Fernando earthquake caused a greater financial loss than the Alaskan earthquake because it occurred in a much larger metropolitan region. Box 16.1 illustrates some other ways that effects of earthquakes can differ in terms of destructive side effects.

If an earthquake can have elements of good fortune, the San Fernando earthquake did. It occurred outside the centers of population in downtown Los Angeles, where it would have done much more damage to old four- and five-story brick buildings. Had it occurred during the peak rush hour of 8 A.M., there would have been more people in buildings, more children in schools, and more cars on overpasses and freeways that gave way.

Similar Quakes Make Different Shakes

BOX 16.1

On December 7th, 1988, in Armenia, S.S.R., one of the deadliest earthquakes of this century killed over 25,000 people. Reports first received from the national seismological reporting services indicated a moderate earthquake with a magnitude range between 6.5 and 7.0 on the Richter scale. The location was pinpointed near the city of Spitak. At first there was no indication that a disaster of major proportions had occurred (fig. 1).

Hundreds of structures over an area of 1,800 square kilometers were totally destroyed, including many buildings of modern design. A survey of the three largest cities most heavily damaged revealed that 87 percent of the buildings in Spitak (pop. 30,000), 24 percent of those in Kirovakan (pop. 150,000), and 52 percent in Leninakan (pop. 290,000) suffered collapse or heavy damage (fig. 2).

Investigations by Soviet, French, and American geologists discovered both reverse thrust faulting and right-lateral faulting in the Lesser Caucasus highlands, about 80 kilometers south of the main range of the Caucasus Mountains. Here convergence of the Arabian and Eurasian plates has created the Caucasus Mountains, which are part of a broad mountainous zone stretching from the Pyrenees and the Alps through southern Europe to Asia and the Himalayas.

Vertical movement along an 8-kilometer (5 mile) zone showed a maximum displacement of 2 meters (6 feet) and a 0.5 meter (1.6 feet) right-lateral horizontal component. This previously unnamed fault strikes parallel to the Caucasus Mountains (fig. 3).

The Armenian earthquake and the Loma Prieta earthquake have a number of things in common (table 1). Both occurred on major plate boundaries with similar magnitudes. Both had strike-slip and vertical thrust components. In both areas, the greatest damage correlated with areas of younger saturated soils, and in the case of San Francisco, a fill area. When shaking occurred, the soils liquefied.

Two factors may explain the higher building failure in Armenia. At Leninakan precast frame panels and composition frame-stone construction was widely used, which failed to hold up. Because the Soviets have a uniform state standard for construction, the buildings were all very similar. Thus all had similar weaknesses.

Secondly, damage patterns and aftershock recordings suggest that the local geology

at Leninakan was a major contributing factor to building failure. Unconsolidated soils and saturated alluvial deposits on a broad lake bed underlain by sedimentary rock was the setting for the Leninakan disaster. This type of geologic setting has been shown to alter the characteristics of seismic waves by increasing the amplitude and duration of shaking.

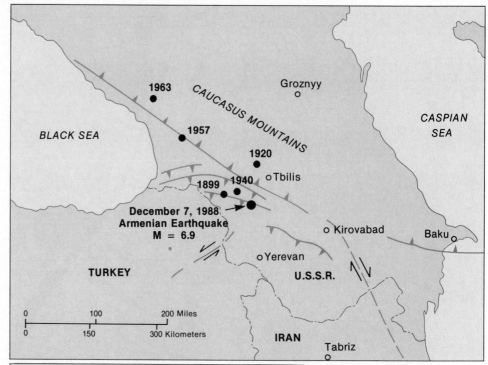

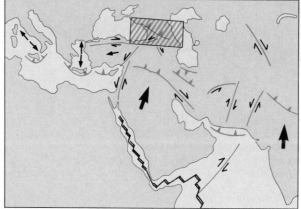

Figure 1

When a 7.0 magnitude earthquake shook Soviet Armenia in 1988, 25,000 people died in less than 45 seconds.

Source: R. Borcherdt (ed.), "Results and Data from Seismologic and Geologic Studies Following Earthquakes on December 7, 1988 near Spitak, Armenia, S.S.R.," U.S. Geological Survey Open File Report 89–163A.

Figure 2

In Spitak, 87 percent of the buildings either collapsed or were heavily damaged.

Photo by R. Borcherdt, U.S. Geological Survey.

Azimuth of fault		292°
Dip of fault	D	55°
Slip	S	2.0m
Strike-slip component	SS	0.5m
Vertical component	V	1.6m
Contractional component	C	1.1m
Horizontal component	H	1.2m
Dip-slip component	DS	1.9m
Plunge of slip	PL	53°
Rake of slip	R	109°

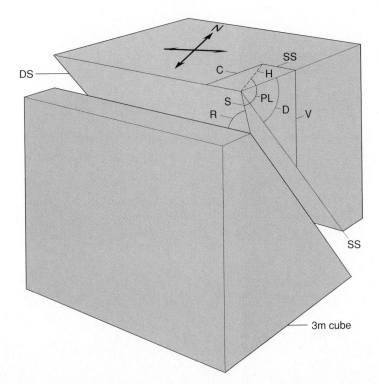

Table 1

Similar Quakes Make Different Shakes

	Armenian Earthquake	Loma Prieta Earthquake
Date	Dec. 7, 1988	Oct. 17, 1989
Duration	25 seconds	15 seconds
Place	45 km (28 mi) north of Mt. Aragats, Lesser Caucasus highland Lat. 40.996° N Long. 44.197° E	6 km (10 mi) northeast of Santa Cruz, CA, in the Coast Range Lat. 37.02° N Long. 121.52° W
Magnitude	6.9–7.0	6.9–7.1
Plates	Eurasian and Arabian	North American and Pacific
Depth of focus	14 km (8.7 mi)	15 km (9.3 mi)
Length of rupture	8 km (5 mi)	14 km (9 mi)
Property damage	Thousands of structures Millions of dollars	Thousands of structures Billions of dollars
Deaths	25,000 people	63 people
Cause of major damage	Shaking and structural design weaknesses Liquefaction	Shaking and structural design weaknesses Liquefaction
Geology of sites most heavily damaged	Alluvial soils and high water table	Bay sands, high water table, and alluvial soils
Aftershocks greater than 4.5 one month after quake	Thousands 21 +	Thousands 13 +
Motion	Right-lateral and reverse thrust	Right-lateral and reverse thrust

Figure 3

Both vertical and horizontal movement occurred along a previously unknown fault. Here both right-lateral and thrust faulting occurred.

Photo by R. Borcherdt, U.S. Geological Survey.

Elastic Rebound Theory

BOX 16.2

According to legend, the Earth was supported by the back of a tortoise. When the tortoise took a step, the Earth shook and trembled. Today's explanations of earthquakes are more scientific and yet have some parallel to the legend of the walking tortoise.

Fault movement starts at a single point several miles below the surface, spreading out and relieving strain over many miles of the fracture zone or fault line. In large earthquakes, such as the 1906 San Francisco temblor, over 10 percent of the plate boundary may be involved in displacement.

According to the **elastic rebound theory**, a fault remains quiet without movement until strain has built up in the rocks on both sides of the fault. The strain rate between the North American and Pacific plates, for example, continues to be about 2 to 4 centimeters (1 to 2 inches) per year. This strain accumulates by a gradual shifting and distorting of the crust, until accumulated stress causes a sudden shear failure, say, on the part of the San Andreas Fault at Loma Prieta (fig. 1). When the accumulated stress finally overcomes the resistance of the rocks, the fault snaps or rebounds back into an unstrained position. The motion of rocks past each other creates the shock waves we call earthquakes.

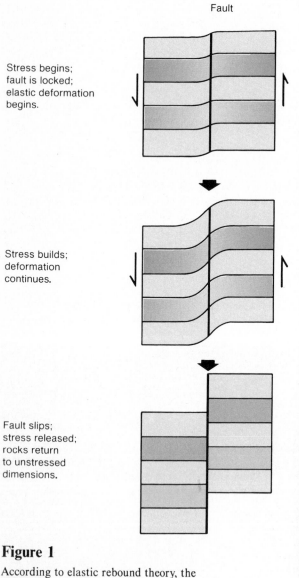

Fault

Stress begins; fault is locked; elastic deformation begins.

Stress builds; deformation continues.

Fault slips; stress released; rocks return to unstressed dimensions.

Figure 1

According to elastic rebound theory, the rocks first undergo stress. Next comes deformation and finally failure, when the rocks snap back into their original undeformed shape, but offset.

Photo by Carol Printice, U.S. Geological Survey.

Figure 16.9

The Madison slide, August 17, 1959, created Hegben Lake along the
Madison River, Montana, as a result of structural instability and
shaking movement triggered by faulting. The earthquake measured
7.1, shattering a dolomite limestone structure that held the schist in
place.

Photo by U.S. Department of Agriculture, U.S. Forest Service.

It has been estimated that 99 percent of the damage in
an earthquake is from shaking or vibratory motion that is an indirect result of the faulting itself. Only 1 percent of the monetary
loss associated with earthquakes is the direct result of surface rupturing and displacement.

GEOLOGIC EFFECTS OF FAULTING

Landslides

Strong shock waves can uproot trees, snap off telephone poles, and
cause massive landslides. Some of the ancient slides, now covered
with vegetation, can only be explained as earthquake induced.

In 1970 a 7.7 magnitude earthquake centered 130 kilometers (80 miles) west of Peru's highest peak, Nevados Huascarán, triggered a massive debris avalanche that buried the towns
of Yungay and Ranrahirca. It took the lives of 18,000 people in
less than one minute. (See box 17.1, p. 384.)

When the Earth shook on August 17, 1959, the Madison
River was modified by a landslide and a lake was formed. Faulting
induces landslides in two ways. First, faulting creates a highly
fractured fault zone. Rocks are weakened in such a way that they
move downhill easily when lubricated by groundwater. Second, the
steep fault-controlled slopes become ideal ramps for landslides (fig.
16.9).

Soil Failure

Earthquake intensities tend to be highest in valleys filled with sand
and gravel or alluvial material, as in the Armenian earthquake.
Filled land and areas reclaimed from swamps and tidal basins suffer
the greatest damage. Structures built on solid rock near the epicenter of an earthquake frequently fare better than distant buildings. In the 1964 Alaska, 1985 Mexico City, and 1989 Loma Prieta
earthquakes, loose unconsolidated alluvial and fill material proved
to be the hardest hit areas.

Soil liquefaction is a condition resulting in soil becoming
semiplastic during severe shaking, as evidenced in the Loma Prieta

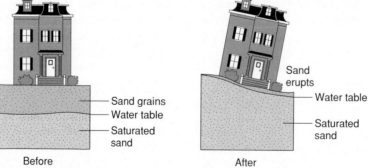

Before   After

Figure 16.10

Soil failure in the San Francisco Marina district. When violent
shaking occurs, water is forced upward between the sand grains. The
entire sand base liquefies. The soil saturates and even erupts in small
liquid volcanoes of sandy water. Buildings sink and tilt in the
liquefied sand.

Photo by Michael Molligan, Lake County Record Bee. Used by permission.

Figure 16.11

At Kodiak Island, Alaska, the 1964 earthquake resulted in a rapid
rise of the ocean in a damaging tsunami.

Photo by the U.S. Coast and Geodetic Survey.

earthquake of 1989. This occurs in fine alluvial soils penetrated
by groundwater. In the Marina district of San Francisco, buildings
swayed and tilted as the bay fill shook violently and weakened as
it liquefied. A structure is no safer than its geologic foundation
(fig. 16.10).

Tsunamis and Changes in Sea Level

Tsunamis, or seismic sea waves, are triggered by ocean floor
faulting, and though they are sometimes called "tidal waves," they
are not related to tidal action. Tsunamis sometimes accompany
large earthquakes; the Pacific coast and Hawaii have been hit on
several occasions.

A wave reaching 15 meters (50 feet) high from an 1812
earthquake struck the Santa Barbara coastline. The Alaskan
earthquake of 1964 triggered a tsunami that caused loss of life
and damage on Kodiak Island (fig. 16.11). Underwater seismic
activity off the southern coast of Mexico during the last week of
June 1990 produced a "tidal wave" that swept away 300 homes
in the fishing village of Cuajimicuilapa. The 1,500 villagers retreated to higher ground before this powerful seismic wave reached
land. It was triggered by motion along an offshore fault system.
Chapter 21, "The Work of Waves," explains the nature of a
tsunami.

Changes in sea level are also one of the risks of living
along the coast in an active fault zone. The Alaskan earthquake
caused the coast in some places to be thrust upward in seconds,
while in other areas downthrows occurred, flooding formerly dry
land.

Marine terraces near the *study area* of chapter 1 are a
testimony to slow-motion changes in sea level over a half-million-
year period.

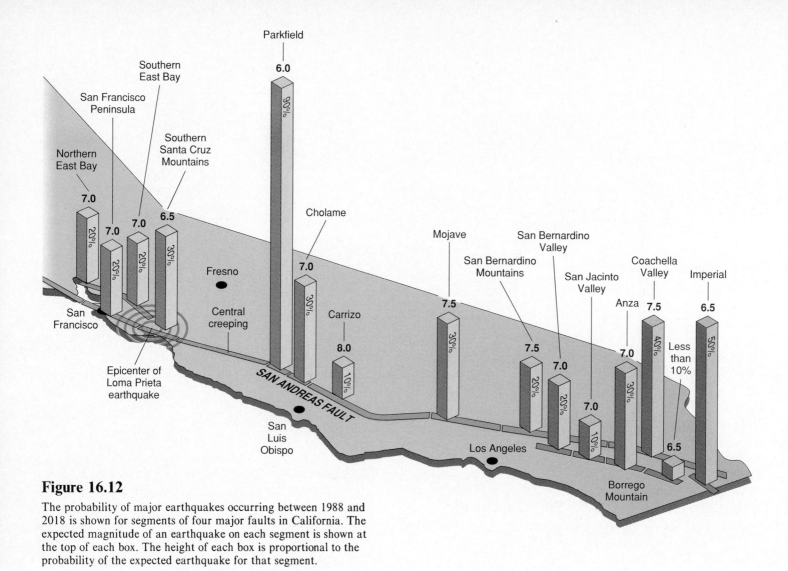

Figure 16.12

The probability of major earthquakes occurring between 1988 and 2018 is shown for segments of four major faults in California. The expected magnitude of an earthquake on each segment is shown at the top of each box. The height of each box is proportional to the probability of the expected earthquake for that segment.

FAULTING IN THE FUTURE

What lies ahead for the quaking earth? What do we do before the next big one hits (fig. 16.12)? Can we learn about past earthquakes in planning for the future? How do landforms influence land use? These are the questions geographers, engineers, regional planners, and government officials are asking around the world and especially in regions of high seismic activity. Looking back at the terrible damage of recent earthquakes, we can learn a number of lessons.

First, most of the loss of life and properties was from shaking, not crustal displacement or rupturing. Shaking affected great areas, while rupture was confined to the zone where the fault broke the surface. Most earthquakes do not break the surface but do their damage by shaking.

Second, planning and legislation must take into account that a great earthquake strikes the rim of the Pacific once every four years on average. Building codes must reflect this high fre-

quency of earthquakes. Buildings, dams, and schools need careful attention in terms of design and location relative to soil, slope, and fault patterns.

Finally, research on earthquake predictions and their effects on people must be accelerated to minimize the damage. The topic is an active and lively area of earth science. Scientists at the U.S. Geological Survey's National Center for Earthquake Research in Menlo Park, California, are mapping faults, making records of past and present large and small earthquakes (fig. 16.13 and box 16.3).

The Loma Prieta quake was expected—though not predicted for any particular date—and areas of probable damage had been mapped and publicized a year prior to the event.

In summary, damage from earthquakes falls into four categories: (1) violent shaking, (2) ground displacement by faulting, (3) landslides and soil deformation, and (4) seismic sea waves or tsunamis and changes in sea level.

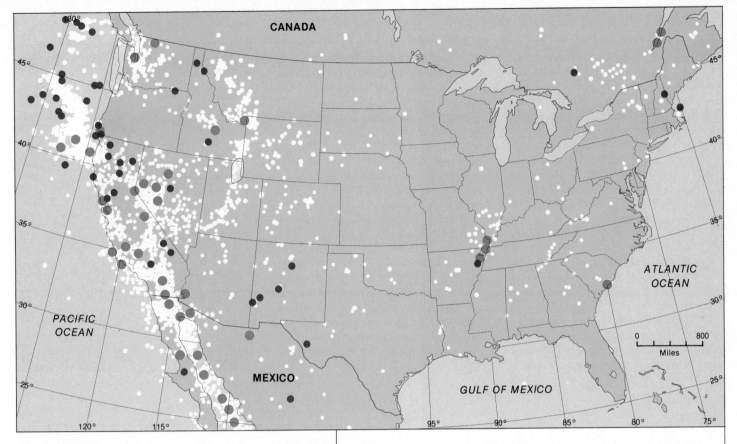

Figure 16.13

Although earthquakes are most common in California and Alaska, they have shaken most states. Earthquakes as large as or larger than the Loma Prieta event (red dots) have occurred in Alaska, California, Hawaii, Idaho, Missouri/Tennessee, Montana, Nevada, South Carolina, Washington, and Quebec. The red pattern indicates where prehistoric events of this size have occurred. This map shows the locations of all historical earthquakes in the 48 conterminous United States of magnitude 5.5 or larger, all earthquakes of magnitude 5 to 5.4 since 1925, all recorded earthquakes of magnitude 4 to 4.9 since 1962, and all recorded earthquakes of magnitude 3.5 to 3.9 since 1975. Clearly, earthquakes are a nationwide problem.

FEATURES OF FAULTING

When faulting occurs, a large number of related geographic features are formed. Fault valleys, scarps, saddles, grabens, domes, sag ponds, and hot springs are just a few of the many fault-related features. Once you learn to recognize these features, you can trace them in your travels.

Virtually all large escarpments and mountain blocks that rise along a fault line are the by-product of millions of years of tectonic activity. Only the very largest earthquakes rupture the surface. In most cases even the most recent features were created by several huge earthquakes occurring hundreds, even thousands of years apart.

Fault Valleys

A fault valley is the by-product of faulting followed by rapid erosion. When the crust fractures along a fault line, the fault zone becomes filled with shattered rock, which is more easily weathered and eroded than the solid rock on either side. Figure 16.14 shows a section of the 1,000 kilometer (600 mile) long San Andreas fault line and accompanying fault valley. Note how the valley runs parallel to fault-formed ridges. Note also how the streams crossing this fault zone are offset. This is because the San Andreas fault system has a **strike-slip fault** motion, which means most of the fault movement is horizontal.

Grabens

Grabens are valleys formed where fault lines run parallel to one another. The block of land between them is pulled downward by gravity while the two adjoining blocks, or *horsts,* rise on either

Surgery on the San Andreas

BOX 16.3

Surgery, especially exploratory surgery on a patient, can be a useful tool in diagnosis of a disease or provide a better understanding of its history and progression. Surgery on the San Andreas fault provides the same type of knowledge. By trenching across a segment of a fault plane, a cross section of the fault can reveal an entire history of recent geologic events. This knowledge not only contributes to the understanding of past events but also provides a tool for predicting future earthquakes.

History seems to repeat itself along the faults of the world. This surgery along the northern portion of the San Andreas fault system has revealed a repeated history of faulting and major earthquakes over a span of several thousand years.

The cross section in figure 1 shows the wall of an excavation dug across the San Andreas fault zone in northern California near Point Arena. These sediments have been broken by at least four earthquakes. Fault A breaks through all the sediments to the surface, and represents the great earthquake of 1906. Fault B does not break to the surface. Because it is truncated by a sand lens, an earthquake must have occurred sometime before the sand lens was deposited. Fault C represents a still earlier earthquake that occurred before the blue sand layer was deposited. Fault D represents a still earlier earthquake. Radiocarbon dating of the red and blue layers gives age control for the three prehistoric earthquakes. The red layer was most likely deposited between 89 B.C. and A.D. 212. The blue layer was probably laid down between A.D. 1040 and A.D. 1384. Two earthquakes have occurred at A and B since the deposition of the blue layer. These and other data collected from similar trenches at this site suggest that the average recurrence interval for large earthquakes along this segment of the San Andreas fault is long—between about 200 and 400 years.

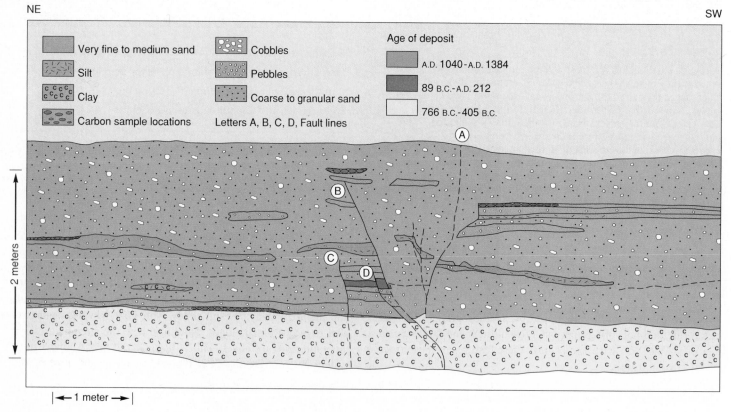

Figure 1

The trench intersects the San Andreas fault in the vicinity of a small alluvial fan. Here mapping of fault history is possible and dates can be determined by using a carbon-14 dating technique.

Photo by Carol Printice, U.S. Geological Survey.

A

B

C

Figure 16.14

The San Andreas fault represents one of the world's most active fault zones. Evidence of faulting can be seen in offset streams, fault scarps, sag pond springs, and offset rock structures. (A) The San Andreas fault is marked by palms, looking northwest along the Indio Hills Springs. (B) The San Andreas fault viewed in false-color infrared near the San Francisco Airport; San Andreas Lake lies in the fault zone. (C) The Carrizo Plain has offset streams

A and C: Photos by R. Wallace, U.S. Geological Survey; B: Photo by NASA.

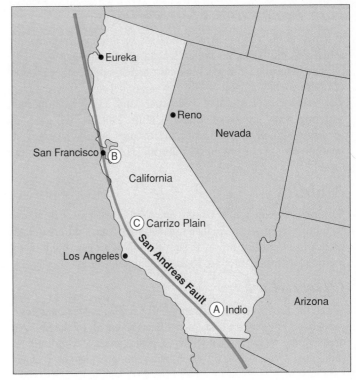

side. The result is a broad, flat valley with steep mountain slopes produced by **normal faulting.** Examples are numerous in the western United States, South America, Asia, and Africa.

Some of the deepest lakes and valleys in the world are located in sunken blocks or grabens. The two lowest points in the Western Hemisphere are located in grabens: Death Valley at −86 meters (−282 feet) and the Salton Sea at −72 meters (−235 feet) in the Imperial Valley of California. Each formed when the crust experienced tensional stress, which caused the sunken block to drop relative to the mountain blocks on each side.

The Dead Sea, the lowest point on the Earth's surface at −392 meters (−1,286 feet), is located in the rift zone and graben of the Jordan Valley. This rift zone is part of the plate boundary between Africa and Arabia. In East Africa, Lakes Victoria and Tanganyika lie in grabens in a major rift zone. The deepest lake in the world (at 1,773 meters or 5,812 feet), Lake Baikal in Siberia, also occupies a rift valley.

Scarps and Steep Mountain Slopes

A common form of evidence of recent faulting is the **scarp,** a steep cliff or ridge formed by fault movement. Scarps represent the actual fault plane or surface of contact between the blocks in motion. The western face of the Wasatch Mountains near Salt Lake City, Utah, and the eastern face of the Grand Tetons near Jackson Hole, Wyoming, are excellent examples of scarps formed along major fault lines (fig. 16.15). Frequently scarps are the sites of waterfalls and rapids, because streams cannot erode fast enough to keep up with recent uplift.

Sag Ponds and Hot Springs

When faulting occurs, stream courses or drainage avenues are sometimes cut off by an impervious layer of rock, causing a lake or pond to form. Green oases in desert regions are often the product of faulting which impounds the flow of groundwater. These springs may support considerable vegetation and serve as important watering holes for livestock and wildlife. Hot springs and soda springs also tend to occur along fault scarps. Many have become health resorts; Palm Springs, California, is one of the more famous ones.

Saddles

Saddles are evidence of a fault line that cuts across a ridge top, forming a mountain pass. The crushed rock from faulting quickly erodes to produce a dip in the profile of the ridge. Many of the major mountain passes are located along fault lines.

Offset Rocks

When the Earth's crust shifts, not only are fences, trees, and highways offset, but so are rock formations. Along all of the San Andreas fault, rocks in contact on opposite sides of the fault are radically different. This offset of formations provides the geologist with information about the amount and timing of movement along the fault. The San Andreas fault has displacements of hundreds of kilometers. For instance, pre-Cretaceous rocks beneath the floor of the San Joaquin Valley near Taft, California, can be matched

Figure 16.15

The Grand Tetons in Wyoming have steep scarps formed along major fault lines that resulted from normal faulting.

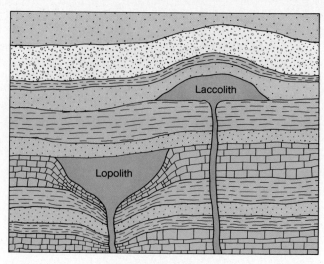

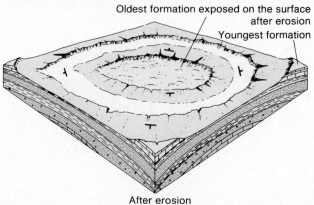

Oldest formation exposed on the surface
after erosion
Youngest formation

After erosion

Figure 16.17

Laccoliths, such as the Black Hills of South Dakota and eastern Wyoming, are formed by a pooling process of magma. Then weathering of a thin veneer of sedimentary rock exposes the crystalline basement rock of the laccolith. The symbols indicate the dip direction of the sedimentary rock beds.

Figure 16.16

The Earth's crust can be deformed into almost any imaginable shape without fracturing. This arched fold or anticline was formed by compressional forces in Calico Hills, California.

with rocks of the same type offshore near Point Arena, California. This represents a displacement of 550 kilometers (350 miles) in 150 million years. Younger rocks are offset by smaller amounts.

Other large-scale offsets have been discovered in the Philippines, New Zealand, and the classic Glenmoor of Scotland. In Scotland, the Glenmoor fault runs in a northeast-southwest direction as evidenced by the linear fault valley and Loch Ness and Loch Lochy lakes connected to the sea by the Caledonian Canal.

Crushed and Deformed Rocks

Rocks in contact with the fault line may be bent against it. This is known as fault drag, the result of friction. Sometimes movement not only deforms but polishes and grooves and even produces physical and chemical changes in the rock. Frequently repeated movement can produce a claylike rock flour, called fault gouge. A contact zone or shear zone may range from a knife-edge line to a region hundreds of yards wide in a complex fault zone.

FEATURES OF FOLDING

Plate collisions not only cause faulting, but also bend and deform the Earth's crust just short of fracture. If compressional forces stop short of fracture or fault, a warp or fold may develop. Solid rock thousands of feet thick may be folded over like so much paper. We all know rock is brittle and can fracture with a good hammer blow, and yet under high temperatures and long-term stress, the Earth's crust can be twisted, stretched, warped, and folded into almost any imaginable shape without fracturing (fig. 16.16).

Domes

Domes are commonly built by pooling of molten rock or magma in much the same way a blister forms on the skin. A magma dome formed primarily by the pooling process is referred to as a **laccolith** (fig. 16.17). The Black Hills of South Dakota and eastern Wyoming are a good example. A thin veneer of sedimentary rock once

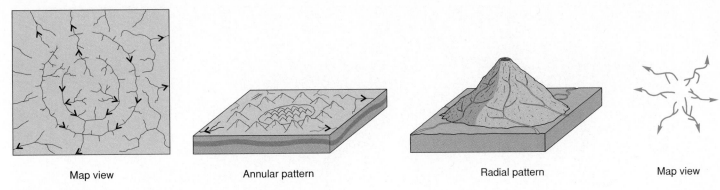

Map view   Annular pattern   Radial pattern   Map view

Figure 16.18

The erosional pattern of an eroded dome changes from a radial
pattern, like spokes of a wheel, to a circular pattern of concentric
rings interconnected into one continuous drainage network.

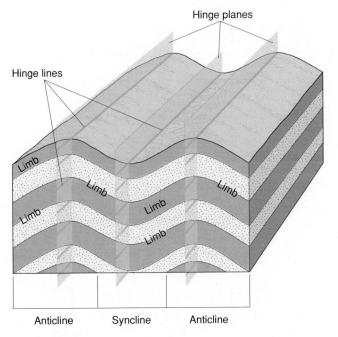

Hinge planes

Hinge lines

Limb

Anticline   Syncline   Anticline

Figure 16.19

Anticlines are upward folds while synclines or geologic troughs are
downward folds.

covered the entire dome, but weathering and erosion never stop.
Currently the eastern central core is eroded to the older crystalline
basement rocks of the continent. Ringing the core are a series of
ridges or **hogbacks** and valleys carved by streams, forming a unique
drainage pattern. In the early stages of development the **radial
drainage pattern** resembled spokes of a wheel (fig. 16.18). As erosion progressed, weaker layers of sedimentary rock eroded more
rapidly, producing an **annular stream pattern.** The streams form
concentric rings in the valleys between the hogback ridges.

Anticlines, Synclines, and Monoclines

Where recent crustal compression takes the form of buckling sedimentary rocks, a distinctive structure is observed. Ridges or
upward folds are called **anticlines** (fig. 16.19), and downward folds
or troughs are known as **synclines.** A monocline is a terrace-like
fold resembling a rounded stairstep. Think of a monocline as a half
of a fold with only one **limb** or side of the fold. It is important to
realize that erosion changes the surface and slope continually. Anticlines and synclines persist as geologic structures, even if the land
surface is eroded to a topographic plain.

The Appalachian Mountains in central Pennsylvania represent a highly folded system that has undergone extensive erosion
and differential weathering. More easily decomposed rocks weather
at greater rates, producing a landform of dissected anticlines and
synclines. Figure 16.20 shows that topographic valleys may occupy
anticlines or synclines; only a close study of the rocks can determine which geologic form is present.

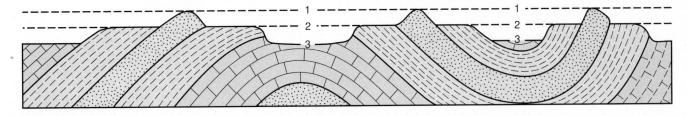

Figure 16.20

Erosion in a final stage of development. Early folding was reduced
by surface erosion in three separate episodes.

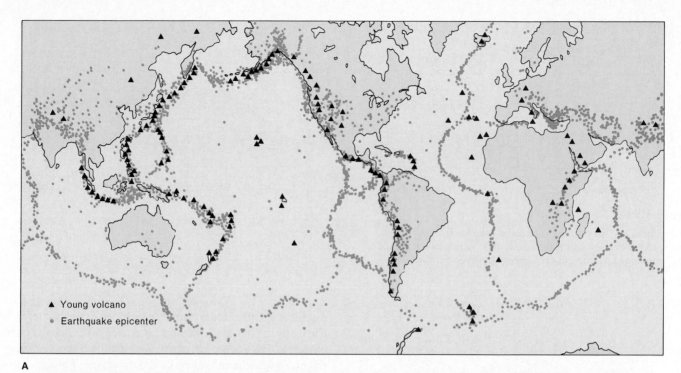

A

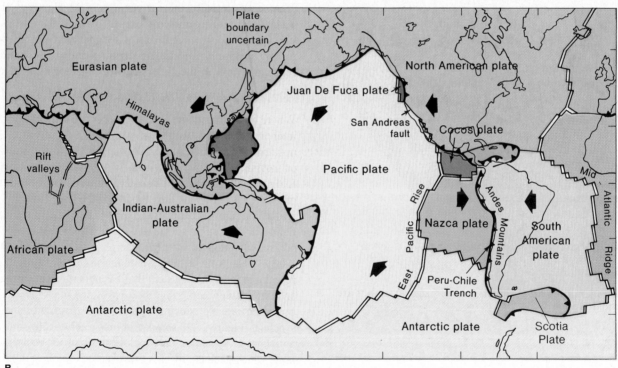

B

Figure 16.21

Volcanic activity and earthquakes, when plotted on the same world map, show a very high correlation. These are the plate boundries and zones of stress and strain.

Anticlines and synclines often occur together as a series of elongated groups like a shower curtain or a wrinkled tablecloth. During the development of anticlines and synclines a number of localized stresses develop. There is tension (stretching) on the outside of the bend and compression (squeezing) on the inside. Erosion works at a greater rate where strain is more severe in these two curved zones. On the limbs a shearing action occurs. Beds tend to slip past one another like cards in a deck. In the process some beds are likely to fracture, while others bend in a plastic manner.

In regions of recent folds, the anticlines are the hill structures and synclines become the valleys, and yet in time anticlines and synclines erode away from the work of weathering and erosion, described in chapter 17.

VOLCANIC LANDFORMS

Volcanic landforms range in scale from the small, symmetrical cinder cones found in volcanic areas the world over to the giant Mauna Loa shield volcano forming the growing island of Hawaii. From the ocean floor to its crest, the island of Hawaii is 9,000 meters (30,000 feet) high, one of the tallest landforms on the Earth, yet only 4,000 meters (13,000 feet) are above sea level.

The Earth's interior energy is the driving force for plate motion and renewal of the ocean floor as well as the energizing mechanism for volcanoes. The eruptions of Mount St. Helens are a classic example of the great energy held under pressure in the Earth's crust. A comparison of earthquake patterns and volcanic activity reveals the close correlation between these two geologic events (fig. 16.21). However, it is misleading to believe that when a strong earthquake occurs, volcanic eruptions will soon occur in the fault zone.

Volcanic features represent some of the Earth's youngest crustal features. Most of the world's major volcanic regions are distributed along the boundaries of the lithospheric plates. As discussed in chapter 15, in some regions the sea floor is spreading, forming major rift zones along the great mid-oceanic ridge systems. In other zones, oceanic plates are being subducted. It is in these zones of spreading and subduction that volcanic activity is prominent. The volcanic forms are represented by cone-shaped islands and mountains and massive lava plateaus and rift zones submerged beneath the sea and dissecting the continents.

We will first examine the processes or causes of volcanism, then relate these processes to a variety of volcanic landforms. Finally, we will focus on this energy potential for human use in the form of geothermal steam and hot water.

Causes of Volcanism

More than 600 active volcanoes and many more dormant ones dot the zones of plate boundaries and crustal stress. Note in figure 16.21 that a major concentration of volcanoes lies along the rim of the Pacific and another extends across the Mediterranean and eastward through Turkey to the mountain chains of central and southeastern Asia. By following the mid-oceanic ridges of the Pacific, Atlantic, and Indian Oceans, one can see a continuous pattern of volcanic features.

A

B

Figure 16.22

Iceland is composed of low-silica, high-temperature mafic magmas, such as occur along zones of seafloor spreading. (A) Note water pipe on roadbed to keep water cool because the ground is warm.
(B) The top of a two-story house and telephone pole are nearly covered by volcanic ash in the town of Heimaey, Iceland, in 1973.
Photos by U.S. Geological Survey.

Landforms resulting from volcanic activity are largely the consequence of two types of plate interaction, consisting of *spreading* and *subduction*. Where plates are spreading apart, magma wells up to fill the voids and forms the mid-oceanic ridges with their occasional small volcanic islands. Although the magma, rich in **mafic** minerals, contains dissolved gases, it is low in viscosity, flowing easily, and producing new ocean floor in the spreading zones (fig. 16.22). Seldom are these oceanic volcanic eruptions violent, but they contribute massive quantities of lava to the rift zones of the oceans and continents.

Figure 16.23

All volcanic forms can be distinguished between central and fissure eruptions. Augustine, a composite volcano, last erupted March 27, 1986, exhibiting a central form of eruption.

Photo by U.S. Geological Survey.

The energy released during an eruption is governed by the nature of the magma originating in the lower crust and mantle. The percentage of silica (SiO_2) in the magma is the key factor determining how violent the eruption will be.

Magmas formed by the recycling of older crustal rocks along the subduction zones often contain more than 65 percent silica. These magmas are highly explosive, being rich in **felsic** minerals and volatile gases. Paradoxically, these magmas are relatively cool, arriving at the surface at a temperature seldom exceeding 900° C (1,600° F). Because of the lower temperatures, the gases begin separating from the fluid well below the surface. Thus the trapped gas builds up pressure and finally overcomes the confining weight of the surrounding rocks. After the first eruption, pressures are reduced. This produces a chain reaction as more gases build up and lift magma again to the surface. The eruption cycle

repeats again and again until all pent-up gas producing the pressure is released. Thus eruptions of felsic magmas with high viscosity are often a series of violent eruptions continuing for days, building up the majestic **stratovolcanoes** such as Mount Augustine, Alaska, Pico de Orizaba, Mexico, and Mount St. Helens, Washington (fig. 16.23). The difference is the magma.

Volcanic Features

Volcanic landforms in their initial stages are determined by the magma and type of plate boundary. Therefore, the features are predictable if we know the nature of the magma and plate contact. Volcanic landforms can be distinguished between central eruptions, where lava and ash are ejected from point sources, and **fissure eruptions,** in which magma reaches the surface along one or more long cracks. Central eruptions generally are violent, with very

Figure 16.24

Paricutín cinder cone eruption at night. Here hot pyroclastic ash and volcanic bombs build the cinder cone.

Photo by R. E. Wilcox, U.S. Geological Survey.

viscous and high-silica felsic and intermediate lavas originating along zones of subduction. Fissure eruptions, with mafic and ultramafic lavas, are quiescent, have low-viscosity lava flows, and are often associated with the seafloor spreading zones.

Central Eruption Forms

Central eruptions, being very explosive, produce three distinctive forms. These include the **cinder cone**, the **composite** or **stratovolcano cone**, and the **volcanic dome**.

Cinder Cones Cinder cones are piles of ash debris from a violent volcanic eruption. On February 20, 1943, in Mexico near the city of Paricutín, a farmer's corn field suddenly began to shake as a major earthquake appeared to be in the making. Actually, it was the birth of a new volcano. Steam and smoke began to escape from the ground, followed by explosions of fire and hot cinders, then lava flowed out of fissures in the field. It grew into a mountain peak of ash and cinders within a few weeks. After a rapid period of growth, Paricutín calmed down to an active, smoldering cinder cone 450 meters (1,500 feet) above the corn field. During its first year it erupted billions of tons of ash, and lava flowed from its base (fig. 16.24).

Cinder Cone peak in Lassen National Park in northern California is a good example of an explosive eruption occurring in the last 100 years in North America. Chemical analysis revealed that its lava and cinders are of continental plate material, rich in silicon and aluminum and highly viscous. These felsic lavas retain very explosive gases that erupt to form cinder cones of symmetrical shape. The straight sides of the cone slope at about 32 degrees, the same angle of repose as that of a sand dune. Cinder cones are very common, being scattered throughout the world's active subduction zones. They represent some of the most explosive types of eruptive activity (fig. 16.24).

Composite Volcanoes **Composite volcanoes** or **stratovolcanoes,** such as Mount St. Helens, are windows into the Earth's bowels. They too are cone-shaped and produce great amounts of lava interbedded with ash. A cross section reveals alternate layers of ash and lava. They are famous for their explosive nature. The composite volcano is one of the most common types found along subductive plate boundaries.

On Sunday, May 18, 1980, at 8:32 A.M., Mount St. Helens erupted on its northwest side, blowing off 4 cubic kilometers (1½ cubic miles) of its crest and lowering the summit by nearly 400

A

Figure 16.25

Central eruptions include cinder cones, composite or strato-volcanoes, and volcanic domes. The forested slope is a shield volcano behind the top rim of Cinder Cone, Lassen National Park, California. (A) Cinder Cone viewed from the top is very symmetrical, resulting from a central eruption. (B) Kilimanjaro is a classic example of a composite volcano.

B: Photo by Robert Jensen.

B

meters (1,300 feet) (fig. 16.26). The force of the explosion has been compared to the largest hydrogen bomb ever exploded.

Before the eruption, the state of the mountain was like a well-shaken can of soda or a bottle of champagne. The gas-rich magma was stirring and welling up inside the mountain, forming a growing dome on the north flank. Like carbonated drinks, dacite magma is capable of holding great quantities of carbon dioxide, as well as other gases, dissolved under pressure. As the pressure increased due to stirring or shaking of the magma, the cap of the mountain's dome slowly rose. Subsequently, when the lid ruptured, the gases, like those that result from shaking a bottle of soda pop, exploded.

Just prior to the eruption, an earthquake, 5.0 on the Richter scale, disturbed the equilibrium in the mountain. Photographs taken at that moment show the dome around its fringes releasing gases and ash. Then seconds later the entire cap blew, and the volcano became a fiery erupting system.

The force of the explosion was so great that every tree was flattened in a 400-square-kilometer (150 square mile) area by the blast (fig. 16.27). The explosion was heard over 32 kilometers (200 miles) away, ash reached the stratosphere, and the jet stream carried the ashes around the world.

Mudflows resulting from melting snow reportedly traveled 130 kilometers per hour (80 mph), clogging streams, rivers, and lakes and causing floods that wiped out roads, bridges, lumber mills, and houses and claimed the lives of scores of people. Timber was destroyed, forest fires flared up, and crops as far away as Montana were damaged by the ash fallout, which measured over 400 kilograms per hectare (350 pounds per acre) in the Yakama Valley located downwind from the eruption. When the rains came it turned the ash into sludge that clogged storm drains, making cleanup almost impossible.

Because the volcano's mud and debris fed into tributaries of the Columbia River, it was necessary for tugboats to guide

Figure 16.26

Mount St. Helens, Washington, a stratovolcano that exploded on
May 18, 1980.

Photo by U.S. Geological Survey.

Faulting, Folding, and Volcanism   *361*

Figure 16.27

The blast flattened trees in a 400 square kilometer (150 square mile) area and rivers were clogged with debris and mud. Numerous fires were started but later snuffed out by the falling ash.

Photo by U.S. Geological Survey.

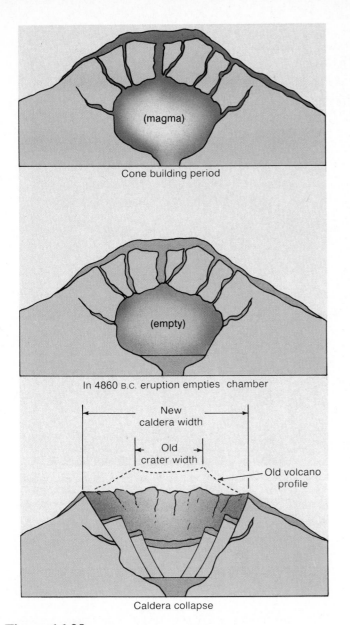

Cone building period

In 4860 B.C. eruption empties chamber

New caldera width

Old crater width

Old volcano profile

Caldera collapse

Figure 16.28

At one time Mount Mazama, the older volcano of Crater Lake, rose 1,200 meters (4,000 feet) higher than the present rim. Like Mount St. Helens, but 42 times more violent, Mount Mazama blew its cork in 4860 B.C. and left a caldera over 13 kilometers (8 miles) wide. Then it filled to become the nation's deepest lake at (1,900 feet). Wizard Island is a small cinder cone and lava flow that followed the earlier eruption.

Photo by NASA.

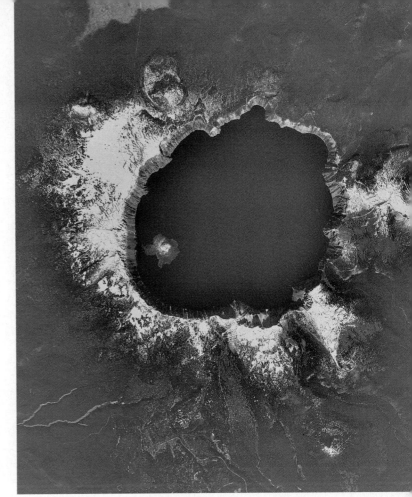

Crater Lake

freighters through mud-swollen waters of the Columbia River near Longview, Washington, for weeks following the eruption.

Emergency dredging was required near Longview when the channel began to clog and water depths reached the minimum for navigation. The silt-filled area of the Columbia River covered a 9 kilometer (5½ mile) stretch. It was estimated that the volcanic silt dredged from the river was enough to make a pile of mud a mile high on a football field size pad.

If Mount St. Helens is true to form, it will experience more eruptions, followed by quiet periods building alternate layers of ash, pumice, and lava, like other composite volcanoes along the "Ring of Fire."

Two geologists from the U.S. Geological Survey, Dwight R. Crandell and Donald R. Mullineaux, predicted two years prior to the eruption, "In the future Mount St. Helens probably will erupt violently and intermittently just as it has in the recent geologic past, and these future eruptions will affect human life and health, property, agriculture, and general economic welfare over a broad area." This same prediction can be applied to every major volcanic peak in the "Ring of Fire." The Cascades were formed by this very process with a similar eruption of this magnitude about every thousand years or so. Following the explosion stage, which builds a layer of ash and pumice, the next phase is cone building by dacite lava flows.

The dacite magma is rich in silica and very viscous, so the process is very slow. Unlike the basaltic, low-silica magmas of the Hawaiian Islands, dacite magma is sluggish and thick, capable of producing a caldera.

Calderas The **caldera** is a special class of volcanic landform represented in the United States by Mount Katmai, Alaska, and Crater Lake, Oregon (fig. 16.28). In these two examples, composite volcanic cone building continued until a violent explosion

Figure 16.29

Shield volcanoes are built by repeated outpouring of low-viscosity lava originating deep in the mantle. Mauna Loa on the island of Hawaii is still actively growing by lateral venting. This photograph shows Mauna Loa's great summit crater Mokuaweoweo. The older shield volcano Mauna Kea is in the background.

Photo by D. W. Peterson, U.S. Geological Survey.

literally blew off the top and left the central portion to subside into the empty cavity. The resultant landform is a **caldera,** meaning kettle. Caldera formation is nature's most violent spectacle, short of an asteroid impact. Mount Mazama, the parent volcano of Crater Lake, once rose several thousand feet higher than the present rim. Following volcanic activity, the caldera has been eroded by streams in older glacial valleys and the crater has filled with water from subterranean percolation and precipitation, to a depth of over 600 meters (2,000 feet). Since the volcano lost its top, cinder cones such as Wizard Island have formed.

The volcanic features we have been examining are the result of subduction with viscous felsic and intermediate lavas. Where spreading zones or plumes are releasing lavas of low viscosity, mafic lavas, produce a unique set of volcanic features. These include shield volcanoes, island chains, volcanic fissures, and flood basalts.

Shield Volcanoes **Shield volcanoes** are massive, gently sloping volcanic domes built of highly fluid mafic magmas. The Hawaiian Islands represent this type of low-viscosity shield: they range from 80 to 160 kilometers (50 to 100 miles) wide at their submerged bases on the ocean floor.

Shield volcanoes are built by repeated outpouring of lava from vents frequently found on a side flank. Volcanic activity is generally nonexplosive and easily observed during periods of eruption. The interior crater walls are generally wide and steep-sided, produced by subsidence as lava is removed through lateral venting (fig. 16.29).

Island Chains Isolated volcanic island chains like the Hawaiian Islands originated in a process different from that which formed the volcanoes along the major plate boundaries. The source of lava is very deep, coming from hot spots beneath the plate. The motion of the plate carries older volcanoes away from the source, and new ones form over the source. The older islands erode and grow smaller, as is apparent in the Hawaiian chain. The length of the island chain continues to grow and is a good indication of how long the source has been active. In the case of the Hawaiian Islands, Midway

Figure 16.30

The Columbia Plateau is an example of a fissure eruption covering thousands of square miles and reaching depths of several thousand feet. Flood basalts from fissure eruptions can spread over an extensive area because of their highly fluid nature.

Photo by U.S. Geological Survey.

Island is the oldest in the chain and the island of Hawaii, the youngest, is still active over the magma source.

Fissure Eruptions and Flood Basalts

Fissure eruptions occur most frequently where the crust is in a state of tension or spreading along the mid-oceanic ridges. They develop two principal forms, the previously discussed **shield volcano** and **flood basalts** covering many hundreds or thousands of square kilometers with lavas thousands of meters thick.

An example of flood basalts is the Columbia Plateau basalts of eastern Washington, Oregon, and southern Idaho (fig. 16.30). There are similar massive basalt plateaus in southern India, eastern Brazil, and Antarctica. They were fed by extensive fissures dispensing highly fluid lavas which flowed for great distances. Fissure eruptions produce the greatest known volumes of volcanic rock.

The Columbia Plateau is the largest outpouring of lava known on North America. Here the lavas cover several thousand square kilometers and measure up to 1½ kilometers (1 mile) deep. The volume of lava may be as much as 21,000 cubic kilometers (5,000 cubic miles).

Outpourings of this type developed along many fissures or cracks in the crust. Each new eruption formed a new layer on top of the older weathered basalts, flooding the region with molten rock. This is why the term flood basalts is used to describe the volcanic accumulations.

Eruptions of this type are almost without exception missing volcanic cones and ash like those associated with the Cascade Range of the Pacific Northwest.

Erosional Forms

Volcanic landforms go through a sequential pattern of development represented by successive stages of buildup, followed by erosion and finally destruction. In the early stages, building processes dominate. After cone building, erosion begins to dissect these features. Calderas often develop following the collapse of a mature composite cone. Some volcanoes such as Ship Rock, New Mexico, become extinct, and others remain dormant, such as Mount Shasta, California, or Mount Rainier in Washington.

Drainage patterns of streams in the mature stage of a volcano resemble spokes radiating outward from the peak of the cone. As erosion continues, the more resistant lava flows become the elevated topographic features and sediments cover the lower valley floors. In the final stages of erosion, all that is left are the most resistant volcanic rocks making up the internal rib structure or dikes that have solidified into resistant crystalline rock. Ship Rock, New Mexico, is a famous example of a volcano that stood much higher above the valley floor several million years ago. Today all that is left of the peak is its resistant skeleton: the throat or volcanic neck of the cone and wall-like dikes radiating from it (fig. 16.31).

Ship Rock exerted such pressure during its eruptive stages that radial cracks formed as the crust expanded under stress. Magma poured into the cracks as fast as they formed.

Ship Rock currently rises 430 meters (1,400 feet) above the surrounding region of horizontal sedimentary rock strata. Given enough time, Mount St. Helens and Lassen Peak will be reduced

Figure 16.31

Erosional Volcanic Forms. Shiprock, New Mexico, became extinct leaving itself exposed to weathering and erosion. At one time it probably stood much higher. Today all that is left are the ribs and throat or volcanic neck of the more resistant dikes and sills.

Photo from the D. A. Rahm Memorial Fund, Western Washington State University.

to a similar form. The evidence that Ship Rock was truly a stratovolcano extending perhaps 4,000 meters (14,000 feet) above the plains is found in the surrounding region. Let's examine these clues:

1. There are more than a dozen similar geologic structures within a radius of 30 kilometers (20 miles). This is typical of a volcanic field.
2. There are fragments of granite buried in the volcanic tuff which was carried upward from sources 2,700 meters (9,000 feet) below the surface. This indicates vigorous upward flow from deep magma chambers. These granite rocks from the basement of the continent must have been torn from the walls or pipe of the volcano.

3. The rock exposed in Ship Rock is volcanic tuff, a by-product of rapidly exploding gases that are released when magma suddenly reaches the surface, reducing the pressure.
4. It is also highly improbable that molten rock under pressure could fill cracks and not break the surface.

Here is a classic example of the ongoing struggle between the forces of mountain-building tectonic processes and leveling forces of weathering and erosion.

Volcanism and Geothermal Energy

Natural steam spouts from the Earth's hot rocks, supplying us with a source of energy. This natural heat has been used since Roman times for heating homes and baths.

Today at The Geysers in northern California, the steam is collected from a number of wells, filtered, and passed through turbines that drive electric generators. Currently, geothermal energy provides more electricity than solar and wind energy combined. Geothermal power plants are also being developed in Mexico, El Salvador, New Zealand, Japan, Iceland, Russia, Turkey, and the Philippines.

Source of Geothermal Energy

When the Earth was born it was a mass of hot liquid and gaseous matter. Perhaps as much as 10 percent of this mass was steam. There is still uncertainty about the processes of early formation, but the original hot mass is still slowly cooling, supplemented by the heat generated by natural radioactivity. It is this heat in the Earth's interior that keeps magma liquid under great pressure. In the mantle zone of the asthenosphere, the molten magma associated with volcanoes is heating the rocks and any water present.

Three conditions are necessary for a geothermal resource: (1) a heat source to provide the energy, (2) a water source, and (3) a reservoir rock that can trap and hold the hot water under pressure. The geothermal energy resource at The Geysers is a large magma chamber 6 to 8 kilometers (4 to 5 miles) below the surface. The water is maintained in the liquid form at temperatures over 177° C (350° F) because of the very high pressure at depths of 2,400 meters (8,000 feet). Because the reservoir rock is fractured shale and filled with cavities and pores, it is an excellent storage container. The lid is an impervious layer of sandstone and serpentine while the walls of this geologic container are bounded by fault planes (fig. 16.32).

When a well is drilled into the reservoir rock, the pent-up high-pressure steam is channeled to the generation plant.

The water can come from several sources. As magmas cool, steam called **magmatic steam** is thrown off. Another source is surface water seeping down into porous rock, which is then heated by magma to later form steam, called **meteoric steam.** It reaches the surface at very high temperatures and pressure. Spectacular **geysers,** such as those found in Yellowstone National Park, can be the result. Water can be trapped in the marine sediments during subduction.

In other regions the steam may be just a **fumarole** or steam vent with sprays of steam like those found at The Geysers of northern California.

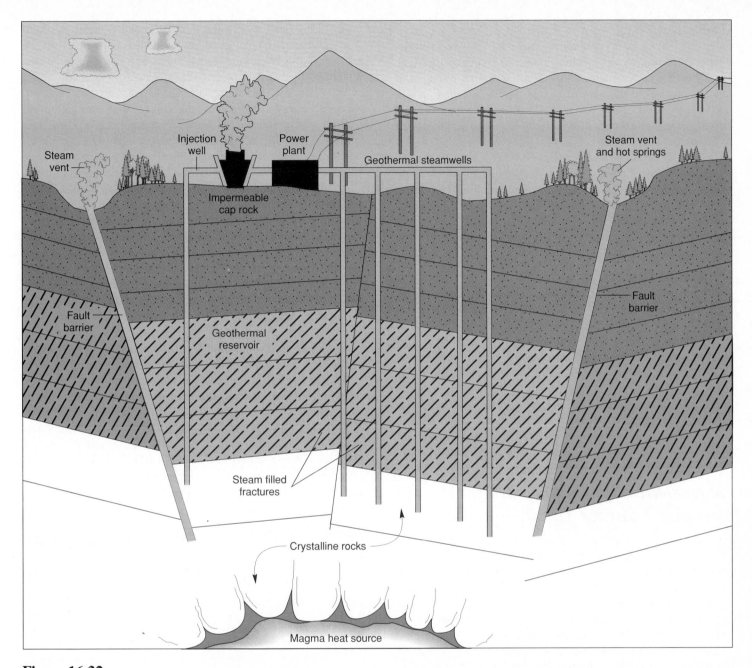

Figure 16.32

Three conditions are necessary for a geothermal resource: a heat source, a water source, and a reservoir rock. Water may have been trapped in the marine sediment of the ocean floor during subduction. This is the largest geothermal power plant in the world. Enough energy is generated at The Geysers to supply the city of San Francisco.

Geothermal Power

Currently The Geysers is generating over a half-million kilowatts of energy, enough to support the city of San Francisco. The U.S. Geological Survey estimates that as much as 1.3 million acres of land, mostly within the western United States, including Hawaii and possibly Alaska, may be potentially attractive for geothermal development. Almost none of this has been tapped and converted to a usable energy form.

It is estimated that there could be at least 2,000 megawatts of geothermal power capacity at The Geysers, which would represent 2 percent of the nation's power capability. This represents the capacity of two Hoover Dams. If a successful research and development program is pursued, it is predicted that geothermal power could supply as much as 13 percent of the U.S. power capability.

Geothermal Heating

Geothermal energy also supplies heat for greenhouses that produce year around. The West has long been known as a geothermal region, but growing evidence now points to large geothermal reserves beneath Delaware, Maryland, Virginia, and North Carolina; others are contemplated beneath New York, Pennsylvania, South Carolina, and Georgia. Temperatures are not hot but warm, ranging from 38° C to 82° C (100° to 180° F). This means the heat cannot power generators of electricity, but it is useful to food processors, greenhouses, and space heaters.

The energy source was located exactly where scientists at Virginia Polytechnic Institute directed the Department of Energy to drill, at depths of 800 to 1,600 meters (0.5 to 1 mile).

Geographic Significance of Volcanic Features

Volcanoes have terrorized humankind for centuries, covering entire cities with fallout of ash, cinders, and bombs. Lava flows have poured down mountain slopes over entire towns. Costa Rica during the 1960s experienced continual volcanic ash fallout, crippling the coffee-oriented economy and adding considerable hardship to those in the fallout zone. Today, coffee is again king. The minerals from volcanic activity have enriched the tropical soils to a new level.

Soils of weathered volcanic rock are among the best in the world. Some of the highest yields of coffee, sugar cane, rice, and corn are found in volcanic soils of the tropics.

Volcanic soils of forested lands produce superior quality trees and higher yields. The wetter regions of the Columbia Plateau support record stands of yellow pine. Over the long term, Mount St. Helens will enrich the productivity of the entire area as fresh mineral deposits add nutrients and help conserve moisture in the soil.

This is nature's way of adding nutrients to an otherwise highly leached tropical soil. Java, the world's most densely populated island, is able to support its population only because of rich volcanic soils.

Mount St. Helens is beautiful and peaceful for the most part these days, yet the tranquility is deceptive. Most of the composite volcanoes that rim the Pacific are termed dormant, which means they have the potential to erupt at any moment. A recent study by the U.S. Geological Survey of the major peaks in the Cascades from Washington to northern California discovered that all had erupted within the last 12,000 years.

The ever-increasing land use of areas around these volcanoes ensures that more lives and property than ever before will be destroyed or endangered. However, putting it in a proper perspective with other environmental hazards such as earthquakes, hurricanes, tornadoes, and floods, the potential hazard is small yet warrants concern, according to the U.S. Geological Survey team.

Mauna Loa on the island of Hawaii, the world's largest volcano, wins first place for a potential eruption. It has been shaking constantly for several years as it tunes up for a blast. If we trace the Pacific "Ring of Fire" around the Aleutian Island chain, geologists count 36 active volcanoes, including several which have had massive eruptions during the last century. Volcanic eruptions in Iceland and the West Indies have ejected poisonous gases and lava, wiping out entire towns in the last decade. An Indonesian volcano had been asleep for 1,000 years before recently erupting again.

The eruption of Sunset Craters near Flagstaff, Arizona, in 1065 A.D., produced an effect similar to Costa Rica by adding fertility to the soil with one added bonus. In the arid zone, corn could not survive until volcanic cinders were discovered to keep the soil moisture from evaporating. For the first time, this region supported heavy settlement. However, after several hundred years, erosional forces finally washed away the thin veneer of volcanic material, and settlements soon disappeared as the arid soil's food-producing capacity declined. Soil infertility of this type may also explain the sudden exodus from the pueblos of Mesa Verde National Park, Colorado.

SUMMARY

Fault motion takes several forms: **normal** and **reverse**, **strike-slip**, and **thrust faulting.** Normal faulting occurs where there is crustal tension. Reverse and thrust faulting result from *compressional forces.* Strike-slip faulting is a *shearing force* where plates are sliding past one another. When the Earth shakes there are a number of geologic effects which include **faulting, landslides, soil failure, tsunamis,** and **changes in sea level.** A variety of fault-formed features are also produced. These include **fault valleys, grabens, fault scarps, sag ponds, hot springs, saddles,** and **offset** and **deformed rocks.**

Folding is another consequence of compressional forces. The crust can be stretched, folded, and warped like a piece of paper. Geologic structures such as **anticlines, synclines,** and **monoclines** are the end product of crustal compression. Erosion then modifies these features to produce a series of parallel ridges and valleys such as those found in the Applachian Mountains of central Pennsylvania.

Volcanism is not the cause of earthquakes, nor do earthquakes cause volcanism, and yet when both are plotted together on a world map, there is a very high correlation. Both are located where the giant plates are separating and colliding. The volcanic landforms or structures are determined by the type of plate motion.

Where plates are in collision, producing *subduction* and oceanic trenches, explosive felsic and intermediate magma is thrust to the surface under pressure to produce **central volcanic eruptions.** Volcanic arcs consisting of **stratovolcanoes, cinder cones, calderas,** and **volcanic domes** are produced. **Fissure eruptions** occur most frequently where the Earth's crust is spreading along the mid-oceanic ridges. Here the lavas are very fluid with minimal violence. Volcanic fissures produce shield volcanoes such as those on the island of Hawaii and massive **flood basalts,** such as those of the Columbia Plateau covering huge areas.

Once eruptions end, volcanic landforms are sequentially eroded producing such features as Ship Rock, New Mexico.

Volcanic regions are of increasing interest because they represent rich soils, mineral resources, and an important source of heat energy found deep in the Earth's crust.

ILLUSTRATED STUDY QUESTIONS

1. Where was the damage from the Loma Prieta earthquake most intense (figs. 16.1 and 16.10, pp. 336, 346)?
2. Describe the primary motion of the Loma Prieta earthquake (fig. 16.2, p. 337).
3. Name the type of fault and direction of motion of the San Andreas and related parallel fault-formed valleys (figs. 16.3 and 16.4, pp. 337, 338).
4. Which of the following types of forces have deformed the rocks in the photograph (box 16.1, fig. 3, p. 342):
   a. tension
   b. compression
   c. shearing
   d. a combination of two of the above

5. Explain elastic rebound theory using box 16.1, figure 1, p. 336.
6. The hills paralleling the valley and trending northwest-southeast were formed by vertical displacement. Name the type of faulting producing the uplift (figs. 16.4 and 16.14B, pp. 338, 351).
7. Describe the type of force producing the Carson Valley (fig. 16.6, p. 340).
8. What type of faulting does compressional force produce (fig. 16.7, p. 340)?
9. Give the magnitude of the earthquake that generated the Madison slide of August 17, 1959. What other geologic factors were involved (fig. 16.9, p. 345)?
10. Describe the fault motion that produced a tsunami (fig. 16.11, p. 349).
11. Name the type of folds shown in the photograph of figure 16.16. Which types of (p. 354)?
12. What term best describes this sunken block form (fig. 16.6, p. 340)?
13. According to plate tectonic theory, volcanism and plate collisions are interrelated. Explain the possible cause of isolated pockets of volcanic rock in the *study area* (fig. 16.21, p. 356).
14. What do hot springs tell us about the area's geologic history (fig. 16.21, p. 356)?
15. Name the following types of volcanic landforms from the eruption processes shown in the photographs. Note the location of each volcanic landform. What does their location tell about their origin (figs. 16.22–6.28, pp. 358–63)?

Leveling of the Land by Weathering and Mass Wasting

17

OBJECTIVES

After completing this chapter you will be able to:

1. Explain the major processes of **physical (mechanical)** and **chemical weathering.**
2. Relate **mass wasting** processes to the *Rock Cycle.*
3. Define **mass wasting** and explain the various types of crustal movement related to this process.
4. Recognize the major landform features produced by weathering and mass wasting.
5. Illustrate the role of human activity in the leveling processes.

Nature's acids dissolve and design the underworld. Carlsbad Caverns National Park, New Mexico.
© Kerrick James

Figure 17.1

The weathering process is a vital part of the rock cycle. The quartz mineral content of the petrified log makes it more resistant than the highly weathered surrounding sedimentary rocks. This varying rate is known as differential weathering.

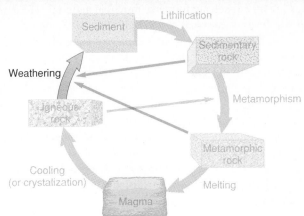

As soon as rocks are formed, the work of the weathering processes and the force of gravity begin leveling the land. Everywhere we look, decay, rusting, and decomposition are weathering the Earth's crust, reducing massive rocks to fine grains of powder. This is **weathering.** All weathering processes can be described as either **chemical** or **physical (mechanical).** Both processes work simultaneously, making it sometimes difficult to separate one process from another. Frequently a chemical process aids a mechanical one, or the reverse can be true. It is a vital part of the entire *rock cycle* (fig. 17.1).

Sometimes weathering is considered destructive because of the damage it does to human works. The weathering that rusts your car, chips the paint off buildings and weakens mortar in bricks, is also building new soils from rocks of the Earth's crust. It has been said that humankind is only three feet away from starvation, the average depth of a soil profile. If weathering suddenly stopped, this profile would soon wear away by agents of erosion, and the earth would lose its basic life resource, the soil. Most plants cannot survive without it; agriculture would become extinct (fig. 17.2).

Our planet is a balanced series of interacting energy systems. **Weathering** is based on energy from the Sun. Part of this energy is converted by photosynthesis to chemical energy; another portion is stored in water vapor as latent energy. Some solar energy enters rocks and minerals as heat to perform the work of weathering, producing expansion, and accelerates chemical reactions of decay or decomposition.

Figure 17.2

The basic life support system of soil is a by-product of weathering. This juniper is being sustained from minerals weathered from this sandstone rock.

The Earth's crust is being weathered more rapidly where heat and moisture are present to chemically dissolve and transport minerals out of the rocks. Water assists the weathering processes in two ways. First it dissolves minerals. Second, gases are also dissolved in water to give the water an acidic character, capable of attacking and dissolving formerly insoluble materials in the rock. This is a form of **chemical weathering.**

Physical (mechanical) weathering is an attack on rocks through a variety of forces that break up rocks and yet leave the rock chemically the same (fig. 17.3). Roots of a large tree, for example, can gradually split a massive rock into smaller and smaller units. When physical weathering occurs, the rock mass is broken into finer particles without chemical changes occurring. Let us examine in more detail a number of **physical (mechanical)** processes and evaluate the basic geometric forms resulting from the disintegration action. These include: **frost wedging, unloading, heating and cooling, mechanical granular disintegration, slaking** and **biological weathering.**

PHYSICAL OR MECHANICAL WEATHERING

Frost Wedging

Frost wedging is a mechanical process that constantly reduces large rocks into smaller ones. When water freezes, its volume increases about 9 percent. If water fills a crack, the expansion process works like a wedge being driven into the crack. The ice crystallization process literally fractures the rock (fig. 17.4). This form of weath-

A

B

Figure 17.3

All weathering processes are either *chemical* or *mechanical*. (A) Mechanical weathering is illustrated by ice forming in a crack and fracturing it. (B) Chemical weathering can result when water dissolves carbon dioxide or sulfur dioxide to become more acidic. Then minerals are dissolved from the rock by acid weakening the rock. Because sulfur dioxide from auto emission is in the atmosphere of Cairo, this Egyptian stone treasure is weathering away very rapidly.

ering is especially active in the middle latitudes of the continents and in high latitudes where below-freezing temperatures are common.

The most striking examples of frost wedging can be seen in the higher mountains, above timberline. When the snows of winter have melted, great fields of broken rock are exposed. Note the base of the cliff in figure 17.5, where rock fragments have collected to form a graded slope of rock dislodged by frost wedging. This rock debris forms a **talus slope,** lying at a stable gradient of 32° to 38°.

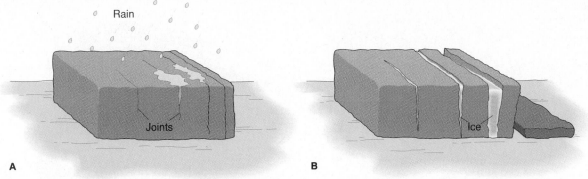

A

B

Figure 17.4

Frost wedging is a mechanical process resulting from water filling a crack, then freezing. Because water expands when it freezes, the ice wedges the rock apart.

Figure 17.5

Talus slopes of rock fragments develop at the base of the cliff from frost wedging above.

Figure 17.6

Unloading is the result of the Earth's continuous adjustment to changing pressures or loads. These rock layers are spalling off as pressure is released. Half Dome, Yosemite National Park, is called an exfoliation dome.

Photo by F. C. Calkins, U.S. Geological Survey.

Unloading

In the mountain belts of the world, where rocks have been uplifted by internal forces and uncovered by leveling processes, the crust has not finished adjusting to the lower pressure resulting from the removal of overlying rock. As the rock is uncovered, it expands and forms peeling sheets. Yosemite National Park is one of the best places to observe the results of this unloading process, known as **exfoliation** (fig. 17.6). The features produced include large exfoliation domes, arches, and rocks stacked like blocks on top of one another.

Heating and Cooling

Energy from sunlight also plays a significant role in the sheeting process. As the Sun warms the rocks, the minerals with higher coefficients of expansion expand more than other minerals. Then as night falls, these same minerals cool and contract. This process, repeated every day, causes the rock to weaken under uneven ex-

pansion rates and fracture along planes of weakness or jointing planes. Sometimes it is difficult to determine whether unloading or heating and cooling is more significant in weathering.

Mechanical Granular Disintegration

A visit to the arid American Southwest will show rocks disintegrating grain by grain. Believe it or not, it is the water in this arid land causing the disintegration. When the Spanish came into the Southwest, they described the rivers in a rather interesting way. They said the rivers run upside down, meaning they flow beneath the surface as groundwater most of the time. Through evaporation this groundwater is brought to the surface, drawn upward by capillary action. When the water evaporates, tiny salt crystals form on the surface, in crevices, and between the minute pores in the rocks. When this form of crystallization occurs, the rocks are broken by the power of crystal growth. It is well known that many of the Indian pueblo settlements were built in caves and niches developed through rock weathering of this type (fig. 17.7).

Figure 17.7
The caves and niches developed by weathering processes provide dwelling sites for early Americans in the Southwest.

Slaking

Sedimentary rocks rich in clay minerals crumble into smaller fragments as a result of alternating between absorption and loss of water. The clay minerals expand as they absorb water, then contract as the water leaves. The effect is similar to heating and cooling of the rock. Changes in particle size cause the rock to crumble, or *slake*.

Biological Weathering

Solar energy is continually transformed into chemical forms through the Earth's biological window, the plant world. As plants through photosynthesis transform and later utilize this energy, they work on the rocks that anchor their root systems. Have you ever failed to repot a plant that grew too large for its clay planter? The result of this forgetfulness can be a "crack pot." Roots have great wedging energy, capable of lifting sidewalks and foundations.

Let's now consider the processes that destroy the rock mass by chemical means. These include: **carbonation solution, hydrolysis,** and **oxidation.**

CHEMICAL WEATHERING

When rocks undergo alteration resulting from exposure to the atmosphere, **chemical weathering** is occurring. Chemical weathering results when minerals are changed chemically. As reshuffling of the atoms occurs, new minerals form. Two ingredients are important for rapid chemical change: moisture and heat energy. Generally, the higher the temperature the more rapid is the change.

Where in the world are these two weathering ingredients most abundant? Geographically, the world's intense chemical weathering is located in the low latitude tropical and subtropical rainy belts, where mean monthly temperatures are well above freezing.

Examine a rock outcropping where the soil-forming process is active, and you will observe changes in the rock mineral structure where weathering is most active. Gradual change in feldspar minerals from the outside surface of a rock inward is a good indication that chemical weathering has progressed from the surface inward.

Carbonation

When it rains, small amounts of carbon dioxide from the atmosphere combine with rainwater to form carbonic acid. The acid is enriched by acid from decaying plant material. As this acidic water enters cracks in rocks, the feldspar minerals which make up the bulk of many igneous rocks are attacked first. When feldspars are dissolved in granite, the by-products are clay, silica, and soluble salts.

Because quartz stubbornly resists chemical change, it remains in rocks such as granite changing it to weak crumbly *decomposed granite*. The effect is similar to loosening bricks in decaying mortar. Thus the soil-forming process begins (fig. 17.8).

The erosional processes then carry quartz crystals away in streams to be deposited in layers and eventually to form beach sand and finally sandstone. Much of the sand on the beaches and sand dunes of the world consists of quartz grains that were set free by weathering of granite or sandstone rocks.

Vast quantities of limestone have been dissolved by water that has taken up carbon dioxide to form a weak acid. As the acid attacks the limestone to form soluble salts, mainly calcium bicarbonate, the groundwater carries the soluble salts into streams and eventually to the sea, to become deposits of future sedimentary rocks.

Solution

Water dissolves more substances than any other solvent. It is commonly referred to as the universal solvent. Salts and many other substances go into a soluble form in water. Some of the dissolved substances never reach the sea but are redeposited in caverns, fissures, and other large openings underground. Chapter 19 is a detailed examination of the processes shaping caverns, and other aspects of groundwater geology.

Hydrolysis

Hydrolysis occurs when water itself combines with a mineral to form a new product. The new mineral is entirely different from the original mineral in all physical and chemical properties. For example, when potassium feldspar undergoes hydrolysis, it becomes kaolinite, a clay mineral with a greasy, slippery feeling. Bauxite is another important altered feldspar product resulting from hydrolysis in the tropical zones of South America, Africa, and Asia. This mineral is a vital aluminum resource.

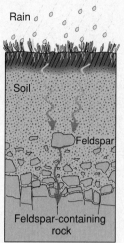

Rain picks up CO_2 from the atmosphere and becomes acidic

Water percolating through the ground picks up more CO_2 from the upper part of the soil, becoming more acidic

A feldspar crystal, loosened from the rock below, slowly alters to a clay mineral as it reacts with the acidic water

The water carries away soluble ions and SiO_2 to the ground water supply or to a stream

Figure 17.8

Carbonation occurs when water with dissolved carbon dioxide forms a weak acid that attacks the feldspar minerals in rock.

Oxidation

The next time you strike a match, you will be witnessing **oxidation.** Remember the rust that appeared on your first bike or car? Oxidation struck again. Oxygen accounts for 21 percent of the atmosphere. It is a very active gas that has a tendency to combine with metals such as iron, magnesium, copper, and aluminum. When this happens, the new oxides formed are weaker and easily removed by weathering and then erosional processes.

Oxidation also affects carbon deposits such as coal in the form of fire. Natural or accidental fires can work into coal beds deep underground and smolder for years.

GEOMETRY OF WEATHERING

The chemical decay of bedrock along jointing planes causes exfoliation or spheroidal forms that peel like onion skins as weathering progresses inward (fig. 17.9). Granular disintegration is another form resulting from the physical weathering of sandstone and chemical weathering of granite along jointing planes. Here grains of minerals are removed in a slow process with quartz remaining as gravel around the base of the rocks, while the feldspars are chemically removed. In chemical weathering, water is an agent for removal, while heat produces the energy for the dynamics of weathering.

GEOGRAPHIC SIGNIFICANCE

Weathering is nature's way of recycling the Earth's crust. Rocks are returned to dust, and the monuments of civilization eventually decay. Weathering is destructive in one sense, reducing materials

Figure 17.9

Weathering produces a variety of geometric forms, including spheroidal layers.

of human creations. Buildings, automobiles, and highways are all in constant decay. On the positive side, much of human pollution is also weathered away. Glass bottles eventually add to the quartz sands of the sea. Paper litter returns to humus. Thus the Earth is gradually cleaned of our garbage. Unfortunately, aluminum products, plastics, and radioactive materials have few chemical enemies and remain intact over long periods of time.

Once mechanical and chemical weathering weakens and breaks the rock into smaller units, gravity begins the leveling process by **mass wasting.**

LEVELING BY GRAVITATIONAL ENERGY: MASS WASTING

The universe contains energy in a variety of forms: light, heat, chemical, nuclear and gravitational, and those are just a few of the more obvious ones. We have seen the role of chemical and mechanical energy in the formation and destruction processes of the earth's crust. If the influence of various forms of energy were rated on a universal scale, gravitational energy (discussed in detail in chapter 2) would win hands down. Although chemical forms dominate our present civilization, gravitational energy is distributed to every single atom in the universe. The most distant galaxy is exerting a gravitational influence on the book you read.

Everywhere on the surface of the Earth gravity is pulling all matter, including the Earth's crust, to the Earth's center of mass. This is **mass wasting.** As we age, gravity causes our flesh to sag and our backs to bend. Just as humans are struggling against this overcoming force, the crust of the Earth is being leveled. Subduction along plate boundaries is the result of this pull.

There are four variables or factors determining the downward movement of materials in this process driven by gravity: (1) the **rate of motion,** (2) the **type of material in motion,** (3) the **water content involved,** and (4) the **type of motion.** The types of movement, or **mass wasting,** fall into three broad categories: flowage, slippage, and rockfall. Each of these types of mass wasting results in distinctive erosional and depositional forms on the landscape (fig. 17.10).

Rate of Motion

The velocity of a landslide under the force of gravity can vary tremendously. Debris can creep downslope at a rate of only one centimeter per year, or free-fall in a rock avalanche or rockfall with velocities in excess of 200 kilometers per hour (125 mph). Slope conditions and relief—the vertical distance between point of origin and final resting position in the valley— are key controls on rate of motion.

Type of Material

The nature of the material can vary depending on such controls as climatic conditions, which in turn affect moisture content in the form of water, ice, or snow. The descending mass may be debris (unconsolidated weathered rock and soil material) or bedrock masses in large boulders weighing several tons.

Figure 17.10

Mass wasting levels the land with the aid of gravity. Four variables determine the nature of the mass wasting process: (1) type of motion, (2) rate of movement, (3) the material in motion, and (4) water content. The Grand Canyon was formed by weathering and mass wasting with the aid of gravity and stream erosion.

Water Content

Water has two important influences on **slides** or **flowage.** First it adds weight to the material and changes its density, increasing its instability. Draining water out of a hill by perforated pipe lessens water pressure buildup and reduces slide potential.

Water is also a lubricant. When clay minerals are saturated in the soil, they have a lower coefficient of friction, hence water "greases the skids."

Type of Motion

Flowage

Soil Creep Some types of mass movement are so slow they are not apparent at all during short periods of observation. Leaning retaining walls, telephone poles, and trees are good indications of **soil creep.** Leaning is a sign that the top profile of soil is slowly creeping downslope under the force of gravity lubricated by moisture (fig. 17.11).

Earthflows and Mudflows Mudflows are rapid, occurring at an observable rate. Water is the lubricant again, and gravity does the work. After a fire has removed the vegetation from a slope, the hillside is a prime target for mudslides. Fires destroy not only the soil layer but also the water quality of streams as channels fill with mud and vegetation debris. If the consistency of the material is that of mud, the moving mass is called a **mudflow.** How fast it moves depends on water content and slope. Mixtures can vary in content from mud so stiff it can hardly move to a fast-moving slurry (fig. 17.12).

The mudflows that continually threaten foothill residences in suburban regions of southern California often follow heavy winter rains on the burned-off mountain slopes (fig. 17.13). Starting as a muddy stream that gains energy from increased runoff, loose material is picked up until the stream begins to move like a slowly migrating dam of mud forced along by impounded water and debris.

Mudflows are very destructive because of their high density and momentum. Large boulders and trees are carried along, leveling everything in the path of the flow. Houses are moved off their foundations, and roads are both covered and torn up.

Earthflows **Earthflows** occur where water is less abundant and movement becomes more sluggish. Where roads have been constructed through weathered sedimentary rock, stability is disturbed. Figure 17.14 shows an earthflow caused by freeway construction through gently dipping beds of sedimentary rock. When the rains came the beds of shale were lubricated, and the beds slipped off one another like a tilted deck of cards.

Slippage

Landslides **Landslides** are generally rapid movements of large masses of rock that have not reached a mud consistency and separate into blocks or masses bonded together in large units. Earthquakes often start the process, setting in motion unstable rock masses. Again, water supplies the lubricant for joints and fault

Figure 17.11

Soil creep is a very slow process that causes leaning poles, trees, and walls. Water is the lubricant in the soil creep process, which involves only the soil profile.

Figure 17.12

Earthflows and mudflows are faster than soil creep. Gravity is the driving force and water is the lubricant.

Figure 17.13

Chaparral fires in California set the stage for winter flows when the rain comes in concentrated periods.

Photo by Gary Bowman.

Figure 17.14

Less water is involved in earthflows, seen in this slide along a recently constructed road cut.

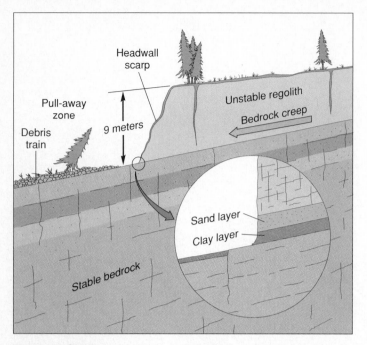

Figure 17.15

Love Creek landslide, January 5, 1982, Santa Cruz County, California. Gently dipping beds of highly fractured and mixed rock debris became unstable when lubricated on a thin layer of sand and clay. It was triggered by 50 centimeters (20 inches) of rain in a 30-hour storm period. The slide buried 9 homes and killed 10 people and ranks as one of the three most destructive landslides in western North America.

Photo by Julie Sowma.

scarps. Most slides are very complex in their movement. Some portions of the slide may have more water and flow while other portions are likely to slip as large blocks. Figure 17.15 illustrates complex slide characteristics of the Love Creek landslide of January 5, 1982, in Santa Cruz County, California.

Landslides come in all sizes and travel at a variety of speeds. The Blackhawk Canyon landslide on the north side of the San Bernardino Mountains of southern California probably reached a speed in excess of 160 kilometers (100 miles) per hour. Velocities can be inferred from the amount of mass and the distance it moved. The toe of the landslide reached 7.2 kilometers (4.5 miles) from the base of the mountains before it came to rest.

Some slides are slow-motion events. Figure 17.16 shows a block separation of well-lubricated beds of sedimentary rock. The beds are almost horizontal, gently dipping in the direction of slippage. Note the backward rotation that occurred in this slump movement.

ROCKFALLS

The **talus** piles that build up at the base of mountain cliffs are the work of gravity. Weathering loosens the material and then gravity takes over (fig. 17.17). Sometimes a slight earthquake, or wind or rain, can start the rockfall. This in turn starts other rocks in motion until a large avalanche occurs (box 17.1).

SUBMARINE LANDSLIDES

Gravity is at work not only above the surface of the sea but below. Recent undersea maps show striking landforms resembling those on dry land. This submarine surface is experiencing the pull of gravity and eroding through landslides.

How do submarine landslides differ from landslides occurring above sea level? First and most important, to have a slide, it takes a force to upset the equilibrium of the geologic structure. The forces involved must overcome the resistance to motion. This upsetting force that triggers the slide may come from earthquakes, waves, or a combination of these with gravity being the underlying driving mechanism.

When Hurricane Camille roared through the Gulf of Mexico in 1969, three oil platforms standing in the Mississippi Delta came tumbling down. At first it was believed that the 122-kilometer-per-hour (76 mph) winds and 15-meter (50 foot) waves pounding the pilings were the cause, but it was discovered that the foundation supports had shifted by landsliding triggered by waves of Camille.

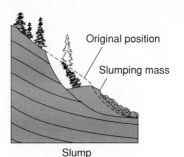

Figure 17.16

Backward rotation as well as forward downslope movement is common in a slump.

Figure 17.17

Rockfalls are without the friction of slides, so velocities can reach free-fall velocities.

Earthquakes are also a major cause of submarine slides. The classic example was the Grand Banks slump of 1929 that got the attention of the world because of the damage it caused to telephone cables linking North America with Europe.

The Good Friday earthquake that hit the Alaskan coast in 1964 caused the waterfront of Valdez to slip into the sea. It was discovered later that sliding soft delta deposits moved seaward in a simple submarine landslide.

Block Slumps

Block slumps occurring on the continental shelf are a common type of submarine landslide. They usually maintain a relatively undisturbed mass whose motion is twofold: downslope and rotational with minimal disturbances of the slide block. Movement may occur suddenly, but more often they move more slowly than other types of slides.

Submarine Flow Slides

Unlike slump block landslides, flow slides are very turbulent, disorderly movements occurring in zones of unstable weak sediments. Motion is usually very rapid and extremely destructive. The triggering mechanism is usually an earthquake coupled with gravity. Some slides cover several square kilometers. The Alaskan earthquake of 1964 created a submarine flow offshore, producing turbulence that engulfed the waterfront.

Submarine mudflows commonly are found at the mouths of rivers, where rapid deposition of fine-grained mud forms a saturated, thick, soft, unstable material. Seismic and sonar studies revealed the profile of the Mississippi River Delta mudflow caused by Hurricane Camille.

The 1970 Earthquake in Yungay, Peru

BOX 17.1

Peru experienced a landslide disaster in 1970 high in an alpine area. It started with an earthquake that triggered a snow avalanche that grew into a **rockfall** and slide, attaining velocities of 400 kilometers (250 miles) per hour. The entire village of Yungay was destroyed, burying an estimated 17,000 people. The rock and ice avalanche traveled 14.5 kilometers (9 miles) and fell 3.7 kilometers (12,000 feet) in less than 4 minutes.

Here is a case where a combination of geologic conditions came into play simultaneously. Heavily glaciated Nevado Huascarán, 6,664 meters (21,860 feet) elevation, was struck by a moderate earthquake centered 100 kilometers (60 miles) away. Second, a slab of glacial ice several hundred meters wide located near the crest was shaken loose. As it came down the steep slope, it plucked and loosened more material in its path. The main mass came down the valley and came to rest in the Santa River Valley, burying 1,800 people, while a smaller portion shot over a curved ridge and became airborne, slamming down on Yungay below (fig. 1).

How can rocks move this fast? Geologists who inspected the region believe that the slide rode on top of a cushion of air trapped beneath it. This produces the same effect as a frictionless plane.

Figure 1

The entire village of Yungay, Peru, was covered by a rock and ice avalanche riding on a cushion of air traveling at frictionless velocities.

Left: Photo by George Plafker, U.S. Geological Survey. Right: U.S. Geological Survey.

Because of offshore oil operations, there is great interest in not only locating these potential hazards but also determining the nature of motion, stability, or instability of the slide area.

Geologic evidence indicates that submarine slides were more common when the continental ice sheets covered the northern latitudes 15,000 to 18,000 years ago. At that time, storage of water as ice lowered sea level by as much as 100 meters. The new shoreline was steeper, and sediments accumulated to produce a thin veneer that easily slid off the steeper slope.

OUR ROLE IN LEVELING THE LAND

We are now capable of moving mountains. Nature works on a greater scale when you consider the Grand Canyon or the Himalayas, but we work with greater speed in most cases. To the physicist, power is defined as work or energy accomplished in a given period of time. Nature's work is usually very slow, often taking millions of years. In contrast to nature's slow erosional and mountain-building processes, we have the ability, with the press of a button, to remove mountains using powerful chemical and nuclear energy.

LANDFORMS: BY-PRODUCTS OF WEATHERING AND EROSION

The present-day mountains, valleys, plains, and plateaus are the by-products of both tectonic forces and slow weathering and sculpturing action that has been continuous since the beginning of the Earth.

All landforms can be described or classified as either depositional or erosional in nature, but both categories are the result of weathering and erosion.

Depositional landforms include beaches, streams, terraces, alluvial fans, sand dunes, and moraines from glacial deposits. Erosional landforms are created by sculpturing or agents of erosion such as streams, rain, ice, and wind, but running water is the most widespread of them all. Chapters 18 through 22 will examine in detail each of these erosional processes.

SUMMARY

The universe has many forms of energy: disintegration of rocks occurs when chemical and mechanical energy through the processes of **weathering** break down the structure of rocks. Gravity is always present to assist.

Mechanical or **physical weathering** produces finer and finer particles but does not alter the rocks chemical structure. **Chemical weathering** on the other hand involves chemical change. **Solution, hydrolysis, carbonation** and **oxidation,** are processes requiring chemical energy. Mechanical weathering breaks down rocks into smaller units by **frost wedging, unloading, slaking, exfoliation,** and **biological weathering** by the **roots of plants.**

Gravitational energy is quietly pulling everything downward. This downward pull involves four variables or factors determining the type of leveling or **mass wasting process.** These include (1) rate of motion, (2) type of motion, (3) water content, and (4) type of material in motion involved in the mass wasting process.

The type of movement falls into three broad categories: **flowage, slippage,** and **rockfall. Soil creep, earthflow, mudflow, landslides, rockfall,** and **talus** development are all examples of mass wasting resulting from the pull of gravity.

ILLUSTRATED STUDY QUESTIONS

1. Explain the role of weathering in the rock cycle (fig. 17.1, p. 371).
2. Explain the role of weathering in the soil formation process (fig. 17.2, p. 372).
3. Contrast chemical and mechanical weathering. Give one illustration of each (fig. 17.3, p. 372).
4. Which form of mechanical weathering is breaking down the rock in this photograph (fig. 17.3A, p. 372)?
5. Give one possible explanation for the cracks in the rocks shown in the photograph (figs. 17.3A and 17.4, pp. 372, 373).
6. Describe the following forms of weathering (figs. 17.3A, 17.5, 17.7, and 17.18, pp. 372–73, 375).
7. Describe the unloading process (fig. 17.6, p. 364).
8. Contrast chemical weathering by oxidation with carbonation (fig. 17.8, p. 376).
9. Define mass wasting (fig. 17.10, p. 378).
10. Contrast soil creep with earthflow (figs. 17.11–12, p. 380).
11. Describe the following mass-wasting processes in each photograph (figs. 17.11, 17.12, 17.14, 17.15, and 17.17, pp. 380, 381, 382, 384).
12. Which forms of energy are at work in each of the photographs (figs. 17.12 and 17.15, pp. 380, 382)?
13. Describe the key control factors causing the Love Creek landslide (fig. 17.15, p. 382).
14. Explain the high-velocity motion of the Yungay, Peru, slide (box 17.1, fig. 1, p. 384).

The Hydrosphere

UNIT V

Water covers nearly three-fourths of the Earth's crustal surface. In the liquid and solid form water scours and sculptures the Earth's varied continental surface. Once its work is complete it returns to the sea with its bounty of eroded materials to begin the process all over again.

Avalanche Falls, Glacier National Park, Montana.
© Barbara von Hoffmann/Tom Stack & Assoc.

The Work of Water in the Hydrologic Cycle

18

Objectives

After completing this chapter you will be able to:

1. Explain the principles of the **hydrologic cycle.**
2. Distinguish between **weathering** and **erosion,** and give examples of each.
3. Explain the key factors that affect the erosional process.
4. Discern between **erosional** and **depositional** features of streams.
5. Explain the work of streams in sculpturing the landscape.
6. Explain the impact of streams on human activity.

Runoff muddies the stream waters after a snowstorm. Arches National Park, Utah.
© Peter Kresan

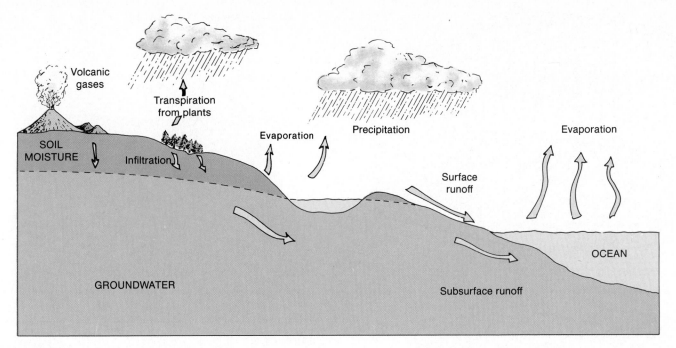

Figure 18.1

The hydrologic cycle consists of evaporation, transpiration, condensation, precipitation, runoff, infiltration, and groundwater movement into streams, rivers, and eventually the sea.

King Solomon, ruler of Israel during thirteenth century B.C., gives us a poetic description of the movement of waters on the Earth, recorded in Ecclesiastes 1:7—"All the streams run into the sea, but the sea is not full; to the place where the streams flow, there they flow again."

There is an endless interchange of water between the oceans, air, and land. Water is the substance that makes the planet Earth unique in the solar system. It is in abundance everywhere on the planet as solid ice, liquid water, and gaseous vapor. It can also be found combined chemically with crystalline rocks. The planet is not only richly endowed with water, but it is at the right temperature. If the planet were colder, most of the water would be locked up as ice and the Earth would be nearly lifeless and cold, perhaps like Mars. At the other extreme, it might have had a hot hostile environment similar to Venus, where the water is primarily in a gaseous state and surface conditions are like a scorching oven. Earth's orbital position and atmospheric filter has allowed the atmosphere to warm enough to keep water in the liquid state over most of the planet.

THE HYDROLOGIC CYCLE

The cycle of water consists of evaporation and transpiration into the atmosphere, followed by condensation and precipitation from the atmosphere back to the surface, and runoff, infiltration, and groundwater movement into streams and rivers on a return journey to the sea. This is the endless **hydrologic cycle** (fig. 18.1). Solar and gravitational energy are the driving forces of the hydrologic

cycle. Solar energy moves great quantities of moisture from the biosphere and lithosphere into the atmosphere, and gravity takes over, returning moisture to the ground and the sea.

Our chief focus in this chapter will be on the movement of water on the surface of the land and back to the sea. Geologists call it **runoff.** The result is a sculptured landscape. Each year solar energy lifts about 400,000 cubic kilometers (100,000 cubic miles) of moisture, and two-thirds of this falls right back into the ocean, but the remainder reaches dry land in the form of rain, snow, sleet, hail, and dew.

A significant portion of the precipitation goes underground by infiltrating porous soil and rocks. Chapter 19 evaluates groundwater movement in the hydrologic cycle. The major portion evaporates back into the air; but an estimated 10 quadrillion gallons, or 38,000 cubic kilometers, runs off to return to the sea. This runoff, year after year for millions of years, is the primary agent responsible for the sculptured landscape on which we build and cultivate the soil.

THE FLOW OF WATER

The next time you take a walk along a stream, note its everchanging velocity and depth. It cuts downward in one place and deposits sand and gravel in another. At one point the stream may be very wide and shallow, hardly moving, and then suddenly it narrows, picks up energy, and hurries along, churning over rocks making a pleasant sound as it cascades to lower levels. On other days this same stream can roar and churn, transporting great rocks

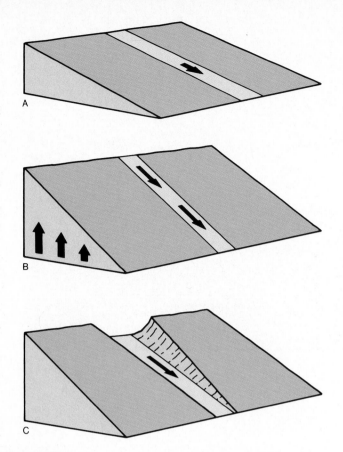

Figure 18.2

Slope or gradient is the key to velocity and resulting erosion. The river in A flows slowly with little erosion. If uplift occurs, as in B, the greater velocity will activate more downcutting, as illustrated in C.

and uprooted trees from along its bank, resculpturing its channel, picking up the land that looked immovable and permanent on high ground. Then it will return to a quiet flow of clean water in newly rearranged channels.

What are the principles of stream flow? How do streams sculpture the earth? Will streams ever finish leveling the land? How do streams affect the way we live? These are some of the questions considered in this chapter.

The work of streams consists of erosion, transportation, and deposition. **Erosion** is simply a process of deepening, widening, and lengthening of its channel. **Transportation** is the process of moving the material eroded, and **deposition** is the unloading or depositing process occurring as the stream loses energy. By considering the hydraulic processes controlling erosion, transportation, and deposition, we can better understand the landforms produced by erosion and deposition. We will first focus on the role of slope and its relationship to velocity.

Slope

The key to motion is **slope,** or gradient. Slope refers to the ratio of vertical fall in a given horizontal distance. Gravitational energy is spent on friction and energy of motion, or kinetic energy. As

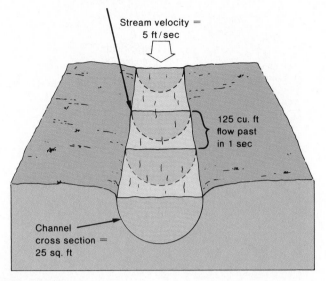

Figure 18.3

Discharge measures the volume of water per second. Stream discharge is measued in a weir. The cross-sectional area A times the velocity V of flow gives the discharge D over a period of time, or $D = A \times V$.

water flows, it loses some of its energy to friction between the water channel and surface of the stream. Water near the sides and bottom of the channel flows slower than water near the center of the channel, where friction is less (fig. 18.2). Friction is the brake, and slope is the accelerator.

Discharge

The amount of water that flows past a point in a stream is referred to by the hydrologist as **discharge.** It can be measured in cubic meters per second or cubic feet per second. As long as no water is lost to groundwater storage, evaporation or transpiration, its discharge rate will be consistent all along the channel.

Discharge can be easily determined by measuring the cross-sectional area and multiplying it by the velocity of the water (fig. 18.3). Velocity will change as the stream channel cross-sectional area changes, but the discharge remains constant. Tubing or rafting down a river is a good way to experience this relationship. As the channel narrows you will pick up speed until you reach a broad portion of the stream where velocities slow to a crawl, and yet the actual discharge is quite constant. Discharge is important to know because it describes the flow rate of the stream.

Turbulence and Velocity

Turbulence is the result of water changing direction and velocity. **Turbulence,** or churning of the water, occurs when the water picks up speed and directional motion becomes very irregular. In turbulent flow, water rapidly changes direction forming eddies and white water. Friction is greatest when water molecules are moving

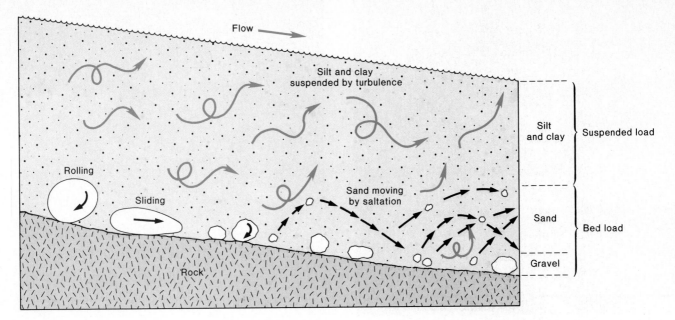

Figure 18.4

Load is equal to the amount of material in transport. This includes boulders, gravel, sand, silt, and dissolved or suspended material such as salts and fine clay particles.

at different speeds, rubbing against one another and against the channel sides. Because of turbulence, only a small amount of gravitational energy is needed to pick up and transport rocks, sand, and silt.

Load

Load is the material being transported by the stream. This includes both dissolved and suspended material as well as silt, sand, gravel, and rocks (fig. 18.4). Velocity and turbulence are influenced by the volume of water and load being carried. More water means greater velocity. Rivers reach their greatest velocities when large volumes of water flow with minimum loads down very steep gradients in restricted channels. By contrast, slowest moving streams are characterized by small volumes carrying heavy loads on broad gentle slopes. The energy of motion or kinetic energy possessed by the stream determines whether the stream erodes or deposits and produces the features related to these processes.

EROSION BY STREAMS

As water moves, it begins to gradually deepen, widen, and enlarge its channel in all directions. A river's erosional potential, or a land's ability to withstand erosion, depends on a number of factors such as rock structure and composition, rates of weathering, soil moisture, precipitation, morphology, and land use in the drainage system.

Fluvial or stream erosion is accomplished by hydraulic action, plucking and scouring particles from the stream bed. The particles range in size from boulders to fine clay sediments. Remember the importance of stream velocity. Members of the famous

Powell expedition down the Colorado River in 1869 could hear rocks pounding against other rocks as they camped along the river. When rocks pound, scrape, and break off additional rock fragments, they undergo an erosional process known as **abrasion** or **corrasion.** The process of continual wearing down of large fragments by abrasion is known as **attrition.**

Water also has chemical solvent properties capable of producing chemical weathering by the stream in the form of solution, hydration, and hydrolysis, referred to collectively as **corrosion.**

TRANSPORTATION OF A STREAM

The processes of erosion, transportation, and deposition are not separate sequential events where the stream first erodes, then transports, and finally deposits. Think of them occurring concurrently, with erosion and transport dominating at one time and place and deposition at another. The dominating process is determined by velocity, and this is determined by slope, stream depth, and the nature of the bed.

Once erosion has begun, streams transport materials they have eroded as well as material introduced by mass wasting and surface runoff. This material, the **load,** consists of suspended particles, sand, gravel, boulders, and chemical solutions.

Clay and silt-size particles move in suspension along with the soluble dissolved chemical load. Sand and gravel-size particles bounce along in the process of **saltation.** Fragments too large to bounce are dragged by stream **traction.**

A

B

Figure 18.5

The velocity and turbulence of water are the key to competence, or the capacity of a stream to carry its load. Turbulence is increased by increasing velocity and friction. The yellow area above the curve shows the velocity that must be maintained to erode various particle sizes. The blue area shows minimum velocities necessary to transport each particle size. The velocities where deposition occurs are shown in salmon color. Sand at point A will not be eroded until it experiences velocities in the yellow zone. (A) The Mattole River is carrying very little load and is not actively eroding. (B) In contrast, the Urubamba River in Peru is actively eroding and carrying a large load of clay, silt, sand, and gravel.

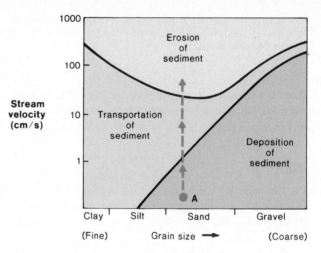

C

The speed and turbulence of the water is the key to the load capacity of a stream. Figure 18.5C shows that the ability of a stream to transport various sized particles is dependent on the velocity of flow. Particles of any size located to the right of the settling velocity curve will settle out because they possess the particle size and velocity value for sedimentation. As you can see, the larger the particle is, the greater velocity the stream must have to suspend and transport it. The erosional velocity in figure 18.5C represents the minimal velocity for particle removal from the stream bed. Transportation by traction occurs for objects larger than a few millimeters. The transportation process requires lower velocities for objects once they are in motion. This is why the traction and erosion velocity curves are at higher values than the settling velocity curve.

A stream is far more complicated than the graph may suggest. Scouring and filling, transportation and sedimentation are constantly in a state of flux. A stream may scour at one place and deposit in another. Weeks or months later, the place that was eroded may begin to fill. Stream channels are not only rivers of water but of gravel, sand, silt, and clay moving from the land to the sea. This material causes either degradation or aggradation, depending on the stream's speed and particle size.

DEPOSITION

As we have seen, a stream's carrying capacity depends on velocity and discharge, or volume per second passing a given point. If either of these factors lessens, the stream begins to unload because it has lost kinetic energy.

If we follow a river downstream to very gentle gradients, it becomes apparent from the flood plain and stream bars and terraces that **deposition** is occurring. Coarse materials settle out first, followed by finer sand, silt, and clay. A dam can cause a depositional process (box 18.1).

FEATURES OF STREAM EROSION

Where does a river begin? Have you ever explored the source of a stream and then followed it downstream to its terminus? Let's return to the small valley described in chapter 1, where weathering

Dams Retain More than Water

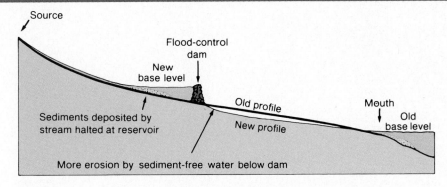

BOX 18.1

Dams across rivers are places where the flow of sediment is interrupted. The reservoir behind the dam in figure 1 has collected most of the sediments because the velocity of the river is greatly reduced as the water enters the lake.

Removal of this sediment and alteration of flow has produced a number of subtle influences downstream. It has been observed that degradation or erosional action has increased, causing a deepening of the channel. Since the stream is carrying less load below the dam, it moves with greater velocity and turbulence. This causes greater undercutting of the banks and deepening of the channel.

The degradation process can also make a negative impact on sand and gravel mining activities downstream. More sand and gravel is removed than deposited.

Figure 1

A dam and reservoir affect a stream profile. The new base level of the reservoir causes deposition in the reservoir and upstream from it. Erosion increases below the dam.

Altering a stream's course for flood control or hydroelectric power generation has both predictable and unpredictable effects. Because our understanding of the process is incomplete, research will continue in order to better predict the results of our impact on the hydrologic cycle.

and erosion rates are among the highest in the United States. The valley lies in faulted and folded sedimentary rocks consisting of sandstone and shale dating back to Mesozoic times, or 70 million years ago, when the rocks were formed. Today these rocks are highly weathered by both chemical and mechanical processes.

Precipitation in this valley is abundant six months out of the year, November through April, often exceeding 250 centimeters (100 inches) annually. Temperatures are mild in the winter, averaging well above freezing, and summer maximum temperatures generally are not hot. The mild temperatures combined with abundant moisture provide for strong **chemical weathering.**

The terrain is very rugged with slopes often exceeding 100 percent, or 45 degrees. Because the terrain is very steep and climatic conditions so moist, settlers have not been attracted to this area for agriculture. This is timberland. Currently over 90 percent of it has been logged at least once and some of it is being logged a second and third time, harvesting a rich forest cover of redwoods and Douglas fir. Logging began in the middle 1800s and continues today. Although logging practices have improved, the whole operation accelerates erosion through removal of the protective vegetation cover. Therefore, in addition to physical factors such as slope, morphology, and geology, human activity is a very significant element in the erosional process. By tracing the stream flowing through the valley, we can observe and examine some of the variety of landform features resulting from the degradation (erosion) and aggradation (deposition) processes of this stream. We will examine the three major segments or courses of the stream, referred to as the upper, middle, and lower course. Figure 18.6 illustrates the main stream and the tributaries that make up the entire **drainage net.** The region drained by the stream and its tributaries

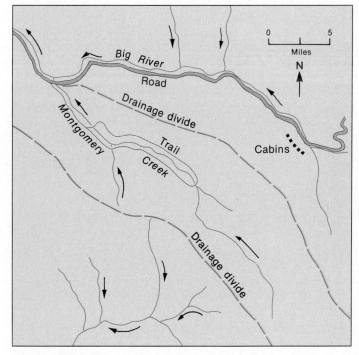

Figure 18.6

Drainage basins vary in size from a few square kilometers, represented by the *study area,* to thousands of square kilometers.

is known as the **drainage basin** or **watershed.** Drainage basins vary in size from a few square kilometers to thousands of square kilometers.

Figure 18.7

Fog drip provides a significant amount of moisture in the water budget of the stream. The Urubamba River in Peru is fed by melting glaciers, springs, precipitation, and fog drip.

The drainage basin of the Mississippi River contains a vast network of tributaries. Most streams have a large number of little streams that feed into successively fewer larger streams.

Robert Horton, an American hydraulic engineer, introduced the concept of stream *order.* A stream with no tributaries or joining branches is called a **first-order stream.** Two first-order streams join to form a larger **second-order stream.** A **third-order stream** is formed by the merging of two second-order streams; thus, third-order streams are considerably larger than first- or second-order streams. Horton demonstrated that the number of streams of a given order drops off sharply with increased order. The drainage net also grows at a consistent rate as orders increase. Size of the system does not seem to affect the validity of these relationships. In most drainage systems, regardless of size, the number of streams in a given order is approximately three times greater than the number in the next higher order.

UPPER COURSE STREAM PROCESSES AND FEATURES

The upper course of the Big River drainage, in our *study area,* includes first-, second-, and third-order streams. By following the hydrologic cycle in this stream valley, we can gain an overview of the stream processes and the features they produce.

Moisture from the atmosphere reaches the Big River drainage basin two ways. First, small amounts resulting from early morning *fog drip* (fig. 18.7), a form of condensation caused by fog and low clouds, affect the area daily during the summer months (chapter 6 explains this process and its influences on the water budget and vegetation). Fog drip represents a very important groundwater source, supplying springs located at the sources of many of the first-order streams. Without these springs, the main river channel would completely dry up and flow only as groundwater.

A second source of moisture in the *study area* is rainfall that begins to fall in late October and early November and continues through April. Because of very high rainfall, 150 to 250 centimeters (60 to 100 inches) annually, the rate of infiltration of rainwater into the soil is insufficient to keep pace with the downpour, and excess runoff begins to develop after a few weeks of storms. Drops of rain are intercepted by the forest cover. Where runoff concentrates, little threadlike rivulets form **sheetwash.** These combine, especially on the higher slopes devoted to pasture and meadow, to concentrate into channels or rills a few centimeters wide and deep. Water from rills then concentrates again into gullies several meters wide and deep, which then feed into streams of first order. In the upper course of Montgomery Creek, the gradient is steepest, averaging about 30 percent, as explained in figure 18.8.

Here, scouring and vertical downcutting is creating a V-shaped series of gullies and first-order streams which feed into the higher order streams downslope.

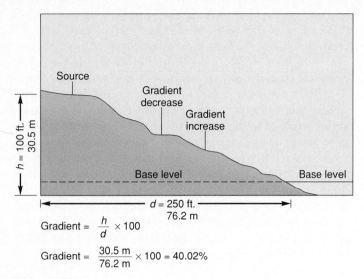

Gradient $= \dfrac{h}{d} \times 100$

Gradient $= \dfrac{30.5 \text{ m}}{76.2 \text{ m}} \times 100 = 40.02\%$

Figure 18.8

The slope of a drainage system can be calculated in terms of percent, by dividing the vertical distance by the horizontal distance times 100. This is the average gradient for the total distance; the actual trace on the graph shows that the gradient can be quite variable within that range.

Because gradients are steep and streams are primarily downcutting deep, narrow channels, they are often referred to as **youthful stream valleys** (fig. 18.9). This concept was developed by the American geographer, William Morris Davis, who organized landforms on the basis of their evolutionary development. The Davis model is more commonly known as the *cycle of erosion.* He examined a series of landforms and classified them according to *structure, process,* and *stages* ranging from youth to old age. Since no one lives long enough to see a landform pass through these stages, we must recognize this as strictly a hypothesis that is valuable in classification and organization of our knowledge about landform development.

The *structure* refers to the rock types or geology. Here in the upper course of the drainage system, the erosional process is not uniform, because the rocks being cut by stream action vary in resistance to weathering and erosion. Where the stream flows over a resistant shale, rapids and waterfalls develop, and the stream picks up energy and velocity with increased scouring and load-carrying ability. These resistant points along the upper course represent *knickpoints,* or points of discontinuity in stream gradient, typical of a youthful stage. Montgomery Creek, the upper course of Big River, has a series of rapids, or small falls, located at several knickpoints.

The Yellowstone River in Yellowstone National Park, Wyoming, represents a classic example of an immense knickpoint. The bed of the Yellowstone River is underlain by very resistant basalt that reduces the ability of the river to downcut (fig. 18.9).

EROSIONAL AND DEPOSITIONAL FEATURES OF THE MIDDLE COURSE

As we move downslope to the middle course of Big River, the most significant change is slope or gradient.

When a stream gradient is reduced to a gentle slope, its sediment load-carrying ability is balanced by erosional processes, and the stream is said to be **graded,** or in equilibrium. There are several indications that this is the case. First, the stream begins to erode laterally, picking up material on one side and depositing it in the form of **point bars** on the other side, where velocities are less. The lateral erosion and deposition produces a meandering, snakelike form (fig. 18.10). However, it is important to remember that this equilibrium condition is only temporary. A long dry period can cause considerable loss of velocity and volume, resulting in deposition or filling. At the peak of the rainy season, this same portion of the stream may straighten its course and increase in velocity, volume, and load. Davis would classify this portion of the stream as a **mature system** once it reaches a graded slope.

Although flooding is always a potential hazard, the rich fertility of the flood plain and the accessibility of the river for both irrigation and transportation has attracted settlers to valleys in this mature stage. The Nile, Yangtze, Mississippi, Ganges, and Mekong rivers are a few examples of productive, densely settled river systems in the mature stages of development.

DEPOSITIONAL FEATURES OF THE LOWER COURSE

Oxbow Lakes and Meander Drainage Patterns

River flood plains may develop **oxbow lakes** and meandering drainage patterns as a testimonial to a river's shifting channels. The oxbow lake forms when the stream meanderings change during flooding. The old meander loops mark the former channel by remaining filled with water, but cut off from the main stream course. The Ucalali River in Peru has produced the oxbow lake Yarina, which was once the channel of the river (fig. 18.11).

Birdsfoot Deltas

A portion of a stream's load eventually reaches a lake, sea, or ocean. Because the stream loses velocity and turbulence in the quiet water, it begins dropping its load. As one channel is clogged, the stream makes another path. Deltas take a variety of shapes. In some cases the stream or river channel divides into several branches, building up its own depositional levees. The Mississippi River has this type of development, referred to as a **birdsfoot delta** (fig. 18.12). The Nile River has a classic triangular shape, like the Greek letter delta,

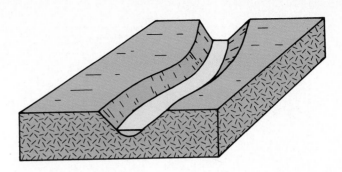

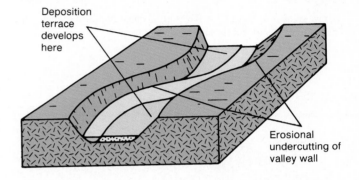

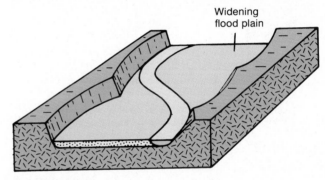

Figure 18.10

When a stream loses gradient or slope, its energy is redirected to lateral erosion and valley widening to produce a mature stream system characterized by meanders. Erosion is greatest on the outside of curves where stream velocity reaches a maximum. Deposition and slower velocities occur on the inside of the curves.

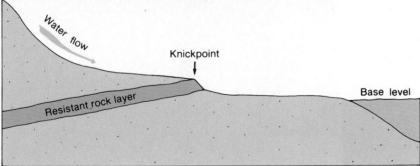

Figure 18.9

Youthful stream patterns, characterized by V-shaped cross sections, are produced by downcutting. The waterfall of the Yellowstone River resulted because of a resistant basalt layer. This interruption is called a knickpoint.

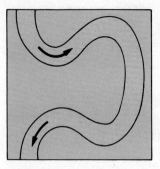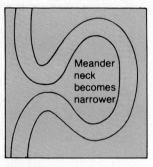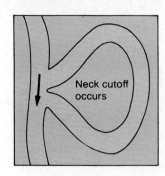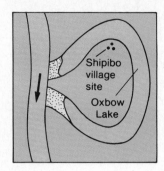

Figure 18.11

Oxbow lakes of the Amazon River are formed in old river channels. They often become village sites.

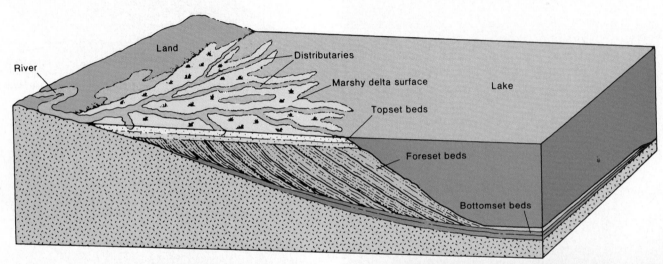

Figure 18.12

The birdsfoot delta of the Mississippi River was photographed by NASA ER-2 aircraft. The photograph, in both visual and infrared wavelengths, enhances the water image. The changing tones reveal sediment load being transported and deposited.

Photo by NASA.

with the city of Cairo at the apex of the triangle. This form is referred to as an **arcuate delta,** because of the fan-shaped, broadly curving shoreline.

When strong wave action and currents intercept a stream depositing its load, sediments are distributed along the coast with a pointed delta form. It is called a **cuspate delta** because of its resemblance to a sharp tooth.

Estuaries, Drowned River Systems, and Rejuvenation

In the lower course of the Alsek River, Alaska, changing meanders broaden the river valley, filling the plain to produce a braided stream pattern. In reality, the cycle may never be completed because of interruptions such as faulting, landsliding, and general uplifting (fig. 18.13).

At the mouth of Big River in the *study area,* it appears as if the lower course has been flooded by the sea. The typical broad flood plains are missing. The ocean mixes with freshwater

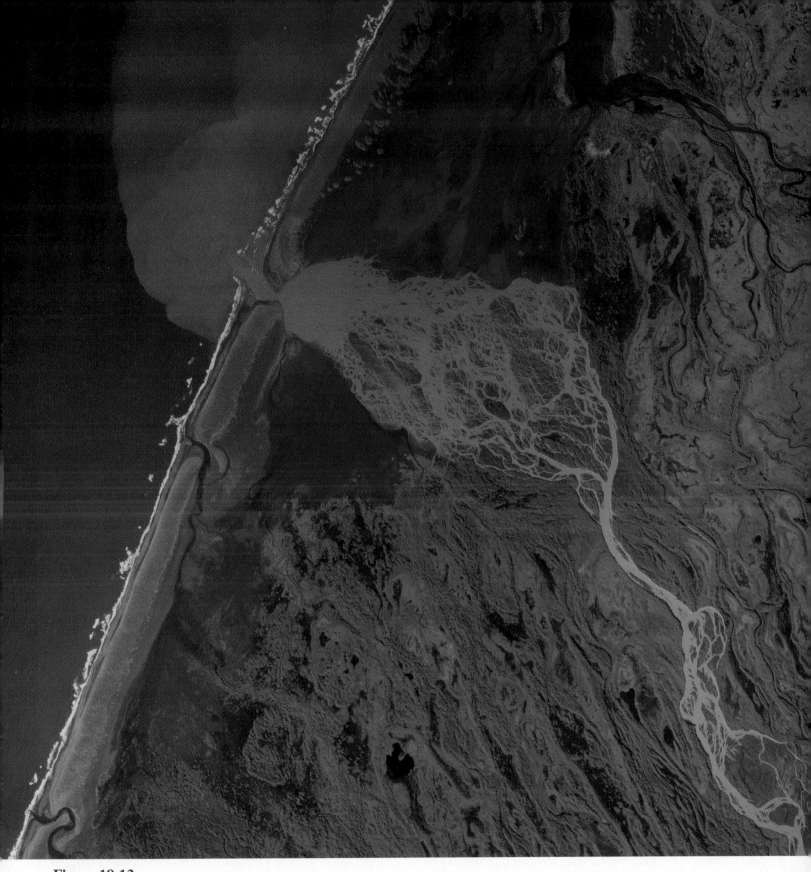

Figure 18.13

The Alsek River empties into Dry Bay in southeastern Alaska.
Along its lower course, the river broadens and develops a flood plain
and a braided stream pattern.

Photo by NASA.

Figure 18.14

A river with entrenched meanders results from rejuvenation of the system. Active downcutting can be reintroduced by a drop in base level or an uplift in the land structure. Damming the Little Colorado River, Arizona, has backed up water into its incised meanders created by rejuvenation.

Photo by NASA.

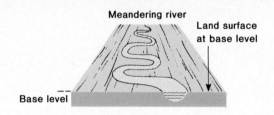

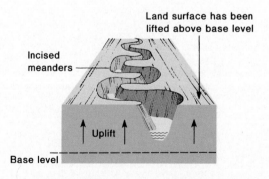

several miles inland during high tide. When sea level was lower, the stream was able to continue downcutting. This would explain why some narrow meandering channels are entrenched and drowned by the sea. During the retreat of the last Pleistocene Ice Age, the addition of water to the sea from melting ice caused sea level to rise rapidly, drowning the lower portions of river systems and producing an estuary environment. Chesapeake Bay, for example, is a major estuary produced by the drowning of the lower course of the Susquehanna River. Because the East coast has a broad extensive coastal plain, Chesapeake Bay covers many square miles of major and minor tidal estuaries, formerly tributaries to the Susquehanna River. Damming of the Colorado River has caused the same effect (fig. 18.14).

As we have mentioned, many typical features of the lower course of rivers are missing because the land has been tectonically active, rising or subsiding relative to sea level, over the last 50 million years. During most of this period, the land has been steadily rising. Streams have never had time to reach equilibrium. Downcutting is the norm. Even as alluvium is deposited and a flood plain begins to develop, **rejuvenation** due to uplift gives the stream new

energy, and downcutting processes dominate the stream's work. In addition to a changing base level, the work of waves, tidal currents, and strong longshore currents constantly excavate the deposited material at the river's mouth. Because the Big River in the *study area* empties into a long, narrow drowned river valley with the

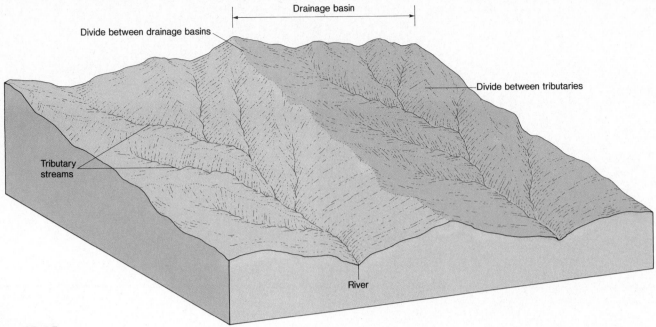

Figure 18.15

Dendritic stream patterns are the most common, resembling the branches of a tree or veins in a leaf. They develop on uniformly erodible rock, such as an uplifted mountain block.

delta confined to the shape of the estuary, this type can be called an **estuarine delta,** similar in form to the Seine River delta of France.

THE GEOMETRIC PATTERNS OF STREAMS

Stream patterns are governed by two opposing forces, gravity and friction. Gravity is the driving force giving energy to all streams to cut through the crust of the Earth. The route a stream takes is the path of least resistance. Internal forces causing tectonic uplift often determine the flow patterns. Rock structures of varying resistance obstruct and direct the downslope movement of water. By examining a variety of stream patterns and the associated geologic structures, we can appreciate the geometry of running water.

Dendritic

Dendritic drainage patterns are characteristic of most drainage systems. Dendritic patterns develop where the geologic structure is relatively uniform in its resistance to erosion. Such rocks as sandstone, shale, and granite often have this erosional pattern (fig. 18.15). The drainage pattern of the small valley of the *study area* is a classic dendritic system.

Trellis

Trellis patterns develop in regions of parallel ridges and valleys. They occur as a result of folding and faulting.

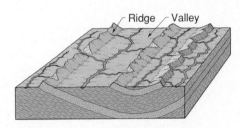

Figure 18.16

Trellis streams consist of parallel main streams with short tributaries meeting at right angles controlled by folded structures.

The pattern is quite conspicuous because the tributaries are short and flow into the main stream at nearly right angles. Here the structure dictates the pattern of flow (fig. 18.16). The ridge and valley system of the Appalachians is a good example of trellis development.

Radial

A **radial pattern** is determined by the resulting runoff in all directions from a central high point (fig. 18.17). Typically, volcanoes such as Mount Shasta, California, or Pico de Orizaba, Mexico, develop this geometric form. A reverse radial pattern develops where drainage occurs into a central basin of circular form.

Annular or Ring

Once a large dome has been partially eroded, the internal structure controls the flow. The once large dome of the Black Hills of

Figure 18.17

Radial streams diverge like spokes of a wheel from Mount Shasta's composite volcano.

Photo by U.S. Department of Agriculture, U.S. Forest Service.

South Dakota, discussed in chapter 16, illustrates the flow of water in the structurally weaker rock (see fig. 16.18). The stream's tributaries meet the main trunk at right angles. In a sense, if we could straighten the ring out we would have a trellis pattern.

Yazoos or Braided Streams

When streams begin clogging their stream channels in the lower courses, they commonly become so overloaded that they develop braided, multiple channels which separate and rejoin continually. A tributary flowing into the flood plain may not join the main channel for miles because of the natural levees preventing the merger of the two systems. These parallel tributaries are called **yazoo streams,** named after the Yazoo River, which runs parallel to the Mississippi River for over 150 kilometers (100 miles) until it finally joins the larger river near Vicksburg, Mississippi.

In summary, a flood plain is commonly found where the stream has lost velocity and energy. Here the river is broad and shallow resulting from repeated deposition when more material is deposited than removed. Braided patterns, oxbow lakes, levees, and yazoos are all typical stream features resulting from this loss of speed (fig. 18.13).

Alluvial Fans

When a fast-moving stream emerges from a narrow, steep, V-shaped canyon into a valley plain, the load material cannot be carried because of the rapid change in gradient. Clogging begins to take place in the stream's channel in the same way described for the formation of deltas and braided streams (fig. 18.18). **Alluvial fans** are commonly found in arid regions, where runoff is extremely variable. A stream may flow only a few weeks each year in arid regions of the Southwest. However, the runoff during those few weeks is often the result of heavy thundershowers in the mountains. Because of little vegetation in this arid zone, runoff and erosion are very active, and the fan receives large contributions of load.

By careful analysis of each runoff period, one can observe the development of a fan. The infrequent rainfall produces clogging of the channel. When the water capacity suddenly drops after the storm, it results in filling of the river channel. When the next torrential rains occur, runoff takes the path of least resistance down another side of the fan that has not been built up by deposition. This repeated process continues to build the fan-shaped system. Along the front of a fault-formed mountain system, fans may merge or coalesce together to form an alluvial apron called a **bajada.**

GEOGRAPHIC SIGNIFICANCE OF STREAM FEATURES

Many towns and cities have located on alluvial landforms for a number of environmentally advantageous reasons. First, water is often available in a large underground flow. Water easily infiltrates the loose coarse sediment of the fan, providing an excellent source of groundwater. Furnace Creek in Death Valley is built on a fan

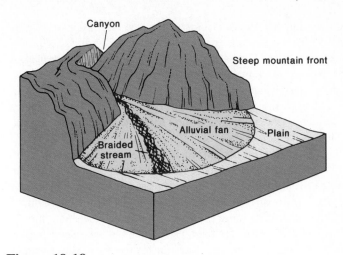

Figure 18.18

Alluvial fans form at the mouths of canyons in desert and semiarid regions.

that has more than enough water for a golf course, date groves, swimming pool, and lodging in the "heat island" of North America, where high temperatures and low annual rainfall are the norm.

Alluvial fans have unique climates. High on an alluvial fan the night air is stirred and mixed as cold, heavy air drains down the stream-formed canyons. When the cold air drains off the slopes at night, mixing with the warm air layers above, the area often remains 5 to 10 degrees warmer than the lower valley floor. This microclimatic condition is common on calm, clear nights in the arid West. Citrus, grapes, and avocados are frost-sensitive crops that grow well on these alluvial fans of California and Arizona because good air drainage keeps the temperatures above freezing during the harvest and blooming seasons.

The alluvial flood plains of Mesopotamia and the Nile were settled long before recorded history. The same can be said of the Indus River of Pakistan and the Yellow River of China. All of these rivers have fertile sandy loam soils easily cultivated with irrigation access. A quick review of a world population map will show that the bulk of the world's population is concentrated on or near major river flood plains in the middle and lower courses.

Herodotus, the ancient Greek who was the first historian, wrote that "Egypt is an acquired country, the gift of the river." Over 95 percent of the population lives in a narrow strip of land 3 to 15 kilometers (2 to 10 miles) wide astride the Nile and in its triangular delta between Cairo and Alexandria. Until the Aswan Dam was built, each year flood waters would begin arriving in late June and reach flood stage in October, following the heavy summer rains in the Ethiopian mountains. The increased flow capacity from these rains enabled a new layer of silt to blanket the lowlands, restoring the fertility of the land.

Soon after the flood waters receded, the cultivators planted seeds in the fresh silt deposits. A dike and tile-lined canal system allowed the ancient Egyptians to control the floodwaters to some degree, impounding some for irrigation and infiltration to leach the salt buildup in the soil. Then in late October, after the soils drained and dried, wheat and barley and other vegetable crops were planted. Harvest occurred in April prior to new flooding. The land was then given a rest, lying fallow until the next flood. The Egyptian culture successfully responded to this river rhythm for thousands of years. Harmony between civilization and nature resulted in sustained occupancy.

In 1960, with Russian assistance, a large dam was constructed at Aswan, the First Cataract. At this point the river flows over resistant granite where rapids prevent river traffic from continuing upstream. The American counterpart for the Nile cataract is Niagara Falls, where the outflow of Lake Erie flows into the Niagara River and over the falls to Lake Ontario. Sudden drops such as these are very good sites for hydroelectric power generation. The newly constructed dam on the Nile backs up water 550 kilometers (350 miles) into the Sudan and makes transportation safer both above and below the dam. Destructive floods have been reduced, but natural soil fertility must be replaced by chemical fertilizers because silt deposits no longer enrich the fields. Since the Aswan Dam was built, new summer cropping is possible along with expanded irrigation land as more water is diverted from the river. Because the Nile carries less volume and load downstream, it is having a negative influence on the delta land. Salt water from the Mediterranean is able to penetrate the groundwater that was once recharged with fresh water from the flooding Nile.

Large rivers like the Nile, Mississippi, Yellow, and Amazon serve as major arteries for inland shipping of bulk cargo and transportation for people. It is hard to visualize what the westward settlement of the United States would have been like without the Mississippi, Ohio, Missouri, Snake, and Columbia Rivers. Rivers throughout the world are not only easy arteries for travel but represent major national and state boundaries. The Rio Grande and Saint Lawrence River mark parts of the United States' boundaries.

Where human activities have accelerated soil erosion, rivers receive the consequences by increased runoff, increased load, and erosion and deposition. The impact of vegetation removal on drainage systems is basically the same, whether it is by logging, fire, mining, subdivisions, road construction, or agricultural clearing. The stream experiences more erosion, transported load, and deposition. A portion of the hydrologic cycle has been modified (fig. 18.19).

Vegetation serves the cycle several ways. First, it forms a canopy or cover, reducing rapid runoff and droplet impact. Second, root systems work like sponges, soaking up water and soil nutrients and reducing the amount of runoff by recycling moisture back into the atmosphere. Third, rocks and the top veneer of soil mantle are held by the web of roots. Plant roots prevent removal by mass wasting and erosional processes. Fourth, suspended load travels downstream and is deposited in river estuaries, causing considerable damage to property, plants, and aquatic life. In places where logging practices have improved, fish are now coming back into these estuaries and streams to resume spawning.

Figure 18.19

Erosion of the land can follow human activity. Logging behind Lake Washington, Washington, exposed the watershed to massive runoff following 29 centimeters (10 inches) of rainfall in a 2-day period. The result was flooding and a debris flow of slash and mud. This house was moved several hundred meters into the lake on a log jam.

Migratory birds are also negatively affected by excessive filling and sedimentation of tidal estuaries. If the estuary vegetation is covered and the waters clogged with new silt, food supplies in their food chain are reduced.

Bays, estuaries, and lagoons along our coast are prime sites for residential, industrial, and commercial development. Many of our great cities and ports are located at the mouths of rivers accessible to trade and water supplies. (In California alone, 75 percent of the estuary lands have been developed for other uses.) All of these important activities have a high price on our natural systems.

SUMMARY

Streams are transportation systems for nature and human activity. Sediments as well as dissolved salts are being carried to lakes and seas by river systems. In these transportation processes, the energy induced by gravity sculptures the land. The work of streams consists of erosion, transportation, and deposition. The key factors that determine these processes are (1) volume or discharge, (2) velocity, and (3) turbulence. As streams continue to work, the stream system and associated landforms are being eroded, passing from youthful to mature and finally old age.

Each of these stages can be identified by erosional and depositional features characteristic of that stage. The stages also often correspond to the courses or parts of the stream. In the upper course, V-shaped valleys, steep gradients, rapids, waterfalls, and lakes are dominant features. Middle course features include a graded stream bed and the beginning of lateral bank erosion and some deposition. In the lower course of a stream, flood plains develop on lower gradient slopes. Stream meanders, braided streams, oxbow lakes, and finally deltas develop at the stream's mouth.

The geometry of a stream is governed by two opposing forces, gravity and friction. The erosional form can be dendritic, trellis, annular, and radial, as dictated by the rock structure.

Braided streams and alluvial fans develop where a stream loses energy and drops its load. The result can vary from a multichannel braided pattern to a fan at the mouth of a V-shaped canyon.

Streams have always been key points of settlement in both arid and humid regions. They bring to a region fertile soils, water for irrigation, transportation routes, and power for industry. They also serve as refuges for migratory wildlife.

ILLUSTRATED STUDY QUESTIONS

1. Name the two variables used to measure discharge (fig. 18.3, p. 390).
2. Rate each of the following streams for turbulence using a scale of 1 to 5, with 5 indicating the highest turbulence (fig. 18.2A, B, C, p. 390).
3. Name the stream gauge that measures discharge. Explain how it is calculated. Give the discharge if the cross-sectional area "A" is 10 square meters and velocity "V" is 2 meters per second (fig. 18.3, p. 390).
4. Describe the various sized particles that make up the load of a stream (fig. 18.4, p. 391).
5. Contrast the Mattole and Urubamba rivers in terms of load and velocity (fig. 18.5, p. 392).
6. Name one additional source of moisture for the Urubamba River (fig. 18.7, p. 394).
7. Define gradient and calculate the gradient of a stream with height "h" 10 meters (30.5 feet) and horizontal distance "d" 100 meters (305 feet) (fig. 18.8, p. 396).
8. Stream erosional processes include erosion, transportation, and deposition. These are not separate processes, but each one will dominate at one time and place in the history and course of a stream. Which stage does this photograph best illustrate (fig. 18.9, p. 397)?
9. What types of erosional patterns develop when a stream loses gradient (fig. 18.10, p. 397).
10. Describe the development of a birdsfoot delta (fig. 18.12, p. 398).
11. Explain the development of an entrenched meander (fig. 18.14, p. 400).
12. Contrast a radial and trellis stream pattern (figs. 18.16 and 18.17, pp. 401, 402).
13. Name the type of stream pattern in these figures. Explain how it was formed (fig. 18.6 and 18.17, pp. 393, 402).

Groundwater and the Hydrologic Cycle

19

Objectives

After completing this chapter you will be able to:

1. Explain the flow of water underground in the hydrologic cycle.
2. Describe the processes shaping the Earth's crust through groundwater activity.
3. Describe groundwater movement through the Earth's crustal zone.
4. Explain the development of springs, aquifers, and geysers.
5. Describe human influences on groundwater quantity and quality.
6. Explain the formation of caverns, sinks, and karst topography.

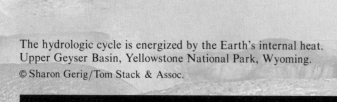

The hydrologic cycle is energized by the Earth's internal heat. Upper Geyser Basin, Yellowstone National Park, Wyoming.
© Sharon Gerig/Tom Stack & Assoc.

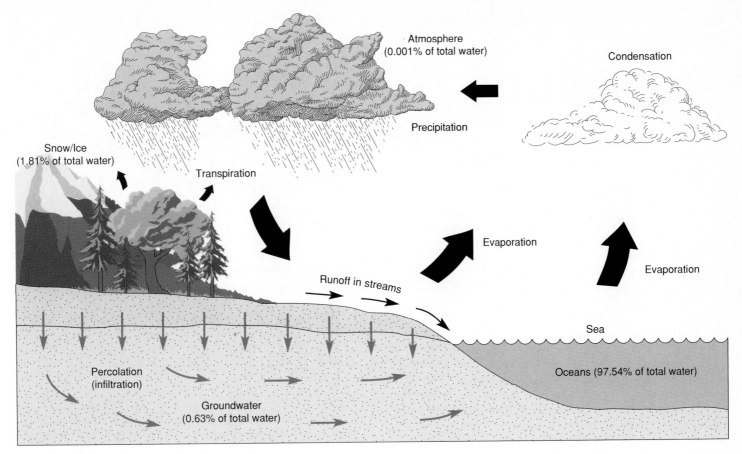

Figure 19.1

Groundwater in the hydrologic cycle. Water falling on the land enters the pores of the soil and percolates into the ground to become groundwater.

In this chapter our focus will be on the portion of the **hydrologic cycle** where water is moving below the ground surface. We will evaluate how it moves and the work it performs.

The cycle is energized by both the Sun and the Earth's gravity (fig. 19.1). Solar energy lifts water from the oceans, lakes, streams, and land into the atmosphere by evaporation. Then gravity returns water to the Earth through precipitation. Gravity keeps the cycle moving by pulling water down to the Earth and back to the oceans. The movement of water within the environment provides the energy and mechanism for many of the geologic processes of weathering and erosion.

Water performs a series of operations on and below the surface of the Earth. As it moves to the ocean, water makes a series of intermediate stops in lakes, ponds, glaciers, and groundwater reservoirs before returning to the sea.

As King Solomon noted, in Ecclesiastes 1:7—"All the streams run to the sea yet they are not full." Solar energy prevents this from happening by maintaining a balance between evaporation and runoff into the sea. Gravity is also a key force. Without it the atmosphere and oceans would boil away under weightless conditions, and the rivers and glaciers would not flow, and precipitation would also cease to fall. Therefore, it is important to realize that the cycle we observe, today functioning in the many environ-

ments of the Earth, is operating because of these principal energy forms. Gravity is a constant over the surface of the Earth, but solar energy varies with latitude and season, producing variations in the cycle through time and space.

The hydrologic cycle in the polar areas functions at a slower rate than in low-latitude environments where solar energy is greater. The middle-latitude zones experience seasonal variations in solar energy and thus produce seasonal variations in the cycle.

The last Ice Age is also a testimonial to changes in the hydrologic cycle. Sea level was lower when ice sheets covered the higher latitudes and upper elevations with several thousand feet of ice. This cooler period locked up much of the Earth's waters and slowed this cycle.

GROUNDWATER VERSUS VEGETATION, CLIMATE, AND GEOLOGY

How much water a given environment stores underground and its rate of movement are dependent on two important factors: climate and geologic conditions. **Groundwater,** "subsurface water," and

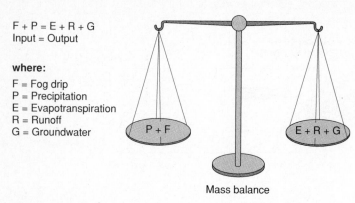

$$F + P = E + R + G$$
Input = Output

where:

F = Fog drip
P = Precipitation
E = Evapotranspiration
R = Runoff
G = Groundwater

P + F

E + R + G

Mass balance

Figure 19.2

The mass balance equation describes the water economy of the hydrologic cycle of a region. No runoff will occur if the plus or input values equal the losses. The global water balance is always a balanced system; however, regional water economies are rarely in balance because of variable climatic conditions and surface and groundwater movement.

"subterranean water" are terms that denote water stored in the pores, cracks, and crevices of the ground we are standing on. Most of this water reaches the surface in the form of precipitation, which varies widely over the Earth's surface. A hard rain rapidly runs off into streams while a drizzle soaks into the soil. Where evaporation is high, large amounts of moisture return to the atmosphere without moving into the groundwater system.

In some coastal locations, summer fog and low clouds have a major impact on groundwater supplies. The protective layer of clouds and fog can reflect the solar rays that dry the land and bake the soil. Condensation from fog drip on plants can increase the soil moisture considerably. This additional moisture supply during the dry summers along the coast of California makes possible forest vegetation such as the coastal redwood (*Sequoia sempervirens*) in areas where annual rainfall would be too low to support these trees. So, cooling fog is a supplier of moisture, and it prevents loss due to evaporation and transpiration.

Vegetation also determines how much groundwater occurs in a given place. Plants not only mine the soil for minerals but function as efficient water pumps. The coastal redwood when it reaches maturity can hold roughly 30,000 liters (8,000 gallons) of water. Each day it draws up hundreds of gallons into the trunk and transpires this moisture into the atmosphere. By the end of summer, transpiration can reduce it to 23,000 liters (6,000 gallons) of water storage, even though the roots are taking up moisture daily.

Where vegetation has been removed in a drainage system, groundwater and surface runoff are significantly increased. A controlled watershed study conducted by the Hopland Field Station, of the University of California, measured the influence of vegetation on runoff and stream flow. After clearing live oaks from a drainage system of known surface runoff, it was discovered that surface runoff increased significantly. Before vegetation removal in the watershed, the stream surface flow ceased during the

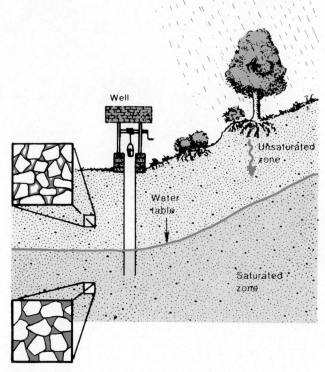

Figure 19.3

The zone of saturation is a porous zone filled by groundwater. The water table marks the upper limit of this zone. Note that the water table corresponds closely to the slope of the land.

summer dry season. However, after clearing the drainage system, the stream became perennial, flowing year-round. Water that normally returned through the plants to the atmosphere now flowed into the stream system as groundwater seepage. When we examine the economy of a water system or water budget of this portion of the hydrologic cycle, we can draw up a "mass balance" (fig. 19.2), an equation which can be expressed as

$$\text{Runoff} = \text{Precipitation} + \text{Fog drip} - \text{Evaporation} - \text{Infiltration} - \text{Transpiration}$$

PROFILE OF UNDERGROUND WATER ZONES

As water infiltrates the ground, some of it is caught by rocks and particles and held or checked in its downward movement. This water is known as "suspended water" and is held above the saturated level in the *zone of aeration*. Two molecular forces hold the water in the zone. First, water particles exert *cohesive* electrical forces on one another. These forces bond water molecules together. Second, the rocks produce attraction forces for water or *adhesion*. The zone of aeration is not saturated most of the time. Here atmospheric gases fill the voids without water. The top layer of the zone of aeration is the belt of soil moisture. Below this belt an intermediate belt, known as the capillary fringe, is just above the water table.

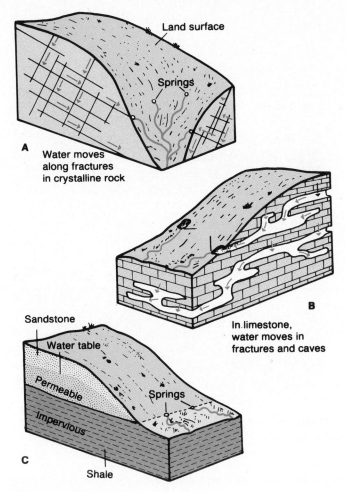

Figure 19.4

Groundwater movement is governed by gravity and the nature of the material it must penetrate. Shale, if highly jointed and fractured, is permeable. Water will travel both vertically and horizontally in the fractures. Sandstone and conglomerate allow water to move freely in the highly permeable rock mass. Limestone is often very impermeable, thus preventing downward movement, unless it is fractured and jointed.

Below the zone of aeration is the *zone of saturation*. Here the cracks, spaces, and pores are filled with water. The water table marks the boundary between the zones of aeration and saturation (fig. 19.3). Let us focus now on the zone of saturation and the work of water there.

Movement Underground in the Zone of Saturation

The molecular properties of water have a remarkable influence on its movement. The liquid state of water permits it to move into any shape regardless of size. However, the nature of water also causes the molecules to behave as though they were covered with an invisible elastic membrane, which acts to contract and pull the liquid back into itself making the least possible surface area. This can be observed when water droplets bead up on a countertop or

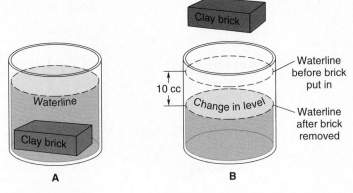

Figure 19.5

Porosity versus permeability. Porosity of a rock determines the water-holding capacity or the volume of air space. Permeability describes the transmission rate through a rock structure. The change in water level in the container measures the water absorbed by the brick, or its porosity. In this example it is 10 cubic centimeters.

cling together in their movement through underground pores and spaces. Its rate of downward movement is governed by both the pull of gravity and the nature of the passageways it must travel through (fig. 19.4).

Porosity versus Permeability

Two characteristics of rock passageways, porosity and permeability, govern movement of water underground.

Porosity of a rock is simply a measure of how much air capacity or volume exists in the rock. Highly porous rock such as volcanic tuff or pumice may even float. To demonstrate to yourself how much void may be in an ordinary brick, take a brick and place it in a known volume of water for 20 minutes. While the brick is soaking, note the bubbles of air leaving the brick. The bubbles of air are being forced out by water moving into the pores. Now remove the brick and you will see two changes. First, the water in the container is lower, and second, the brick is heavier because of increased water content in the pore spaces (fig. 19.5).

The brick demonstration clearly illustrates water's ability to penetrate the very small pores, and it also shows how even a dense clay material can hold substantial amounts of water. This does not mean water can be easily transmitted through a porous material. Take the same brick and try to remove the water or pass water through it. Bricks are nearly impermeable to water.

Whether or not groundwater is available in a region depends on not only the porosity of the materials but the ability of water to move through a rock zone, or its **permeability.** This is determined not only by the number of pores, but also by the size and shape of the water passages. Water molecules tend to not only cling together but adhere to surfaces with tremendous strength. Therefore, water passes more easily through sandy materials than porous clay simply because the molecular attraction holds the water better in the small clay spaces (fig. 19.6). Larger spaces reduce the amount of water surface tension and encourage penetration and movement.

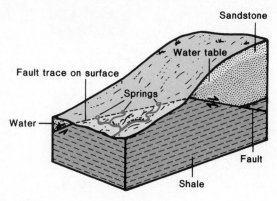

Figure 19.6

Permeability describes the ability of water to move through a rock. The shape and size of the water passages are the controlling factors. In this example, water moves easily through sandstone but meets resistance in shale. Springs are frequently found along rock boundaries or fault planes.

Remember, the energy that causes water to flow downslope is **gravity;** it tugs on each water molecule and sets it in motion downward to the water table. Solar energy causes capillary movement upward in direct opposition to the pull of gravity. The net direction and velocity is determined by the strength of these two forces.

The Water Table

The downward movement of water brings it to the water table, the top of the zone of saturation. This zone is not a pool or lake like a body of water or flowing underground stream. Some limestone areas have underground streams such as in the Mammoth Caves region of Kentucky, but in most cases water occupies pores in the rock. The rock is saturated, much like the brick in the demonstration.

The water table may lie hundreds or thousands of feet down in arid regions. It may be near the surface in humid or poorly drained regions. Note that this is the condition in the *study area* on the valley floor. Here, giant coastal redwoods strike water at two meters below the surface. The water in the nonsaturated zone of aeration is known as **vadose water.** Its content varies from day to day depending on weather conditions (fig. 19.3).

Where fairly homogeneous geologic conditions exist, the water table corresponds roughly with the slope of the land, higher in elevation under the hills and lower under valleys, but its relief is far less pronounced. It reflects only the major surface slope changes, as figure 19.7 shows. During a dry period the water table not only lowers but tends to become more and more horizontal and reflecting less the irregularities of the surface contours. Groundwater flow stops when the table flattens out completely.

When a well is drilled and water is drawn out, a depression or dimple is created in the water table at the well site. The rate of pumping determines how pronounced the depression becomes. Geologists refer to this well-induced dimple as a **cone of depression** (fig. 19.8).

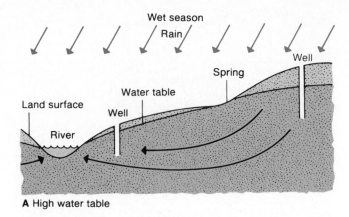

A High water table

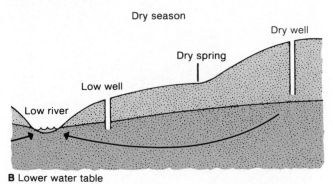

B Lower water table

Figure 19.7

Where homogeneous geologic conditions exist, the water table corresponds closely to the slope of the land. The amount of base flow in a stream is a function of the seasonal patterns of rainfall and evapotranspiration. This determines the height of the regional water table.

Aquifers and Artesian Wells

Along the eastern slopes of the Rocky Mountains from North Dakota south to New Mexico, groundwater flows under the force of gravity for many miles eastward in permeable sedimentary rock consisting of sandstone and limestone. Bodies of water-transmitting rocks are known as **aquifers,** from the Latin for "water" and "to bear." The best aquifers are unconsolidated sand and gravel, sandstone, and limestone such as the Ogallala aquifer of the Great Plains. The aquifers of Artesia, New Mexico, owe their existence to the permeability of limestone which has chemically weathered, enlarging the fractures and bedding planes into water channels (fig. 19.9).

Movement in an aquifer is usually slow and lacks the turbulence of surface water. Just as in the case of running water, there must be a slope or gradient. This is the water table's **hydraulic gradient.** It can be computed by simply dividing the vertical distance the water flows underground by the horizontal distance measured from the point of intake to the point of discharge. This vertical distance is known as the **head.** If the horizontal distance is 1,000 meters and the head is 100 meters, the hydraulic gradient is 0.1

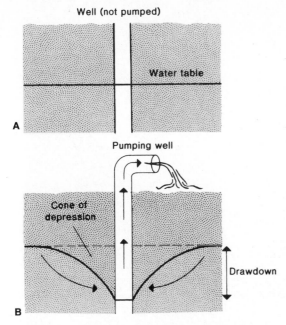

Figure 19.8

A cone of depression is the result of well development. When water is drawn from a well, a dimple or depression results.

or 10 percent. The greater the hydraulic gradient or slope, the greater the water flow. Typical aquifers flow at 1.5 meters per day and slower.

Artesian Systems

The Roswell, New Mexico, artesian system occurs in a region of gently sloping limestone and sandstone beds, chiefly rocks of Permian and Pennsylvanian age that occupy most of eastern New Mexico and western Texas and parts of Oklahoma and Kansas (fig. 19.9). The geologic structure is rather simple with Permian limestone dipping eastward from the crest of the Guadalupe and Sacramento Mountains on the west, to the Texas–New Mexico border on the east. Most of the groundwater begins as rain and snow that falls on the extensive drainage basins of the streams that are tributaries to the Pecos River. Not all precipitation contributes to the groundwater supply; some evaporates, and a large portion runs off in the surface channels of the Pecos River. Plants also utilize smaller amounts. The rest sinks into the ground and is drawn down by gravity to the main zone of saturation. The water in the artesian basin moves on downslope in an eastward and southwestward direction from the mountains on the west to the Pecos River valley, where it is discharged at the surface through wells and springs.

The artesian condition is produced by the following geologic conditions. Precipitation that falls in the mountains and limestone uplands enters the permeable Picacho Limestone aquifer. The water moves downslope in these beds eastward through solution passages, fractures, and joints until it reaches the western edge of the alluvial basin, where the impermeable cover formed by the

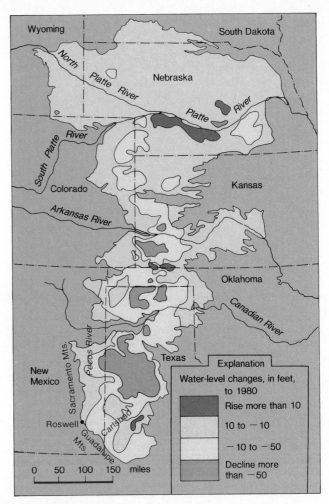

Figure 19.9

The Roswell, New Mexico, artesian system occurs in a geologic setting of gently eastward dipping limestone and sandstone beds. The water enters the system in the Sacramento Mountains and flows downslope to the Pecos River Valley. It represents the southern portion of the High Plains aquifer system or Ogallala aquifer.

Pecos Formation and alluvial fill prevents most of the water from passing into the overlying formations and confines it in the Picacho Limestone. Because of the hydraulic pressure, some of the water surfaces as springs, finding its way through various cracks in the impermeable layer. Most, however, is trapped by the cap of the Pecos Formation above and by relatively impervious underlying beds of limestone in the Picacho and Nogal formations (fig. 19.10). Being under artesian pressure, the water rises approximately to the level at which it was first confined by the overlying beds, but not exactly to that level because of friction between the moving water and rocks.

Environmental Concerns

The development of artesian water for agriculture brought about numerous problems. As the agricultural base expanded in the Pecos River valley, the water supply began to decline. Originally artesian

Figure 19.10

The water is trapped between a limestone cap rock of the Pecos Formation and an impervious layer of limestone underlying a permeable aquifer rock of the Picacho Limestone. This permeable sandstone is the conduit for underground transport.

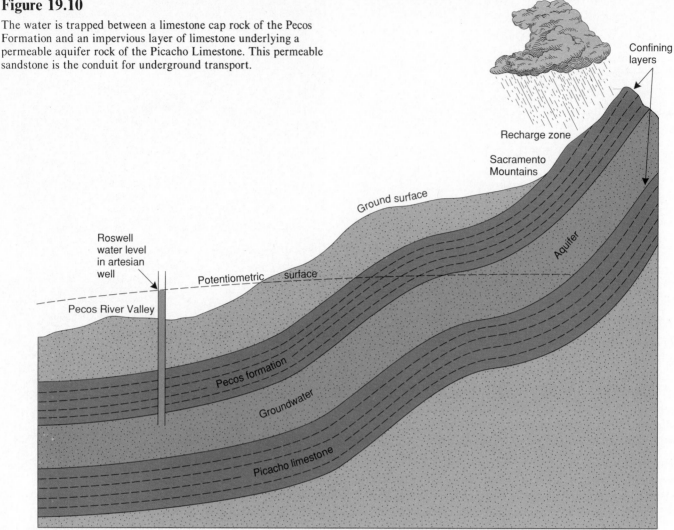

flow occurred over 1,717 square kilometers (663 square miles), but because of heavy draft upon the artesian reservoir, it decreased to 1,292 square kilometers (499 square miles) in 1916 and less than 1,100 square kilometers (425 square miles) today. Farther northeast of the Rockies, the Ogallala aquifer is being depleted by overdrawing by wells in Nebraska and adjacent states underlain by this artesian system.

Legislation currently regulates use of this resource to preserve it for continuous agricultural settlement. An artesian system will recharge itself if consumption is not greater than infiltration.

THERMAL SPRINGS AND GEYSERS

The Pomo Indian people for thousands of years have sought relief from the aches and pains of life in therapeutic hot springs. Warm Springs, Georgia, long before the Civil War, was used as a health and bathing spa. Here the groundwater is heated by natural heat of the Earth's interior, which warms with depth at the rate of 1.8° C per 100 meters (1° F per 100 feet). Water infiltrates the surface as cool water, around 16.5° C (62° F), in the Hollis Formation at a depth of over 1,100 meters (3,600 feet). It warms as it descends until reaching an impervious layer that forces the water back up. It emerges at a nice warm temperature of 36° C (97° F), ideal for soaking.

Palm Springs, California, is another place first made famous for its spas. However, the waters of Palm Springs are probably heated by hot rocks in contact with a mass of magma not too far below the surface.

Hot springs are scattered throughout the world along zones of tectonic activity. Where the Earth's crust is under stress, heat probably caused by friction of the rocks grinding past one another within the fault zone can heat the groundwater.

The Icelanders gave us the name geysers for hot springs that periodically eject water with great force. The action takes place in a natural boiler room. Groundwater moves downward into the zone of saturation where the water is heated by rocks in contact

Figure 19.11

Geysers result from superheated water under high pressure. Old Faithful in Yellowstone National Park, Wyoming, is a classic example of a geyser eruption resulting from superheated water. Old Faithful sends water 50 meters into the air for over 4 minutes until the pressure is relieved.

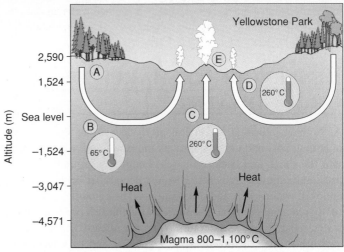

with magmas that have risen into the upper crust. Heat is then slowly conducted to the water-saturated rock. The water temperature can rise well above the normal boiling point because it is under so much pressure. The higher the pressure, the higher is the boiling point temperature. Eventually the water rushes upward, followed by a pressure drop, and the boiling point is reached. At this point the water at the base of the natural conduit turns to steam, blowing water and steam out of the ground.

Old Faithful in Yellowstone National Park, Wyoming, is the most widely known erupting geyser. It erupts in the following manner. First, a quiet flow of water bubbles to the surface, then a few seconds later it subsides. This seems to relieve enough pressure to lower the boiling point of water in the conduit. Then the eruption of steam and water occurs, abruptly sending water 50

meters (165 feet) into the air for nearly 4 minutes. Once the pressure is relieved and the water in the underground reservoir has emptied, the groundwater heats up again to repeat the cycle in about one hour. It is not as faithful as the name suggests, for it may erupt again in 30 minutes or as late as 90 minutes (fig. 19.11).

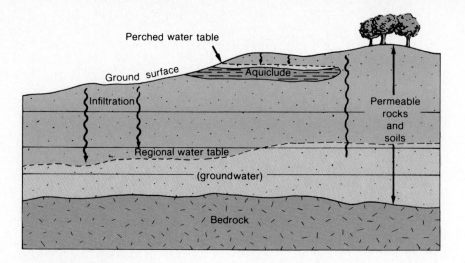

Figure 19.12

A perched water table often results when a spring develops above the main water table. It results from groundwater collection over an impervious rock layer or aquiclude above the main water table.

SPRINGS AND PERCHED WATER TABLES

Springs represent a quiet surfacing of groundwater. As water infiltrates the surface and moves downward, it may reach an impervious layer of rock on the water table. At this point water begins to spread out laterally until it surfaces in an opening known as a **spring.** Springs develop wherever the flow of groundwater is redirected to the surface.

Occasionally when a spring develops above the main water table, the water table is referred to as a **perched water table.** The spring results when groundwater collects over an impermeable rock layer (fig. 19.12).

RESERVOIR RECHARGE OF GROUNDWATER

As we have indicated, the hydrologic cycle is driven by the forces of both gravity and the Sun. Gravity moves water downslope, back into the Earth and back to the sea. Long-term groundwater and precipitation measurements show a close parallel between spring and well water levels and precipitation. The amount of water actually reaching the zone of saturation to replenish or **recharge** the water table varies considerably. We have already suggested the variables of fog drip, precipitation, vegetation, and seasonal change in determining the amount of groundwater recharge. In many locations large reservoirs and settling basins or spreading grounds are used to put water into underground storage. The entire San Fernando and San Gabriel Valleys of southern California are recharged with local runoff and imported water from the Sierra Nevada watershed and Colorado River. However, most southern

Californians drink water coming from wells tapping this underground supply that was imported from the Colorado River and Mono Basin east of the Sierra Nevada via aqueducts. In a sense it is "water in the bank."

CAVERNS AND GROUNDWATER

Get on an elevator and travel 250 meters (800 feet) down underground to an environment that is always cool, moist, and dark. You are in one of 300 known caves in the Carlsbad Caverns group, located on the northeastern slope of the Guadalupe Mountains in southern New Mexico (fig. 19.13). The caves were formed in a reef created by algae and other marine life that secreted a lime ooze during Permian times about 250 million years ago. Following a period of reef development, sediments buried the reef deposits, then over long periods of time **joints** or cracks formed in the fossilized limestone reef, which set the next stage for cave development.

As we go into the dimly lit rooms you can hear water dripping from the ceiling. This is quite a contrast from the hot, dry desert above. Rainfall comes in heavy downpour but in small amounts in this country. Because the ground is very porous limestone, it seeps quickly downward through cracks and pores. As gravity pulls it down to the water table, the water undergoes chemical change to become a weak acid, corroding the limestone. Carbon dioxide in the soil from decaying plants and organisms combines with water to form carbonic acid. This slowly dissolves the rock to create large rooms below the water table in the saturated zone.

In time, mountain-building forces raised the Guadalupe Mountains and the water table lowered, exposing the etched cavities to the atmosphere. At this time the caverns were bare, without decorations. But soon **stalactites** (fig. 19.14) began to hang from

Figure 19.13

The Carlsbad Caverns, New Mexico, were formed in limestone from a Permian coral reef over 600 kilometers (400 miles) long. The next stage produced joints and cracks allowing groundwater to penetrate and etch out the cavern. Water rich in carbon dioxide forms a weak acid that dissolves the limestone. Many geologists believe a stronger acid was needed to form such large caves. From deposits of gypsum (calcium sulfate) came sulfur to form sulfuric acid, a powerful corrosive acid, that may account for these large caves. After uplift and lowering of the water table, the beautiful decorations of stalactites and stalagmites began to form and are still forming.

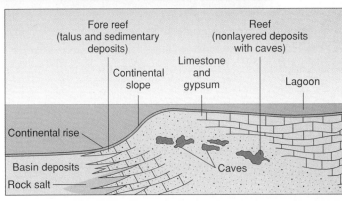

Figure 19.14

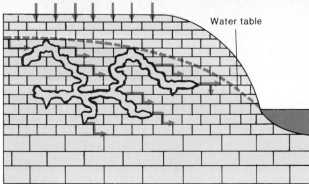 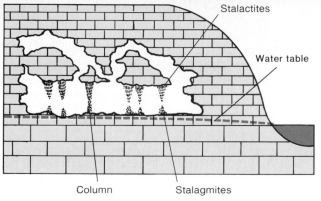

Stalactites and stalagmites form with each drop of mineral-laden water dripping from the ceiling. When they join, a column is formed. Notice in the top center of the photograph a stalactite is growing without regard to gravity, thrusting upward. This upward motion is governed by crystal shapes, impurities, and the force of water under pressure.

Photo by Thomas Clabe Campbell.

the ceiling. These form as groundwater drips slowly, leaving a film of gypsum (calcium sulfate) and calcium carbonate that has precipitated out of solution in the cool cavern environment. The deposits grow with each drop but ever so slowly. The water solution that drips off of a stalactite strikes the floor of the cavern and builds a pedestal or **stalagmite.** When a stalactite and stalagmite meet, a **column** is formed reaching from the floor to the ceiling of the cavern. Over the centuries the cave decorations grow as groundwater seeps down through the zone of aeration. The excess may collect into underground streams and ponds and either exit as a spring or continue downward into the zone of saturation below the water table. Where cavern ponds have evaporated, beautiful calcite crystals have been formed in a variety of geometric patterns.

It is believed that not all limestone caverns have formed in the same manner. Some caves, such as the Mammoth Cave system of Kentucky, are believed to have formed above the water table. The Mammoth Cave network consists of long, winding channels with many signs of underground stream erosion being the primary formation process.

KARST TOPOGRAPHY

Let's leave this cool, humid environment and examine the influence of groundwater on surface features. The word **karst** is a term associated with limestone and dolomite regions where surface features and underground drainage is modified by chemical weathering (fig. 19.15). The word comes from the name of a limestone plateau bordering the Adriatic Sea in Italy and Yugoslavia.

Karst landforms develop primarily where rainfall is more abundant. For example, the entire region of southeast New Mexico is underlain by limestone, but karst features are hard to find. They can only be found in the Pecos River valley. The central Sahara is also underlain by limestone, but there are few signs of karst topography in that arid region where evaporation is high and moisture is limited to scattered oases.

Where rainfall is abundant, limestone quickly erodes as it dissolves into solution. Therefore, some of the best examples of karst landscapes are found in the humid subtropics such as central Florida and wet equatorial regions of the world.

Depressions on the surface or **sinkholes** characterize these regions. Large circular lakes may develop where the water table is near the surface.

World maps of limestone deposits located in the humid climates do not tell the whole story. In some regions the rock is too permeable. The chalk cliffs of Dover, England, soak up water too rapidly for karst features to form. Where dolomite limestone is too dense, water cannot percolate through it to etch and dissolve minerals into solution. The best geologic conditions occur when the rock is highly jointed and has bedding plains that enable water to pass through it. The limestone regions of central Kentucky and Missouri and the limestone belts of the Appalachians all have limestone beds that are highly jointed. Where these conditions exist limestone just below the surface may dissolve as water seeps downward through the joints, and eventually a depression or sink is formed. Sinks also form where the surface collapses into a cavity formed below. In both cases caves are common.

Karst regions are scattered throughout the world. Some lesser known regions include the Nullarbor Plain in coastal southern Australia, a coastal plateau over 650 kilometers (400 miles) long and 140 kilometers (90 miles) wide. The Yucatan Peninsula of Mexico is dotted with sinkholes, known locally as *cenotes.* Throughout southern China and Southeast Asia, on the Malayan Peninsula, Java, Celebes, and the Moluccas there are a variety of karst features (fig. 19.16).

In the rainy tradewind belts of the Caribbean Islands, rounded, asymmetrical hills called pepino hills in Puerto Rico and *mogotes* in Cuba have formed as a result of heavier precipitation and runoff erosion on the windward sides of these limestone islands. Headward stream erosion coupled with chemical weathering have caused the limestone ridges to migrate in the downwind direction, producing asymmetrical hills.

GROUNDWATER AND PEOPLE

The work of groundwater not only has created beautiful landscapes through the sculpturing process, but has also carved niches for human occupancy.

The Pueblo Indians of the Southwest, well known for their house construction, agriculture, and pottery, developed many of their housing sites where groundwater created large caves and sheltered ledges. A good example of cliff dwellings is located in the Walnut Canyon National Monument, Arizona (fig. 19.17). Here, the Sinagua Indians occupied the canyon roughly from 1120 to 1250. They found an ideal house site on sheltered ledges in the Kaibab Limestone. Here the beds are almost horizontal, covering many hundreds of square miles. Most of the ledges are shallow, extending back into the rock no more than 3 or 4 meters (10 to 12 feet). The Indians built masonry walls in front and partitioned them into rooms. Each house had a view of the valley and opposite canyon wall. The outer walls were set back far enough under the ledge to avoid the weather. Many of the original walls are still intact, and all without the help of cement mortar.

Similar cliff dwellings are located throughout northern Arizona, New Mexico, and southwestern Colorado. Mesa Verde National Park near Durango, Colorado, and Tanti National Monument near Globe, Arizona, represent two examples of the importance of groundwater in shaping a dwelling site for the Pueblo Indians of the Southwest.

Water from the spring rains moves down through the porous sandstone, volcanic tuff, and limestone cap rock until it reaches denser limestone at the cave floor level. Gradually groundwater has etched a concave depression into the side of the mountain where the groundwater sits on impervious lower rock. At this point springs emerge, providing an ideal water source in the dry country.

Groundwater action can shape both excellent and bad house sites. The Mesa Verde sites encouraged settlement because of the natural shelter etched by groundwater. However, in many areas of the world, groundwater is the hidden culprit of landslides and cave-ins. Where sedimentary rock is stacked like a tilting deck of cards, the top cards, when well lubricated with groundwater, will slide. That is what happened in the Love Creek slide of 1982

Photo scale 1:127,000

Scale   1:24,000
Contour interval 5 feet
Latitude   27° 52′ 30″ N
               to 27° 55′ to 30″ N
Longitude   81° 33′ W
               to 81° 37′ W

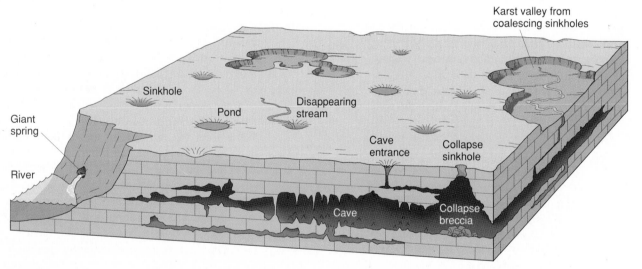

Figure 19.15

Karst topography results from chemical weathering of limestone formations, especially in humid climatic regions, such as Lake Wales in Florida.

Photo by NASA.

Figure 19.16

Limestone pillars at Yangshou Guangxi, China, are karst features created by weathering and erosion of limestone in a humid subtropical monsoon climate.

Photo by Roger Foote. Used by permission.

Figure 19.17

Cliff dwellings in Walnut Canyon National Monument, Arizona, sit in niches that result from gradational weathering of less resistant limestone.

near Santa Cruz, California, discussed in chapter 17. It took several homes resting on landslide-prone rock. Geologists believe that the water from heavy winter rains may have taken four to five months to percolate to a depth where the moisture could saturate and lubricate the gently dipping beds.

Groundwater may also be a problem if you live in a relatively low lying area of poor drainage and a high water table. Homes and buildings may settle, sink, and have seepage. Mexico City is a classic example of a building site in these conditions. When Aztecs came down from the arid north invading and defeating the Toltecs around 1300, they built their new capital city of Tenochtitlan on an island in the middle of Lake Tescoco in the Basin of Mexico. Then, in 1607 a Spanish engineer, Enrico Martinez, partly drained the basin by digging a ditch and tunnel across a low divide to the headwaters of the Panuco River system, which flows to the Gulf of Mexico. As a result of draining the lake, the water table was lowered, resulting in settling of the land. Urban structures began to sink. Buildings lean and sink as the dry lake bed adjusts to the falling water table. Today wells are illegal, and tall buildings must be engineered to float and adjust to the spongy ground.

Arizona also has a sinking land problem from a falling water table. Geologists estimate that 6 million years ago, groundwater started building up beneath the desert floor between the Santa Rosa and Picacho mountains in a vast sand and gravel aquifer. Then about 100 years ago, settlers began to irrigate their crops from this groundwater reservoir. Since 1915, over 35 trillion gallons of water has been pumped from beneath this central Arizona basin.

As a result of this heavy drain on the water table, the land has sunk, in some places as much as 4 meters (12 feet) in the last 40 years. At least 200 fissures have been discovered, ranging up to 14 kilometers (9 miles) in length. Some geologists consider this region to have the world's largest number of fissures resulting from subsidence. Cracks are also beginning to damage roads, buildings, sewage facilities, and irrigation channels. The expansion of Tucson and Phoenix into this region is causing considerable concern to planners for two reasons. First, increased population means a greater draw of water from the basin. Second, expanding urban sprawl raises the potential for more structural damage.

Below the surface, even more dramatic changes are occurring. The top of the water table has dropped as much as 120 meters (400 feet) in some places. As the aquifers dry up, alluvial deposits consolidate into less permeable alluvium.

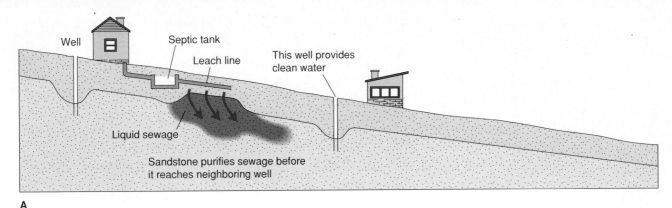

A

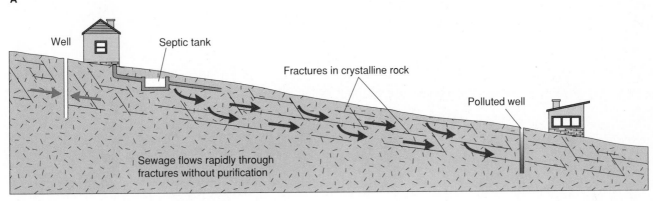

B

Figure 19.18

Toxic chemicals can travel many miles in the groundwater system by aquifers. The septic tank symbolizes the place of intake of toxic substances. Far downstream, a well will pick up polluted water in the aquifer.

Studies continue to monitor new cracks that first appear as hairline fissures, small holes in a series, and slight depressions. Then when summer thundershowers pound the area, runoff causes the cracks to widen and deepen.

Water Quality

Poor water quality is almost always an emblem of the poor nations of the world. Waterborne diseases—cholera, dysentery, typhoid, malaria and others—weaken the vitality of nearly one in seven people in the developing nations. Toxic chemical wastes released by the industrialized nations have created water quality problems, too. Here the problem is not bacterial but chemical in nature. Because most usable water comes from groundwater, quality is of prime concern.

The quality of groundwater is determined primarily by the path water takes as it moves through the hydrologic cycle. As soon as moisture condenses on hygroscopic nuclei in the atmosphere, its quality is affected by these condensation nuclei. On the descent to ground, more qualitative changes occur. Gases such as sulfur dioxide, nitrogen oxides, and carbon dioxide become part of the water's chemistry. Because water is a universal solvent, dissolving a great variety of elements and compounds, it takes on many characteristics of the environment it passes through.

Once the droplet reaches the ground, gravity then continues to pull water downward through the pores and spaces in the 65 rocks. In this section of the hydrologic cycle, water may become *hard*. By this term we mean that calcium, iron, magnesium, and other elements are dissolved in high concentrations. The water may also become either acidic or basic, depending on the ratio of hydrogen to hydroxyl ions in the water.

In the northeastern United States, surface and groundwater located near major industrial centers are becoming acidic resulting from increasing amounts of sulfur dioxide in the atmosphere, which is readily dissolved in the rainwater and introduced into groundwater and soils. The sulfur source is coal used as a fuel. In many areas, levels toxic to plants and aquatic life are being reached.

Human activity influences groundwater quality not only through air pollution, but also through our many activities on the land's surface. It has been well established that nitrate levels are higher in wells near heavily fertilized agricultural land and also dairy farms. The location of industrial and agricultural activities is of paramount importance in determining the quality of groundwater in a region. If toxic elements or bacteria are released in an aquifer, they have the potential of traveling many miles in the aquifer, causing water pollution in the wells tapping the aquifer (fig. 19.18).

Another water quality challenge is developing in irrigated lands, where poor drainage exists. When Alexander the Great's troops moved into the Indus River basin, they came upon an old desert civilization on the banks of the greatest river they had ever seen.

Figure 19.19

The United States-Mexican border is a sharp contrast in soil conditions. The bright red squares are fields of low-salinity irrigation on the California side of the border. Mexico is using higher salinity water resulting in low yields of cotton. Bright red is an indicator of healthy green fields in false-color infrared photography. Currently the United States is developing a desalinization plant to improve water quality for Mexico.

Photo by NASA.

The Indus River and its five tributaries represent a major natural resource. Here, over 40 million persons live on the vast flat plain, and over 25 million make their living from agriculture. The river carries more than twice the flow of the Nile. Half of this water is used for irrigation of over 9.3 million hectares (23 million acres), the largest single irrigated region on Earth. Beneath the northern portion of this region lies a huge reservoir of fresh groundwater equal to 10 times the annual surface runoff to the Indus and its tributaries. In spite of this great resource, poverty is the norm, not well-fed affluent people. Food must be imported just to keep the population alive.

The problem of productivity is both cultural and physical. It is a problem of land, water, and people interacting to produce the current economic situation. Because the land is nearly flat, waterlogging and salt accumulation are slowly destroying the fertility. Nearly 20 percent of the irrigated land is saline. Salinization occurs when groundwater and diverted stream water becomes overloaded with dissolved salts from chemical fertilizers, and high evaporation causes increasing concentrations of salts.

Saline problems are not uncommon in the United States and Mexico. The Imperial Valley of southern California and nearby Mexico face the same poor drainage and waterlogging conditions. Solutions require vast amounts of fresh water to leach the fields and wash the salts from the soil into the water table (fig. 19.19). In these dry lands, fresh water is seldom close by. However, a cure in Pakistan might be possible by drilling fields of large wells to pump groundwater and spread it on the irrigated saline land. This spreading process would lower the water table, reducing waterlogging caused by salt buildup and leaching the sodium salts downward into the deeper groundwater.

Seawater incursion is another major problem where fresh well water is overdrawn. If fresh water is overdrafted, hydraulic pressure from the salt water adjacent to the shoreline can invade the porous strata. When the Aswan Dam began restricting and diverting water from the delta region, the salty Mediterranean began claiming the coastal delta lands and wells became salty. Drought can produce the same effect. During the two-year drought in California in 1976–1977, wells in the delta region of the Sacramento and San Joaquin Rivers became saline in many locations. Fresh water that normally recharged these wells was not available.

Water Quantity

Not only is quality of prime concern, but availability is often a problem in many parts of the Earth. Water is plentiful or abundant in many locations while it is often in short supply where demand is great. More than one-third of all fresh water, for example, is in Canada and the Soviet Union.

Much of the water that falls on the land flows back into the sea overland and in underground systems by transpiration and evaporates into the air. Even agriculture, the principal consumer of water, takes little of the continent's water. Less than 4 percent of the total river flow is used for irrigation.

The problem is one of distribution and conservation. About half of the water provided for irrigation is lost in transpiration, and less than half the water that reaches the fields is utilized by plants.

In a recent year, India sank more than 78,000 new wells. This may relieve the short-term water shortage and help this needy land meet its food balance, but it means an ever-increasing drain on its water resource bank. It is estimated that by the end of the century, India will use all of its available water resources. India is symbolic of rapidly declining water resources in the developing nations of the subtropical and arid lands.

SUMMARY

Groundwater represents the portion of the hydrologic cycle where water moves below the surface. Gravity drives this portion of the cycle, delivering water from the clouds to the ground and back to the sea. As this motion occurs, traces of its subsurface path are left in the rocks. Karst landscapes with depressions or sinks, disappearing streams, springs, aquifers, and caves are all marks of groundwater processes.

The amount of groundwater and runoff in a given region is a function of climate, vegetation, soil, geologic character, and human activity. Runoff can be expressed as a mass balance equation, which states that runoff equals precipitation, plus fog drip, minus evaporation, infiltration, and transpiration. The movement of groundwater is governed by the molecular properties of water and the porosity and permeability of the Earth's crust.

Water quality and quantity vary over the surface of the Earth. Quality is a function of the path the water takes. Groundwater takes on many of the environmental characteristics of its place. Water passing through calcium-rich rock becomes rich in calcium. The same is true of regions rich in silica, iron, and magnesium.

Water is abundant in supply in many locations, yet in short supply for many of the world's thirsty nations. It is a problem more of distribution and conservation. Many of the Earth's thirsty regions are rapidly overdrawing their groundwater supplies to meet the needs of their expanding populations. India and Pakistan are prime examples.

ILLUSTRATED STUDY QUESTIONS

1. On which slope is solar energy pumping the most water back into the atmosphere (fig. 19.1, p. 407)?
2. Explain the role of plants in cycling the water (fig. 19.3, p. 408).
3. Explain the forces that are holding the water to the plant leaf surface (figs. 19.4 and 19.10, pp. 409, 412).
4. What is the force giving water the pressure in this photograph (figs. 19.5 and 19.6, pp. 409, 410)?
5. Which photograph illustrates porosity? Permeability (fig. 19.10, p. 412)?
6. Which of the following types of rocks serve as aquifers (fig. 19.11, p. 413)?
7. What is the probable heat source of the hot springs shown in the photograph (fig. 19.12, p. 414)?
8. Explain the formation process of the stalactites shown in the photograph (fig. 19.14, p. 416).
9. Describe the geologic conditions for sinkhole formation shown in the map view (fig. 19.15, p. 418).
10. Explain the recharge processes that keep this fog belt river flowing during the long, dry season lasting from May through October (fig. 19.18, p. 421).

The Work of Ice in the Hydrologic Cycle

20

Objectives

After completing this chapter you will be able to:

1. Identify and locate glacial landscapes.
2. Explain the processes and causes of glacial formation for **alpine** or **valley** and **continental glaciers.**
3. Explain the mechanisms of glacial movement as a portion of the **hydrologic cycle.**
4. Describe the formation processes of erosional and depositional glacial features.
5. Identify possible direct effects of the last Ice Age on the Earth's surface features as well as changes in sea level, and extinction of species.
6. Describe possible causes of the Ice Age in the light of current knowledge of Earth-Sun relationships and plate tectonic theory.

Glacial highways to the coast driven by gravity, Kennicott Glacier, Alaska.
© Thomas Kitchin/Tom Stack & Assoc.

Ice Ages come and go, leaving their mark on the Earth as the driving force of gravity moves ice in the form of glaciers over the land. As snowfall gradually builds up a deep layer of snowpack, it slowly hardens into ice under its own weight. Again we see *gravitational energy* playing a major role in compaction. Motion of the glacier is also triggered by gravity. When a critical depth of ice is reached, the glacier starts its slide. Sliding occurs when instability of the layer is enough to overcome the force of friction. Gravity not only pulls the glacier down but also produces pressure to melt the ice at subzero conditions, thereby lubricating the moving mass.

The death of a glacier is mainly the result of *solar energy* or insolation. The melting rate is regulated not only by the amount of solar energy striking the surface but the reflectivity or **albedo** of the surface. If the winter snowpack is unusually extensive, the highly reflective expanded snow surface will absorb less energy and reduce the melting and evaporation processes. The energy balance, then, is the key to glacial life and death. Thus a glacier is really a regulated energy system driven by the force of gravity and solar rays.

When glaciers move over the Earth's surface they create a variety of landforms resulting from *erosion, transportation,* and *deposition.* These masses of moving ice have produced some of our finest vacation parklands and wilderness scenery. Glaciers have also deposited on North America and northern Europe great quantities of rock, gravel, and soil in the form of a thin veneer or blanket over the land.

By studying present glaciers and the depositional and erosional features produced by past glacial periods, geologists, glaciologists, and geographers are interpreting these features that today mantle much of the Northern Hemisphere's land surface. Many scientists question whether this Ice Age has reached its end. Perhaps we are still in an interim period between glacial cycles. More glacial and meteorological research are needed to unlock the secrets of the ice.

In this chapter, we will examine the anatomy of glaciers and then describe the variety of landscape features caused by glaciation. Finally, we will examine possible causes of the Ice Age.

GLACIERS AND THE HYDROLOGIC CYCLE

Glaciers, snowfields and ice combined, make up the portion of the **hydrologic cycle** which is in a solid state. Eighty-nine percent lies on the Antarctic continent. The total ice reserve consists of 27 million cubic kilometers (6.5 million cubic miles) of water, which sounds like a lot, until we realize it represents only 1.7 percent of the world's water.

Although glaciers are reservoirs for global fresh water, they are very active, receiving a fresh supply in the form of snow during the winter and releasing water by melting in the summer. Water moves through this cooler hydrologic system, first falling as snow, changing to glacial ice, and then melting.

The oldest glacier, in Greenland, spans only 25,000 years, a rather short period on the geologic time clock. The life expectancy of a glacier depends on how fast the ice melts relative to the replacement by new snow. These rates are influenced by changes in the hydrologic cycle, principally moisture in the snow form, temperature, and evaporation. What will happen to the ocean level if a warming trend develops? How will this affect our shorelines? According to scientists, if all of the glaciers were to melt today, sea level would rise 60 meters (200 feet) and flood many of the world's populated lands, located along the perimeters of the continents.

If we were to move back into another Ice Age as we had 20,000 years ago, Seattle, Philadelphia, New York, Boston, and Chicago would be choking in glacial ice and the shoreline of North America would be much lower than it is today. Glaciers have left scars on the continents where today the land is ice free. Through climatic changes in the hydrologic cycle, glaciers could return to middle-latitude cities of North America and Europe. So, even though we may think of glaciers as distant things because of their northern latitude or high elevation, they are very much a factor that has shaped our local environment and could possibly affect it again.

ANATOMY OF A GLACIER

A **glacier** is a body of ice lying entirely or largely on land, which was formed chiefly by compaction and recrystallization of snow. It also flows or at some time has flowed over the Earth's surface. Gravity is its driving force. There are two fundamental glacial forms: (1) **alpine glaciers,** also called **valley** and **piedmont** glaciers, and (2) **ice sheets** or **continental glaciers** (fig. 20.1).

Alpine glaciers are formed and confined to mountainous terrain. Where the glacier descends from high mountain valleys and merges with two or more valley glaciers at the base of a mountain, a piedmont glacier is formed. A single glacier may also become a piedmont glacier if it spreads out over a plain at the base of a mountain.

Ice sheets cover extensive areas and can move outward in all directions, like a pancake on the griddle. Currently, both Greenland and Antarctica are covered by ice sheets. During the past Ice Age, glaciation left its imprint on some 10 million square kilometers (4 million square miles) of North America (fig. 20.2) and an area half that size in Europe and Asia.

Glaciers may also be classified by the type of motion they exhibit. Valley and piedmont glaciers are considered ice streams, flowing largely under gravity down valleys, and their motion is in one direction highly controlled by local topography. Movement of ice sheets or continental glaciers is produced by compressional forces within the ice mass moving in multiple directions and not highly controlled by surface or underlying topography.

GLACIAL FORMATION

Glaciers form above the snowline where two basic requirements are found, abundant snow and low temperatures. Where are these two climatic conditions most commonly found? Figure 20.2 shows the two principal regions: high latitudes and high elevations, especially near coastal regions where the prevailing winds are on-shore. The two largest ice sheets located in Greenland and Antarctica, and also Alaska's largest glaciers are all found close

A. Mount Rainier Alpine glacier, Washington

Figure 20.1

There are two fundamental glacial forms: (A) valley or alpine
glaciers, and (B) ice sheets or continental glaciers.
B: Photo by Felder, U.S. Geological Survey.

B. Byrd glacier, Antartica

to the oceans on the windward sides. Here moist winds rise up the
mountain slopes, cooling and condensing moisture into snowflakes,
which fall to the ground to begin the transformation process into
glacial ice.

  These beautiful hexagonal ice crystals gradually lose their
geometric form on the ground and grow denser, smaller, and thicker
as the spaces between them fill up. The snow mass becomes granular in texture. The granular snow, called **firn,** begins to change
into ice, a form like a sedimentary rock. When its specific gravity
reaches 0.8, the ice is so compact that it is impermeable to air and
is similar to a metamorphic rock because the crystal grains are
changed and now interlock (fig. 20.3). Pressure due to the weight
of overlying ice continues until the maximum specific gravity of
0.92 is reached. Since the specific gravity of water is 1.00, the ice

Figure 20.2

Glacial formation occurs where two basic requirements are found, abundant snow and low temperatures. These conditions are found in high latitudes and high elevations, especially where storm tracks deliver moisture in the form of snow. When more snow accumulates than melts each season, a glacier will form. The last ice sheet covered the region shown.

floats. The icebergs that break off in Antarctica, along the Alaskan coast, and in the North Atlantic are of glacial origin. Because its density is slightly less than water, only one-tenth of the iceberg is visible above the water line. Water's higher density creates a buoyancy force to float the iceberg.

GLACIAL MOVEMENT AND ECONOMY

Glacial movement is determined by two opposing forces, *gravity* and *friction*. Glaciers move downslope very slowly, but the movement can be observed by placing stakes in the ice and measuring their change in position over a period of weeks or months. Velocities usually run from a few millimeters to a meter per day. A row of stakes placed straight across the ice of a valley glacier will gradually form an arc because ice movement is greatest in the center of the ice where friction is less than the contact surfaces. Ice flows with plastic characteristics near the bottom of the glacier. The upper part of the ice is brittle because it is not under pressure, as brittle as an ice cube dropped on the floor. You can demonstrate to yourself the plastic characteristic of ice that allows it to flow under pressure. Just freeze a long bar of ice (18 by 2 by 1 inches). Support it at both ends and place a weight in the middle of the bar. After a few hours, the bar of ice will be bent but will not spring back to its original form after the weight is removed.

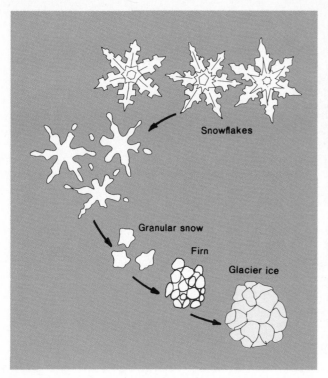

Figure 20.3

Snow passes through a transformation process from snowflakes to a granular snow called firn. Then even more compact ice results from the pressure of burial under more snow. The photograph shows glacial ice at the terminus of a Swiss glacier.

Photo by Tim Wallen.

Whether ice shatters or bends, force is applied, but the rate or speed of the pressure makes the difference in the results. The more pressure there is on the ice occurring over a period of time, the more plastic it becomes. Brittle fracturing occurs only near the surface of the glacier where pressures are lower.

Crevasses, or cracks in glaciers, never extend too deep because ice becomes increasingly plastic with depth. Pressure works like antifreeze. Ice tends to melt at lower temperatures when under higher pressures. Deep in the glacier, melting provides lubrication for movement. The process is called *regelation*. You can see this principle work by taking two ice cubes, placing a weight on one, and watching the rate of melting of each cube. The loaded cube will have a faster rate of melting.

At some point in time as a snowfield builds up, just as in mass wasting, the slope gradient allows gravity-driven flowage to take place. Gravity is the driving force, assisted by the plastic and lubrication properties of ice under pressure.

A glacier is like a bank account that is sometimes balanced and at other times is either overdrawn or increasing. A glacier consists of an input net **accumulation zone** from snowfall, which is separated from the net loss zone. The dividing boundary is the **snow line,** where losses occur by melting and evaporation at a greater rate than can be replenished (Fig. 20.4).

As the weight of the extra snow upstream increases, gravity-driven downslope movement occurs, replenishing the loss downstream. If it is a wet year and snowfall is abundant, the snowline lowers and the downstream zone grows larger. During a dry period the reverse happens. Climate is the chief control: when the climate changes, the economy of the glacier is modified.

It must be pointed out that melting of snow and ice is more than a matter of temperature. The melt rate is a function of the heat balance of the glacier. Heat gains to the ice come from insolation, reradiation from clouds, warm maritime air masses, and latent heat released in the formation of frost or dew. When frost is formed, 680 calories of heat are released per gram of water. Dew formation releases 540 calories per gram.

Heat losses from the ice result from reradiation back into space as well as absorption for sublimation and melting. The energy required for sublimation is 680 calories per gram of ice. Melting of ice amounts to 540 calories per gram. The greater the precipitation as snow during the winter, the less runoff a glacial stream will carry. The key factor is really reflectivity, or albedo. If the winter snowpack is unusually high and persists into the summer, more solar energy is reflected back into space and less is available to melt the ice. Thus streams fed by glaciers are regulated by the energy budget of the ice field.

A

B

Figure 20.4

A glacier has an input zone or zone of accumulation (white ice field) resulting from snowfall and a net loss zone where losses are greater than gains. The two zones meet at the snowline or firn limit. Climate is the chief control of growth or shrinkage and the location of

snowline. These two photographs of a Canadian glacier were taken 23 years apart. Photo A was taken in 1957. In photo B, taken in 1980, the glacier had shrunk as well as retreated and icebergs are missing.

Photos by U.S. Geological Survey.

We have observed during the first half of this century a general global shrinkage of glaciers. Glaciers have receded in the European Alps, Sierra Nevada of California, and the mountains of Alaska. The ice of these glaciers still moves downslope, but the thickness is less and the terminus shifts upstream because melting takes place at a greater rate than downstream movement.

Whether or not glaciers advance or retreat, the rate of motion reflects the environmental conditions on both a short-term and long-term basis. Therefore, glaciers are important indicators of climatic environmental change. In many locations, glaciers have completely disappeared since the last Ice Age, but they have left their mark on the land. Let's examine some of these scars on the higher latitudes and elevations of the continents.

GLACIAL LANDFORMS

Glacial ice at one time covered most of Canada, Alaska, and the northern United States as well as northern Europe and the northern Soviet Union. The higher mountains of the continents also were blanketed by ice. Then the climate changed and a warming trend set in and glaciers began receding, leaving uncovered a landscape quite different from those outside the ice-covered zones of the lower latitudes and elevations (fig. 20.5).

The work of glaciation, like stream and wind action, includes *erosion, transportation,* and *deposition.*

Glacial Erosion

The work of a glacier resembles simultaneously that of a bulldozer, scraper, and conveyor system. It plucks, gouges, and scrapes the surface. In this process the glacier carries away material that has fallen onto it from nearby cliffs, along with the load of rock material it has scoured. The rock material it carries gives the ice scour power necessary to cut, groove, pluck, and polish the surface over which it flows. These materials serve as abrasion tools. You can demonstrate to yourself the abrasive character of an ice-rock mixture. First, take a clean ice cube and rub it across clay or plastic or any easily scratched surface. Then put sand on it and note how the ice can now scratch the surface once it has a cutting tool.

On a grand scale, glaciers are producing a variety of landforms by erosion. After the glacier disappears, these unique topographic features are exposed. Figure 20.6 shows both striations, or scratches and grooves, and a polished surface. The striations were caused by scraping rocks embedded in the ice while the polished surface resulted from finer particles of silt and sand in the base of the glacier acting like sandpaper on a smooth wood finish. Weathering has since removed some of the polished surface.

Roche Moutonnée

While scouring is going on, the ice is plucking fragments from a rock outcropping, creating an asymmetrical form with a cliff face

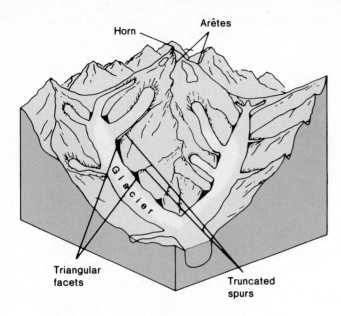

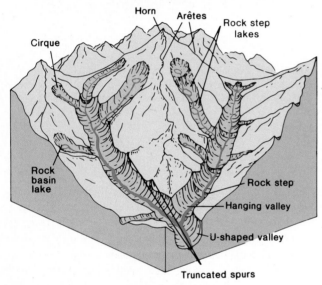

Figure 20.5

Once the glaciers recede from the continents, a new landscape is created with new lakes, streams, valleys, and hills, as well as many deposits of rock and gravel.

on the downstream side and a smooth upstream side, creating an erosional feature known as a **roche moutonnée** (from French, meaning roughly "sheep rock").

Cirques

High in such alpine regions as the European Alps, Rockies, and Sierra Nevada is one of the most characteristic glacial features, the **cirque** (pronounced "sirk"). Figure 20.7 shows a typical cirque basin, a steep-walled amphitheater or half-bowl cut into the head of a glacial valley in the Sierra Nevadas. It has been excavated by glacial plucking in combination with **frost wedging.** When ice forms in the cracks of rocks, the rocks are wedged or pried apart and often fall down the steep walls of the cirque basin.

Figure 20.6

Evidence of scouring includes grooves and scratches in the rocks as well as polished surfaces.

Figure 20.7

Cirques are natural amphitheater basins sculptured by ice through the frost wedging and glacial erosion processes. These cirques hold the Cottonwood Lakes, Sierra Nevada mountains of California.

Figure 20.8

The Lagunas Llanganuco are glacially formed lakes in the Peruvian Andes. Small lakes or tarns arranged in stairstep sequence are known as paternoster lakes.

Photo by Mike Gibson.

This continuous process of frost wedging is also responsible for sharp sawtoothed ridges known as **arêtes** and horn-shaped peaks, similar to the Matterhorn of Switzerland. The **horn** develops as cirques form on all sides of a mountain.

The floors of many cirques are rock basins, while others contain small lakes known as **tarns.** When a series of glacially formed lakes develop in stairstep fashion in a cirque basin, they are referred to as **paternoster lakes** (fig. 20.8). The divides located between the horns are known as **cols.** These serve as mountain passes between cirques, funneling population movement between mountain systems. Major train and highway routes as well as trails are located on cols.

Figure 20.9

Glacial valleys are U-shaped and poorly drained with meadows,
lakes, and meandering streams. Tributaries in hanging valleys are
often left high above the main valley floor. Waterfalls and rapids are
the result, such as Yosemite Falls in Yosemite National Park.

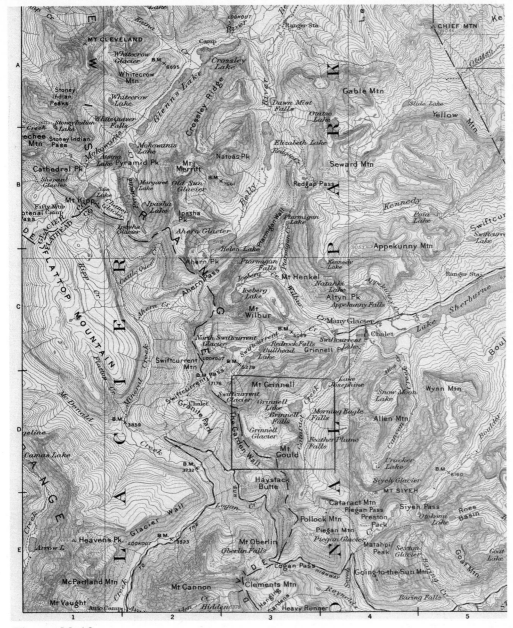

Figure 20.10

Map of Grinnell Glacier, Glacier National Park, Montana.

Scale 1:125,000
Contour interval 100 ft.
Latitude 48° 41′ N
to 48° 57′ N
Longitude 113° 34′ W
to 113° 54′ W

Glacial Valleys and Plains

Glacial valleys are generally troughlike and U-shaped in cross section. Glaciers usually follow the path of least resistance. Therefore, existing stream valleys become the avenues converting the valley from a V-shape to a broad-bottomed, smoothed, **U-shaped valley.** The valley floor or plain is often poorly drained with a series of meadows, lakes, and meandering streams. Smaller tributaries feeding the main valley stream often enter the valley by waterfalls and cascading streams from **hanging valleys.** Glacial erosion produces the main U-shaped trough of the main valley below the floors of the tributaries, leaving them hanging high above the major valley floor (fig. 20.9).

As glaciers move they leave plucked, steep steps instead of a graded slope as streams often do. Plucking makes the risers, and frost wedging and gouging develop the treads or benches of the steps. Figure 20.10 shows cirques, U-shaped valleys, and many other glacial features in Glacier National Park, Montana. Note that Lake Sherburne, a trough lake, fills a U-shaped valley, and each tributary has a cirque.

Locate Grinnell Glacier on the map in figure 20.10 and notice the waterfalls and stairstep lakes that drain into Lake Sherburne. Small tarns are located in a cirque basin with moraines, creating a natural dam separating each lake.

Figure 20.11

Fiords are glacially formed U-shaped valleys that have been drowned by rising sea level. Millford Sound, New Zealand, with a hanging valley and waterfall.

Photo by Michael Stammer, Dept. of Scientific and Industrial Research Vessel, New Zealand.

Numerous alpine glaciers are scattered throughout the world wherever cool and snowy conditions exist most of the year. There are over 1,000 glaciers in the United States outside of Alaska, most located in the Cascade Range of Washington. They grow in regions where snowfall more than offsets melting. They favor high altitudes and latitudes where most precipitation comes as snow. Glaciers can be found in the high South American mountains near the equator in Ecuador and Peru, and also in Tanzania, East Africa, on Kilimanjaro above the 4,500-meter (15,000 foot) level.

Valley glaciers today also cover portions of Alaska, Antarctica, Greenland, New Zealand, eastern North America, northern Europe, central Asia, and the mountain peaks of the tropical island of New Guinea.

Fiords

Where glacial valleys extend to the sea in the coastal regions of Norway, Alaska, Maine, British Columbia, Chile, and New Zealand, the valleys appear to be drowned by the sea, but actually they are deeply scoured glacial troughs known as **fiords** (fig. 20.11). Ice in some fiords may have accumulated to more than 1,000 meters (3,300 feet) below current sea level. Only a massive glacial ice movement could continue to carve at this depth. Somes Sound, Maine, is an example of a fiord formed by an ice sheet of the last Ice Age. Today glaciers are currently forming fiords in southeastern Alaska and Greenland.

Figure 20.12

Glacial deposition occurs downstream in the zone of ablation or wastage to build ridges of rock debris known as moraines. If the deposits are scattered along the lateral valley walls, the deposits are known as lateral moraines. Terminal moraine deposits form at the base of the glacier.

Photos by James F. Seitz, U.S. Geological Survey.

Glacial Transport Depositional Features

As we have mentioned, glaciers resemble simultaneously a bulldozer, scraper, and conveyor system. In the process of erosion the glacier transports large quantities of material, but unlike streams, its load is neither sorted or layered. Huge rocks are moved right along with **rock flour** or fine sand and silt.

As the ice melts in the downstream **zone of wasting,** the rock load is released from the melting ice. The landforms resulting from this glacial deposition give the landscape a unique character. This debris is generally referred to as drift or **till.**

A ridgelike accumulation of till deposited by a glacier along its lower margin is known as a **terminal moraine.** The glacier piles it up in bulldozer fashion. Where glacial deposits are scattered along the lateral valley walls, they are known as **lateral moraines** (fig. 20.12).

Some glacial deposits are streamlined by the sculpture of the moving glacier. When the till resembles elongated parallel hills, they are referred to as **drumlins.** The orientation of the drumlins gives a clue to the direction of the moving ice sheet.

Glacial activity results in poor drainage, creating swamps and ponds. Steep-sided knolls or **kames** are formed where glacial deposits have piled up against or on stagnant ice, as it melts. Kame terraces also form where streams have deposited their load in a

Lateral moraines   Medial moraines   End moraines

Terminus of glacier   Recessional moraine   Ground moraine   Terminal moraine

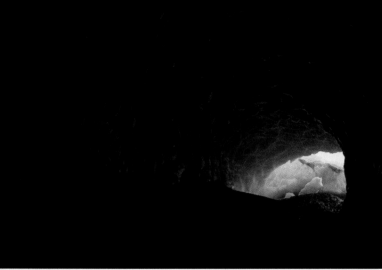

Figure 20.13

Eskers result from deposition of material carried by streams fed by the melting ice. This ice cave was carved by a stream fed by melting ice from Mount Rainer's glaciers. The stream-carried material forms the esker.

glacial valley. If a large iceberg is trapped in the terrace deposits, a **kettle** basin is formed as the ice mold melts in the terrace deposits.

Glaciers not only produce a variety of depositional features, but in the processes of deposition, drainage is often modified. **Eskers,** stream deposits formed by streams under the glacier, leave a landform resembling the stream pattern before the ice melted (figs. 20.13 and 20.14). They consist of depositional ridges of sand and gravel that meander over the glacial flood plain. Figure 20.14 shows glacial depositional features near Jackson, Michigan, including kettles and a terminal moraine. A classic esker is identifiable as Blue Ridge, running in a northeasterly direction across the Michigan Meridian and Highway 127.

Outwash Sediment and Loess Deposits

From the terminal moraine, downstream deposits of gravel, sand, and silt are carried by the streams of meltwater as they flow out of the glacier. The sediments spread out to form an outwash plain. The great **loess** or windblown sediment deposits of the Mississippi and Missouri River valleys were brought into the valleys by winds,

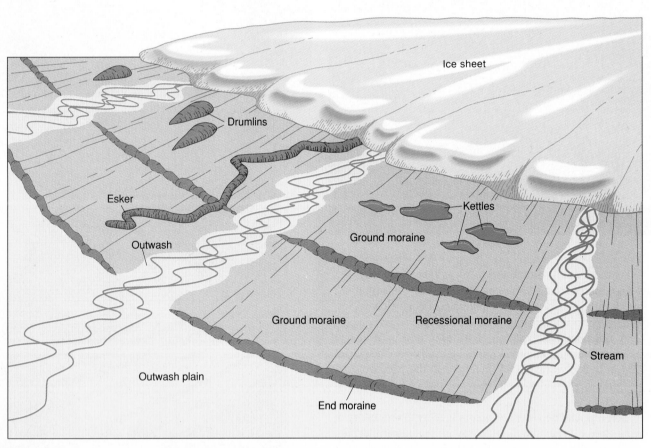

Figure 20.14

Retreating ice sheets left depositional features near Jackson, Michigan, with classic eskers, kettles, and terminal moraines. Blue Ridge is an esker. Many of the small ponds were formed from ice blocks that melted to create cavities or kettles.

Scale   1:62,500
Contour interval   10 feet
Latitude   42° 3′ N
    to 42° 11′ N
Longitude   84° 18′ W
    to 84° 27′ W

Figure 20.15

Outwash plains and loess deposits of the Mississippi and Missouri River valleys form a rich veneer up to 30 meters (100 feet) thick over bedrock.

Photo by L. B. Buck, U.S. Geological Survey.

Figure 20.16

Erratics are commonly referred to as haystack boulders because of their appearance on the Missouri plains. They are geologically and geographically foreign to their present location. When arranged in a fan-shaped pattern or boulder train, their place of origin can be easily traced.

which picked up the silt of the outwash plains of the great ice sheets. These deposits make up some of the best agricultural soils in North America, forming a veneer up to 30 meters (100 feet) thick over large regions (fig. 20.15).

Erratics

At the other end of the particle size spectrum, glaciers carry very large rocks known as haystack boulders. When the rock composition is different from that of the bedrock beneath it, the rock is referred to as an **erratic** (fig. 20.16). Some weigh thousands of tons and serve as proof that the region was glaciated because only continental ice sheets or valley glaciers could import them into the area. Erratics containing copper, found in Missouri, can be traced back to their northern Michigan source several hundred miles to the north. When a group of erratics spreads out in a fan shape downslope from the parent bedrock source, the pattern is called a *boulder train* (fig. 20.16).

ANTARCTICA AND GREENLAND: REMNANTS OF THE LAST ICE AGE

Today most of the planet's ice is locked up in Greenland and Antarctica's ice sheets.

The ice sheet of Greenland covers 1,700,000 square kilometers (670,000 square miles) or seven-eighths of its surface, and reaches an elevation of over 2,700 meters (9,000 feet) above sea level. By making use of seismic wave measurements across the surface, profiles were developed showing that in places the land beneath the ice is below sea level and the ice is over 3 kilometers (10,000 feet) thick. Only the mountainous margins are comparatively barren while the continental ice spills through the deep valleys like white tongues stretching to the sea.

Australia-size Antarctica is far larger, covering over 15 million square kilometers (6 million square miles), with over five-sixths of it covered by ice. This represents 89 percent of the world's ice supply and two-thirds of the fresh water on the planet. Because Antarctica is centered on the South Pole, it is much colder than Greenland. Antarctica's floor is covered by 3,000 to 4,000 meters (10,000–13,000 feet) of ice, which forms a dome with a border fringe of mountains and the great Transantarctic Range dividing the Continent.

Seismic wave measurements reveal ice thickness up to 3,600 meters (12,000 feet) near Vostok Station at the South Geomagnetic Pole. Scientists near Byrd Station have drilled 2 kilometers (1.3 miles) down into the ice to discover its age, at that depth, to be 50,000 years old. They also located ice in Marie Byrd Land to be resting on a surface 2,000 meters (6,500 feet) below sea level producing a thickness of over 4,000 meters (13,000 feet) (fig. 20.17).

Clues to Plate Motion

Not all of Antarctica is covered with ice. Victoria Valley, for example, is as barren and dry as the Sahara. The surface is subdivided by thawing and freezing cracks. About 600 kilometers (400

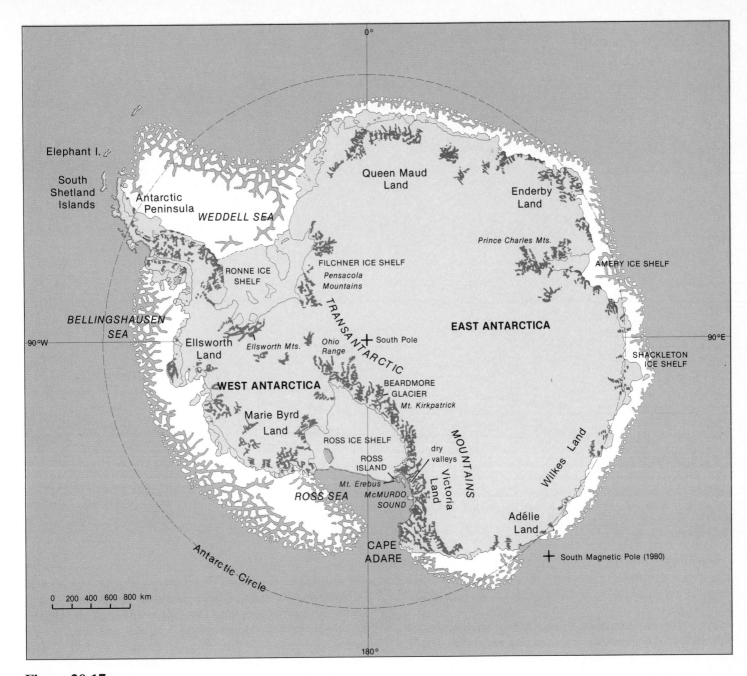

Figure 20.17

The ice caps of West and East Antarctica rest on a surface 2,000 meters (6,500 feet) below sea level in some places, with ice thickness over 4,000 meters (13,000 feet).

miles) to the south a fossilized bit of jawbone of an ancient amphibian gives new evidence that Antarctica was once part of a supercontinent called Gondwanaland. Paleontologists have found similar specimens in Australia and South Africa. They believe that this labyrinthodont, which lived 200 million years ago, could not swim in salt water. The land masses of Antarctica, Africa, and Australia must at one time have split off and slowly drifted apart. The fossil was found only 520 kilometers (325 miles) from the South Pole, near the head of Beardmore Glacier.

The White Desert

The snowfall on the center of Antarctica is not only the whitest and cleanest but the lightest anywhere because of low temperature. Cold air holds less moisture. At Byrd Station, 25 to 50 centimeters (10 to 20 inches) of snow falls each year. The season's total will be squeezed down into 10 to 20 centimeters (4–8 inches) of solid ice. By taking core drilling samples, glaciologists can study climatic change. The upper core sections of ice layers resemble the

Figure 20.18

Ice sheets upon reaching the ocean break up into large and small icebergs. One unique feature of Antarctica is its vast floating ice shelves. Lower Beardmore Glacier, Antarctica, reaches the sea, breaking up and delivering a load of icebergs.

Photo by New Zealand DSIR, Antarctica, December 1977.

rings of a tree, marking the seasons. At deeper levels the ice can be dated using carbon-14, which is trapped in air bubbles in the ice.

One very important feature missing in Greenland but prominent in Antarctica is the vast floating ice shelves. The Ross Ice Shelf covers 5,000 square kilometers (2,000 square miles) and a thickness of 70 meters (230 feet). Ice shelves are fed by glaciers and new snowfall, which compacts into ice in the quieter waters and bays (fig. 20.18).

EFFECTS OF THE LAST ICE AGE
Changes in Sea Level

Along the coasts of North America there are old shorelines standing high and dry at heights ranging from a few feet to more than 150 meters (500 feet) above sea level. As we indicated earlier in the chapter, sea level dropped when ice sheets and alpine glaciers began to grow during the last Ice Age. Probably these old shorelines were formed during one of the past *interglacial* ages, when ice on Greenland and Antarctica was less abundant than today. We cannot be sure this is the case, but it is a good possibility. General uplift could have also produced the same effect, elevating shorelines, but a universal global uplift is difficult to explain. We find elevated shorelines along portions of all of the continents.

The submarine canyons of New York's Hudson River, the sediment-buried canyon of the Sacramento–San Joaquin delta region, and many other rivers flowing to the sea from the continent's interior can be explained by the work of the river when sea level was lower. Some of the world's best harbors are found along both the West and East coast, resulting from a rise in sea level from the melting ice from the last glacial period. The retreat of Hubbard Glacier is contributing to this rise in sea level (fig. 20.19).

The ice of the Pleistocene is believed to have reached a thickness up to 2 kilometers (1.2 miles). The force of gravity created an enormous pressure on the land, causing the crust to sag. It is estimated that ice pressure may have reached 1,100 tonnes per square meter (1,000 tons per square yard). There is evidence that parts of Scandinavia and Canada are slowly rising since this

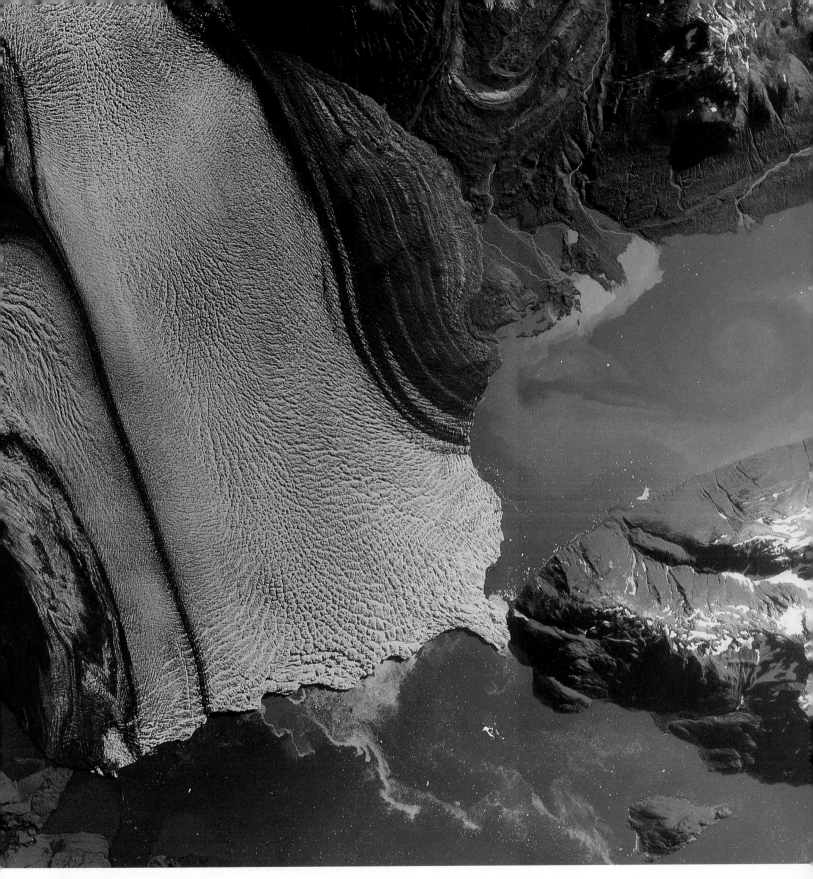

Figure 20.19

Drowned shorelines and raised or elevated marine terraces are clues to the past ice ages. Tectonic activity can produce the same effect. Melting or shrinking ice caps change the volume of the oceans markedly. Over the last nine centuries, Hubbard Glacier has retreated out of Yakutat Bay, Alaska, leaving its bed behind as a fiord.

Photo by James F. Seitz, U.S. Geological Survey.

The Work of Ice in the Hydrologic Cycle **441**

Figure 20.20

Mono Lake, California, is an example of a once much larger glacial lake, which is now being depleted of its water to serve the growing urban development of Los Angeles by the Owens Valley Aqueduct. These tufa towers are formed when calcium-bearing freshwater springs well up through alkaline lake water rich in carbonates that precipitate out as calcium carbonate around the mouth of a spring underwater.

pressure has been released after the ice melted. This may explain why parts of Greenland and Antarctica now lie below sea level under the enormous pressure of their ice sheets.

Today not only is the land rising but sea level also appears to be on the rise, averaging about 1 millimeter per year. This has been attributed to melting of the ice sheets on Antarctica and Greenland. Where uplift occurs at a slower rate than rising sea level, drowned coastlines develop. If uplift is more rapid, marine terraces often develop. These coastal features are discussed in chapter 21.

Pluvial Lakes

During the last glacial age, lakes were more numerous than they are today outside of the glaciated regions. The basin of Great Salt Lake, Utah, was formerly occupied by Lake Bonneville with an estimated volume equaling that of Lake Michigan, covering 50,000 square kilometers (20,000 square miles) and over 300 meters (1,000 feet) deep. Today it is less than 15 meters (50 feet) deep at its maximum depth and covers just over 5,200 square kilometers (2,000 square miles). However, Great Salt Lake is currently increasing again in size.

Mono Lake in California is a much smaller example of a once larger glacially fed lake. Because of a growing southern California demand for water, spring formed tufa towers are now exposed above the current receding lake level. At one time, past shorelines were 270 meters (900 feet) higher (fig. 20.20). The glacial lake bed deposits, Ice Age beaches, and deltas are all there to give their glacial history. These lakes are **pluvial,** meaning they existed under a formerly wetter climate. They were fed by both meltwater and increased precipitation, which accounted for the greatest increase. Cooler weather with lower evaporation helped maintain the level. Many are located outside today's glacial belts. The largest of the late Pleistocene ice marginal lakes in North America was Lake Agassiz, now known as Lake Winnipeg in southern Manitoba, Canada (box 20.1). Lake Agassiz at different times occupied large portions of the Red River Valley of North Dakota, northwest Minnesota, and extensive areas of Manitoba, Saskatchewan, and Ontario. A lacustrine plain (former lake bed) of over 250,000 square kilometers (100,000 square miles) occupies its former site and forms some of the richest chernozem soils of this area of North America and also some of the flattest terrain.

The Great Lakes were at one time much larger than they are today. The lakes lie in former river valleys that were enlarged and deepened by glacial erosion. Advances and retreats of the ice sheets have produced a very complex history still being interpreted by geomorphologists.

In the Great Basin country of California and Nevada, a number of large lakes formed during more humid conditions. Nevada's Pyramid and Winnemucca Lakes near Reno were part of the large irregular pattern of ancient Lahonton Lake. It reached a depth of over 150 meters (500 feet). Mono Lake, California, just east of Yosemite National Park, was also much larger, reaching a depth of over 270 meters (900 feet).

Pluvial lakes were also abundant in North Africa, Australia, Chile, Argentina, Bolivia, and Asia. The Dead Sea of Israel and Jordan extended over a much larger region, and both the Caspian and Aral Seas were much larger during the last Ice Age.

Permafrost

Another significant feature of the last Ice Age is the development of **permafrost,** soil that is frozen and remains frozen below the top layer. When the water in the soil freezes, this creates a impermeable surface reaching depths up to 300 meters (1,000 feet). Today permafrost, dating back to the last ice advance, covers over 20 percent of the land area of North America, Asia, and Europe. During the summer only a thin surface thaws, about 1 meter (3 feet) thick. At one time permafrost extended deep into North America. When the ice sheet began to retreat, this frozen land retreated too. Climate is the primary factor determining its distribution, but soil characteristics and vegetation cover are also important in determining permafrost location.

There are three main types of permafrost: *continuous, discontinuous,* and *sporadic.* In the extreme northern areas, continuous permafrost is present everywhere, reaching a depth of more than 180 meters (600 feet). In Siberia it is 600 meters (2,000 feet) thick in some places. Farther south it is thinner and in some places may vanish. Under these conditions it is referred to as discontinuous. Where only small islands of it exist, it is called sporadic. Sporadic patterns develop in marginal frozen conditions and in regions of rugged terrain where many microclimatic variations exist. Slope exposure, air, drainage, and vegetation cover all affect the sporadic pattern.

Today, because Arctic climates are generally warmer, permafrost is slowly shrinking poleward in pursuit of the ice sheets, and yet because it is buried so deep, permafrost recedes very slowly.

One of the most difficult problems posed by permafrost is constructing buildings and roads in it. These structures cause thawing of the ground, which then induces settling and sliding and cracking. The structure can be its own worst environmental enemy.

THE ICE AGES OF THE PAST

The records in the rocks indicate that extensive glaciation has covered the land again and again throughout the Earth's history. Rocks known as tillite (till converted into solid rock) in strata of many different ages testify to these earlier periods of glacial activity covering much of North America and Europe. Both fossils and tillite, collected in Antarctica and Greenland, indicate that these regions have been frozen for the past 20 million years. Even older *tillite* indicates glaciation has been going on since Precambrian times, more than 600 million years ago.

The last Ice Age ended 10,000 to 15,000 years ago, leaving fresh erosional and depositional imprints on much of North America and Europe. The Laurentide ice sheet of North America centered on Quebec and spread in all directions, reaching south to the Missouri and Ohio Rivers. In Europe, the Scandinavian ice sheet centered on the Baltic Sea. The British Isles had a smaller ice sheet that merged with the Scandinavian ice cap. Ice also spread into central Germany and across the northern European plains from the Alps; others came down from the Rockies, Sierra Nevada, Andes, and Himalayas. All experienced extensive alpine glaciation at the time of ice sheet expansion. Today, only small glaciers exist in the mountains of the world, and ice sheets cover only Greenland and Antarctica.

The Pleistocene Epoch, or Ice Age, began about 2 million years ago, representing four, possibly five advances, or stages. The last, known as the Wisconsin Stage, formed most of the major glacial features of today and erased most of those from the older stages.

ICE AGE THEORIES

No event of prehistoric times has given rise to more speculation than the Pleistocene Ice Age. The only exception might be the theories attempting to explain the origin of life and diversity of living things. The two questions have one thing in common: they cannot be verified by eyewitness accounts. The only testimony we have is recorded in the rocks and glacial ice. Therefore, until we can collect more data and actually observe similar events to test our hypothesis, we are still in the realm of a scientific hypothesis.

Theories began to be proposed in the seventeenth century, and new ones are still being formulated. A recent tabulation lists no less than 54 explanations for causes of the Ice Ages. Newer hypotheses are, on the whole, more reasonable and yet more complicated, because they must account for many facts that were unknown to earlier students. In general, glacial theories are divided into two categories: (1) those based on conditions on or within the Earth are **terrestrial,** and (2) those that appeal to conditions entirely outside and independent of the Earth are **astronomical** in nature.

Lake Agassiz, an Ice Age Remnant

BOX 20.1 The photograph shows the beach ridges of Lake Agassiz about 30 kilometers (20 miles) west of Grand Forks, North Dakota. The view is toward the northwest. This is the Campbell Beach, which is the most continuous of the many beaches surrounding the old lake plain. It can be followed for a distance of about 4,800 kilometers (3,000 miles) along the former shoreline. To the east of the beach (to the right) the surface materials are mainly sandy lake sediment (the heavy lake clays begin still farther to the east and are not bordered by any beach). The beaches have been mined for gravel and sand. They constitute the only source of aggregate in the Red River Valley. Where it has not been disturbed, the Campbell Beach is about 4 to 6 meters (15 to 20 feet) deep and about 90 meters (300 feet) wide.

Figure 1

Beach ridges about 30 kilometers (20 miles) west of Grand Forks, North Dakota.

Photo by John Bluemle, North Dakota Geological Survey. Used by permission.

Consider the following examples of significant hypotheses that could possibly explain an upset in "normal" climatic patterns. First a sample of terrestrial hypotheses:

Terrestrial Hypotheses

1. Plate tectonics and mountain-building processes caused continents to rise in elevation and drift poleward into a colder climatic zone. These movements brought the land into the snowline where glaciers form.
2. The CO_2 content of the atmosphere was reduced. Carbon dioxide absorbs the long-wave radiation given off by the cooling Earth and then reradiates much of this energy back to the surface, preventing it from rapid cooling. A reduction by 50 percent of CO_2 in the atmosphere could cause a drop of 4° C (7° F), enough to bring on a major ice age.
3. The Ice Ages correlate with active periods of volcanism. Great quantities of volcanic dust in the atmosphere could filter and reflect the solar rays back into space.
4. Volcanic dust served as nuclei for the condensation of moisture, thereby encouraging precipitation in the form of snow in the higher latitudes and altitudes.

Here are some examples of astronomical hypotheses.

Astronomical Hypotheses

1. There was an actual variation in solar energy reaching the Earth. A variation in solar energy producing a drop in mean global temperature of only 0.5° C (1° F) can start a rapid buildup of glaciers in the higher latitudes and elevations.
2. Energy reaching the surface can be altered in a variety of ways resulting in an Ice Age. Some have suggested energy interference by a meteoric dust cloud, affecting solar energy in much the same way high cirrus clouds filter the rays of the Sun causing a general cooling trend.
3. As we discussed in chapter 2, the Earth's axis is in a *precession* or wobble, completing one cycle every 26,000 years as it orbits the Sun in its elliptical orbit. Some believe changes in the wobble could cause changes in the amount of solar energy received by the Earth. Currently we are closest to the sun (perihelion) during winter in the Northern Hemisphere. In 12,000 years we will be at aphelion during this same winter season. This variation of distance can cause an insolation variation of more than 5 percent, thus perhaps triggering an Ice Age.
4. Finally, it is suggested that the Sun itself varies in energy output, but there is no firm evidence for this conclusion. It was first supposed that decreased solar radiation led to glaciation. However, today this view is being questioned by many meteorologists and glaciologists studying the relationship between the Earth and Sun. It is now argued by some climatologists that the opposite is true. A warming trend induced by increased solar energy, coupled with mountain-building processes during the late Pliocene and early Pleistocene epochs, could have caused increased evaporation over the land and sea, resulting in more moisture in the atmosphere and greater atmospheric circulation. This increase in moisture content and greater circulation of the atmosphere pumps moist air into the higher latitudes and altitudes above the freezing level, resulting in greater snowfall.

Glacial periods correspond to an expanded polar vortex (polar air circulation) and a vigorous atmospheric circulation with greater temperature gradients between higher and lower latitudes. It has also been noted that solar variation causes shifts in the global winds. Polar easterlies, westerlies, and trade winds shift toward the equator once the Ice Age is in full swing and then shift poleward during interglacial times. These shifts can be accounted for in both major cycles and interglacial variations. When the westerlies shift poleward, greater moisture is pumped into the polar lands to fall as snow, which eventually compacts into ice.

According to the solar variation hypothesis, the Pleistocene Ice Age had four solar maximums. Each ice sheet developed in concert with a rise in solar radiation. Temperatures and solar energy first experienced a gentle decline prior to the onset of increased energy. This was favorable for a cooling of the oceans and formation of a small ice cap in higher elevations and latitude. The ice cap would further cool the atmosphere, producing favorable conditions for the oncoming expanded ice sheet when solar energy and snowfall began to increase. As the ice field grew, more energy was reflected off the ice cap. Increased cloudiness in the polar areas then resulted in further contrasts in temperature between the equator and poles. As solar radiation increased further, the ice caps began to melt and the glaciers started their retreat. Thus, a warm, humid interglacial period followed. This cycle was repeated with a gentle decline in solar energy and temperature followed by increases in solar radiation with more evaporation, condensation, and precipitation worldwide.

Ice Ages and the accompanying glacial ice sheets need one important element: increased moisture in the snow form. Low temperature does cause water to freeze, but unless more moisture is available in the atmosphere an ice cap, glacier, or ice sheet will not grow. Without increased precipitation the land will remain cold and the air stagnant. Solar energy again is the driving force of this cooler hydrologic cycle, lifting water out of the sea, followed by precipitation where gravitation takes over, first by pulling snow to the surface and then by pressing out the snow into ice and moving it downslope or across the plains to sculpture the Earth.

Perhaps the validity of this solar energy model will be revealed in the work of astronomy and glaciology. Astronomy has recently given us a great deal of information about the Sun's variation in energy output. We know that the Sun pulsates on an 11-year cycle resulting in very minor fluctuation in solar energy. Current upper atmospheric studies are attempting to discover the affects of this 11-year cycle on the weather. It appears there are some indirect effects, but to what extent is not clear.

As astronomers survey the universe, they find stars throughout our galaxy and neighboring galaxies similar in structure and composition to ours. Some pulsate rapidly with great bursts

Figure 20.21

Ice cap expansion on Mars is a cyclic event due to variation in solar energy reaching the surface, in turn the result of its orbital pattern. Mars is 4½ million kilometers (3 million miles) farther from the Sun during its Southern Hemisphere winter, when the south polar ice cap extends halfway to the equator. Note the frozen carbon dioxide and water vapor in the shadows of the rocky terrain fractured by frostwedging.

Photo by NASA.

of energy. Others, such as the star Mira in the bright constellation Orion, pulsate over longer periods, on the order of months and years. Perhaps the Sun and other stars have periods extending millions of years. Because recorded human history is so short, we can only speculate that the Earth's surface has been receiving solar energy irregularly throughout its long history but only at specific points in time. Solar variation alone may not account for all of the vari-

able factors. Mountain-building, volcanism, lower carbon dioxide content, and even precessional change in the Earth's orbit synchronized with the Sun's energy fluctuations may be part of the Ice Age model or hypothesis.

Since continents appear to shift great distances over long periods of time, Ice Age features in lower latitudes can be easily accounted for. According to plate tectonic theory, Australia was at one time in a colder, higher latitude near the south geographic pole. Some scientists think that coupling plate movement with general uplift, folding, faulting, and volcanism are essential to the makings of an ice age. Increased volcanic activity produces great quantities of atmospheric dust, causing more reflection of solar energy back into space and hence lowering the temperature on the Earth's surface. As we have mentioned, more condensation nuclei in the volcanic dust form will favor increased precipitation. Volcanism also releases carbon dioxide and water vapor into the atmosphere. Both of these gases are key regulators of the hydrologic cycle.

General uplift, folding, and faulting would bring mountain ranges and highlands into the zone of snowfall above the snowline, causing glaciers to form.

A correct hypothesis of the causes of the Ice Age cannot ignore the following known facts about glacial ages:

1. Fluctuations of climate are global but varying only a few degrees and are very irregular in period.
2. Glaciations began on the highlands and high latitudes where the land is above the snowline and where not only cold but moist conditions exist.
3. Glaciers migrate to lower lands and sometimes cover great areas until stopped by warmer climates or oceans.
4. During the last 2 million years glaciers have reformed in their former locations.

Modern theories become more complex as our information and understanding about the Earth increases. Perhaps expanded knowledge of plate tectonic theory coupled with more information about extraterrestrial Ice Ages on Mars and perhaps other planets will bring a unified view that can account for the known fact about the Pleistocene (fig. 20.21).

SUMMARY

Probably no event of prehistoric times has given rise to more speculation than the Pleistocene Ice Age. We are left with only the record of the rock and current Ice Age conditions located in the alpine mountain regions of the continents and the frozen polar regions of Greenland and Antarctica.

Our focus in this chapter has been on the formation and economics of two fundamental glacial forms: **alpine glaciers** and **continental glaciers.** Glaciers make up the portion of the hydrologic cycle in the frozen state, receiving and releasing water through the cycle. The driving force of this portion of the cycle is gravity. The Earth's gravity first compacts the freshly fallen snow into firn, a granular form of snow. As compaction continues, the specific gravity increases to 0.92, the density of ice. Gravity then moves this newly formed ice downslope at a creeping pace ranging from a few millimeters up to a meter per day. The ice flow is plastic, especially near the bottom of the glacier.

A glacier has an economy resembling a bank account that is sometimes overdrawn during lean years and summer drought periods and then well endowed with more income than outgo during periods of greater precipitation. Glacial periods represent the latter condition.

When ice moves over the continent or down a valley, a set of unique glacial features are produced. They represent erosional scars where ice impregnated with abrasive rock material scours, scrapes, and gouges the Earth's surface.

Glaciation, like the erosional work of water and wind, includes erosional and depositional features. Examples of erosional features are **roches moutonnées, cirques, arêtes, horns, cols, tarns, paternoster lakes, U-shaped valleys,** and **hanging valleys. Fiords** are glacial valleys that have been flooded by rising sea level resulting from the melting of the major ice sheets that covered the northern continents.

Glaciers not only scrape the land but, like streams, deposit vast quantities of rock, sand, and gravel. However, unlike stream deposits, the material—**till**—is neither layered or sorted. The depositional features take the form of ridgelike accumulations of till known as **moraines.** When till takes special forms, specific names such as **drumlins, eskers,** and **kames** are applied.

The past Ice Age produced a variety of effects, including changes in sea level which created drowned coastlines and some of the world's finest harbors. Many of the great basins and pluvial lake beds of the world are products of cooler, more humid times. These basins are located outside glacial regions but can be attributed to the last Ice Age.

Permafrost represents another Ice Age feature. Today permafrost is slowly retreating to the poles as the land continues to thaw.

The geologic record preserved in the rocks indicates that extensive glaciation has covered the land again and again throughout the history of the Earth. Glacial features can be found on all continents dating back more than 600 million years. The Pleistocene Epoch or last Ice Age began about 2 million years ago, representing perhaps four or five advances. Current theories can be divided into two categories: (1) **astronomical**—those that appeal to conditions entirely outside and independent of the Earth, and (2) **terrestrial**—those based on conditions on or within the Earth.

ILLUSTRATED STUDY QUESTIONS

1. Explain the connection this fiord may have with glacial periods of the past (fig. 20.19, p. 441).
2. The drowned channel in the photo continues offshore. Explain the possible effects of the Ice Age on this channel (fig. 20.11, p. 434).
3. Identify the glacial features shown in this illustration and indicate how they are formed (fig. 20.14, p. 436).
4. What process formed this ice cave (fig. 20.13, p. 436)?
5. Explain the correct process for each feature shown in the illustration (fig. 20.12), p. 434).
6. Identify three erosional features on this map (fig. 20.10, p. 433).
7. Explain how Yosemite Falls was formed (fig. 20.9, p. 432).
8. Describe how paternoster lakes are formed (fig. 20.8, p. 431).
9. Explain the role of ice in the development of a cirque basin (fig. 20.7, p. 430).
10. What evidence has Lake Agassiz left behind that identifies it as an Ice Age remnant (box 20.1, fig.1, p. 444)?
11. Contrast terrestrial and astronomical theories of the last Ice Age. Which view does Mars tend to support (fig. 20.21, p. 446)?

The Work of Waves

21

Objectives

After completing this chapter, you will be able to:

1. Explain the interrelationships between *wind* and *waves*.
2. Describe the processes of **wave refraction** along the shoreline.
3. Describe longshore currents, their role in moving sand along the coast, and the effects of dams and construction on the sand supply.
4. Explain the effects of tides on **estuaries, harbors,** and **beaches,** and the impact of people on these features.
5. Explain the causes of a **tsunami** and its potential destructive effects.
6. Describe the processes of **wave erosion** and **deposition** and the features produced.
7. Cite two processes responsible for changes in sea level.
8. Identify coastal features experiencing changes in sea level.

Wave energy from the wind, Cape Kiwanda, Oregon.
© David Muench Photography

Figure 21.1

Change is constant along the shorelines of the world as wave energy erodes headlands and deposits fresh sand along the quieter waters. This headland is retreating. Bedding planes exposed at low tide can be matched to those exposed in the eroding cliff.

As you watch the restless waves build and crest, then break on a rocky shoreline or sandy beach, you are witnessing another example of nature's energy in the form of wind-driven waves and currents reshaping this important environmental boundary. The shape of this shoreline boundary and the nature of the materials being deposited and sculptured by the wind and waves all affect human activities. Our attention in this chapter will focus on the processes shaping coastal landforms and the features being produced by the energy systems of wind and waves. We will also consider human impact on these systems.

Along the coastlines of the continents and islands of the world, change is constant (fig. 21.1). Beaches are the most readily modified coastal features: easily moved by wind and waves, changing from season to season. Some beaches have disappeared completely within the short span of a storm season, while others have grown in size.

Many shorelines consist of steep cliffs, while others may be drowned tidal flats and estuarian lagoons. Let's examine the work of waves and currents as they continually create or modify our shoreline habitat.

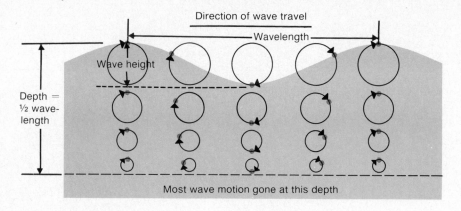

Figure 21.2

Wind is the driving force of waves. Waves travel as a series of looplike oscillatory motions. This orbital motion of water in the waves weakens with depth and is gone at over half the wavelength.

The shoreline boundary is three-dimensional, with interfacing surfaces of the *atmosphere, hydrosphere, biosphere,* and *lithosphere.* At this boundary, solar energy is energizing two fluids, the atmosphere and the ocean. Through differences in heating, both the air and water stir to create large convectional motions. As the air moves over the ocean surface, friction exerts a force on the water to set it in motion. It can be compared to two gears engaging, with the wind driving the water to produce global ocean currents that reflect the paths of planetary wind patterns. Remember that all of this wind and subsequent ocean movement originates in the sun.

WAVE ENERGY FROM THE WIND

Most waves receive their energy from the wind. Perhaps you have observed that on a windy day at the beach, the waves are larger and hit the beach with greater force. Bays are usually much quieter because they are protected behind the **headlands** extending seaward on either side of the bay.

When the wind blows across the surface of the sea, friction between air and water results in a looplike motion of water particles in a wave of oscillatory motion. The depth of the oscillation equals exactly one-half the wavelength (fig. 21.2). The distance between the crest of one wave and the crest of another is the wavelength. Because the driving force of air friction is at the surface, there is a rapid decrease of energy downward. Thus, smaller and smaller loops develop below the largest surface loop until no motion occurs below a depth of one-half the wavelength. Wave action, then, is a surface event. This is why you can dive under a large breaker and feel only minor force from it. This also means that waves do little erosional work below the depth of one-half the wavelength.

As the wave approaches land, we must also remember that the motion of the wave is very different from the actual water particle motion. Wind blowing across a field can give the same optical impression of wave motion. The tall grass bends forward and then back, but the eye sees only a forward wavelike motion. Water also moves forward and back, but we see only the advancing wave. This is an illusion of forward water motion.

When a wave is affected by shallow water, the oscillation loop is disturbed. At the depth of one-half the wavelength, the wave feels bottom, or bottoms out, and the wave's oscillatory motion and form are altered. The loop form begins to change into an elliptical shape, leaning forward as the bottom drags, and the wave begins to increase in height as interference with the bottom increases. Friction on the bottom not only heightens, but also shortens the wavelength, because the water piles up as the wave slows to move onshore. On a windy day the wave can double in size and greatly increase in steepness, especially on the front side. The wave breaks at the point when the steep front can no longer support itself. The rear portion of the wave overrides the front, and the wave breaks.

Surf

A surfer is home free with a good ride if he or she can catch the wave just as it is in the breaking stage and can move inside the curl of the unstable breaking wave (fig. 21.3). At this point, wave energy is assisted by gravity, which comes into the picture, pulling the water down on the beach. The water is now turbulent and muddy, churning and mixing with **backwash,** or the returning water. This turbulent water between the breakers and the beach is known as the surf. The water moves forward until all forward energy is spent. Because its momentum carries it high on the shore, gravity carries it back to the sea both in narrow channels eroded by the returning water and in broad sheets. When the surf is very turbulent, the returning water can produce very strong currents know as **rips** or **undertow** (fig. 21.4). This returning water transports silt, sand, and gravel on the beach back into the sea.

The energy of waves in the surf zone is quickly consumed by turbulence, friction on the floor, and in transporting rock materials. Sediments constantly migrate back and forth between the shore and sea. The finest particles are carried into deeper water before they settle out, at depths greater than one-half the wavelength.

Figure 21.3

When a wave feels bottom, its progress slows, its height grows, and its wavelength shortens. The rear portion of the wave rises up and overtakes the front portion, then the wave breaks and spills on the beach.

Photo by Sheryl Wallen.

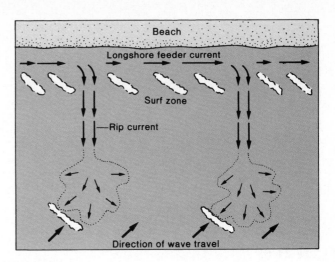

Figure 21.4

Undertows, or riptides, driven by gravity, are strong seaward currents from the backwash of the sloping shorelines.

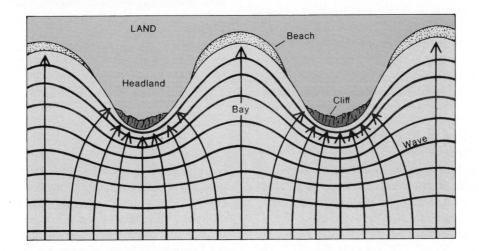

Figure 21.5

Wave refraction concentrates wave energy on headlands and other seaward extensions of the coastline. A wave always bends to approach the shore nearly perpendicular to it.

As you walk along a rocky coast or sandy beach, you can determine for yourself the depth at which surf is eroding. The key is in the height of the breaking wave. Most waves break in depths ranging between their wave height and 1.5 times their height. If the wave is 4 meters (13 feet) high, then the depth of erosion should not be more than 1.5 times that height, or 6 meters (20 feet). Thus, the surf zone is a narrow, vertical zone which fluctuates with local tides and the heights of waves. Since waves are rarely over 6 meters in height, erosion is limited to about 9 meters below mean sea level. The surf cuts like a knife edge in a horizontal plane, producing a wave-cut bench in the tidal zone.

Wave Refraction

A wave is a carrier of energy and cannot disappear until all of its energy is spent or removed. Some energy is consumed by oscillations and friction of the water, but most is lost when the wave encounters an obstruction such as the shallow ocean floor. When this happens along a coastline, the wave slows, crests, and refracts (bends) to compensate for loss of energy in the shallow water. An obstacle, such as a headland, causes refraction of waves moving onshore, concentrating wave energy (fig. 21.5). **Refraction** causes waves to line up to the portion of coast they are approaching, focusing wave energy on the headland. Energy is dissipated, instead

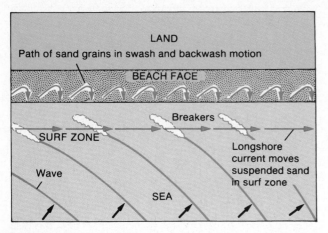

Figure 21.6

Longshore currents carry a river of sand parallel to the shoreline inside the breaker zone when waves strike the beach at an angle other than perpendicular. The beach also moves in the same direction. This is referred to as beach drift. Swash and backwash motion move the sand in and out with each wave, but the net motion is along the beach.

of concentrated, in bays because the energy is refracted away from them. Refraction focuses wave energy on the obstruction in much the same way a magnifying glass focuses refracted light. Accordingly, headlands erode more rapidly and bays become depositional zones. For this reason, over long periods of time coastlines become smoother and less irregular as the headlands are eroded and the material is deposited into the bays in the form of sand and gravel.

LONGSHORE CURRENTS

Have you ever been swimming in the surf and suddenly discovered you were swimming several hundred yards down the beach? Sand is being carried in the same manner by **longshore currents,** or currents parallel to the coast inside the surf zone (fig. 21.6).

When waves approach the beach, refraction causes the waves to line up more parallel to the shoreline as they reach the breaker stage. However, because wave approach is seldom parallel to the beach, water and sediment move onto the beach in a motion called **swash.** Once all forward energy is spent, gravity pulls water and sand back into the surf zone in a **backwash** motion. This same motion is repeated continuously as swash deposits sand and backwash erodes it. The angle of approach of the advancing wave determines the direction of the longshore current along the beach, transporting sand and pebbles in the surf in a zig-zag movement down the beach known as **beach drift.**

The general, prevailing direction of migrating sediment along the coast is determined by the prevailing wave angle approaching the beach. On both the east and west coasts of North America, longshore currents and beach drift carry material southward. Glacial sand from Maine, for example, eventually reaches the coast of Virginia.

Where losses due to wave action are greater than deposition, beaches begin to disappear. Along both the East and West

Figure 21.7

The nature of the source of sand determines not only the quantity but the quality of sand. The source may be hundreds of miles inland or along the coast from its point of deposition. The beach is never static. Once beach sand is deposited, wind or waves or both will pick it up and move it to a new location. Small waves in this protected bay deposit sand in the quiet water.

coast, dam construction for hydroelectric power and flood control purposes is not only holding up the flow of water in rivers, but also cutting off sand supplies for our beaches. Streams transport the sediments that make up our sandy coasts. If you examine closely the sediment of the beach, you will find not only pebbles and sand from the local cliffs and marshes, but material originating perhaps hundreds of miles inland. Beach sand begins its journey in the headwaters of rivers and travels to the sea by the conveyer system of water (fig. 21.7).

Once streams carry sediments to the ocean, longshore currents and beach drift, coupled with small waves, build beaches and bars along the coast. If the sand supply is reduced or altered, beaches can decline from wave action and beach drift.

The beaches of Israel, southern California, and the resort Black Sea coast of the Ukraine are currently losing sand from their beaches more rapidly than the replacement processes. This loss is caused by the large-scale inland dam construction of the 1940s and 1950s for flood control. Recent harbor and resort development also has removed considerable quantities of sand from beach areas, causing interference with natural processes. Longshore movement is altered by manmade harbors where breakwaters extend into the sea. Breakwaters encourage deposition by removing the energy from the system. When the waves and surf become underloaded

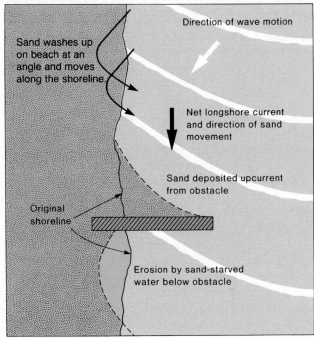

Figure 21.8

Longshore currents are altered by breakwaters, piers, and port development. When this happens, deposition results in the quiet water zones, and erosion is enhanced where the sediment load has been reduced and wave action is unaltered. Note the sand deposit upcurrent from the groin along the shoreline near Lima, Peru.

(Diagram labels: Direction of wave motion; Sand washes up on beach at an angle and moves along the shoreline; Net longshore current and direction of sand movement; Sand deposited upcurrent from obstacle; Original shoreline; Erosion by sand-starved water below obstacle)

following deposition, the wave compensates by eroding sand from along the beach until it becomes loaded again (fig. 21.8). Thus docks, dams, and sea walls all can interfere with the movement of sand along the coast and can alter the beach in a permanent way.

TIDES—GRAVITY AT WORK

The dynamic nature of the shoreline is governed not only by winds and waves, but also by celestial gravitational forces. The gravity of the Moon and Sun continually pulls on the Earth, causing the tides (fig. 21.9). In most places, twice each day, lunar and solar gravity moves water onshore, creating currents that keep estuaries moving and open like a natural dredge. Then, when a high tide is reached, the waters move back to the sea, carrying wastes, sediments, nutrients, and oxygen to the marine world.

When the tides come in, not only do estuaries fill, and plants and mud flats disappear under the water, but waves break higher on the beach, eroding different parts of the coast. Thus, changing tides allow waves to work at different levels on the shore.

Not only are shoreline features influenced by tides, but the features of the shoreline determine the pattern of the tide. An island far from shore generally has two high and low tides in a 24-hour period. Other places, where the land constricts the flow, may have only one major high and low tide. Where the water is funneled into a long, narrow channel, tides may peak once daily with

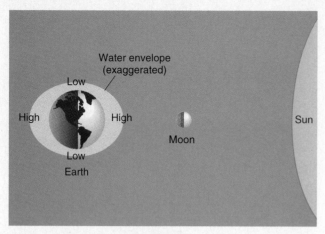

Spring Tide

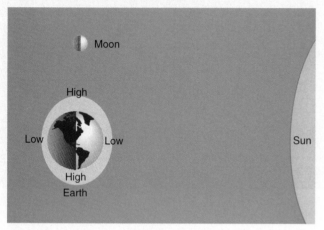

Neap Tide

Figure 21.9

Tides are caused primarily by the gravitational pull of the Moon and, secondarily, the Sun. Most places have two high tides twelve hours apart and two low tides falling in between, as each spot on Earth rotates through the changing gravitational fields due to the three bodies. The highest tides of the month, called **spring tides,** occur when the Earth, Sun, and Moon are in a straight line and the forces are complementing one another. This occurs at full and new moon phases. The most moderate tides, known as **neap tides,** occur when the Sun and Moon are at right angles, thus partially canceling each other's influence. Neap tides are associated with first and third quarter lunar phases.

ranges of over 12 meters (40 feet). The Bay of Fundy in eastern Canada is an example of a long channel which reinforces tidal movement, producing some tides of over 15 meters (50 feet).

The height of tides varies from day to day along all coasts. The highest and lowest tides of the month occur when the Earth, Moon, and Sun are aligned, at new and full moon. Alignment causes their gravity to work in the same direction. Tides during this period are called **spring tides.** When it happens, flooding is not uncommon in low-lying areas, and boats can be stranded on dry land if they don't move into deeper water as the tide recedes. Al-

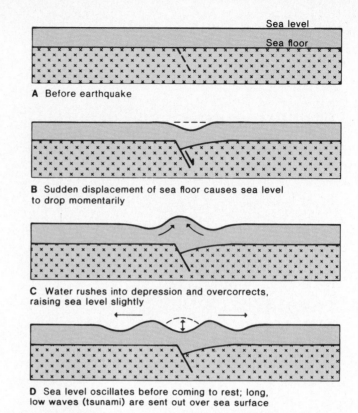

A Before earthquake

B Sudden displacement of sea floor causes sea level to drop momentarily

C Water rushes into depression and overcorrects, raising sea level slightly

D Sea level oscillates before coming to rest; long, low waves (tsunami) are sent out over sea surface

Figure 21.10

The tsunami or seismic wave is generated by seismic energy. Tsunamis originate at the epicenter and travel over 720 kilometers per hour (450 miles per hour). In deep water even the largest go almost unnoticed, but as they touch bottom near land, friction can cause them to rise to terrifying height. Whole coastal towns have been known to be wiped out without warning. Today, worldwide alert networks provide some protection against loss of life.

though this alignment is repeated monthly, in late December and January spring tides reach an annual maximum daily range because the Earth is at perihelion in its orbit. The Sun, being 4.8 million kilometers (3 million miles) closer, exerts a greater gravitational tug on the Earth's surface.

Lowest high tides, or **neap tides,** occur when the Moon, Earth, and Sun are at a right angle, creating a weaker net gravitational effect. This occurs at first and third quarters of the Moon.

Increased runoff can also raise the tidal level, especially if it is coupled with strong onshore wave action. Many harbors have been badly damaged when high tides, enhanced by high ocean waves, are coupled with flood stage of a river. The river water becomes impounded as it approaches oncoming waves at the mouth of the bay. The tide rises, and water backs up in the bay from the swollen river and incoming salt water. Back in 1900, Galveston, Texas, experienced 4.6-meter (15-foot) high tide due partly to a hurricane in the Gulf of Mexico. Gale winds pushed the sea toward shore, compounding the tidal effect. The sea wall completely failed, flooding the city and drowning nearly 6,000 people.

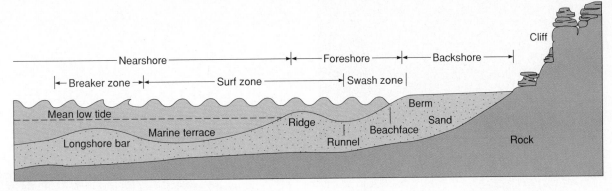

Figure 21.11

A profile of a typical shoreline reveals a wave-built sand berm, a marine terrace offshore, and rocky sea cliff at the landward limit of wave action.

TSUNAMIS, SEISMIC SEA WAVES

The most destructive waves of all are caused not by tides or winds, but by earthquakes under the sea. These waves have long been called, incorrectly by the general public, "tidal waves." Scientists have adopted a Japanese word **tsunami,** meaning "harbor wave." These are seismic in origin (fig. 21.10). Tsunamis begin at the epicenter of an earthquake and travel at speeds in excess of 720 kilometers per hour (450 mph). As with wind-generated waves, the water itself does not advance. Crests of individual waves may be 15 minutes apart and may not be noticed by a ship as they pass beneath it, because in deep water they maintain a very low profile. However, when they reach shallow water, tsunamis suddenly rise to great heights and destroy everything in their path along the shoreline. The shoreline profile helps determine the height. On a low-lying shore they have been known to rise to 15 meters (50 feet), the height of a five-story building, and to more than 30 meters (100 feet) at the head of a narrow harbor or inlet where the water piles up with no place to go. These waves build and break according to the same principles described earlier in this chapter.

Because the Pacific Ocean is ringed by major earthquake belts, it has more tsunamis than any other water body. Japan has had 15 major ones since records have been kept, dating back to 1496. One in 1896 killed over 27,000 people.

When Krakatoa, in the East Indies, erupted in 1883, it generated a tsunami that killed 36,380 people and destroyed many lowland villages on island after island as the waves rolled by. Not only did its smoke and ash travel around the world, but its tsunami also left its mark on tidal gauges as far away as the English Channel. The Alaskan earthquake of March 27, 1964, caused a major tsunami along the northern California coast at Crescent City, where it caused flooding and loss of life and property.

PROFILE OF A SHORELINE

If we could drain all the water away from a shoreline to examine the profile of an undisturbed zone, we would see a series of ero-

sional and depositional features produced by wave and tidal action. Figure 21.11 shows a cross section from the shore, the landward limit of effective wave action, to the **marine-built terrace** offshore.

If we assume that our profile is of a recently submerged coast of gentle slopes, wave action under these conditions cuts a notch that evolves into a sea cliff. This is the start of a **wave-cut bench** or abrasion platform. The material eroded by the waves is deposited seaward as a wave-built terrace.

The **wave-built terrace** consists of materials removed from the wave-cut bench. Together they make up the marine-built terrace. At the shoreline is the beach and rocky sea cliff. This is the upper landward limit of wave action. This beach, then, is a depositional feature upon a wave-cut bench and is not permanent because the materials are in constant transport. A wave-cut bench can end suddenly or grade gently into a sloping platform.

Given enough time, erosion widens the wave-cut bench and cuts deeper into the cliff. As waves continue cutting inland, they spend energy as they cross wider expanses of shallow water, eroding the wave-cut bench. When the wave energy has created a concave slope just steep enough to move eroded material seaward, the shoreline is said to have a profile of equilibrium. At this point beaches form. Waves no longer can remove sandy beach material as efficiently, so the beach is considered part of a mature stage. The sandy beach widens during the times of less wave energy, but may be destroyed when a major storm hits. Large waves erode beaches, while small waves build them.

Frictional Energy Shapes the Shore

Wave-cut sea cliffs and wave-cut benches and marine terraces are the two most abundant erosional features produced by the work of waves.

Wave-Cut Cliff

When the waves attack a cliff, the surf cuts most actively at the base of the cliff, producing a horizontal, sawlike action on the cliff. This causes undermining, resulting in mass wasting of the cliff. If the cliff does not erode easily, a notch forms at the base. The notch may develop first into sea caves, the **sea arches,** and tunnels. Where a sea arch has collapsed, offshore **stacks** remain in its place (fig. 21.12).

Figure 21.12

Erosional features include wave-cut cliffs, sea caves, arches, and stacks. These stacks are located along the southern coast of Victoria, Australia.

Photo by Holly Brackmann. Used by permission.

Wave-Cut Bench and Marine Terrace

The wave-cut bench is a product of wave and surf activity. It slopes gently to the sea and is partly covered by sediments from cliff erosion. At low tide, some benches are entirely exposed. If the coast has been uplifted, a series of marine terraces, often resembling giant stairsteps, will be exposed higher upon the land as described in chapter 11. Each step, or marine terrace, represents a former wave-cut bench with its associated cliff (fig. 21.13). The highest terrace represents the earliest marine erosion. Vegetation and soils may be altered gradually by time on each terrace, so a series of terraces represents a sequential change in the environment, starting with a marine environment at the shoreline and proceeding to a stable climax plant community on the highest terrace.

Gravitational Energy Makes Deposits

Beaches

The most frequently changeable depositional features along the coast are beaches. They form in zones of more abundant sediment and in bays where quiet water is protected from winds.

Figure 21.13

A wave-cut bench is the product of wave and surf action. Changes in sea level can expose a series of marine terraces, cut by the eroding surf. At one time each step was at sea level and spaced 100,000 years apart.

Photo by Carol Printice, U.S. Geological Survey.

We can define a beach as a wave-washed zone of sediment along a coast, extending through the surf zone. The material can range in size from small rocks and pebbles to fine sand and silt. In this zone, sediment or beach material is on the move. The sediment material making up the beach comes partly from wave erosion of the nearby cliffs, but mostly from alluvial material washed into the sea from streams and rivers. The beach does not really end at the low tide level, but the sediment forms a carpet on the wave-cut bench and seaward from the breakers into quieter, deeper water. If the beach is defined in terms of sediment that can be transported onshore by waves, beaches extend from the surf zone to a depth of approximately 10 meters (33 feet). A depth greater than that is generally beyond the driving force of the wave.

Beaches are always on the move. Figure 21.14 shows a California beach in the winter and the summer at the same location. The larger waves of winter are destructive erosional agents to the beach. Smaller summer waves push sand upon the beach, reconstructing it each year.

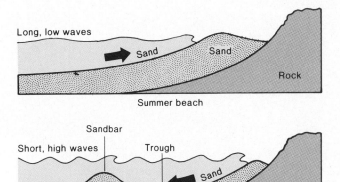

Beach sediment is never static. Energy from the waves moves sediment onto and along the beach, forming a variety of depositional features. By tracing the movement of sediment along the coast by longshore transport, one can observe depositional features growing in the quiet, protected waters of bays and coves. Here in the quiet water, a **spit,** or sand bar, projects from the coast where the shoreline curves abruptly. It is being fed by longshore currents. The northern tip of Cape Cod, Massachusetts, is a classic example (box 21.1).

When a spit completely closes a bay, it is referred to as a **bay-mouth bar** (fig. 21.15). Strong tidal or river currents can prevent spits from becoming a bay-mouth bar. Sometimes a barrier island will develop as tidal and river currents erode gaps in the long, narrow beach deposits. Coney Island, New York, and Atlantic City, New Jersey, are both located on barrier islands. Other extensive barrier islands include the Cape Hatteras chain in North Carolina and Padre Island, Texas, which is 130 kilometers (80 miles) long. It is an oversimplification to suggest that all barrier islands have formed in the same manner. Some are thought to be the result of submergence of the coastline as sea level rose after the last Ice Age. This submergence drowns valleys and leaves exposed small barrier reefs and islands, which trap sediments in the backwaters behind them.

Figure 21.14

Beaches are temporary features, especially the individual sand particles. Streams move sand down to the sea, then longshore currents transport the sediment to quiet water zones and deposit it as beach material. Later, storm waves may move the sand again to a new location. This California beach scene was photographed in January and again in July. Winter storms produce larger waves that erode the beach. Smaller summer waves deposit more than they erode.

Cape Cod from the Sea

BOX 21.1

When the last major glacial ice sheet began retreating from the North American continent, sea level was about 120 meters (400 feet) lower along the Cape Cod coastline. The continental shelf was exposed and served as a rich habitat for the mastodon, mammoth, and other now-extinct mammals of the Ice Age period.

As the ice melted, sea level rose first very rapidly, about 30 centimeters (1 foot) per 100 years, and slowed to one-third that rate 2,000 years ago. At each successive level, wave action eroded the disappearing shelf area. The eroded material was not just washed into the ocean, but was transported by longshore currents and redeposited by waves on the beach to form a growing sand bar of new land.

The largest spits around Cape Cod are located at Provincetown, Monomoy Island, and Sandy Neck (fig. 1). There are also numerous small bars and spits along the Cape Cod shoreline, serving as testimonials to the work of waves and currents, followed by wind-deposited dunes. The Sandy Neck spit grew as longshore currents carried sediment and deposited their load at the outer end. Sometimes, as in this case, the spit curved to form a hook and produced a protected quiet water zone where intertidal marshes could develop. In the last 3,000 years the Sandy Neck spit at Barnstable has grown from a small spit a little over 1.5 kilometers (1 mile) long to its current length of 10 kilometers (6 miles) and a large lagoon. As long as there is a source of sediment, the spit will continue to grow.

Where the work of waves stops, wind picks up the load and moves the sand inland to build coastal dunes reaching 30 meters (100 feet) above sea level.

Eventually the work of waves will take its toll on the cape, and the land will be reduced to not much more than a few sandy dune deposits surrounded by a rocky shoal. This will happen when the sand supply is reduced by dams across streams or by gravel extraction. Currently the signs are pointing toward this fate. On the ocean-facing cliffs of lower Cape Cod, about 2 hectares (5 acres) per year is lost to the waves. At the tip of the spit, less than 1 hectare (about 2 acres) per year is being added. Putting it simply, for every acre lost by erosion, only half an acre is replaced by spit development.

Source: US Geological Survey

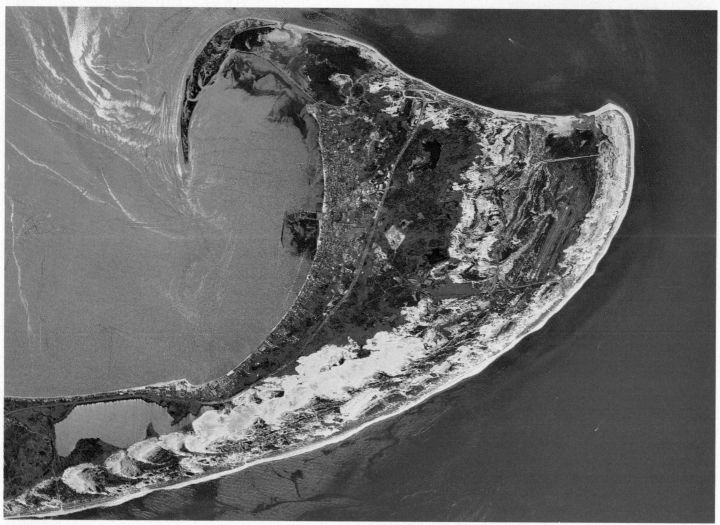

Figure 1

The Sandy Neck spit, located on upper Cape Cod.
Photo by NASA.

Figure 21.15
Depositional features include bay-mouth bars, or sandbars. They develop as a result of a spit closing the mouth of a bay. The spit at Bodega Head, northern California, has nearly closed Bodega Bay. Wave refraction can be observed in the lower righthand corner of the photograph.

Photo by NASA.

Figure 21.16

Bays and estuaries are formed by bay-mouth bars and spits, as well as changes in sea level creating drowned river systems. When waters of different salinities mix, a rich nutrient trap is formed. These backwater regions are the nurseries for a wide variety of fish and resting places and wintering grounds for migratory birds. Kakadu National Park, Northern Territory, Australia, is drained by the South Alligator River. This river system is a refuge for migratory birds and freshwater and saltwater crocodiles. In the wet season, a sea of shallow freshwater spreads over the flood plains, triggering an explosion of aquatic plant and animal life.

The **tombolo** represents a unique depositional bar that connects a stack with the beach. The offshore stack reduces the energy in the longshore current and causes sand to be deposited, building a bar from the stack to the shore. Wave refraction focuses most of the wave energy on the island, and in the quieter waters behind it, deposition occurs to build the bar connection from the island.

Estuaries

Behind bay-mouth bars, spits, and tombolos are some of the world's richest habitats, full of life. This rich habitat results from a mix of fresh and salt water in the estuary. An **estuary** is simply all or part of the mouth of a river, stream, or bay, having connections with the sea, where salt and fresh water from surface runoff and groundwater mix to produce a diluted or brackish water environment. Estuaries represent some of the most fertile regions on Earth. When waters of different salinities mix, a nutrient trap is created. Nutrients are not swept out to sea, but flow in and out with the tide. This ebb and flow of the tide continuously provides food, nutrients, and oxygen to the living system. Waste products are also flushed or utilized by the estuarine ecosystem. Generally speaking, a flowing system is much more fertile and productive than a standing system.

The ebb and flow of tidal water and runoff from streams carries plankton from the marsh throughout the estuary. This primary food supports clams, oysters, and other plankton feeders which are the next links in the food chain. Clams and oysters also perform a filter function, removing wastes in the estuary. The quiet waters also provide a niche for algae and eelgrass plant communities to develop. These plant communities produce an environment for a host of organisms such as snails, worms, insects, and crustaceans, which are fed upon in turn by a variety of fish, shorebirds, and marsh birds in the Alligator River estuary of the Northern Territory of Australia (fig. 21.16).

Bays and estuaries are also important nurseries for many species of freshwater and marine fish. Sixty-one major fish species have been recorded as spending part of their life cycle in a coastal estuary in California alone. They include Pacific herring, white sea bass, corbina, and surf perch. Ten square kilometers (4 square miles) of estuary in Rhode Island are the breeding waters for flounder, accounting for up to one-quarter of the region's catch.

Estuaries are also essential resting places, feeding areas and wintering grounds for migratory birds. The coastal estuaries of North America are visited each winter by a distinguished list of globe hoppers including the American widgeon, mallard, great blue heron, and snowy and American egrets.

Figure 21.17

Shorelines of submergence are the result of drowning when the coast experiences rising sea level. A classic example can be found along the coast of Mount Desert Island, Maine. Scouring by glaciers and drowning by a rise in sea level created this rugged shoreline of submergence.

Photo by NASA.

The real danger facing this habitat today is urban sprawl. On both the east and west coasts of North America, estuaries are being converted into harbors, home sites, and industrial development. In California, just 25 percent of the estuary habitat in existence in 1900 remains today.

SHORELINES OF THE WORLD

All along the coastlines of the world, erosion, deposition, and changes in sea level are continually occurring. Coasts are continually changing on both a seasonal cycle and a long-term geologic span of time. Examination of a series of coastal regions reveals a gradual evolution. Let's first examine shorelines that are experiencing rising sea level resulting from tectonic sinking or actual increases in ocean level. We refer to these as submerged or drowned shorelines.

Shorelines of Submergence

When the Pleistocene glaciers melted, the oceans of the world increased in size, rising higher on the land, flooding coastal plains and valleys. Mount Desert Island, Maine, shown in figure 21.17, is a classic example of a shoreline of **submergence.** The shape of this new shoreline is closely related to the landforms prior to rising sea level. Note the many bays, coves, and drowned valleys represented by Somes Sound and associated bays and coves. Because this area has also been sculptured by continental glaciation, the shoreline produced is the product not only of waves, currents, and streams, but also of moving ice of the last Ice Age. Somes Sound, Goose Cove, and Pretty Marsh Harbor are all drowned glacial valleys or **fiords.** Glaciers that once covered the region scooped out the stream-formed valleys into U-shaped troughs, then changing sea level drowned them. Chapter 20 examines the glacial processes responsible for the many finger lakes and ponds, poor drainage, and glacially scoured bedrock and narrows.

Figure 21.18

Fiords mark the coast of southeastern Alaska, Chile, Greenland, and New Zealand. This fiord in Milford Sound, New Zealand, is an extensive inland waterway on the west coast of South Island. This photograph was taken aboard the New Zealand marine research vessel *Rapuhir*.

Photo by Michael Stammer, Dept. of Scientific and Industrial Research Vessel, New Zealand.

Fiord shorelines are common at latitudes between 50° and 70°. The best examples of fiords are located in five regions. Those on Norway's west coast are among the deepest and largest in the world. Sogne Fiord is the largest, extending inland 248 kilometers (155 miles). Another major fiord region is along the beautiful shoreline of British Columbia and the Panhandle of Alaska. At the same latitude is the Southern Hemisphere, fiords are found in southern Chile, beginning at Puerto Montt and extending south to Tierra del Fuego. A fourth region of fiords forms the west coast of South Island, New Zealand (fig. 21.18). Greenland is the largest region of glacial submergent coastline. The entire island's shoreline is made of fiords.

Given enough time and wave erosion, fiord coasts such as Maine's will be modified. The cliffs will retreat, producing a wider wave-cut bench, and the shoreline will become smooth. The net result is a continual distribution of wave energy over a wider coastal zone. Eventually Mount Desert Island will reach a condition similar to Martha's Vineyard, Massachusetts, an island where barrier bars have closed the bays to the sea. Figure 21.19 shows a typical evolutionary cycle of coastal erosion. Where does wave action stop or reach equilibrium? Can the oceans eventually reclaim the land? The record in the rocks indicates that the continents date back to as far as we can read Earth history. The abundant evidence indicates that sea level is never static. Therefore, the cycle of coastal erosion never completes its task of eroding a continent-sized land mass.

Figure 21.19

Coasts evolve through time, as erosion takes off the rough edges and fills in the valleys. The Roman fortress at Ceasarea, Israel, is now under wave erosion attack. The shoreline was over 100 meters offshore when the fortress was built some 2,000 years ago.

A **ria** shoreline is another form of submergence of the land with respect to sea level. Rias are formed when the ocean drowns valleys and coastal plains in nonglacial regions. Again, there is a relative rise in sea level; the ocean can increase in volume, or the coastal land can sink. Submergence resulting from subsidence of the land is a local phenomenon affecting one specific region, while submergence caused by increased ocean water is a universal phenomenon affecting all coasts. Much of the East coast, from Maryland and the Potomac River southward to Florida and the Gulf of Mexico, represents a ria coast. San Francisco Bay is an

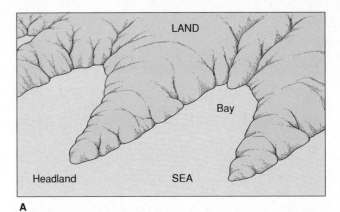

A

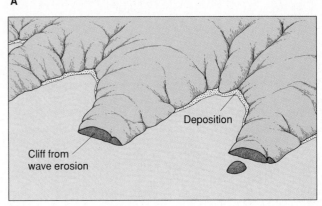

B

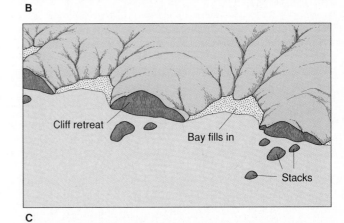

C

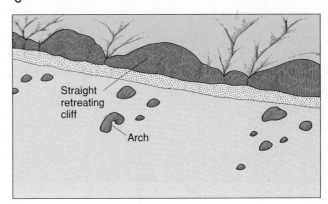

D

example of a ria shoreline on the West coast where the delta of the Sacramento and San Joaquin Rivers has been drowned, giving it the ria characteristic.

Shorelines of Emergence

Shorelines produced by a drop in sea level or uplift of the land can elevate coastal shoreline features well above sea level. The causes for shoreline emergence are again related to either climatic change or local tectonic forces generated by plate tectonic motion. Uplift in Scandinavia is climatic in origin, occurring at a rate of 1 centimeter per year in those areas once covered by thick ice sheets during the Ice Age. As the weight of the ice was removed, the crust began slowly rebounding to its former position.

Figure 21.20

Uplift is also occurring where tectonic forces are thrusting the earth upward, as evidenced by an abandoned shoreline near Mobile, Alabama. Note also a drowned river in a region of marine terraces. The Earth never stands still.

Photo by NASA.

Along the coast of California, another process is causing shorelines to rise. Here the Pacific crustal plate is slowly grinding and buckling the North American plate. Marine terraces, carved by the ocean waves, are notched in the mountain slopes at heights of nearly 150 meters (500 feet) above the current sea level. In some places, over eight terraces have been cut by wave action when the land was at a lower level. On each terrace, old beach sands, dunes and wave-cut eroded cliffs can be found. The oldest terraces are farthest inland and highest in elevation. Currently, at sea level, waves are starting a new terrace which will form the next step as the land rises in escalator fashion.

On the East coast, from Virginia to Florida, there are elevated spits, bars, beaches, and terraces reaching elevations of over 50 meters (160 feet). Here the causes are similar to those on the West coast. The land has risen due to crustal uplift. Figure 21.20 shows an abandoned shoreline 9 meters (30 feet) above current sea level near Mobile, Alabama.

Many shorelines of the world are not easily cataloged as simply emergent or submergent. One example is where volcanism creates more land on a volcanic island through eruptions and lava flows to the sea. Currently, the island of Hawaii is adding several acres per year as lava flows from Kilauea reach the sea. Coral reefs, built on the shores of tropical islands, and deltas are other examples of coasts that cannot be simply classified as emergent or submergent. Along the East and West coasts of the United States, shorelines of all kinds have some evidence of both submergence or emergence. Therefore, in reality, it is easy to oversimplify nature's work. Figure 21.20 also shows a drowned river mouth located in a region of marine terraces near Mobile, Alabama. Here is a case of both emergence and submergence. This condition is referred to as a compound shoreline.

SUMMARY

Our attention in this chapter has focused on the work of waves, currents, and tides as they sculpture and reshape our coastlines. Erosion begins with the work of waves energized by terrestrial winds and by the celestial forces of the Earth, Moon, and Sun. When the wind blows across the water, friction between the wind and water sets in motion wave action, ranging from ripples to walls of water 15 meters (50 feet) high generated by hurricane-force winds.

When waves reach shallow water equaling one-half the wavelength, the wave "feels" bottom and begins to change form, increasing in height and shortening in wavelength. Gravity then takes over, pulling the water down on the beach, producing swash and then backwash, as energy is consumed by the turbulent surf.

A wave is an energy system which spends itself by oscillations and friction with the water and the shallow ocean floor. When a wave touches bottom, it slows down, crests, and refracts parallel to the shore. Refraction concentrates wave energy on the seaward extensions of the coastline. These headlands erode more rapidly than the more protected bays, so, over long periods of time, coastlines become smoother as the headlands are eroded and deposited as sand in the bays.

Sand moves not only onshore toward the beach, but parallel to the coast by longshore currents. The direction of migration of the sediment-carrying current is determined by the prevailing wave angle approaching the beach, which is ultimately determined by prevailing winds.

Along both coasts of North America, sand is being carried southward, building beaches and other depositional features.

Tides affect the coastline in a variety of ways. When the tide comes in, estuaries fill, covering mud flats and marsh grass along the sandy beach and rocky shoreline, and waves break higher on the land, eroding and depositing new sediments. Tides influence the shape of the shore, yet the shape of the shore determines the pattern of the tide. Narrow funnel-shaped estuaries or bays can experience very high tides. The highest tides of the north, or spring tides, occur when the Sun, Moon, and Earth are lined up in a straight line. The most moderate tides, or neap tides, occur when the Moon is in the first or third quarter, resulting in opposing forces between the Moon and Sun.

Tsunamis, or seismic sea waves, represent the greatest sea waves. They are produced by earthquakes. They travel outward from the epicenter in all directions at speeds in excess of 700 kilometers per hour (450 mph). Wave heights generated by tsunamis have been known to reach 30 meters (100 feet) above the sea's normal height at the head of a narrow harbor. Tsunamis are extremely destructive in low-lying areas.

All the shorelines of the world are continually experiencing erosion, deposition, and changes in sea level in both daily and long-term cycles. Shorelines can be classified by their erosional and depositional features and also by recent emergence or submergence. Shoreline depositional features include beaches, spits, bay-mouth bars, tombolos, estuaries, and lagoons. Erosional features develop where wave action is cutting away at exposed headlands and beaches. These features include wave-cut cliffs, sea caves and arches, tunnels, and stacks. The wave-cut bench is a product of wave and surf activity in the surf zone. Where the land has risen or sea level has dropped, the benches become an elevated marine terrace.

Shorelines of submergence can be classified as glacial in origin, or fiords, while those formed by rising sea level in nonglacial areas are called ria shorelines. Many shorelines are not easily cataloged as emergent or submergent. Volcanic islands grow when eruptions occur. Coral reefs along a shore and deltas at the mouths of rivers cannot be simply classified as emergent or submergent. Some shores show signs of both emergence and submergence, illustrating again that nature is not easy to classify.

ILLUSTRATED STUDY QUESTIONS

1. Name each of the erosional features of the shoreline in the mouth of Big River (figs. 21.1 and 21.12, pp. 449, 456).
2. Identify each depositional feature in the photograph (fig. 21.15, p. 449).
3. Explain the tectonic significance of the marine terrace in this photograph (fig. 21.13, p. 456).
4. How many terraces can you count in the photograph (fig. 21.13, p. 456)?
5. How are the estuaries of Cape Cod being formed by waves in the photograph (box 2.1, fig. 1, p. 458)?
6. What evidence is there that this is a drowned river system (figs. 21.17 and 21.18, pp. 461, 462)?
7. What is the name of the current moving sand from the mouth of rivers (figs. 21.6 and 21.7, p. 452)?
8. Explain the role of wind in developing waves (fig. 21.3, p. 451).
9. Describe the influence of sea bottom in wave development (fig. 21.3, p. 451).
10. Describe the astronomical positions of the Sun and Moon during spring and neap tides (fig. 21.9, p. 454).
11. Explain why the tsunami is mainly a Pacific Ocean wave. Describe the type of fault motion associated with it (fig. 21.10, p. 454).

The Work of Wind and Water in Arid and Wind-Formed Landforms

22

Objectives

After completing this chapter you will be able to:

1. Explain the role of solar energy in moving the Earth's atmosphere and its load of sand, silt, and clay-sized particles.
2. Describe wind and water **erosion, transportation,** and **deposition** in arid climatic regions.
3. Explain the significance of **climate, sand supply,** and surface conditions, such as **vegetation** as environmental factors governing the patterns of arid and wind-formed landforms.
4. Describe geomorphic features produced by wind and water erosion in arid regions.
5. Give examples of human impact on arid and wind-formed landforms.

Wind and water sculpture the landscape. Monument Valley Tribal Park, Arizona.
© Kerrick James

Figure 22.1

The air is a fluid capable of carrying great quantities of suspended particles. This 1930s dust storm in Prowers County, Colorado, had wind speeds of 50 kilometers per hour (30 mph).

Photo by U.S. Soil Conservation Service.

Energy from the Sun warms the planet's atmosphere and surface, setting up convection currents of wind and water. Without solar energy, the atmosphere and oceans would be lifeless and still. Consider the air and water as fluids moving about the planet and driven by components of solar power and gravity. Warm air expands and rises, to be replaced by cooler, heavier air. This convectional circulation occurs on a seasonal cycle as well as a local daily circulation. The atmosphere can be thought of as a conveyer system. As this gaseous fluid moves it carries suspended particles, which sandblast, scour, and finally settle in a quiet spot on the land (fig. 22.1). Water in arid regions is also a major force developing a series of unique arid landforms. In this chapter, we will first examine the work of wind and then water in arid regions.

DEFLATION—WIND PICKUP

Wind by itself is nearly frictionless, and yet when loaded with particles it is capable of sculpturing the land into a variety of forms. Sculpturing occurs in two principal ways. First, the wind removes particles by a process known as **deflation** (fig. 22.2). This is similar to water's hydraulic action. The wind sweeps away loose material.

Figure 22.2

Deflation is a removal process in which the wind sweeps away loose material. It can produce a blowout in fine-grained silt and sand, as illustrated near Harrison, Nebraska. The pillar top is the original level of the land.

Photo by N. H. Darton, U.S. Geological Survey.

Remember the dust in your eye when a gust of wind suddenly blew across the school yard? Some of these particles can sting your skin, while others are so small they can only be seen with a microscope.

The grit in the air attacks the surface of the Earth and sandblasts it. This is the second way sculpturing occurs. We refer to this process of erosion as **abrasion** or **corrasion.**

While the particles are in motion they not only act as abrasive agents but undergo some wear and tear, or **attrition.** This causes a reduction in particle size through bumping, rubbing, and scraping. The same process occurs in a stream bed reducing rock to smaller sizes.

Wind transports particles in three ways: traction, saltation, and suspension. **Traction** is a rolling, pushing, and dragging of particles that are too large to be lifted off the ground. The wind velocities must be high and the surface conditions must be conducive to particle movement. An unusual example of traction can be seen in Death Valley National Monument on a dry lake at the northern end of the monument known as the Devil's Race Track. After a rain, the lakebed sediments become very slick and well lubricated. At one edge of the lake, rocks from a cliff frequently fall onto the lake bed, where the wind moves the rocks about like rafts, leaving tracks of their irregular path. These rocks weigh several pounds and yet move about when the wind and surface conditions are just right. For many years no one could explain the tracks; movement is so infrequent and the location is so isolated that this movement has never been witnessed or photographed.

When movement is in the form of skips and leaps, it is referred to as **saltation** (fig. 22.3). Wind speeds must generally be in excess of 18 kilometers per hour (11 mph) to move particles of sand by saltation. Sand generally lifts off the ground about a meter (3 feet). The wind sorts the particles by size, just as if we were to sort them with a sieve. The finest particles travel highest off the ground and remain suspended in the air the longest. Transportation by **suspension** is the temporary support of rock particles by the air. This requires turbulence in the air, just as it does in water. Suspension occurs with fine clay-sized particles, which may take days to settle after a storm. In order to have abrasion, deflation, and attrition, three environmental factors are necessary. These include the right climate conditions, sand supply or source, and proper vegetation cover. Let's examine these factors to better understand their role in the work of wind.

CLIMATE AND THE WORK OF WIND

The distribution of **eolian** or wind-formed landforms gives several clues to their location pattern. First, most of the sand dunes found on the Earth are related to specific locations, and climate is paramount. In most locations, active sand dune areas are typically in the most arid zones or major desert belts. Second, sand deposits are widely distributed along the world's coastlines of strong onshore winds, in climates ranging from the tropical areas to the polar

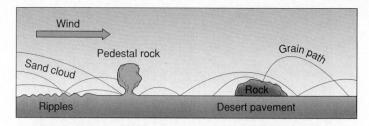

Figure 22.3

Saltation moves particles over the surface when wind speeds exceed 18 kilometers per hour (11 mph). Wind velocities are lowest at the surface where friction is greatest. Friction plays a major role in reducing wind erosion.

land of Asia and Alaska. Therefore, it is apparent that climate is a key factor in explaining the eolian pattern, but not the only one. Sand supply and vegetation cover are also important variables.

SAND SUPPLY

Most of the sand found in desert locations and along coastal plains comes from several principal sources. These include weathered sandstone, depositional river materials, delta deposits, flood plains, beaches, lake shores, and glacial deposits. Sand and silt must be available in large continuous supply for eolian deposits to form. Therefore, exposure coupled with weathering and erosion are essential for this continuous supply.

VEGETATION

Vegetation scarcity, or lack of ground cover, is another important factor contributing to the distribution of wind deposits and erosional features. Where vegetation is sparse or missing, the surface is wide open for wind and water erosion. Generally speaking, the arid regions of the subtropical and tropical deserts lack substantial vegetation. The important thing to remember is that all three conditions are necessary concurrently for extensive eolian features. Many places have one or two of these environmental factors, such as vegetation scarcity and arid conditions; yet, without a sand supply, erosional and depositional features will not arise. In regions where wind depositional features are confined to coastal areas there is an abundance of sand, favorable ocean currents and tides, and prevailing winds that move the sand onshore. However, if dune vegetation is abundant, sand ceases migrating onshore and dunes become stabilized.

A good example of vegetation takeover of dunes can be found between Florence and Coos Bay, Oregon, where the 68-kilometer (42 mile) stretch of sand extends up to 4 kilometers (2.5 miles) inland at some points. Now the dunes appear to be dying or are in the process of being stabilized by a new invader, European beach grass (fig. 22.4). The grass was imported several years ago to stabilize the sand at the mouths of coastal rivers and navigation channels. It has since migrated into the dune region, covering the

Figure 22.4

These Oregon dune deposits are being taken over and arrested by a creeping invasion of grasses recently imported to stabilize the sands at the mouths of rivers. The vegetation cover is now cutting off the sand source of the dune field.

Photo by Beth Horn, U.S. Department of Agriculture, U.S. Forest Service, Pacific Northwest Region.

sandy surfaces, at the rate of 6 to 18 meters (20 to 60 feet) per year, and starving the dunes of sand. The dunes that have been around for centuries will be stabilized in less than 100 years. As the beach grass stabilizes the dunes, native sedges, willow, pine, and spruce begin taking a foothold. Aerial photographs reveal that the dunes are losing about 55 hectares (140 acres) of open sand per year.

In most parts of the world, a loss of dune sand would be a welcome sight. Every year over 100,000 hectares (250,000 acres) of farming land is reclaimed by the desert. Recently the United Nations calculated that around the world a total of 630 million people were in danger of losing their land through this **desertification process.**

With the help of local people and the Algerian government, Windy Campbell Purdie, a New Zealand forester, planted 100,000 trees in a belt 1 kilometer wide and irrigated them with waste water. Five years later the forest was thriving and the roots had penetrated deep enough to find their own water. Between the rows of trees, shrubs and grasses, fruit trees, and cereal crops have also been planted.

The Algerian government's next step was a 20-year project to create 1,400 kilometers (900 miles) of green belt from the Tunisian border in the east to Morocco in the west. Miss Campbell hopes to someday plant a green belt from the Nile to the Atlantic along the northern edge of the Sahara desert. If the work is not done, she says the desert will continue to grow until it reaches the

Figure 22.5

The wind's sculpturing ability creates a variety of unique landforms in the form of pedestal rocks.

shores of the Mediterranean Sea. This experiment is a graphic example of the effectiveness of plants in binding the sand and ending the devastating process of wind erosion.

WIND EROSIONAL FEATURES

Let's examine the types of erosional features associated with eolian deposits.

Wind abrasion produces several unique features, yet the only topographic form that can be attributed primarily to wind abrasion is the **yardang.** The name is applied to ridges between erosional grooves or furrows in weak, soft deposits of silt or clay hills. According to geomorphologists, wind abrasion manifests itself through (1) polishing and pitting, (2) grooving, and (3) shaping and faceting. All of these are widespread throughout the arid lands. When there is strong wind, an abundant sand source, and the absence of vegetation, abrasion can produce **ventifacts,** or polished and faceted surfaces.

Pedestal rocks are sculptured by the blowing sand to form a pedestal or narrow base capped by a less eroded mantle. The abrasion process is greatest at the height where sand supply and

wind velocity combine to produce the greatest scouring. Wind velocity increases with height above the surface, but the sand supply in the air decreases rapidly above the ground. Therefore, maximum abrasion occurs at about the 60-centimeter (2-foot) height, where wind velocity is high and sand moving by saltation is most active (fig. 22.5).

Deflation is continually at work removing loosened weathered material and is responsible for the formation of many depressions, known as **blowouts.** An example of a large deflation basin is located in the Laramie Basin of Wyoming. It is 14.5 kilometers long, 9.7 kilometers wide, and 45 meters deep (9 miles long, 6 miles wide, and 150 feet deep). Here the strong **chinook winds** descend from the Rockies, eat the snow, and scour the landscape creating these basins.

LAG DEPOSITS (DESERT PAVEMENT)

As the wind blows across the desert floor, a sorting process occurs. The finer materials are removed to be deposited downwind, while

Buggies on the Pavement

Dune areas have been modified not only by vegetation but by vehicular traffic. Today's automotive technology has developed a variety of off-road vehicles ranging from dune buggies to trail bikes. As more people look for open space, the desert dunes and desert pavement become the highways for the weekend recreationist.

The desert appears rugged enough, and yet as data come in we are beginning to see significant permanent changes caused by off-road vehicle travel. The major environmental impact by recreational vehicles is not only vegetation damage but desert pavement breakup (fig. 1). The desert floor is very fragile, sealed by the thin pebble and sand crust. Wind can blow across an undisturbed desert floor and find very little sand supply. However, when the desert pavement is broken, fine sand and silt particles become deflated more easily by the wind. The vegetation is then threatened as roots are exposed from deflation, and soil moisture is lost due to increased evaporation from the broken surface. Desert pavement not only caps the sand but also traps moisture below the crust, so badly needed by plants in this thirsty land. Today the recreationist, conservationist, and preservationist are in a struggle over the best human use for the arid regions of the West.

Source: U.S. Geological Survey

Figure 1

Desert pavement is fragile, like the desert ecosystem itself. Once the surface is disturbed by vehicle tracks, winds deflate the ruts, exposing the surface to more wind and water erosion.

Photo by R. Wilshire, U.S. Geological Survey.

Figure 22.6

Desert pavement is not something to drive on; it is actually a thin covering over loose sand. The finer material is sorted out by the wind leaving behind coarser rock particles polished and cemented together by evaporative salts and coarse sand.

Photo by R. Wilshire, U.S. Geological Survey.

coarser rock particles are left behind as **lag deposits** or **desert pavement** (fig. 22.6). The term "desert pavement" is very descriptive of its appearance. The stones that make up this lag deposit are often shiny and varnished in appearance. They are cemented together by a thin coarse crust of sand and salts brought to the surface by capillary action to form a cement after the moisture is evaporated into the air (see box 22.1).

The varnished, smooth appearance is the result of two processes. Wind abrasion shapes the stone in concert with chemical weathering, which colors the stones with iron and manganese oxides. The enamel-like appearance is known as **desert varnish.** The sand and silt-sized particles can be found downwind from the desert varnish and pavement. Let's examine these downwind deposits, removed by deflation.

WIND DEPOSITS

Loess

Wind deposits are formed by a sorting or winnowing process: the coarser sand falls out first, while finer silt and clay-sized particles may travel in suspension many hundreds of miles. Variation in wind velocity can produce a certain amount of interbedding of sand, silt,

Figure 22.7

Loess is a mineral-rich silt and clay deposited by the wind. These loess deposits of the Mississippi Valley near Vicksburg, Mississippi, are some 1,200 kilometers (750 miles) downwind from the Sand Hills source region in western Nebraska.

Photo by E. W. Shaw, U.S. Geological Survey.

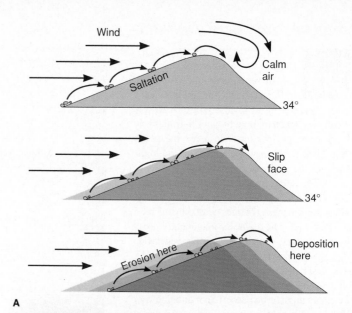

A

B

C

Figure 22.8

Sand dunes come in many colors, shapes, and sizes, but they all can be defined as heaps of sand that are mobile. Notice the advancement between 1988 and 1990 in B and C.

and clay, but the general rule is for silt and clay-sized particles to deposit downwind from the sand dune region. The resulting **loess** material is found in both humid and arid regions but always leeward or downwind of the sources, which may be located along shorelines, stream courses, weathered sandstone, or glacial outwash. Figure 22.7 shows loess far leeward and east from its source in the Sand Hills of western Nebraska. The vast Sand Hills region, covering 62,000 square kilometers (24,000 square miles), is an example of a sand supply produced from the weathering of sandstone deposits on the Great Plains.

Many areas of North America, Europe, and Asia have vast sand and loess deposits from the Pleistocene glacial period. These deposits originated from glacial outwash and delta regions. In North America, loess was probably derived from glacial outwash along the lower Mississippi, Missouri, Ohio, and Wabash River valleys. Similar loess deposits are found in Europe along the Rhine, Rhone, and Danube valleys and very extensively in the Ukraine region of the USSR.

The loess of northeastern China probably originated in the Gobi Desert, and loess deposits in the steppe region of Siberia and Turkestan in the USSR are also of desert origin. Loess is usually 40 to 50 centimeters (16 to 20 inches) thick, forming a mantle over existing older bedrock resembling a soil profile. It is especially thick, up to 30 meters (100 feet), in the Palouse region of eastern Washington and Oregon and southern Idaho. However, the loess deposits of northeastern China win the thickness prize, reaching 60 meters (200 feet) in thickness over much of the area.

The loess deposits of the Argentine pampas, which extend into Uruguay and southern Brazil, are from the arid leeward side of the Andes, similar to the desert source region of northeastern China. Loess is a valuable soil resource rich in nutrients and of excellent physical properties, discussed in chapter 10.

Figure 22.9

White Sands National Monument, Malpais, New Mexico, satellite view. Pure white gypsum is scooped up by the wind as it blows over nearby dry Lake Lucero.

Photo by the Technical Application Center, NASA.

Dunes

In the wind sorting process, sand **dunes** are found upwind from loess silt deposits. A dune can be defined as a mobile heap of sand accumulated by the wind (fig. 22.8). One of the myths held about dunes is the idea that they dominate hot, dry regions. The Australian, Sahara, and Arabian deserts are closest to that perception, but even there, sand deposits represent a small percentage of the surface area. The most extensive coverage of sand dunes is in Saudi Arabia. One-third of the nation, or 1 million square kilometers (400,000 square miles), is dune covered. There dunes may reach 200 meters (700 feet) in height. In Africa, the Sahara has about 800,000 square kilometers (300,000 square miles) of dune region, commonly called **ergs** (sand regions). Most deserts are **regs,** or rock regions. Our own North American deserts have very limited active ergs, confined mainly to coastal zones and interior continental areas in the Great Basin region of Nevada, Colorado, and New Mexico. Large dune fields are also found in Mexico, Arizona, and southern California near the Colorado River flood plain and delta.

Groups of dunes are referred to as dune colonies, complexes, or chains. Active dunes change their shape and location as the wind moves them about. One form can evolve into another. One of the most unusually beautiful dune sites is White Sands National Monument, in the Tularosa Basin near Alamogordo, New Mexico. Here 1,300 square kilometers (500 square miles) of pure white gypsum (calcium sulfate) dunes cover the land. The gypsum is washed into nearby Lake Lucero as a soluble salt from the nearby mountains. Evaporation transforms the solution into sand-sized crystals to be deflated and deposited downwind in the dunes region by the prevailing southwest winds (fig. 22.9).

Figure 22.10

The barchan or crescentic dune has its tips pointing downwind. The leeward side is concave and steep.

Photo by E. D. McKee, U.S. Geological Survey.

To the north of the white gypsum deposits is another dune area almost as large, but not as well known because the sand is made up of more common gray quartz with a blend of feldspar and mica minerals from the granite mountains upwind. Whether sand dunes are of granite or gypsum origin, their forms can be classified by characteristic shapes. These shapes are the result of sand supply, wind velocity and direction, and vegetation cover.

Barchans or Crescent Dunes

The **barchan** or crescent-shaped dune is a curved dune with its tips pointing downwind (fig. 22.10). The windward slope is gentle, 5° to 10°, while the leeward side is concave and steep, forming an angle of 31° to 34°. Barchan dunes line up in chains, closely packed where the sand supply is abundant and more dispersed and smaller where sand sources are scarce. Sand leaves the barchan dune from the tips of the horns. Note in figure 22.10 how this determines the position of the next dune downwind, which draws its sand supply from the tips of those behind it.

Transverse Dunes

Transverse dunes, when viewed from the air, resemble a large sand sea. The dunes form a series of waves parallel to one another with the wind blowing at right angles on the ridges (fig. 22.11). Like the barchan or crescent dune, the windward face is gentle and the leeward slip face is steep, reaching again the stable angle of 31° to 34°. The troughs are deep furrows, often 60 meters (200 feet) below the crests. Sand seas exist only where there are abundant sand supplies.

Parabolic Dunes

Along the coastline of California and Oregon and the shoreline of Lake Michigan are some of the best examples of parabolic dunes. These are long, scoop-shaped hollows or parabolas with points facing into the wind. This type of dune is formed where the wind removes sand from the windward hollow by deflation and deposits it on the leeward slopes.

A

B

Figure 22.11

(A) An aerial view of transverse dunes in eastern Iran gives the illusion of a choppy sea. (B) On the ground, a profile of a transverse dune is similar to a barchan dune with a steep slipface and gentle windward slope. Those seas of sand develop where the sand source is abundant.

A: Photo by NASA; B: Photo by Wilshire, U.S. Geological Survey.

A

The **coastal blowout** dune is a good example of the parabolic class. These dunes form adjacent to beaches with a large sand supply and strong prevailing onshore winds. The sand is piled high in a horseshoe shape. On the windward side, a saucerlike depression is formed as the wind blows out the sand to form the ridges. On the leeward slip face, the sand advances over the ground, burying trees and shrubs in its path (fig. 22.12).

These dunes develop in semiarid regions downwind from deflation hollows. Sand is caught by brush, and it accumulates in a horseshoe series of ridges. In some places parabolic dunes grade into long hairpin forms stabilized by vegetation. Parabolic dunes seem to always be associated with vegetation.

B

Figure 22.12

Parabolic dunes are common forms along the shorelines of oceans and large lakes. Deflation produces a hollow blowout on the windward side and steep slipfaces on the leeward slopes. (A) The close-up shows bedding planes in the slipface of a coastal parabolic dune near the *study area*. (B) The aerial view shows this beach dune in 1952. Today the dunes have reached the road at several locations. The prevailing winds are northwesterly.

B: Photo by U.S. Department of Agriculture, Soil Conservation Service.

Figure 22.13

These Algerian longitudinal dunes on this Landsat image consist of
parallel ridges that can stretch out over 300 kilometers (200 miles)
where strong prevailing winds blow.

Photo by the Technical Applications Center, NASA.

Longitudinal or Seif Dunes

Dunes that form elongated ridges parallel to the prevailing wind are referred to as **longitudinal dunes** or **seif dunes.** They attain heights of over 100 meters (330 feet) in Egypt, and some in Iran are as high as 200 meters (650 feet).

Longitudinal dunes can often stretch out over 300 kilometers (200 miles) long. Their crests form knifelike ridges with gaps or corridors between adjacent dunes where the rocky desert floor is exposed. Some geomorphologists believe longitudinal dunes are a modification of a barchan form, produced by strong cross winds at right angles to the long ridges. They grow in height and width during cross winds and elongate during prevailing wind periods. Good examples of longitudinal dunes are located in the Algerian desert (fig. 22.13).

Star Dunes

In North Africa, Arabia, and Iran a distinctive star-shaped dune exists. It consists of a multiple pointed base and reaches over 100 meters (300 feet) in height above the base. Star dunes are rather fixed, serving as important landmarks and reference points for desert travelers.

GEOGRAPHIC SIGNIFICANCE

How does human activity influence wind-formed landforms? As we have stated, eolian activity can be divided into three components: **erosion, transportation,** and **deposition.** All three are interrelated and dependent on **sand supply, wind velocity,** and **surface conditions** such as vegetation and landform morphology or shape. Wherever human activity modifies any of the three processes, or influences the flow of air, sand source, and vegetation features or surface characteristics, change will occur within the system.

The Dust Bowl during the drought of the early 1930s, on the plains of eastern Colorado, western Oklahoma and Kansas, and neighboring regions, was created by the western expansion of agriculture. Dust was picked up by dry winds from freshly plowed croplands that had failed to produce a vegetation cover (fig. 22.14). If the land had been left undisturbed, the drought and wind would have had far less erosional influences. In the 1960s, very similar effects occurred in Siberia resulting from the Virgin Lands programs of the USSR. The short-grass steppe land of Central Asia was stripped by the plow and rains failed to bring a vegetation cover. The land was then ripe for erosion.

The transportation of sand has been greatly reduced by the use of trees because sand travels close to the ground. Windbreaks can effectively reduce the wind velocity and impair the ability of air to drive and suspend particles. In the southwestern United States, the Australian eucalyptus and other drought-resistant trees capable of growing in windy conditions serve as wind screens. Where they have been introduced, not only has wind erosion been reduced, but evaporation of soil moisture has also been slowed.

Figure 22.14

The Dust Bowl of the 1930s occurred when a severe drought followed removal and overgrazing. Large areas of short-grass steppe had been plowed under or browsed upon by cattle and sheep. Here windblown silt deposits bury a crop.

Photo by U.S. Department of Agriculture, Soil Conservation Service.

DESERT FEATURES AND THEIR CONTROLS

Climate is the chief control of weathering rates, as discussed in chapter 17. In arid climates where limestone is less likely to weather and erode, features are likely to be prominent ridges with angular features, like those made of other more weather-resistant igneous and metamorphic rocks.

Sparse vegetation also contributes to both wind and water erosion. Vegetation shields the soil, and bonds the soils and geologic structures into a more stable mass. Hence, low-rainfall regions with sparse vegetation are some of the most eroded areas on Earth.

Rock structures are also significant controls on desert feature formation. A visit to the Sun Belt of the southwestern United States reveals two distinct geologic provinces or regions: the Colorado Plateau and the Basin and Range province (fig. 22.15).

The Colorado Plateau centers on the intersection of Utah, Colorado, New Mexico, and Arizona. The geology is primarily horizontal beds of sedimentary rock like those exposed at the Grand Canyon of Arizona. This high plains country has been eroded by rivers and streams into plateaus, mesas, buttes, hogbacks, and cuestas (fig. 22.16).

Where flat-lying and erosion-resistant sedimentary or volcanic rocks form a cap or **plateau** bounded by cliffs, erosion takes its course, and the cliff is gradually eroded back. A plateau can evolve into a **mesa,** or flat-topped hill bounded by cliffs. Continued

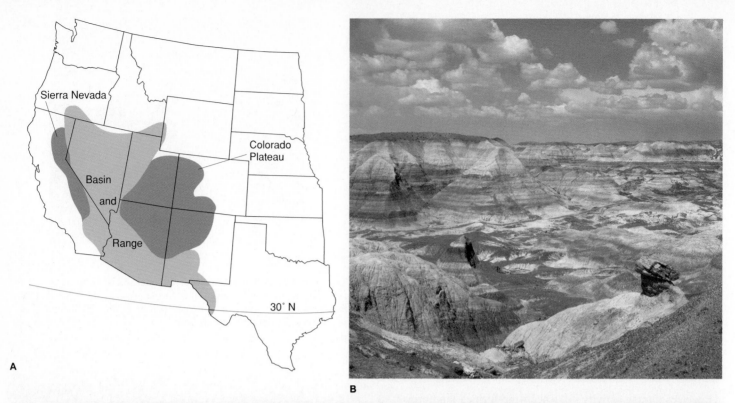

Figure 22.15

(A) The Colorado Plateau and the Basin and Range province in the southwestern United States. (B) Petrified Forest National Park in northern Arizona. The more resistant petrified log is last to be weathered. (C) Death Valley in eastern California is a fault-formed basin located between two uplifted blocks.

A

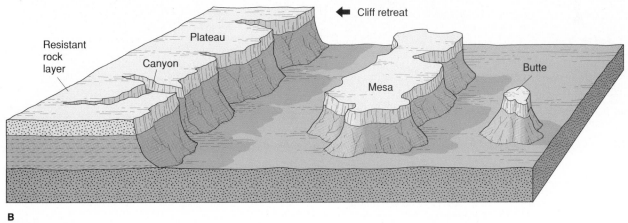

B

Figure 22.16

(A) The butte was photographed at Monument Valley, Arizona.
(B) Characteristic landforms of the Colorado Plateau. Erosional
retreat of a cliff at the edge of a plateau can leave behind mesas and
buttes as erosional remnants of the plateau.

Photo by Carol Printice, U.S. Geological Survey.

A

Figure 22.17

(A) Monocline near Mexican Hat, Utah. Rocks on far left and far right are horizontal; rocks in the center are steeply tilted downward to the left. (B) Steplike monocline folds often erode so that resistant rock layers form hogbacks and cuestas.

Photo by Jack MacMillan.

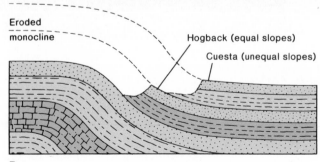

Eroded monocline

Hogback (equal slopes)

Cuesta (unequal slopes)

B

Figure 22.18

Basin and range topography in Death Valley, southeastern California, looking northwest. The Panamint Range is on the west side of the valley; the Black Mountains are in the foreground. Faults separate the valley floor from the mountains. Large alluvial fans are building out from the base of the mountains onto the flat, salt-encrusted valley floor.

flooding is common. Thus rock debris and mudflows from flash floods build **alluvial fans** at the base of the mountain ranges. As described in chapter 18, alluvial fans develop when rapid runoff with a large load is suddenly dropped where the stream loses gradient and energy, and makes a wider channel. Once the channel is clogged, the next runoff will take another path of least resistance, thus building a fan-shaped sediment pile at the mouth of the canyon. Where fans merge, a **bajada** forms.

The finer clay and dissolved salts are carried onto the flat valley floor. When there is no exterior drainage, a shallow lake forms. If it evaporates, a salt flat or **playa** forms (fig. 22.18).

Erosion of the mountain eventually produces a *pediment,* which is a stable, gently sloping erosional surface underlain by bedrock whereas the bajada is a depositional surface.

Summary

Solar energy drives the convectional processes that circulate the air over the surface of the Earth. The wind deflates or picks up dust and sand and transports this material to new locations. On the way, erosional processes of abrasion sandblast the Earth's surface, and **attrition** breaks larger particles into smaller ones.

The **transportation** process includes **traction, saltation,** and **suspension.** Traction is a rolling, pushing action of the particles, while saltation is more of a skipping motion, and suspension is temporary support of rock particles by the air. Large particles move by traction while suspension carries fine clay-sized particles.

Climate, sand supply, and **vegetation** are the three major environmental factors governing the patterns of wind-formed landforms. The primary distribution of sand deposits reflects the arid regions of the world. However, aridity alone is not enough, as many dune regions are located along the coasts from the tropics to the Arctic. Where vegetation has been modified in eolian regions, changes occur in sand supply and in the configuration of the eolian deposits.

erosion will result in a **butte** or narrow, flat-topped pinnacle. Where the Colorado plateau is folded, monoclines and synclines will erode to form steeply lifted sharp ridges or **hogbacks** and less steeply tilted **cuestas** with one steep slope (fig. 22.17).

Basin and Range country, making up most of Nevada, northern Utah, and southern Arizona and New Mexico, is characterized by parallel fault-formed mountain ranges and basins resulting from normal faulting, described in chapter 16. Because the mountain catches most of the heavy precipitation, rapid runoff and

Erosional features include **abrasion-formed ventifacts, pedestal rocks,** and **yardangs.** Deflation produces blowouts or deflation basins. **Desert pavement** or lag deposits result from a process of cementation of a thin layer of salt, sand, and pebbles which are polished smooth by the blowing grains of sand to form a natural cement.

Depositional features are formed by a sorting process. The coarser sand settles out first, while fine silt and clay-sized particles, or **loess,** may travel in suspension many hundreds of miles and then blanket a broad region. Loess is a valuable soil resource with excellent chemical and physical properties.

Dunes make up a small percentage of the surface area of deserts. The world's largest dune region is in Saudi Arabia, where one-third of the nation is dune covered. This type of desert is referred to as an erg desert; rocky deserts are known as **regs.**

Active dunes change their shape and location as the wind moves them about. The **barchan** or crescent dune is shaped like an arc with its tips pointing downwind. **Transverse dunes,** when viewed from the air, resemble a large sand sea. The dunes form a series of waves in parallel patterns with the wind blowing at right angles to the sandy ridges. **Parabolic dunes** often develop along coastlines; they have a parabolic shape with points facing into the wind. These dunes have long scoop-shaped hollows or parabolas produced by deflation on the windward side. The **coastal blowout dune** is a good example of this parabolic class.

Longitudinal or **seif dunes** form elongated parallel ridges aligned in the direction of the prevailing wind, often stretching out over 300 kilometers (200 miles) downwind. It is believed that the knife-edge ridges are produced by strong crosswinds that blow intermittently.

Star dunes display a distinctive star-shaped base that builds to a large sand hill often reaching 100 meters (300 feet) in height above the base.

Eolian activity can be divided into three processes: **erosion, transportation,** and **deposition.** When human action modifies any or all of these three components, change will occur within the system. Examples are many, but the American Dust Bowl of the 1930s and the soil losses of the 1960s in the Virgin Lands program of the USSR represent two case studies of human impact on these processes.

Arid features are controlled by **climate, vegetation,** and **geologic structure.** Arid conditions influence slower weathering rates, hence more angular features. Precipitation is erratic and low but sometimes torrential, creating rapid runoff and erosion with flooding and mudflows. In the arid southwestern United States, there are two distinct geologic provinces: Basin and Range and the Colorado Plateau.

Basin and Range country is a series of faulted mountain and valley systems. Here one can observe **alluvial fans, bajadas, playas,** and **pediments.**

The Colorado Plateau consists of massive, nearly horizontal beds of sedimentary rock hundreds of meters thick. Here mass wasting and erosion by rivers, chiefly the Colorado River, have created arid features which include **plateaus, mesas, buttes, cuestas,** and **hogbacks.**

ILLUSTRATED STUDY QUESTIONS

1. Match the terms with each of the following figures (figs. 22.1 to 22.3, p. 468).
   a. deflation
   b. suspension
   c. saltation
2. Explain the role of vegetation in reducing wind erosion on this beach (fig. 22.4, p. 469).
3. Identify the wind erosional feature in this photograph (fig. 22.5, p. 470).
4. Explain the processes that form desert pavement. Why is it so fragile and yet important in the survival of desert plant and animal communities (box 22.1 and fig. 22.6, p. 471)?
5. What is the source of the loess deposit in this figure (fig. 22.7, p. 472)?
6. What evidence of dune movement is present between 1988 and 1990? Approximately how many meters (feet) of movement has taken place (fig. 22.8, p. 472)?
7. Name the dunes in each figure and explain the favorable conditions for the development of each dune form. (figs. 22.10 to 22.13, p. 476).
8. The trees invaded by dunes show the strain that wind puts on their branches. Why does the wind blow consistently on shore on both a daily and seasonal basis in the *study area*? Note: persistent cold ocean current maintains water temperatures around 10° C (50° F) while the interior land warms to over 38° C (100° F) (fig. 22.12, p. 475).
9. Climate, sand supply, and vegetation are important factors determining the potential development of wind-formed landforms. From the photograph, explain the source of sand, favorable climatic conditions, and role of vegetation (fig. 22.12, p. 475).
10. Describe the location of the Colorado Plateau and Basin and Range country (fig. 22.15, p. 478).
11. Compare each of the following desert features: butte, mesa, and plateau (fig. 22.16, p. 479).
12. Contrast cuestas and hogbacks in terms of slope (fig. 22.17, p. 480).
13. Explain the one possible origin of mountains and valleys in the Basin and Range province (fig. 22.15, p. 478).
14. Explain the origin of each (figs. 22.15 and 22.18, pp. 478, 481):
    a. alluvial fan
    b. playa
    c. desert pavement

The Earth's Systems

VI

How long the Earth's systems can support the current quality of life will depend not only on the natural systems but upon our treatment of these systems. The design features are truly a miracle of marvelous proportions, delicately balanced, enduring, yet fragile. This new perspective of a unique integrated system of air, water, land, and life is in your hands.

Landscape Arch, Arches National Park, Utah.
© Doug Sherman/Geofile

The Miracle of Design

23

Moon rise over the giant saguaro cactus.
© John Gerlach/Tom Stack & Assoc.

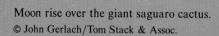

484

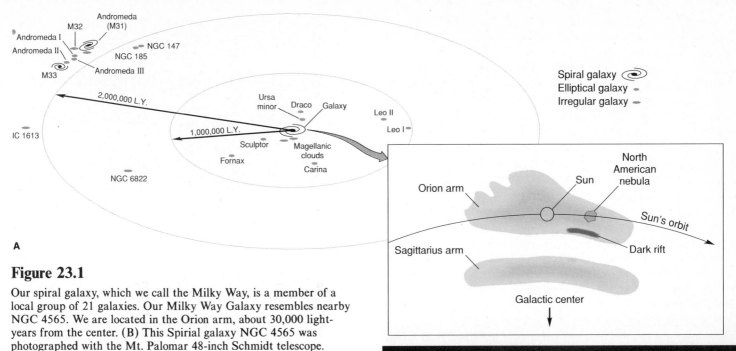

A

Figure 23.1

Our spiral galaxy, which we call the Milky Way, is a member of a local group of 21 galaxies. Our Milky Way Galaxy resembles nearby NGC 4565. We are located in the Orion arm, about 30,000 light-years from the center. (B) This Spirial galaxy NGC 4565 was photographed with the Mt. Palomar 48-inch Schmidt telescope.

B: A Palomar Observatory Photograph. Used by permission.

B

The design of the Earth is truly mind-stretching. Each of the major spheres of the Earth covered in this text is amazingly integrated into one system capable of supporting life.

In this chapter, we will review the major themes of the text by focusing on the uniquely integrated aspects of Earth's design. For example, its position in the universe, astronomical properties, geologic processes, climatic conditions, and finally, life forms testify to this unique system. From the smallest single-celled organism to the largest giant coastal redwood, all reveal unique design features. This final chapter is meant to instill in you a profound appreciation and respect for our home, the planet Earth.

Our Earth is truly an oasis in a desert of space, not unlike the oasis of a desert community, where life abounds because of important life support characteristics. These include air, water, soils, and proper temperature and radiation qualities. Let's take another look at our universe to see where these characteristics are likely to occur to produce life as we know it.

THE UNIVERSE

The field of astronomy has given us many interesting insights into the unique astronomical aspects of the Earth. On a clear evening, take a small telescope, a pair of binoculars, or even a scope on a rifle, and gaze into the heavens toward the star-rich field of the Milky Way. You will be amazed to discover how many stars fill your view.

Can life exist out there in other places in the universe? What conditions in our part of the universe make life possible? Why is there only one Earth in our solar system? Because of our lack of knowledge about the vast realm beyond, few answers are available to these questions. We can, however, find the evidence of design for life in our little corner of space.

Our galaxy we call the Milky Way is a member of a local group of 21 galaxies. Each of these galaxies consists of billions of stars, hydrogen gas, and interstellar dust forming galactic shapes ranging from spirals to spherical forms. Figure 23.1B shows an edge-on view of a galaxy like ours, and also our position in the spiral arms. What we don't see in this picture are the magnetic and gravitational fields that form the system's hidden linkage. The gravitational field of the galaxy holds the system together and prevents it from flying apart. Magnetic forces give each star, and corporately each galaxy, a giant magnetic field capable of deflecting and trapping cosmic rays.

These two energy fields are so intense in most places in our galaxy that life on our planet could not survive except in two doughnut-shaped regions located between 32,000 and 40,000 light-years from the center of the galaxy. Magnetometers, capable of measuring the magnetic field strength of our galaxy, show that even a change of 0.25 light-year at right angles to the plane of the spiral would be a fatal move, causing radical magnetic changes to our Earth environment. A similar calculation shows an even greater environmental change in our gravitational field. Therefore, we can say there are only two safe doughnut-shaped harbors in space, representing about 1/150 of the total volume of our galaxy. This is a very conservative figure, but it gives us a probability of roughly 1 in 150 of being located at the proper place in the galaxy to have life even possible.

The Sun

Can we hitch the Earth's wagon to just any star? Our nearest star, the Sun, provides just the right amounts of light, heat, and energy that drive the water, wind, and energize the biosphere.

Within the range of the largest telescopes one can observe billions of other stars. Each belongs to a specific spectral class or family of stars. Of all the star groups, only a small percentage radiate electromagnetic energy in the range beneficial to life forms on the Earth.

If we briefly examine the *Hertzsprung-Russell diagram* showing spectral classes of stars in our galaxy (fig. 23.2), the diagram tells us the Sun is a class G star, of medium temperature, producing primarily yellow light. It belongs to the main sequence and the middle age class of stars. Stars in the main sequence located in classes O, B, A, and F are very hot, bright blue and white stars and are rapidly consuming their hydrogen fuel. Spectral classes K and M are cool, dim, slow-burning stars, producing mostly red, infrared, and orange light. Red giants and white dwarfs, located off the main sequence, are dying stars. Therefore, only the stellar class G stars on the main sequence are capable of producing proper radiation for life as we know it. The white and blue stars are too hot and produce too much deadly x-ray and ultraviolet, while the K and M stars are too cool and don't generate enough visible energy waves for photosynthesis. If we assume there are 100 billion stars in this galaxy and there are a million stars like ours, the ratio of the two numbers reduces to 1 to 100,000, meaning that a probability of having a sun like ours is only one in 100,000.

Solar Distance

The nature of our galactic position in the spiral is critical. However, just as critical is the precise location of our planet's orbit in the solar system. Let's examine the critical nature of this orbital pattern as it relates to the Earth's environment and life support system. As discussed in chapter 2, the amount of energy reaching the Earth's surface varies inversely with the square of distance from

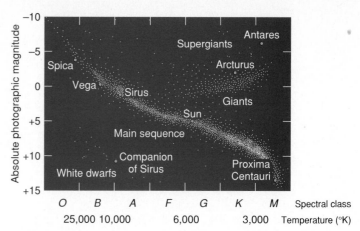

Figure 23.2

The Sun is located on the main sequence at the Hertzsprung-Russell graph, a middle-aged class G star of the fifth magnitude. It became a star about 6 billion years ago.

the energy source, assuming no atmospheric interference. Therefore, each planet has a predictable solar constant or energy input based on the inverse square law of light.

If we were located in any other orbital pattern, life as we know it could not exist because temperatures would be either too hot or too cold. Chapter 2 discusses the mean temperatures of the nine known planets. Only the Earth has a mean temperature that allows life dependent on liquid water to exist. Since both plants and animals are between 99 and 50 percent by weight of water, it is critical that temperatures remain in very narrow range between freezing and boiling. Our distance from the Sun is a key factor in maintaining this narrow range. All one needs to do is examine our planetary neighbors, and note that water is either missing or in the frozen or gaseous state.

The Earth's Motions

The Earth's orbital velocity in space is also critical. If we were to slow down or speed up a few miles per second, our path around the Sun would be altered. It has been calculated that an increase in velocity of just 13 kilometers (8 miles) per second would send us on a journey beyond the solar system. If we were to slow down by the same amount, the Sun's gravity would cause us to spiral into the Sun's atmosphere. Our current orbit is the result of a balance of forces between the Sun's gravitational pull and our orbital motion. Fortunately, when we do slow down slightly at **aphelion,** the Earth is accelerated again by the gravitational component of the Sun.

The rotation rate is also critical for life. The Earth can be compared to a giant rotisserie. The rotation rate averaging 360 degrees in 24 hours ensures even heating during the diurnal (daily) pattern. Consider the Moon's rotation rate of 360 degrees in four weeks. This long period of two weeks of daylight and two weeks of darkness is a major factor in causing daytime temperature to soar over 93° C (200° F) and nighttime temperatures to −130° C (−200° F). If we had a rotation rate faster or slower than the current 24-hour period, tidal patterns would also be affected. Wind and ocean currents would all be altered as ranges in new daily temperatures would change. In summary, the entire heat budget would be altered by a change in the rotation rate. A slow rate would produce temperature extremes between night and day and a fast rate would tend to eliminate any diurnal temperature cycle. How these changes would affect life is uncertain, but the planet Earth would certainly have a very different temperature environment.

The Earth's Tilt

Consider the Earth's inclination or tilt of 23 1/2° with the normal or perpendicular to the plane of its orbit. If the tilt did not exist and the Earth's spin was perpendicular to the plane of its axis of rotation, seasons would not exist: each latitude would receive solar energy without seasonal variation. The Sun's rays at noon would strike the Earth at the same angle each day, and the path of the Sun across the sky would be faithfully the same each day.

Climate, vegetation, and soil patterns would be more in line with latitudinal lines. The seasonal rhythm of plant and animal life we observe and experience would not exist. Migrations, hibernations, fall colors, budding of spring, and tree ring development are just a few changes that would disappear or be altered by an axis without a tilt.

THE UNIQUE ATMOSPHERE

The gases of the Earth's atmosphere are a blend of ingredients found in just the right proportions to support life. Recall from chapter 3 that only the Earth has oxygen and carbon dioxide in the right proportions for life. If oxygen did not exist or existed at a higher or lower percentage or as ozone (O_3), respiration could not occur in the life forms we observe for both plants and animals.

It is difficult to speculate on the origin of the atmosphere, but we can safely state that it is currently maintained in the correct proportions by the natural processes of the Earth's systems. Through photosynthesis, green plants convert solar energy into chemical forms, and oxygen is given back to the air. When a tree adds a ton of wood in the growth process, it also generates a ton of oxygen. If plants could grow on Mars, this gas would soon escape into space because its gravitational hold is too weak. The gases we have are still with us because gravity keeps a hold on this envelope of air. This gravitational hold has some interesting additional side effects that make life possible on this planet. First, and most obvious, is atmospheric pressure. Air exerts a pressure of approximately 1,000 millibars (15 pounds per square inch) at sea level. Because this pressure is not felt, it is taken for granted. The weight of the atmosphere, like a lid, keeps the oceans from boiling away at sea level temperatures. Thus, the Earth's atmosphere and hydrosphere are uniquely linked together by gravity.

Our atmosphere also serves as a cosmic shield, protecting us from deadly radiation in the form of low-energy cosmic rays, solar wind particles, and excessive ultraviolet radiation. Ozone in

Figure 23.3

The atmosphere is our protection shield. It filters harmful ultraviolet and X-ray radiation and maintains an even temperature through a greenhouse effect.

Photo by Jacqueline (Pearce) Standley.

the stratosphere forms a first line of defense by intercepting ultraviolet radiation. By the time solar energy reaches the surface, most of the harmful portions of the electromagnetic spectrum have been diluted or filtered to safer levels (fig. 23.3).

Carbon dioxide, as we have mentioned, also performs a major role in the Earth's life support system. This gas comes to us currently from processes of decomposition, respiration, and combustion. Like oxygen, it is currently recycled in the biosphere. Although it makes up less than 1 percent of the atmosphere, its role is nearly as remarkable as oxygen's (fig. 23.4). Green plants could not grow without it. **Photosynthesis** requires carbon dioxide to combine with water to form starches, sugars, and oxygen.

Figure 23.4
Fire is a rejuvenating force. It also returns carbon dioxide to the
atmosphere in a *carbon cycle* that is necessary for life.
Photo by Gary Bowman.

Carbon dioxide, along with water vapor, is also woven into our Earth's heat budget or terrestrial thermostat. Without these gases, the Earth would have a poor **greenhouse effect.** Long-wave infrared radiation, given off by the Earth, is partially absorbed by these gases and reradiated. Thus, carbon dioxide and water vapor together form a blanket for the planet. Because of this blanket we have moderated temperatures and an environment more conducive to life, that sprang forth in a very narrow planetary temperature range.

If the atmosphere were thinner, we could also expect increased destructive meteoric bombardment from outer space. Every day several tons of meteoric debris is intercepted by the Earth and atmosphere without harmful impact. As meteors race through the Earth's atmospheric shield, most are consumed to ash and gases or deflected back into space like a rock skipping on a pond. Only the very largest ones are capable of impacting the Earth and doing any damage. Fortunately these are very rare.

The percentages of elements in the Earth's crust, atmosphere, and oceans show that the unique distribution of ocean, land, and sea results from the final compound formations. If vast quantities of hydrogen had not escaped the gravity of the Earth before it combined with oxygen to form water, we would have a planet totally under water or nearly so.

If oxygen were present in a larger percentage, say 50 percent, lightning bolts would have exploded our forests or eliminated them by fire. Most of the green belts of the world, and again our source of oxygen renewal, would be lost by fire. If oxygen were reduced to 10 percent or less, life might have adjusted perhaps without the tool of fire, a key element in the development of civilization. Fire has left its impact on all living systems. It is used to clear land, drive wild animals into traps, and, of course, cook, process metals, and refine clay pottery and adobe bricks. Plant communities are modified and renewed by this environmental factor.

Nitrogen fills the air we breathe. It represents 78 percent of the atmosphere, compared to 21 percent for oxygen. After hydrogen, oxygen, and carbon, the three most plentiful elements in all living things, nitrogen is next in abundance.

Nitrogen gives the air its body and pressure. It dilutes oxygen to a percentage conducive to life. Although all living systems need nitrogen to build new cells and conduct metabolic processes, most nitrogen is unavailable in the biologically useful nitrate form. Before it can be used it must be in the ionized state. Nature can extract this gas from the air several ways. Lightning strokes are a significant method of extraction, but the primary source of nitrogen comes from nitrogen-fixing bacteria. Nitrogen gas is utilized directly by these bacteria in the soil and in root nodules of plants, especially the legumes. When these organisms die, they release nitrogen compounds that can be utilized by other plants. There is also good evidence that legumes release nitrogen through the root system while still alive. Thus nitrogen, the building block of life, is cycled through the biosphere in the food chain so that life can continue to grow and reproduce (fig. 23.5).

Therefore, this nearly inert gas that gives the air we breathe body is of paramount importance in the life support system of the planet. Hence, we see the balance of design in the makeup of gases and percentages of elements in the earth, sea, and sky.

The Miracle of Design   **489**

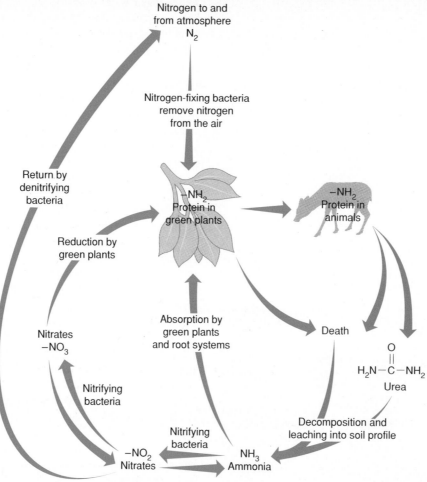

Nitrogen to and
from atmosphere
N_2

Nitrogen-fixing bacteria
remove nitrogen
from the air

Return by
denitrifying
bacteria

Reduction by
green plants

$-NH_2$
Protein in
green plants

$-NH_2$
Protein in
animals

Nitrates
$-NO_3$

Absorption by
green plants
and root systems

Death

$$H_2N-\overset{\displaystyle \overset{O}{\|}}{C}-NH_2$$
Urea

Nitrifying
bacteria

Decomposition and
leaching into soil profile

Nitrifying
bacteria

$-NO_2$
Nitrates

NH_3
Ammonia

Figure 23.5

The *nitrogen cycle* is essential to plant growth. Nitrogen can be lost
by interrupting this cycle through deforestation and excessive
leaching.

Photo by Gary Bowman.

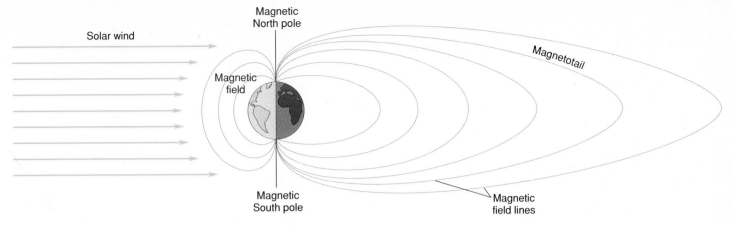

Figure 23.6

Our electrical shield, the magnetosphere, protects us from harmful cosmic particles and solar wind.

The Magnetosphere

As we examine our planet, we can observe how it is uniquely blessed with not one but two cosmic shields. The first is our atmosphere, while the second is in the form of the magnetosphere, a magnetic web woven with lines of a magnetic force field that encompasses the Earth's spherical shape (fig. 23.6). It filters and deflects the most lethal cosmic rays and the particles of the solar wind. The magnetic field is also a navigational guide for pigeons, bees, and even aquatic organisms. These two cosmic shields alone are essential for life. All other positive life support factors would be nullified if this shield were not present. During stormy periods in the Sun's atmosphere, solar flares unleash high-energy cosmic particles and deadly X-ray energy capable of destroying any life forms in its path.

THE LITHOSPHERIC DESIGN

Those stones in a creek, giant granite cliffs of Yosemite, or snow-covered lava slopes of Mount St. Helens, Washington, all testify to a unique lithosphere that provides the nutrients and platform or foundation for all living systems and their activity. The general properties of the lithosphere can be inferred from the recording of seismic waves that arrive at seismographs around the globe during earthquakes.

Seismologists infer that the Earth's core probably consists of high-density iron and nickel with a liquid outer core and solid inner core. The upper mantle layer beneath the crust is in a dynamic plastic state that causes the surface crustal plates of the ocean floor and continents to drift about like thawing ice on a lake in spring. In this process, fresh crustal material emerges along spreading zones and regions of subduction. Recall from chapter 15 that it is in these regions of spreading and plate collisions that mountains are formed.

Once crustal material is exposed to the atmosphere, gravity, weathering, and erosion begin reducing the surface to sea level. The mountains are slowly broken down into transportable particles and carried away to the quiet waters of lakes, seas, and ocean basins to be deposited, cemented together, and formed into sedimentary rock. This crustal sedimentary rock accounts for over 70 percent of the continental surface. When these rocks are reheated in zones of subduction located along the plate boundaries, continental basement rocks rich in silicon are formed. Because of this **rock cycle,** our planet's crust is subdivided into lower-density **felsic** or continental material granite and higher-density ocean-floor plates of **mafic** basalt rich in heavier minerals of magnesium and iron (fig. 23.7).

Thus the very foundations of the major continental land masses bob above the heavier ocean plates like floating ice. Without the rock cycle, the Earth would be a very different place. Continents would either be missing or lacking the lower-density rocks. Oceans would cover nearly the entire surface, and perhaps only marine life would exist.

The plate boundaries where the Earth's major slabs are sutured together are very dynamic zones. Here are the Earth's major earthquake zones, volcanic eruptions, and geothermal activity.

If it were not for the volcanic activity, our Earth would be very hostile to the development of life. Let's examine the reasons why. First, volcanic activity is a major source and contributor to the gases of our atmosphere, which in turn influence the Earth's heat budget.

Throughout the Earth's history, volcanoes have supplied the planet with many of its gases. Each volcanic eruption releases great quantities of water vapor, sulfur, nitrogen, and carbon dioxide, compounds that make up the major atmospheric components. This heat budget is basically controlled by dust particles, carbon dioxide, water vapor, and reflective and absorbent surface characteristics.

Fallout from volcanic activity in the form of dust and ash is another of nature's ways of replenishing soil nutrients (fig. 23.8).

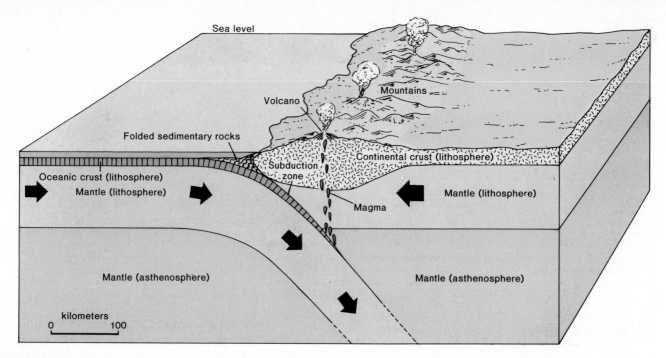

Figure 23.7

The *rock cycle* is part of the *plate tectonic cycle:* seafloor spreading and moving over an eruptive center brings new crust to the surface, and subduction returns old crust into the mantle where temperatures are high enough to produce metamorphic rocks and melt the crust to a magma form. When magma returns to the surface, new volcanic rocks are formed and exposed to weathering and erosion in preparation for sedimentary rock development.

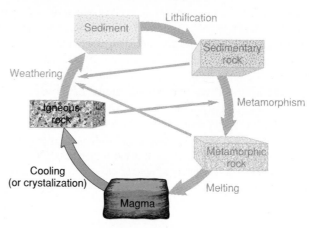

After volcanic fallout, vegetation is invigorated. Farmland soils become more productive. Where volcanic pumice and ash falls on marginal semiarid lands, the surface receives a fresh supply of nutrients and a blanket of ash with moisture retention characteristics. Recall from chapter 16 how the Sunset Craters near Flagstaff, Arizona, produced an ash cover that made the soil capable of supporting a corn-based culture on land once too arid for farming.

Erosion, Weathering, and Soils

Wind

The Earth's winds sift and sort the silt and sand and sculpture the rocks in their path. The work of wind takes several forms, which include depositional and erosional features. These features, which include pedestal rocks, dunes, loess, blowout basins, and desert pavement, are not only nature's sculptured art forms but provide a clue to the stability of the desert. Wind has a sealing effect. **Desert pavement** is not a blacktop highway to Las Vegas or Reno but a natural armor on the desert surface. It forms as wind erosion and evaporation selectively remove the finer particles and moisture and leave behind salt precipitates to bond the coarse surface together and shield the underlying material from more erosion (fig. 23.9).

Wind-borne sediment particles are deposited downwind in a selective manner to form dunes and **loess** deposits many feet

thick. Loess is a well-drained, fertile, fine silt soil making up some of the richest farmland in the world. Because it consists of fine, dust-size particles, loess is found deposited hundreds of miles downwind in quieter zones of the atmosphere. Here is an example of the selective grading processes of nature's air transportation system at work.

Water Levels the Land

Not only is the Earth's crust in a dynamic state of tectonic activity but, once exposed above sea level, the force of gravity, with the aid of water, begins the leveling process. As streams and rivers flow to the sea, tons of sediment and dissolved minerals are continually removed each day. Not all becomes sedimentary rock. Some falls out as fresh flood plain alluvial deposits on the land to create some of the most productive lands on the Earth. The world's major population settlements are located on plains and river valleys where alluvial soils are abundant.

Figure 23.8

Nature's fertilizer is volcanic ash. Shown here is Mount St. Helens during its major eruption of May 18, 1980.

Photo by Jim Hughes, U.S. Department of Agriculture, U.S. Forest Service.

Soil Development Unique to this Planet

The entire soil-forming process is unique to the Earth. In order for plants to grow, they need to be anchored in a nutrient-rich medium. Bedrock alone does not satisfy most plant needs. The minerals must be in a soluble form. For this to happen, nature must break up the rock through the **weathering** process described in chapter 17. As a result of weathering, the top few inches of the Earth's lithosphere is a finely ground mixture of rock partially blended with organic

Figure 23.9

Desert pavement seals the fragile surface from wind erosion.

Photo by R. Wilshire, U.S. Geological Survey.

Figure 23.10

Like this oasis, our planet is also an oasis in space.

DESIGN OF THE HYDROSPHERE

matter, organisms, air, and water. This thin veneer covers the Earth in varying thicknesses, making possible a fertile environment for the biosphere. Again, one can see how interrelated the Earth's systems are to one another. The lithosphere, atmosphere, and hydrosphere create these conditions for the dynamic life of the biosphere. These spheres are linked together through the flow of energy and materials in the **rock cycle** and **hydrologic cycle.**

It has been said that humankind is just three feet away from starvation, or the average depth of the Earth's soil. This fragile zone of interface between the Earth's major spheres is the primary life support base for the planet. Water and weathering make it happen.

The planet Earth was really misnamed if one considers that over two-thirds of its surface is covered by water, and even the dry land areas are great groundwater reservoirs, not to mention the invisible water vapor in the air.

Truly the **hydrologic cycle** is the most unique feature of the Earth's planetary characteristics. Water and life are without question inseparable (fig. 23.10). The nature of water's molecular structure is the key to its physical and chemical properties. These unusual properties are of prime importance to the development and existence of life. Water holds the title of "universal solvent," which means it dissolves more substances than any other dissolving agent. This property alone is essential for life to exist. Water is the medium of nutrient transport (fig. 23.11). Minerals would not be available to plants and animals if water did not have this solvent characteristic.

Water dissolves not only solids but also life-supporting oxygen for the marine world. Gases such as carbon dioxide, which result from respiration, decomposition, and combustion, are also dissolved by water. If this did not happen, we would have found long ago the planet's atmosphere warming up, due to an ever-increasing greenhouse effect caused by a carbon dioxide buildup. Currently this greenhouse effect may be our greatest environmental threat as carbon dioxide continues to grow in the atmosphere and in the oceans.

Figure 23.11

Water moves nutrients in the form of silt and makes deposits on the flood plains of the earth. Flooding of the Red River during the 1975 flood in North Dakota is nature's way of fertilizing the soil. A number of factors contribute to floods on the Red River. Typically the ground is still frozen after a rapid snowmelt, so almost all the water runs off. Drainage is slow due to the gentle northerly gradient, about 4 to 6 inches to a mile near Oslo. Sometimes the areas to the north are still frozen and blocked by ice even after areas to the south have thawed. The floods carry away huge amounts of topsoil, but because the floodwater spreads out so much, especially in the downstream areas, they also tend to build up a layer of relatively fertile silt on the lake plain. In fact, the surface materials throughout much of the flood plain area are entirely flood silts; the actual Lake Agassiz deposits are generally buried from 3 to 10 feet deep.

Photo by John Bluemle, North Dakota Geological Survey. Used by permission.

Water's physical properties also make possible life on this planet. As water temperatures fall, the density increases to a maximum at 4° C (39.2° F), then decreases again as the freezing point is reached. When ice forms at 0° C (32° F), ice has a density lower than water and floats with one-tenth of its volume exposed above the water level. As a result, lakes are insulated by the ice cap, and water at 4° C, with its higher density, remains at a lower level. Fish and plant life have a place to live in the warmer, sheltered waters below the ice even though surface temperature may be well below freezing. If water reached its maximum density in the solid form, as most substances do, rivers and lakes would fill up with ice, and aquatic life would not be able to survive.

Because ice expands when it freezes, it is also a very important weathering agent in cooler climates. Water freezing in rock cracks breaks down rock into smaller pieces in the soil-forming process, thus releasing life-giving minerals to plants in these cold environments.

Water's high **specific heat** is also an important feature. Because it takes five times as much energy to heat a gram of water when compared to a gram of rock, the oceans are a heat sink, storing energy from the Sun. The temperature of the oceans remains moderate and the planet in general is temperate because of this unique characteristic of water.

Water also has boiling-point properties that are rather unusual when compared to most solvents. If we compare water to alcohol, gasoline, and ether, water takes first place with the highest boiling point. This may not seem significant at first, but consider what would happen if water's boiling point was more typical of

A

B

Figure 23.12

Mars (A) and Earth (B) are two environment systems in contrast: life versus no life. What is missing environmentally in the Mars picture?

A: Photo by NASA.

other substances, such as alcohol. The oceans, lakes, and rivers would not exist. The Earth's water would be primarily in the vapor form like the atmosphere of Venus. Thus, life as we observe it could not exist. All biological systems receive their nourishment by water-based circulatory systems. Each living cell is a reservoir of water and an aquarium for life's chemical processes. Life and water are inseparable.

Water in motion also transports heat. We can see this on a global scale as great ocean currents redistribute energy from the lower to the higher latitudes. On a human scale, our own body circulation, mostly water, performs the same functions of heat redistribution and nutrient transport.

THE BIOSPHERE

The Miracle of Community

You are alive! The Earth is alive. Is there a more important fact? None of us would be alive if we were not woven and integrated into the complex web of the **biosphere.** We are a family, community, one unique system of plants and animals, called the biosphere. Nothing exists alone. The whole cosmos is involved in keeping you alive. John Muir said, "When we try to pick out anything by itself, we find it hitched to everything else in the universe."

Adaptation is a relationship between living things and their surroundings. The surroundings are designed to fit the organism just as much as the organism is fit for the environment. It is a hand-in-glove relationship; otherwise, no life in that niche, the small valley, the entire Earth.

In other words, the Earth has evolved to be well fitted for life, and life is well fitted for the Earth. We have a good match. If the Earth, the small valley, the place were not fit for living systems, life would not be here but perhaps somewhere else.

This seems so simple, and yet the prospect of it happening and producing life as we know it somewhere else, is hard to imagine unless the environment were like our Earth. Was it a rare accident that our planet was sculptured for a biosphere? If it was chance, the likelihood of it occurring elsewhere is slim. Or was it a design of common occurrence in the universe? If this is the case, other suns in the vast reaches of space are niches with the same life support system capable of supporting biological communities much like our own planet (fig. 23.12).

We began our study of physical geography in a *study area* valley in chapter 1, a microcosm of the Earth. We saw how the geographer views the Earth by applying the geographic perspective in a small valley in nothern coastal California. Let's return now to this microcosm and refocus our attention on the biosphere, or life symbolized by the redwood plant community, in order to more fully appreciate the miracle of design.

As one enters the valley, one is impressed with the size and majestic stature of these red giants. They appear to need no other support. All other plants shrink at their feet, like servants paying tribute to their king.

Symbiosis in Design—"Togetherness"

No redwood stands alone! What design features give the coastal redwood its ability to survive for thousands of years as a plant and

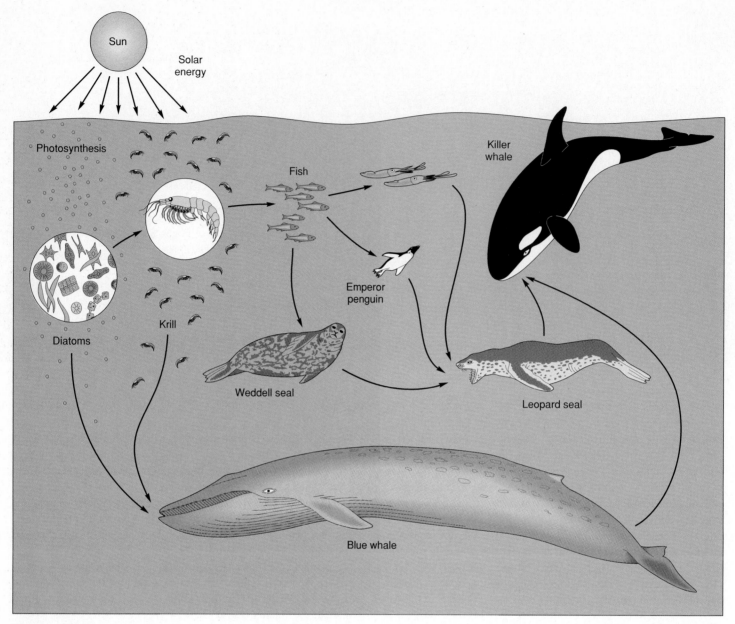

Figure 23.13

The *food chain* is really a flow of energy from the Sun. The top member of the food chain in Antarctica is the whale. The major energy-trappers are the minute diatoms that soak up energy from the sun using photosynthesis to convert water, carbon dioxide, and minerals into living tissue.

maintain its community? If all other members of this plant community were removed, the redwood would quickly die. All plants and animals share *symbiotic* relationships. In symbiosis—literally, "life together"—organisms of two different species live in a close association. Fleas on a dog, barnacles on a whale, algae and fungus in a lichen, and a hermit crab in a seashell are all examples of this kind of "togetherness." Without symbiosis, extermination would quickly occur for most organisms even though all other environmental factors could be considered positive.

The diversity, abundance, and interdependence of living things are impressive. The entire forest community is involved in the transformation of energy and the flow of nutrients and energy. From the smallest bacteria to the giant redwoods, each is related, and the cohesive bond is the flow of energy and materials through this living system.

Energy Flow

As we look at a forest system, the food chain in Antarctic waters, or any biological community, the circuit flow of energy provides

Figure 23.14

Fire is a stimulant to the forest.

Photo by Gary Bowman.

integration and gives meaning to the biosphere (fig. 23.13). If the flow of energy is cut off for even a short period, death and decay result, and the potential energy in the system is either stored or redirected to other living units. Utilization of energy and reproduction are the two primary functions of every living organism, and without energy disorganization and death result.

Form and Function

The coastal redwood is alive where it is because its form and function fit the environmental spectrum where it is growing. The great American architect, Louis Sullivan, said, "form follows function." This statement applies to biological design as well as architectural form. The coastal redwood community is a prime example of this design. Its height and density distribution, as described in chapter 12, all testify to its role in the plant community.

Each plant of the forest has its specific environmental requirements, or narrow spectrum of conditions favorable for its life cycle. As this narrow spectrum changes from place to place, the form and functions of the community also change. There are really two coastal redwood regions, one located on the river terraces and one on the steep mountain slopes. Each is unique in form and design, reflecting different environments. Redwoods on the steep slopes are smaller and more homogeneously blended with other conifers. The river terrace site produces environmental conditions that result in taller trees of uniform age with few competitors. Thus, the trees on the river flats are giants standing alone without competition.

The life of the forest is carried out by its individual units: trees, ferns, insects, birds, reptiles, fish, and many other units. What gives meaning to it all is the form and function. It is not just a department store of unrelated items under a common roof of the atmosphere. There is organization and activity, or *internal coherence* and *spatial interaction*. The fact that the redwoods have survived an Ice Age, floods, fires, and now humans, is illustrated in its miracle of design.

The forces that attack and kill most plants appear to be the very reasons the redwoods have endured for thousands of years. Essential to the longevity of the redwoods are these catastrophic events. These trees are well equipped to endure months without rain, and then months of heavy downpours and floodwater that completely destroy most plants by burying them in mud. Only the redwoods thrive under the stress of landslides, stream erosion, and silt deposits. Fires kill the competition of redwoods but rejuvenate and fertilize the redwood forest itself with a fresh supply of ash nutrients (fig. 23.14). Even where logging is carried on, redwoods reseed or resprout.

Take a deep breath and fill your lungs and sense of smell with the forest air. Stoop down and pick up the duff, or forest litter. Scoop up some soil and massage it between you fingers. Listen for a few minutes to the many sounds that express the life and vigor around you. Look at the forest, then its trees, shrubs, ferns, flowers, birds, reptiles, and variety of other animals. Soon you will begin to totally sense this community powered by the Sun and nourished by soil, air, and water.

The Miracle of Growth

Life is a quest for growth, dictated by the genetic code of the plant. Some plants grow ever so slowly, but the design is predictable and begins with seed germination and ends at death.

The rate of growth of the redwood is phenomenal. Under laboratory conditions, redwoods have been observed to grow 11 centimeters (4 1/2 inches) per day. However, in a natural environment, rates vary widely from almost none to over 1.2 meters (4 feet) per year.

Two principal energy forms, gravity and light, direct its movement during growth. If a seeding is grown inside a horizontal tube, its stem will turn up toward the light in the response known as *phototropism*. Its roots will turn down as soon as it is able to escape the container. This downward motion in response to gravity is known as *geotropism*. In the weightlessness of a space laboratory, plants show no orientation at all. A leaning redwood will grow more rapidly on the side facing the ground, building a buttress known as a *flatiron* to give it more support (fig. 23.15). A hormone, auxin, is the growth stimulator which collects along the leaning or lower side. Exactly how auxin concentrates in areas to correct nonvertical growth is still a mystery.

Although a tree increases in height each year, a nail placed at a given height will remain at that height until the tree dies. The only growth in a tree trunk is in its girth or diameter. In this category as well as in height, the redwood is amazing. Girth growth averages about 1 1/4 centimeters (1/2 inch) in diameter each year to about age 40, when it begins to slow down. A typical redwood reaches about 70 centimeters (28 inches) in diameter in 100 years; it is one of the world's fastest growing cone-bearers.

Contrary to popular belief, not all redwoods are large and old. The forest population represents all ages and sizes, much like the human population of a large city. There are redwoods near the *study area* several hundred years old, but only 6 to 9 meters (20 to 30 feet) tall. These trees survive in very poor soil conditions of the pygmy forest. Typically, in 2.59 square kilometers (1 square mile), only one tree may be over 2.4 meters (8 feet) in diameter and more than 1,000 years old. These are generally located on the better soils of the alluvial flats where flooding is common.

Drought-Resistant Design

To stand next to these giants is a humbling experience. These survivors of floods, fires, insects, drought, and human activity live on because of their unique adaptability. Each giant is really a water storage tank that it draws upon beginning in June. A tree 1.5 meters (5 feet) in diameter and 75 meters (250 feet) tall holds roughly 30,000 liters (8,000 gallons) of water to help it survive the long, dry summer. By late summer it has lost about one-fourth of that water. Because its root system is shallow and inefficient, this reservoir is essential to its survival. During the fall, winter, and spring, it is recharged for the next dry period.

Flood Insurance

On many alluvial flats in redwood country, floods periodically deposit silt as much as 1.2 meters (4 feet) thick. When this happens, the tree is capable of sending up vertical and horizontal roots into the fresh layer of deposits. With each deposit a new root system develops, and the old one dies.

This flood insurance design has allowed the redwood to remain and thrive while many of the plants typically associated with it are removed. Flooding reduces its competition. Tan oak, madrone, and Douglas fir, commonly associated with the redwood, drown during flooding. Because redwood can survive extremely low quantities of oxygen and moist conditions, it thrives like a swamp plant.

Fireproof Design

The remarkable redwood is not only waterproof but also fire resistant. Its thick bark shields its life support system of the cambium and sapwood. If it is seriously burned, it can sprout again from its base from dormant buds, or if its leaves or twigs are burned off, in a few days a new crown will begin to grow. Fire scars are common in the forest as a testimony to the redwood's ability to survive this killer. Some trees are hollowed out at the base by fire to produce cavernous hollows called "goose pens" and "catfaces," probably because they were used by the early settlers to keep geese in and the fancied resemblance to a cat's face. Sometimes a tree will be burned hollow in the entire lower trunk to produce a telescope tree. Thus the redwoods live on, while their competition is reduced by fire and floods.

Animal Resistance

While insects and fungus are melting down the dead cores and injured branches of most trees, the redwood has no tree-killing organisms. A number of insects and animals find the redwood a nice home for their life-style, but none can kill it. However, some animals and fungi in time can do considerable damage to the tree.

It is bothered by the gray squirrel to some degree. This tree lover strips away bark for its nest and girdles the tree some distance from the top. Deer are also heavy browsers of young seedlings and crown sprouts. Bears rip off the bark to chew on the nutrient-rich cambium layer beneath. In general, though, animals do not play a major direct ecological role. Their primary function is one of attacking most of the redwoods' plant competition. The redwood forest is rich in animal life, but the tree itself is not a major part of the food chain, while its plant competitors are very much affected. Again we see a plant with few enemies, accounting for its long endurance.

Resistance to Shade and Tolerance of Light

All plant life requires light to grow. When you stand in a redwood forest, the sky is hidden except for small peepholes where beams of light stream to the ground. The redwood is capable of photosynthesis beyond the range of most plants. It can carry on when its competitors are starving for energy.

The redwood can endure shade and yet thrive in light of the brightest intensities. Present-day logging practices of thinning by selective cutting open up the forest to light. The remaining trees

Figure 23.15

Two forces, gravity and solar energy, determine how and where the redwoods grow. As the tree leans, additional growth builds a buttress on the underside. The left side is nearly vertical where rapid compensating growth has stabilized this leaning giant.

Figure 23.16

Reproduction by root sprouts is the last method of reproduction in a changing climate.

rapidly accelerate their growth rate from well under 1 percent volume change to over 10 percent per year. This rapid rate change continues for several years until light and nutrients are again in reduced supply.

Death in Design

A species has only three choices when faced by a hostile environment: to migrate, mutate, or die. Let's consider each of these alternatives forced on the coastal redwood. The coastal redwood's days of migration are over. Its design is uniquely adapted to its boxed-in location along an 800-kilometer (500 mile) foggy coastal strip of North America, extending from Curry County, Oregon, to Monterey County, California. Very few occur more than 50 kilometers (30 miles) from the coast. It is believed the tree has been here over 40 million years, making it a very old resident. It has occurred in other parts of the world, with fossil remains being discovered in Europe, Japan, China, and various parts of the United States. There is even a petrified forest of redwood in Sonoma County, California.

Mutation is unlikely, since it has changed so little in its 40 million years of fossil record. Nature's third choice, death and extinction, appear inevitable. The tree appears doomed. This death verdict is strange when one considers its time on the Earth, but a number of environmental factors seem to be on a collision course to speed up the extinction prediction.

Botanists put the primary blame on a changing climate. The climate here for thousands of years has been slowly becoming more arid. According to local foresters, redwoods in Sonoma County and southward no longer reproduce by seeds (fig. 23.16). The summers are too long and dry for seeds to germinate. They are able to reproduce through root sprouts. This same trend is true on the higher slopes above the fog layer, where coastal redwoods need three to four times more moisture during the month of August than their chief competitors.

Human activity is also a major factor causing the environment to change at an accelerated rate. When we alter any of the trees' environmental controls or elements, it sets off a whole chain of events. Excessive logging destroys the soil base, increases runoff, and causes excessive flooding on the alluvial flats where big trees are located. Dam construction for flood control can lower water tables below the dam and raise them above. Both are detrimental. Occasional floods replenish the mineral supply.

Overprotection in a park setting can discourage occasional fires that are important in the survival design. It does not follow that by setting fires we can be assured of preservation. Per-

Figure 23.17

Death and decay bring life through nutrient recycling.

haps this forest will continue in spite of our predictions because it has a historical track record far beyond the level of human experience. We just don't know what the redwood has been able to adjust to during its 40 million years of varied history.

Death to individual trees is normal and part of the design. Death and decay bring life through nutrient recycling. In our *study area,* some mighty redwood giants have fallen to the ground to pave the way for new life by opening the forest to light (fig. 23.17).

Death comes mostly by windthrow. A few trees start leaning on their boggy foundation and just fall over without a breeze. Their small root systems, disproportionate to their size,

cannot hold them up when either high winds or lean sets them off balance. Compensation for lean is a slow process, and it only works when the tree is in a young, healthy, rapid state of growth, producing large volumes of wood. Thus, gravity takes its toll and humbles these giants.

As we take leave of the redwoods in the *study area,* we are struck with the amazing strength and cohesive nature of the Earth's living systems. Change and death are certain, but life's regenerative powers are also just as certain. We may be optimistic about the Earth's survival ability if we respect the miracle of design.

Appendix A
Climatic Classification

THE KÖPPEN CLIMATIC SYSTEM

The main purpose of any classification of climates is to provide a useful tool to evaluate a variety of climatic types in a comprehensive systematic manner. The Köppen climatic system used in this text defines climates according to the major climatic elements: precipitation and temperature. Since this system was first published in 1918 it has been modified by R. Geiger and W. Pohl, two of Köppen's students. This present version was published in 1953.

Both temperature and precipitation are viewed in terms of their effect on plant growth, resulting in climatic boundaries that correlate closely with vegetation patterns.

Köppen identifies five major groups, which are further subdivided into eleven principal climatic types. The five major groups are denoted by capital letters A through E.

Tropical Rainy Climates (A)

The mean monthly temperature of the coolest month is greater than 64.4° F (18° C). Precipitation potential exceeds evaporation. These climates have no winter season.

Dry Climates (B)

Evaporation exceeds precipitation on the average. Therefore, there are no water surpluses or permanent streams originating in the B climates.

Warm Temperate Rainy Climates (C)

The mean monthly temperature of the coldest month is below 64.4° F (18° C) but above 26.6° F (−3° C). Precipitation exceeds potential evaporation, and at least one month has a mean temperature above 50° F (10° C). Unlike the A climates, summer and winter seasons are the norm.

Snow Forest Climates (D)

The mean monthly temperature of the coolest month is below 26.6° F (−3° C). However, the average temperature of the warmest month is above 50° F (10° C). This boundary corresponds closely to the poleward limit or upper elevation of forest growth, the "timberline."

Polar Climates (E)

The mean monthly temperature of the warmest month is below 50° F (10° C). These climates have no true summer season, although seasonal variation does occur, especially over Antarctica and the northern borders of North America and most of Asia.

The selection of temperature and precipitation as the basis for this system stems from the facts that both are among the most important climatic elements and that these data are among the most available throughout the world.

The five major groups are divided into subgroups according to variations in temperature and precipitation. A second and third letter are used to indicate variations in precipitation and temperature conditions that produce distinct climatic types.

DRY CLIMATES SUBDIVISION

The capital letters S and W are used to designate semiarid or steppe (S) and arid or desert (W) climates. These letters apply to only the dry B climates. The boundary of the B climates can be determined by using the formulas listed in table A.2.

Köppen uses the annual precipitation, in inches, as the deciding factor defining the limits between dry and rainy climates using the formula $r = 0.44t - 8.5$ where t is annual mean temperature in degrees Fahrenheit and r is the annual precipitation. This empirically derived formula resulted from the relationships between temperature, precipitation, and vegetation patterns. He

Table A.1

Climate Data for Cities throughout the World

Climate/ Place	Jan.	Feb.	Mar.	Apr.	May	June	July	Aug.	Sept.	Oct.	Nov.	Dec.	Year
Am	Jakarta, Indonesia (elev. 26 ft.; lat. 6° S, long. 107° E)												
t (° F)	79	79	81	81	81	81	80	81	81	81	80	80	80
p (in)	11.8	11.8	8.3	5.8	4.5	3.8	2.5	1.7	2.6	4.4	5.6	8.0	70.8
Aw	Darwin, Australia (elev. 104 ft.; lat. 12.5° S, long. 131° E)												
t (° F)	84	83	84	84	82	79	77	79	83	85	86	85	83
p (in)	15.2	12.3	10.0	3.9	0.6	0.1	0.0	0.1	0.5	2.0	4.7	9.4	58.7
BSk	Albuquerque, NM (elev. 5,311 ft.; lat. 35° N, long. 106.5° W)												
t (° F)	35	40	46	56	65	75	78	76	70	58	44	37	57
p (in)	0.4	0.4	0.5	0.5	0.8	0.6	1.2	1.3	0.9	0.8	0.4	0.5	8.1
BSk	San Diego, CA (elev. 12 ft.; lat. 33° N, long. 117° W)												
t (° F)	55	56	59	62	64	66	70	72	70	66	62	57	63
p (in)	2.0	2.2	1.6	0.8	0.2	0.1	0.0	0.1	0.2	0.5	0.9	2.0	10.4
BWh	Cairo, Egypt (elev. 381 ft.; lat. 30° N, long. 31.5° E)												
t (° F)	55	57	63	70	76	80	82	82	78	74	65	58	70
p (in)	0.2	0.2	0.2	0.1	0.1	0.0	0.0	0.0	0.0	0.0	0.1	0.2	1.1
BWh	Lima, Peru (elev. 395 ft.; lat. 12° S, long. 77° W)												
t (° F)	70	72	72	68	64	61	60	59	59	61	63	67	65
p (in)	0.1	0.0	0.0	0.0	0.2	0.2	0.3	0.3	0.3	0.1	0.1	0.0	1.6
Cfa	Atlanta, GA (elev. 1,010 ft.; lat. 33.5° N, long. 84.5° W)												
t (° F)	45	46	51	60	69	77	79	78	73	62	51	45	61
p (in)	4.4	4.5	5.4	4.5	3.2	3.8	4.7	3.6	3.3	2.4	3.0	4.4	47.1
Cfb	Bergen, Norway (elev. 144 ft.; lat. 60.5° N, long. 5.5° E)												
t (° F)	34	34	36	42	49	55	58	57	52	45	39	36	45
p (in)	7.9	6.0	5.3	4.4	3.9	4.2	5.2	7.3	9.4	9.2	8.1	8.2	79.2
Cfb	Paris, France (elev. 246 ft.; lat. 49° N, long. 2.5° E)												
t (° F)	38	40	46	52	58	64	67	66	62	54	45	40	53
p (in)	2.2	1.8	1.4	0.7	2.2	2.1	2.3	2.5	2.2	2.0	2.0	2.0	24.4
Cfc	Reykjavik, Iceland (elev. 92 ft.; lat. 64° N, long. 22° W)												
t (° F)	31	32	34	38	44	50	52	51	48	42	35	32	41
p (in)	4.0	3.1	3.0	2.1	1.6	1.7	2.0	2.6	3.1	3.4	3.6	3.7	33.9
Csb	Portland, OR (elev. 30 ft.; lat. 45.5° N, long. 124° W)												
t (° F)	40	44	48	54	59	64	68	68	64	56	47	42	55
p (in)	5.4	4.9	4.2	2.4	1.9	0.6	0.4	0.6	1.8	3.5	6.0	7.1	39.9
Csb	Santiago, Chile (elev. 1,706 ft.; lat. 33.5° S, long. 71° W)												
t (° F)	67	66	62	56	51	46	46	48	52	56	61	66	56
p (in)	0.1	0.1	0.2	0.5	2.5	3.3	3.0	2.2	1.2	0.6	0.3	0.2	14.2
Cwa	Hong Kong, China (elev. 109 ft.; lat. 22° N, long. 114° E)												
t (° F)	60	59	64	71	78	82	82	82	81	77	70	64	72
p (in)	1.3	1.8	2.9	5.4	11.5	15.8	15.0	14.2	10.1	4.5	1.7	1.2	85.4
Cwb	Mexico City, Mexico (elev. 7,340 ft.; lat. 19.5° N, long. 99° W)												
t (° F)	54	57	61	64	65	64	62	62	61	59	56	54	60
p (in)	0.2	0.3	0.5	0.7	1.9	4.1	4.5	4.3	4.1	1.6	0.5	0.3	23.0
Dfa	Chicago, IL (elev. 607 ft.; lat. 42° N, long. 88° W)												
t (° F)	26	28	36	49	60	70	76	74	66	55	40	29	51
p (in)	1.9	1.6	2.7	3.0	3.7	4.1	3.4	3.2	2.7	2.8	2.2	1.9	33.2
Dfb	Calgary, Canada (elev. 3,540 ft.; lat. 51° N, long. 144° W)												
t (° F)	13	17	26	40	50	56	62	60	51	42	28	19	39
p (in)	0.5	0.5	0.8	1.0	2.3	3.1	2.5	2.3	1.5	0.7	0.7	0.6	16.7
Dfb	Portland, ME (elev. 43 ft.; lat. 44° N, long. 74° W)												
t (° F)	22	23	31	42	53	62	68	67	59	49	38	26	45
p (in)	4.4	3.8	4.3	3.7	3.4	3.2	2.9	2.4	3.5	3.2	4.2	3.8	42.8
ET	Angmagssalik, Greenland (elev. 95 ft.; lat. 65.5° N, long. 37.5° W)												
t (° F)	19	19	21	27	36	43	45	45	39	32	27	23	31
p (in)	2.9	2.4	2.6	2.1	2.0	1.8	1.5	2.1	3.3	4.7	3.0	2.7	31.1

Source: *Tables of Temperature, Relative Humidity, and Precipitation for the World, 1958,* Meteorological Office, Air Ministry of Great Britain.

Table A.2
Boundaries for B Climates

Precipitation	Boundary between *BS* and Rainy Climates	Boundary between *BW* and *BS*
Maximum in summer	$r \leqq 0.44t - 3$	$r \leqq \dfrac{0.44t - 3}{2}$
Maximum in winter	$r \leqq 0.44t - 14$	$r \leqq \dfrac{0.44t - 14}{2}$

found that the vegetation changed from forest to steppe forms where the rainfall is less than the value $0.44t - 8.5$.

The B climates can be further subdivided between steppe (BS) and desert (BW), using the above formula and dividing by two:

$$\text{Desert (BW)} \quad r \leqq \frac{0.44t - 8.5}{2}$$

The desert (BW) exists where the precipitation (r) drops to one half or less than the limit between the dry (B) and moist (C and D) climates. The boundaries between B climates and rainy climates are also affected by the seasonal patterns of precipitation. If the rainfall has a winter maximum, evaporation takes a smaller toll on the water available to plants, so the formula is adjusted accordingly. Summer maximum rainfall zones experience higher water loss due to evaporation, so Köppen formulas are modified accordingly.

The third letters used in Köppen's B climates describe temperature.

h—indicates hot, dry conditions where the mean annual temperature is above 64.4° F (18° C).

k—indicates that the mean annual temperature is below 64.4° F (18° C). Therefore, a BWh and BWk indicate whether the desert is hot and dry or cold and dry. BSh or BSk gives the same differentiation for the steppe semiarid climates.

f—means that sufficient precipitation falls in all seasons; it is used with the A, C, and D climates.

w—means that the dry season occurs in the winter; it is used with A, C, and D climates.

s—means that the dry season occurs in the summer; it is used with A, C and D.

m—is used only with the A climate to indicate a monsoon rainfall pattern. This means there is sufficient precipitation to support a rainforest, but a distinct dry season occurs during the period of high sun.

The C and D climates use the third letters a, b, and c to indicate temperature variations; the letter d is used with D climates only.

a—hot summers—the mean temperature of the warmest month is above 71.6° F (22° C).

b—warm summers—the mean temperature of the warmest month is below 71.6° F (22° C) but at least four months are above 50° F (10° C).

c—cool summers—summers are cool and short, less than four months have mean temperatures above 50° F (10° C).

d—very short summers with cold winters; the coldest month is below −36.4° F (−38° C).

The above letters produce the following climates, described in the text. See figure A.1 for a global pattern of climate.

GLOBAL CLIMATES

Tropical rainforest (Af). Precipitation of the driest month is at least 2.4 inches (6 cm).

Tropical monsoon (Am) (with distinct rainy seasons). At least one month has less than 2.4 inches (6 cm).

Tropical savanna (Aw). Like the Am variety, at least one month has precipitation less than 2.4 inches (6 cm). The annual rainfall is insufficient to support a tropical rainforest. The vegetation pattern is usually a parkland of grasses and groves of trees.

Steppe (BSh). A tropical semiarid grassland environment where the mean annual temperature is above 64.4° F (18° C).

Middle-latitude steppe (BSk). A middle-latitude semiarid grassland environment where the mean annual temperature is below 64.4° F (18° C).

Tropical desert (BWh). A tropical arid climate with insufficient precipitation to support a grassland and a mean annual temperature above 64.4° F (18° C).

Figure A.1
World climatic regions.

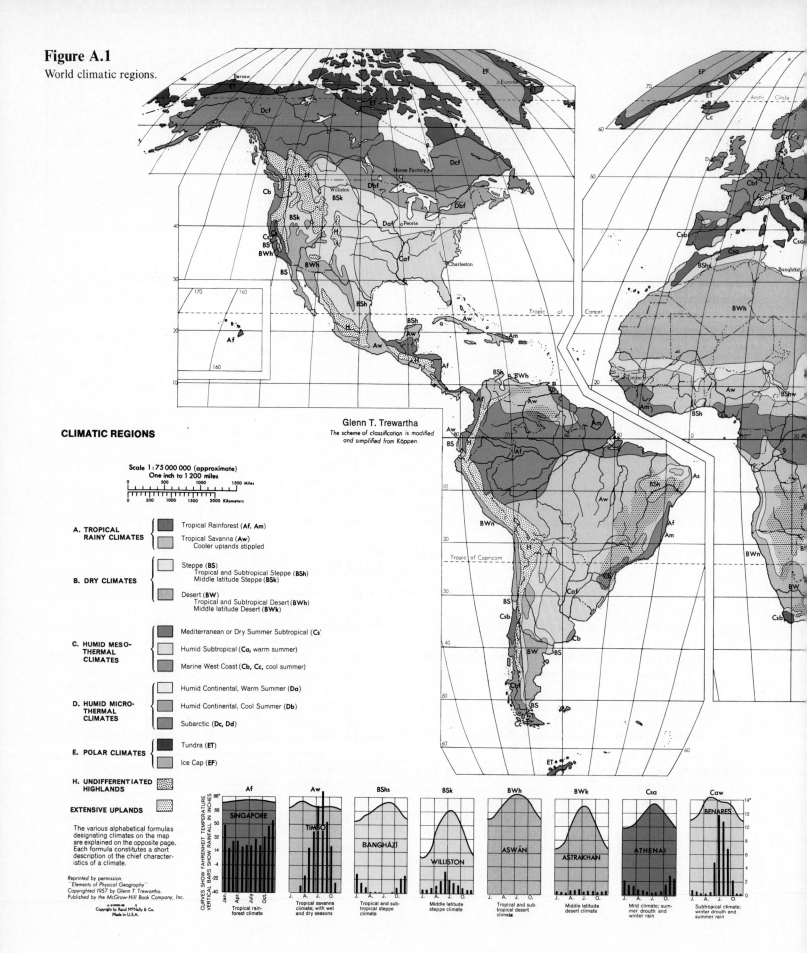

Glenn T. Trewartha
The scheme of classification is modified and simplified from Köppen.

CLIMATIC REGIONS

Scale 1:75 000 000 (approximate)
One inch to 1 200 miles

0	500	1000		1500 Miles	
0	500	1000	1500	2000 Kilometers	

A. TROPICAL RAINY CLIMATES
 Tropical Rainforest (**Af. Am**)
 Tropical Savanna (**Aw**)
  Cooler uplands stippled

B. DRY CLIMATES
 Steppe (**BS**)
  Tropical and Subtropical Steppe (**BSh**)
  Middle latitude Steppe (**BSk**)
 Desert (**BW**)
  Tropical and Subtropical Desert (**BWh**)
  Middle latitude Desert (**BWk**)

C. HUMID MESO-THERMAL CLIMATES
 Mediterranean or Dry Summer Subtropical (**Cs**)
 Humid Subtropical (**Ca**, warm summer)
 Marine West Coast (**Cb, Cc**, cool summer)

D. HUMID MICRO-THERMAL CLIMATES
 Humid Continental, Warm Summer (**Da**)
 Humid Continental, Cool Summer (**Db**)
 Subarctic (**Dc, Dd**)

E. POLAR CLIMATES
 Tundra (**ET**)
 Ice Cap (**EF**)

H. UNDIFFERENTIATED HIGHLANDS

EXTENSIVE UPLANDS

The various alphabetical formulas designating climates on the map are explained on the opposite page. Each formula constitutes a short description of the chief characteristics of a climate.

Reprinted by permission:
"Elements of Physical Geography"
Copyrighted 1957 by Glenn T. Trewartha.
Published by the McGraw-Hill Book Company, Inc.

A-510000-88
Copyright by Rand M°Nally & Co.
Made in U.S.A.

CURVES SHOW FAHRENHEIT TEMPERATURE
VERTICAL BARS SHOW RAINFALL IN INCHES

Af — SINGAPORE
Tropical rainforest climate

Aw — TIMBO
Tropical savanna climate; with wet and dry seasons

BShs — BANGHÁZI
Tropical and subtropical steppe climate

BSk — WILLISTON
Middle latitude steppe climate

BWh — ASWÂN
Tropical and subtropical desert climate

BWk — ASTRAKHAN
Middle latitude desert climate

Csa — ATHENAI
Mild climate; summer drouth and winter rain

Caw — BENARES
Subtropical climate; winter drouth and summer rain

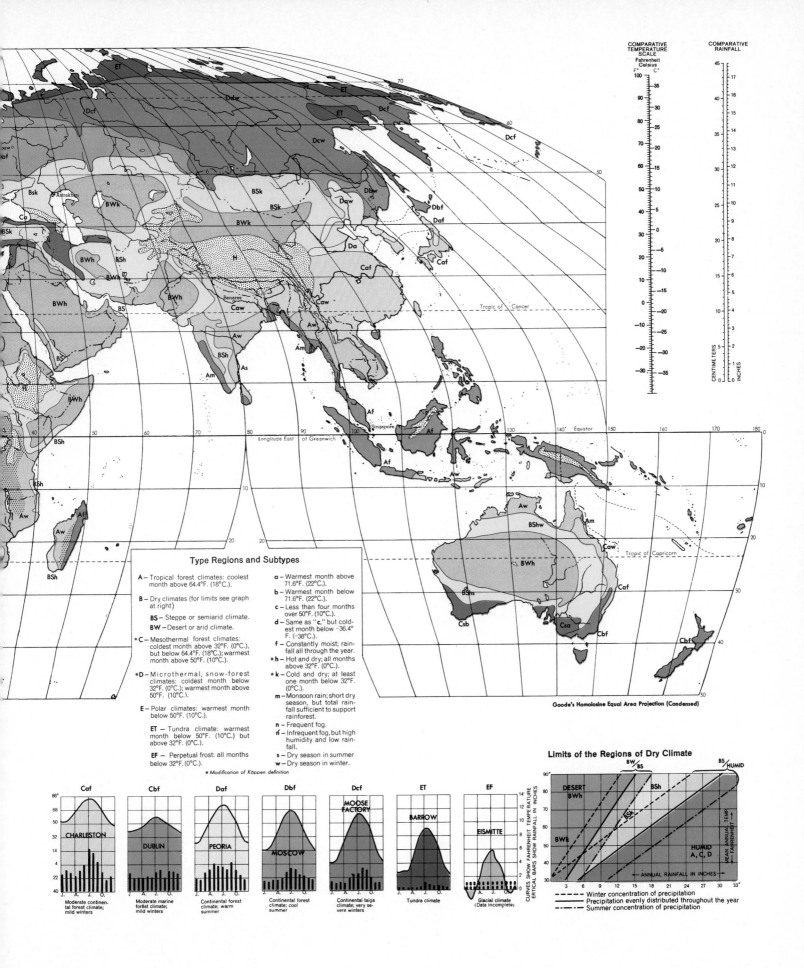

COMPARATIVE TEMPERATURE SCALE
Fahrenheit Celsius

COMPARATIVE RAINFALL

Goode's Homolosine Equal Area Projection (Condensed)

Type Regions and Subtypes

A – Tropical forest climates: coolest month above 64.4°F. (18°C.).

B – Dry climates (for limits see graph at right)

BS – Steppe or semiarid climate.

BW – Desert or arid climate.

*C – Mesothermal forest climates: coldest month above 32°F. (0°C.), but below 64.4°F. (18°C.); warmest month above 50°F. (10°C.).

*D – Microthermal, snow-forest climates: coldest month below 32°F. (0°C.); warmest month above 50°F. (10°C.).

E – Polar climates: warmest month below 50°F. (10°C.).

ET – Tundra climate: warmest month below 50°F. (10°C.) but above 32°F. (0°C.).

EF – Perpetual frost: all months below 32°F. (0°C.).

a – Warmest month above 71.6°F. (22°C.).

b – Warmest month below 71.6°F. (22°C.).

c – Less than four months over 50°F. (10°C.).

d – Same as "c," but coldest month below –36.4° F. (–38°C.).

f – Constantly moist; rainfall all through the year.

*h – Hot and dry; all months above 32°F. (0°C.).

*k – Cold and dry; at least one month below 32°F. (0°C.).

m – Monsoon rain; short dry season, but total rainfall sufficient to support rainforest.

n – Frequent fog.

ń – Infrequent fog, but high humidity and low rainfall.

s – Dry season in summer.

w – Dry season in winter.

* Modification of Köppen definition

Caf
CHARLESTON
Moderate continental forest climate; mild winters

Cbf
DUBLIN
Moderate marine forest climate; mild winters

Daf
PEORIA
Continental forest climate; warm summer

Dbf
MOSCOW
Continental forest climate; cool summer

Dcf
MOOSE FACTORY
Continental taiga climate; very severe winters

ET
BARROW
Tundra climate

EF
EISMITTE
Glacial climate (Data incomplete)

CURVES SHOW FAHRENHEIT TEMPERATURE
VERTICAL BARS SHOW RAINFALL IN INCHES

Limits of the Regions of Dry Climate

DESERT BWh

BWk

BSh

HUMID A, C, D

MEAN ANNUAL TEMP. FAHRENHEIT

ANNUAL RAINFALL IN INCHES

- - - - Winter concentration of precipitation
———— Precipitation evenly distributed throughout the year
–·–·– Summer concentration of precipitation

Middle-latitude desert (BWk). A middle-latitude desert climate where the mean annual temperature is below 64.4° F (18° C).

Humid subtropical (Cfa). Hot, humid summers with no dry season. The warmest month is over 71.6° F (22° C). Precipitation of the driest month exceeds 1.2 inches (3 cm).

Marine west coast (Cfb & Cfc). Mild (Cfb) to cool (Cfc) summers and winters with precipitation in all seasons. The average temperature of the summer months is above 50° F (10° C) but below 71.6° F (22° C).

Subtropical monsoon (Cwa). Mild, dry winters with hot, humid summers with average temperature of the warmest month at least 71.6° F (22° C) or higher.

Mediterranean (Csa & Csb). Warm (Csb) to hot (Csa) dry summers and mild, rainy winters with precipitation for the driest month less than one-third the amount that falls during the wettest month of the winter rainy season.

Snow forest continental (Dfa & Dfb). Summers are humid and range from mild (Dfb) to hot (Dfa) with the average temperature of the warmest month above 50° F (10° C) and winter months 32° F (0° C) and below with snow in the winter and rain in the summer.

Subarctic (Dfc, Dwc, Dfd, Dwd). Winters are long and cold and summers are mild and short. At least one month averages above 50° F (10° C) but not more than three, and none reach 71.6° F (22° C). If the average temperature of the coldest month is below −36.4° F (−38° C), the letter d is used.

Tundra (ET). This is a summerless climate. The average temperature of the warmest month is below 50° F (10° C). Precipitation is very low.

Icecap (EF). Average temperature of warmest month is 32° F (0° C) or below. This is a land of permanent frost or permafrost.

Highland Polar (EH). Icecap climate located in the high mountains outside the arctic region.

Appendix B
Soil Classification Systems

THE U.S. COMPREHENSIVE SOIL CLASSIFICATION SYSTEM: 7TH APPROXIMATION

The U.S. Comprehensive Soil Classification System, completed in 1960 by the Soil Conservation Service of the U.S. Department of Agriculture, is commonly referred to as the 7th Approximation because it represents the seventh revision of an original proposal developed in the 1950s. In this new system the physical and chemical properties as they now exist are the chief focus, rather than formations, processes, and factors.

The system's categories include 10 orders, 40 suborders, 120 great groups, 400 subgroups, 1,500 families, and finally thousands of soil series named after the local place in which a homogeneous soil type is found. The orders and suborders represent global patterns. See table B.1 for a comparison of equivalents between the 1938 U.S. Soil Classification System and the 7th Approximation.

After the properties of a soil's horizons are diagnosed, the soil is assigned to a series and family, which is a member of a group and global order. Surface horizons, referred to as epipedons, are darkened by organic matter and include the eluvial horizons. There are six of these horizons and 13 subsurface horizons. Three other horizons are the *duripan,* a horizon cemented by silica or aluminum silicate; the *fragipan,* a loamy subsurface horizon with platy structure and high bulk density; and *plinthite,* rich in sesquioxides, highly weathered and poor in humus.

These diagnostic horizons combined together are known as the *pedon,* a three-dimensional unit of soil with a minimum area between 1 and 10 square meters. This is the basic unit of soil studied and classified into a soil series.

We shall briefly summarize the characteristics of each soil order in the Comprehensive System; table B.1 gives the corresponding suborders. The ten orders will be described in order of increasing soil activity and profile development.

Entisols

Entisols are soils of recent development without horizons, with the following suborders. They can be found in all climatic environments.

> Fluvents—on alluvial deposits (alluvial soils).
> Psamments—sand or loamy texture.
> Arents—strong artificial disturbance.
> Orthents—loam or clay organic content decreases regularly with depth.

Histosols (Greek **histos,** *tissue)*

Composed primarily of organic matter, Histosols develop in watery environments. Here organic matter breaks down slowly due to low oxygen content in the water. They occur from the arctic to the tropics. Peat and muck grow in this environment. Suborders are not finalized.

Vertisols (from Latin, to turn)

Horizon development of Vertisols is hindered by constant churning of the soil caused by repeated wetting and drying. They tend to have clay minerals capable of great water-holding capacity. When dry, these soils may form deep cracks which swell closed when wet.

> Uderts—usually moist cracks open only short periods.
> Usterts—dry for short periods.
> Xererts—dry for long periods, often called adobe in the West.
> Torrerts—usually dry, wide, deep cracks open most of the time.

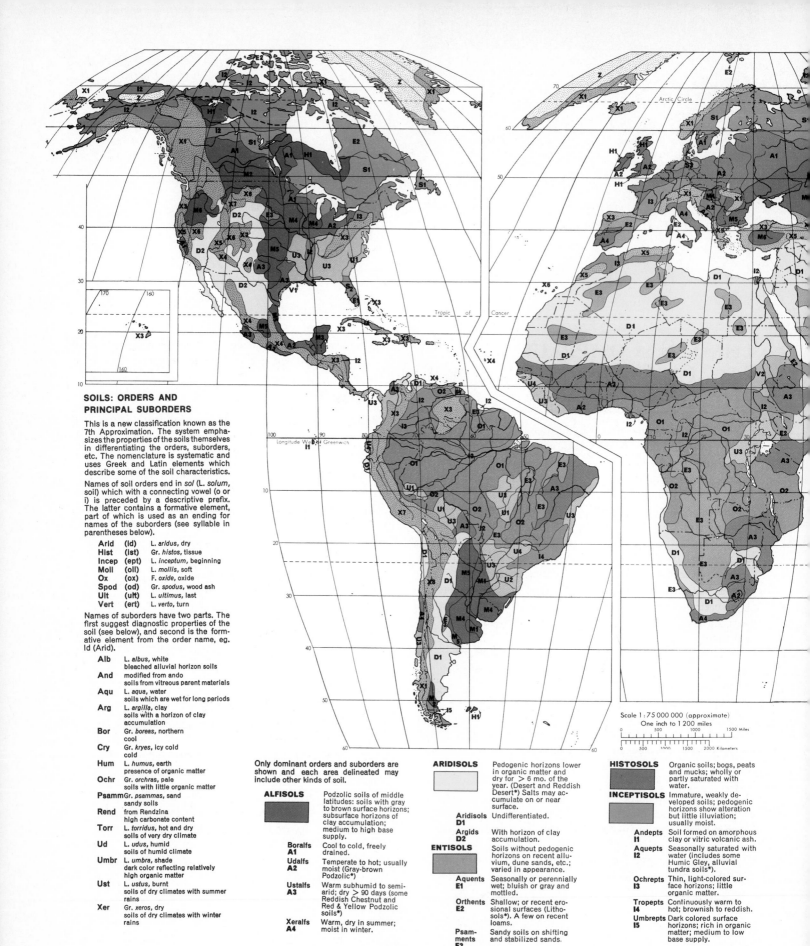

SOILS: ORDERS AND PRINCIPAL SUBORDERS

This is a new classification known as the 7th Approximation. The system emphasizes the properties of the soils themselves in differentiating the orders, suborders, etc. The nomenclature is systematic and uses Greek and Latin elements which describe some of the soil characteristics.

Names of soil orders end in *sol* (L. *solum*, soil) which with a connecting vowel (o or i) is preceded by a descriptive prefix. The latter contains a formative element, part of which is used as an ending for names of the suborders (see syllable in parentheses below).

Arid	(Id)	L. *aridus*, dry
Hist	(Ist)	Gr. *histos*, tissue
Incep	(ept)	L. *inceptum*, beginning
Moll	(oll)	L. *mollis*, soft
Ox	(ox)	F. *oxide*, oxide
Spod	(od)	Gr. *spodus*, wood ash
Ult	(ult)	L. *ultimus*, last
Vert	(ert)	L. *verto*, turn

Names of suborders have two parts. The first suggest diagnostic properties of the soil (see below), and second is the formative element from the order name, eg. Id (Arid).

Alb	L. *albus*, white bleached alluvial horizon soils
And	modified from ando soils from vitreous parent materials
Aqu	L. *aqua*, water soils which are wet for long periods
Arg	L. *argilla*, clay soils with a horizon of clay accumulation
Bor	Gr. *boreas*, northern cool
Cry	Gr. *kryes*, icy cold cold
Hum	L. *humus*, earth presence of organic matter
Ochr	Gr. *orchras*, pale soils with little organic matter
Psamm	Gr. *psammas*, sand sandy soils
Rend	from Rendzina high carbonate content
Torr	L. *torridus*, hot and dry soils of very dry climate
Ud	L. *udus*, humid soils of humid climate
Umbr	L. *umbra*, shade dark color reflecting relatively high organic matter
Ust	L. *ustus*, burnt soils of dry climates with summer rains
Xer	Gr. *xeros*, dry soils of dry climates with winter rains

Only dominant orders and suborders are shown and each area delineated may include other kinds of soil.

ALFISOLS
Podzolic soils of middle latitudes: soils with gray to brown surface horizons; subsurface horizons of clay accumulation; medium to high base supply.

Boralfs A1 Cool to cold, freely drained.

Udalfs A2 Temperate to hot; usually moist (Gray-brown Podzolic*)

Ustalfs A3 Warm subhumid to semi-arid; dry > 90 days (some Reddish Chestnut and Red & Yellow Podzolic soils*)

Xeralfs A4 Warm, dry in summer; moist in winter.

ARIDISOLS
Pedogenic horizons lower in organic matter and dry for > 6 mo. of the year. (Desert and Reddish Desert*) Salts may accumulate on or near surface.

Aridisols D1 Undifferentiated.

Argids D2 With horizon of clay accumulation.

ENTISOLS
Soils without pedogenic horizons on recent alluvium, dune sands, etc.; varied in appearance.

Aquents E1 Seasonally or perennially wet; bluish or gray and mottled.

Orthents E2 Shallow; or recent erosional surfaces (Lithosols*). A few on recent loams.

Psamments E3 Sandy soils on shifting and stabilized sands.

HISTOSOLS
Organic soils; bogs, peats and mucks; wholly or partly saturated with water.

INCEPTISOLS
Immature, weakly developed soils; pedogenic horizons show alteration but little illuviation; usually moist.

Andepts I1 Soil formed on amorphous clay or vitric volcanic ash.

Aquepts I2 Seasonally saturated with water (includes some Humic Gley, alluvial tundra soils*).

Ochrepts I3 Thin, light-colored surface horizons; little organic matter.

Tropepts I4 Continuously warm to hot; brownish to reddish.

Umbrepts I5 Dark colored surface horizons; rich in organic matter; medium to low base supply.

Scale 1 : 75 000 000 (approximate)
One inch to 1 200 miles

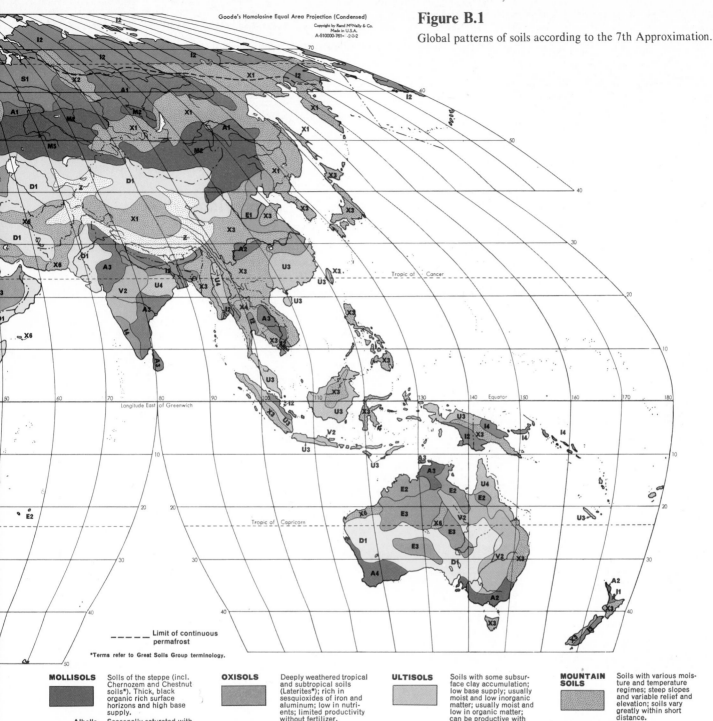

Global patterns of soils according to the 7th Approximation.

Goode's Homolosine Equal Area Projection (Condensed)
Copyright by Rand McNally & Co.
Made in U.S.A.
A-510000-761- -2-2-2

— — — — Limit of continuous permafrost

*Terms refer to Great Soils Group terminology.

MOLLISOLS Soils of the steppe (incl. Chernozem and Chestnut soils*). Thick, black organic rich surface horizons and high base supply.

Albolls M1 Seasonally saturated with water; light gray subsurface horizon.

Borolls M2 Cool or cold (incl. some Chernozem, Chestnut and Brown soils*).

Rendolls M3 Formed on highly calcareous parent materials (Rendzina*).

Udolls M4 Temperate to warm; usually moist (Prairie soils*).

Ustolls M5 Temperate to hot; dry for > 90 days (incl. some Chestnut and Brown soils*).

Xerolls M6 Cool to warm; dry in summer; moist in winter.

OXISOLS Deeply weathered tropical and subtropical soils (Laterites*); rich in sesquioxides of iron and aluminum; low in nutrients; limited productivity without fertilizer.

Orthox O1 Hot and nearly always moist.

Ustox O2 Warm or hot; dry for long periods but moist > 90 consecutive days.

SPODOSOLS Soils with a subsurface accumulation of amorphous materials overlaid by a light colored, leached sandy horizon.

Spodosols S1 Undifferentiated (mostly high latitudes).

Aquods S2 Seasonally saturated with water; sandy parent materials.

Humods S3 Considerable accumulations of organic matter in subsurface horizon.

Orthods S4 With subsurface accumulations of iron, aluminum and organic matter (Podzols*).

ULTISOLS Soils with some subsurface clay accumulation; low base supply; usually moist and low inorganic matter; usually moist and low in organic matter; can be productive with fertilization.

Aquults U1 Seasonally saturated with water; subsurface gray or mottled horizon.

Humults U2 High in organic matter; dark colored; moist, warm to temperate all year.

Udults U3 Low in organic matter; moist, temperate to hot (Red-Yellow Podzolic; some Reddish-Brown Lateritic soils*).

Ustults U4 Warm to hot; dry > 90 days.

VERTISOLS Soils with high content of swelling clays; deep, wide cracks in dry periods dark colored.

Uderts V1 Usually moist; cracks open < 90 days.

Usterts V2 Cracks open > 90 days; difficult to till (Black tropical soils*).

MOUNTAIN SOILS Soils with various moisture and temperature regimes; steep slopes and variable relief and elevation; soils vary greatly within short distance.

X1 Cryic great groups of Entisols, Inceptisols and Spodosols.

X2 Boralfs and Cryic groups of Entisols and Inceptisols.

X3 Udic great groups of Alfisols, Entisols and Ultisols; Inceptisols.

X4 Ustic great groups of Alfisols, Entisols, Inceptisols, Mollisols and Ultisols.

X5 Xeric great groups of Alfisols, Entisols, Inceptisols, Mollisols and Ultisols.

X6 Torric great groups of Entisols; Aridisols.

X7 Ustic and cryic great groups of Alfisols, Entisols; Inceptisols and Mollisols; ustic great groups of Ultisols; cryic great groups of Spodosols.

X8 Aridisols; torric and cryic great groups of Entisols, and cryic great groups of Spodosols and Inceptisols.

Z Areas with little or no soil; icefields, and rugged mountain.

Soil Classification Systems   **513**

Table B.1

1938 U.S. Soil Classification System and Approximate Equivalents of the 1960 U.S. Comprehensive Classification

Soil Types		1938 System: Great Soil Groups Order	1960 System Order
Zonal	Pedalfer	Laterite Yellow-red podzolic Latosol	Oxisol Ultisol Alfisol Inceptisol
		Podzol Gray-brown podzolic	Ultisol Spodosol Alfisol Inceptisol
		Prairie Degraded chernozem Noncalcic brown	Alfisol Mollisol Inceptisol
		Tundra Alpine turf Polar desert	Inceptisol
	Pedocal	Chernozem Chestnut	Mollisol Alfisol
		Brown Sierozem Desert	Aridisol Mollisol Alfisol
Intrazonal		Solonchak Solonetz Soloth	Inceptisol Aridisol Mollisol Alfisol
		Humic gley Bog Planosol	Inceptisol Mollisol Alfisol Spodosol Ultisol Entisol
		Brown forest Rendzina	Inceptisol Mollisol
Azonal		Lithosols Regosol Alluvial	Entisol Inceptisol Mollisol

Source: USDA, U.S. Government Printing Office.

Inceptisols (Latin inceptum, beginning)

Inceptisols are moderately developed soils on recent alluvium, glacial or eolian deposits, or volcanic ash.

> Aquepts—with gleying features.
> Andepts—on volcanic ash.
> Tropepts—in tropical climates.
> Umbrepts—in crystalline clay minerals, thick black surface horizon.
> Orchrepts—in crystalline clay minerals, light-colored surface horizons.

Aridisols (Latin aridus, dry)

Aridisols have very shallow profiles resulting from infrequent downward movement of water. These soils are low in organic matter and rich in salts and bases. Salty iron and silicate crusts form on the surface, producing a stable layer known as desert pavement. Aridisols are pedocals in the older system consisting of desert soils.

> Argids—with an argillic or clay A horizon.
> Orthids—accumulation of salts such as calcium carbonate and gypsum.

Mollisols (Latin mollis, soft)

Mollisols have a dark, rich humus layer in the A horizon and a high base content. Grasses form the primary plant cover in this transitional zone between the more humid forest and arid steppe and desert. Mollisols do not harden and crack extensively when dry but maintain favorable physical properties for tillage. Calcification is the soil-forming process, producing some of the most fertile soils in the world for grain crops.

> Albolls—seasonal perched water table, bleached B horizon.
> Aquolls—seasonally wet, gray subsurface horizons, and gleying characteristics.
> Rendolls—subsurface B horizon with large amounts of calcium carbonate.
> Borolls (in cool regions)—North Central United States.
> Udolls—in mid-latitude humid climates without large amounts of calcium carbonate or gypsum.
> Ustolls—mostly in semiarid regions, dry for long periods of time.
> Xerolls—in Mediterranean climates with wet winters and long, dry summers.

Spodosols (Greek spodos, wood ash)

Spodosols are leached in the A horizon (albic), producing an ash color. The B horizon (spodic) is iron-colored, sometimes forming a clay hardpan. The podzolization process dominates this order, producing acidic, low-fertility soils where leaching is severe. These are the cooler midlatitude soils.

Aquods—features of gleying.
Ferrods—with little humus in spodic horizon.
Humods—with little iron in spodic horizon.
Orthods—with iron and humus.

Alfisols (Derived from aluminum, Al, and iron, Fe)

The A horizons or orchric epipedons of Alfisols are yellowish brown, being colored by iron and aluminum oxides. Bases have been moderately leached, producing a clay hardpan in the B horizon. Alfisols are in a zone of transition between the drier Mollisols and the humid Spodosols and Ultisols. Alfisols range from the southern Great Lakes region to the tropics. These soils can be very productive where leaching is moderate. The American "corn belt" is partially located on Alfisols. These soils are formed by podzolization.

Aqualfs—gleying features, seasonally wet.
Boralfs—in cool to cold regions, forests.
Udalfs—in middle latitudes to tropical regions, usually moist, dry only during short periods.
Ustalfs—in middle latitudes to tropical regions, reddish brown, dry for long periods.
Xeralfs—in areas of rainy winters and dry, long summers.

Ultisols (Latin ultimos, ultimate)

Ultisols are formed by the podzolization process. They are similar to Alfisols but represent a greater degree of leaching. They are generally in warmer, more humid regions than the Alfisols and have redder, more acidic horizons due to iron and aluminum oxides that have experienced significant leaching of bases. The forest cover is very important in maintaining the nutrient cycle. Reforestation ensures rapid leaching of nutrients. These soils are found in the subtropics along the southern Atlantic states and the lower Mississippi valley. These soils represent a transition between the podzolization and laterization processes, being equivalent to the red and yellow podzols of the 1938 system.

Aquults—seasonally wet, gleying features.
Humults—in warm, humid regions, high in humus content.
Udults—in continuously moist regions, low in organic content.
Ustults—in seasonally wet-dry climates of the subtropics and tropics, low in organic content.
Xerults—in regions of rainy winters and dry summers; dry over long periods, low in organic content; vegetation is grassland and woodland.

Oxisols

Oxisols represent the highest level of leaching characteristic of the humid tropics and the laterization process. The oxisols have a sub-surface B horizon (oxic horizon) consisting of iron, aluminum, and clay oxides and hydroxides with bases absent. Oxisols are more widely distributed than the climate responsible for producing them. Some currently exist in rather arid regions, indicating that climates have changed since they were formed. Vegetation cover is a key factor in recycling the nutrients, as in the Ultisols.

Aquox—gleying features, bogs and meadows.
Humox—high in organic matter, low in bases and moist.
Orthox (latosols)—moist, low in organic matter and alkalinity.
Ustox (latosols)—partly dry for long periods of time.

THE 1938 UNITED STATES SOILS CLASSIFICATION SYSTEM

The 1938 Soil Classification System evolved from the Classification of World Soils developed by V. V. Dokuchaiev in 1900. He is credited with producing the first natural soil classification, based on observable features in the soil. Dokuchaiev observed that many soil types had a definite environmental association. This led to the concept of zonal soils, the primary feature of 1900 and 1938 soil classification systems.

The 1938 system developed by C. F. Marbut, Chief of the U.S. Soil Survey, stressed not only the zonal concept but also azonal and intrazonal soils. These represent the three main orders of soils in this system. *Zonal soils* have well-developed horizons and reflect broad climatic and vegetation belts of the world.

The zonal soils can be divided into two categories that reflect the soil-forming process. Drier zonal soils, rich in bases such as calcium and magnesium, produced by calcification processes, are called *pedocals*. Soils formed in the more humid climates, where forest dominates the plant cover, are referred to as *pedalfers*. Pedalfers are formed by the laterization and podzolization processes, which leach the soil of bases leaving iron and aluminum in the solum or topsoil. Soils that lack horizon development are classified as *azonal*. They have not had time to develop in a given environment. Examples are recent glacial, alluvial, and windblown deposits. Volcanic ash and geologically disturbed areas fall into this order. Intrazonal soils make up a third order consisting of soils resulting from local site factors. Soils formed in marshes and tidal flats, where drainage is poor, fall into this category along with soils rich in a given mineral such as calcium.

The soil orders are divided into suborders reflecting environmental influences such as vegetation and climate and the other soil-forming factors. The suborders are further divided into 40 great soil groups. The groups are divided further into families and families into series, bearing local place names corresponding to the location in which the soil type occurs (table B.2).

Table B.2

Suborders and Great Groups of the Zonal Order of Soils: Traditional USDA System

Suborders	Great Soil Groups	General Traits
Soils of cold regions	Tundra soils	Soils of the active layer in permafrost regions; poorly drained, often waterlogged with heavy organic accumulation in nine horizons
Light-colored soils of arid regions	Desert soils Red desert soils Sicrozem Brown soils Reddish brown soils	Soil-forming regime is salinization or calcification; salt accumulation usually heavy in the upper soil and may form a caliche (or petrocalcic layer); organic content very low; color usually light, but red and brown are common because of iron and magnesium oxide staining of particles
Dark-colored soils of semiarid, subhumid, and humid grasslands	Chestnut soils Reddish chestnut soils Chernozem soils Prairie soils Reddish prairie soils	Soils of the calcification regime; substantial organic content in O and A horizons (mollic horizon); B horizon rich in calcium carbonate; color ranges from light brown (chestnut) to dark brown (chernozem and prairie)
Soils of the forest-grassland transition	Degraded chernozem soils Noncalcic brown or shantung brown soils	Soils in midlatitude prairie-forest transition where climate is subhumid; strong organic layer, calcium carbonate weak or absent in B horizon
Light-colored podzolized soils of forested regions	Podzol soils Brown podzol soils Gray-brown podzolic soils	Soils of the humid midlatitudes characterized by a strong to moderate organic layer, leached and light-colored A horizon, and a B horizon with distinct concentrations of iron and aluminum oxides
Lateritic soils of forested subtropical and tropical regions	Yellow podzolic soils Red podzolic soils (and terra rossa) Yellowish-brown lateritic soils Reddish-brown lateritic soils Laterite soils	Heavily weathered (leached) soils of the humid tropics; light organic layer over a deep zone of clays rich in iron and aluminum oxides; heavy accumulations of iron oxides form laterite layers, and aluminum oxides may give rise to bauxite reserves in some areas

Source: USDA, U.S. Government Printing Office.

Appendix C
Soil Nutrients

Plant nutrients are classified either as macronutrients or micronutrients, based on relative amounts that are found normally in plants. Let's examine the functions of 18 of these important nutrients.

MACRONUTRIENTS

Carbon. The brick in cell construction and a component of sugar. It influences color and fragrance of blossoms. Carbon is constantly recycled from the atmosphere to the plant's tissues and back into the atmosphere through decomposition and respiration.

Hydrogen. Used in the plant cell in the manufacture of sugars and starches. Carbon and hydrogen make up the greatest percentage of all elements in plants. Hydrogen is the most abundant element in the universe, being the primary fuel for the stars.

Oxygen. Used by plants to breathe, and in the respiration process, oxides and complex organic compounds are formed. Plants also release oxygen in photosynthesis and provide the atmosphere with continual replacement.

Nitrogen. A key building material in the plant. It stimulates vegetation growth and gives plants a healthy, green appearance. A lack of nitrogen causes low protein content, stunted growth, and a pale, yellowish color. The primary source of soil nitrogen is the atmosphere, of which nearly four-fifths is nitrogen. However, higher green plants cannot utilize gaseous nitrogen directly. It must first be combined with other elements. The process of producing usable nitrogen is called **nitrogen fixation.** This process can occur in several ways, including lightning, industrial chemical processes, air pollution, and several aspects of microorganisms. Legumes such as alfalfa, peas, and beans are favorite hosts for nitrogen-fixing bacteria. Nitrogen is also provided by decaying plants and animals. There are thus two interrelated subcycles. Nitrogen in one cycle moves from the soluble nitrate form to plants and then to animals and back again into the soil. In the other cycle, free atmospheric nitrogen is fixed primarily by bacteria and lightning into the soil and eventually released to the atmosphere as nitrogen gas.

Phosphorus. Essential to all plant growth as an active element of the cell's nucleus. It stimulates early growth and root formation and hastens seed production and maturity. It is necessary for photosynthesis; without it, plants stop growing.

Potassium. A key element needed in making starches and sugars, and it enhances the size, taste, and color of fruits. It is one of the three major fertilizer elements. The primary commercial source is potash, or potassium oxide, coming primarily from the mines near Carlsbad, New Mexico. Symptoms of a potassium deficiency include slow plant growth, weak stalks, and shriveled seeds or fruit. Plants easily succumb to drought, low temperatures, and poor soil texture conditions.

Calcium. Corrects soil acidity and can be used to improve permeability in alkaline or saline soils. As a nutrient, it is considered essential to healthy cell walls and aids in the development of root structure. Symptoms of calcium-starved plants are early shed of blossoms and buds and weak stems. Structures and terminal buds die under severe deficiency.

Magnesium. An important component of chlorophyll and may aid in the movement of starch within a plant. It is also believed to be essential in the formation of oils and fats. When a plant lacks magnesium, leaves lose their color at the tips and between the veins. Leaves are abnormally thin and may even dry up and die. Twigs on trees are weak and subject to fungus infections.

Sulfur. Essential for life, being a constituent of all plant protein and some plant hormones. The plant hormones are key growth regulators for plants and are also important in animal nutrition. Symptoms of sulfur deficiency are like those of too little nitrogen. The plant is stunted and pale green to yellow in color. Too much sulfur can be toxic to plants. This can occur in highly polluted industrial areas where sulfur dioxide in high concentrations kills the vegetation.

Carbon, hydrogen, oxygen, nitrogen, phosphorus, and potassium are considered major nutrients as they are used in the greatest amounts. Calcium, magnesium, and sulfur are needed to a lesser extent, so they are called secondary plant food elements. The next nine elements are micronutrients. Plants use them in very small amounts. However, the fact they are needed in smaller amounts should not imply they are less essential for plant vigor.

MICRONUTRIENTS

Manganese. With the aid of iron, assists in the synthesis of chlorophyll. It is also closely associated with copper and zinc, acting together as a catalyst in plant growth. Manganese deficiency leads to chlorosis, a yellowing of the leaves while the veins remain green.

Boron. Is associated with calcium utilization. Whenever an imbalance develops between the proportions of calcium and boron in a plant because of a boron deficiency, the ends or buds of a plant fail to develop properly. The amount needed ranges between 5 and 30 parts per million in solution. Any excess can produce severe toxicity. Boron for fertilizers comes from mineral deposits in Death Valley and the Mojave Desert in California.

Symptoms of a boron deficiency include not only tip deformities, but fruits such as apples develop corky texture, and root crops develop "brown heart."

Copper. Acting as a catalyst of other elements within the plant, it promotes the formation of vitamin A. It also regulates the consumption of nitrogen when too much nitrogen is available in the soil. An excess of copper is very toxic.

Soils high in organic matter and weathered sandy soils are often low in copper. A severe deficiency may cause serious stunting of growth, while moderate deficiency may merely reduce yields.

Iron. Serves as a catalyst in the production of chlorophyll. It also enters into oxidation processes which release energy from sugars and starches.

A symptom of an iron-starved plant is chlorosis, affecting the youngest leaves first. Iron deficiency is especially common in arid soils of the West with high pH values. By lowering the pH, iron in the soil can be released.

Iron is abundant in most soils, but many factors inhibit its availability to the plant. If there is an accumulation in acid soils of heavy metals such as copper, manganese, zinc, and nickel, iron deficiency may result. Soils rich in lime are most likely to contain too little iron. It is hard to supply iron in a form available to plants, but it can be added in the form of soluble compounds called **iron chelates.** Also, through the addition of ammonium sulfate, soil acidity will increase, favoring natural releases of iron.

Molybdenum. Supplies the nitrogen-fixing organisms with molybdenum-containing enzymes. This same nutrient not only stimulates nitrogen fixation, but reduces plant diseases associated with low molybdenum levels.

Plants suffering from molybdenum deficiency are stunted and yellow, resembling nitrogen-deficient plants. This micronutrient has a rags to riches story. Vast acreages of land in Australia near Adelaide were found to be so low in molybdenum that no nitrogen-fixing legumes could be grown, and most plants in the area were in poor health. By applying only 2 ounces per acre, the land was converted to a highly productive rangeland.

Zinc. Is closely associated with iron and manganese in the formation of chlorophyll. Zinc is abundant in most soils, and yet it is a nutrient unavailable to plants if the pH is at the wrong level. It has been observed that plants suffer zinc deficiency when the pH of the soil is increased. It is readily available below values of 5.5, but reaches lowest levels between 5.5 and 7.0

Zinc deficiency in plants causes small terminal leaves, abnormal plant structure, and poor fruit bud formation.

Chlorine. Is a negatively charged anion, unlike the other nutrients, which are positively charged cations. It is needed in large supply when compared with other micronutrients. Chlorine-starved plants produce low yields because this nutrient is important in the metabolism of carbohydrates, production of chlorophyll, and the water-holding capacity of plant tissue. Rainwater is a continuing source of chlorine. Except in very humid regions, it is not deficient. It is easily washed out of the atmosphere, as it is very soluble in water.

Cobalt. Like molybdenum, is important for nitrogen fixation. It has been established that cobalt is necessary for the survival of *Rhizobia,* the nitrogen-fixing organisms in legumes. Without cobalt, nitrogen fixation ceases. Therefore, it has an indirect role. Plants may not need it, but microorganisms and higher animals such as cattle and sheep need it in their diet to gain weight, thrive, and multiply. Actually, cobalt deficiency is a vitamin B deficiency, as cobalt is a key element in the vitamin complex.

Sodium. Is necessary for maximum yields. Celery, sugar beets, turnips, and table beets show increased growth when this nutrient is present in the proper proportions. It can also improve soils low in potassium.

There are many other trace elements that are believed to play an important role in the diet of plants. Elements such as silicon and aluminum are known to be absorbed by plants, but whether or not they are needed has not been confirmed. This survey of 18 nutrients provides us with a chemistry that is essential for plant health and life. Perhaps other elements will be discovered to be essential to the chemistry of plants, and yet, it is important to remember that no single element is a complete plant food, capable of magic guarantees to achieve wonder growth. All nutrients work in concert, blending to produce the complex alphabet soup of soil that gives life to the planet.

Glossary

A

abrasion The "sandblasting" process in which wind drives sand and silt particles against exposed rock or soil causing it to be worn away.

absolute humidity The mass of water vapor per unit volume of air (grams per cubic centimeter or meter).

abyssal plains Vast flat plains on the floor of the oceans. In the Atlantic, the average depth is 5,500 meters (18,000 feet) below sea level.

adiabatic Occurring without gain or loss of heat. Applied to the change in temperature of a mass of gas which is undergoing expansion (cooling) or compression (heating) without actual loss or gain of heat from outside. This often occurs within an ascending or descending air mass. On expansion, the lowering of the temperature may cause the air to reach the **dew point.**

adiabatic lapse rate The cooling or heating of air in a rising or descending parcel at a rate of approximately 5.5° F/1,000 feet or 1° C/100 meters.

advection fog A fog that forms when a warm, moist air mass moves horizontally over a cooler land or sea surface, thus reducing the temperature of the air below the dew point. San Francisco is often foggy, resulting from warm, mild air moving east onshore over cold ocean currents.

aftershocks Smaller earthquakes that follow the primary earthquake. They can last for several months.

air mass A uniform body of air usually in regions of high pressure in which temperature and moisture are uniform over a large area. The border region, called a frontal zone, serves as a transition between air masses.

air mass lifting A general uplift of a large body of air, generally occurring in major cyclonic storm activity. It is the result of general instability of the atmosphere.

albedo The reflectivity of a body, usually expressed in percent. An albedo of 100 percent indicates total reflection and no absorption. Snow and ice range from 45 to 85 percent, while fields and forests range from 3 to 25 percent.

alfisols Soils with a yellowish-brown A horizon colored by aluminum and iron compounds.

alluvial fan A cone of alluvium located at the mouth of a canyon and resembling in outline an open Oriental fan.

alluvial soils Soils that are **azonal** unless very old. These soils consist of unconsolidated sand, silt, clay, and gravel deposited by streams on flood plains and deltas. They are usually very rich and frequently replenished by floods. The Mississippi, Nile, Yangtse, and Mekong are good examples of places where these soils are in production and supporting large-scale agriculture economies.

alluvium (alluvial) All unconsolidated fragmental material laid down by a stream in its bed, on its flood plain, and in lakes, deltas, and estuaries, comprising clay, silt, sand, and gravel.

alpine glacier **Glaciers** that form in the mountains. These long and narrow valley glaciers occupy previously formed valleys and bring the ice from small collecting grounds or cirques down to lower elevations to the zone of ablation.

altocumulus (Ac) An intermediate height type of individual clouds massed closely together. It appears fleecy and cellular.

altostratus (As) An intermediate height cloud of grayish, uniform sheets usually of wide extent, through which it is possible to see, as through ground glass.

andesite Igneous volcanic rock that is intermediate between the most acidic and basic types.

aneroid barometer An instrument for measuring atmospheric pressure, consisting of a metal box almost exhausted of air, whose flexible sides expand and contract with the changing air pressure.

annular stream pattern A ringed pattern of drainage, occurring in dome regions such as the Black Hills of South Dakota.

anticline A fold whose sides slope apart, caused by compressive forces in the Earth's crust.

anticyclone A high-pressure center of stationary or rapidly moving pressure centers.

aphanitic Fine grain texture.

aphelion The position in Earth's orbit that is farthest from the Sun, occurring on July 4.

aquatic community A plant and animal community located in a water environment.

aquifer A porous and permeable conduit of layered rock, which holds and allows movement of water within it.

arcuate delta A **delta** form that is shaped like a fan.

arête A steep-sided rocky ridge or crest between two adjacent **cirques.**

aridisols Soils of desert regions, which are rich in soluble minerals and low in organic compounds.

artesian well Water that flows to the surface under gradient pressure.

asthenosphere Plastic layer in the Earth's upper mantle, which moves in response to convectionlike currents and generates movements in the Earth's lithospheric plates.

atmosphere The blanket of air that surrounds the Earth or any planet, composed of a mixture of gases.

atmospheric pressure The weight of the air at sea level that exerts a pressure of about 1,000 millibars (15 pounds per square inch). Think of this pressure as the weight of a column of air extending to space.

autumnal equinox See **equinox.**

azonal soils Soils that have poorly developed profiles either because of insufficient time or because slopes are too steep to allow for soil profile development.

B

backwash The return flow of waves down a beach slope. The volume of water is less than the **swash** because of infiltration into the sand.

bajada Series of interconnecting **alluvial fans** along the base of a mountain range.

Ballot's Law Stand with your back to the wind and the low pressure will be on your left, the high on your right.

barchan dune An isolated arc-shaped dune. The horns of the crescent point downwind and indicate the direction of dune migration and prevailing wind.

barometric slope A pressure gradient created by differences in pressure between two locations.

base level The lowest level to which a river can erode its bed, normally assumed to be sea level. A lake is a temporary or local base level.

batholith A massive igneous intrusive rock.

bay-mouth bar Low mound of sediment deposited by waves and currents at the mouth of bays.

beach Unconsolidated sand and gravel at the edge of the sea between the low-tide line and upper limit of wave action.

bed load Large fragments that move close to the channel floor by rolling, sliding, and skipping in low leaps (**saltation**).

big bang The original event that started the universe.

biological weathering Weathering resulting from biological activity of plants and animals.

biomass The total organic mass of all living organisms; also, fuel derived from modern (unfossilized) organisms.

biomes A region or habitat with a distinctive plant and animal community or ecosystem. The term *biochore* can also be used in this same sense. Each biome represents a response of plants and animals to global climatic controls.

biosphere The living sphere of the planet or zone where biological activity occurs.

birdsfoot deltas A delta in which sediments form deposits of **alluvium** along major distributaries that are separated by arms of the sea.

block separation Where a rock has numerous joints, the common form of breakup is by block separation.

body waves Seismic waves that travel through the Earth's interior (P waves and S waves). P waves travel through solids and liquids. S waves cannot travel through liquids, such as the Earth's outer core or large magma sources.

bora Regional name for cold air movement from the interior to the Adriatic coast of Yugoslavia.

braided stream A stream whose course consists of a tangled network of interconnected diverging and converging shallow channels.

breccia A sedimentary rock composed of angular fragments cemented together in a fine matrix.

broadleaf evergreen A broadleaf tree that remains green with foliage in all seasons.

butte Small, flat-topped, erosional remnant in semiarid or arid regions.

C

calcification A soil-forming process where evaporation exceeds precipitation for most months of the year, in dry climatic regions of the lower middle latitudes. Rainfall is not sufficient to adequately leach bases and salts. Capillary action brings calcium carbonate to the upper horizons in dry periods. Calcification can produce a very fertile soil, rich in bases and humus, if rainfall is sufficient to support tall grasses.

calcium carbonate ($CaCO_3$) with two crystalline forms, calcite and aragonite, occurring commonly in limestone, marble, chalk, and coral. It dissolves in water containing carbon dioxide to form calcium bicarbonate, $Ca(HCO_3)_2$, which is present in most natural waters. When calcium bicarbonate decomposes, calcite is deposited at temperatures below 30° C and aragonite above 30° C.

caldera Large bowl-shaped crater caused by volcanic collapse, apparently because of the withdrawal of large quantities of magma from beneath the surface during volcanic eruptions.

caliche A layer of calcium carbonate deposited in arid soils at the surface because of the evaporation of capillary water.

capillary attraction A force that is the result of cohesion of water and its adhesion to soil particles. Evaporation causes capillary attraction upward in the soil resulting in the deposition of dissolved salts in the top layer.

caprock A resistant stratum; it may be responsible for a waterfall or rapid in a stream.

carbonation A specific type of solution process where carbon dioxide in water forms a weak acid capable of dissolving limestone rock. The calcium carbonate in the limestone is slowly dissolved in the form of calcium bicarbonate.

carbon dioxide (CO_2) A colorless, odorless gas whose concentration is about 0.0345 percent (345 ppm) in a volume of air near sea level. It is a selective absorber of longwave infrared radiation and, consequently, it is important in the Earth's atmospheric **greenhouse effect.**

carnivore An animal with primarily a meat diet.

cation exchange capacity Exchange of positive ions in the soil solution.

cementation The process by which sediments adhere together through the deposition or crystallization of mineral material between grains.

centrifugal force The apparent force that appears to pull an object away from the center of rotation.

chaparral A type of evergreen shrub vegetation capable of withstanding long summer drought. It consists of broadleaf evergreen oaks and shrub plants with small leathery leaves (**xerophytes**). It is common throughout the Mediterranean climate.

chemical energy Energy resulting from a chemical reaction.

chemical sedimentary rock Rocks formed from chemical precipitates or compounds precipitated directly from water. Limestone is the most abundant of the chemical sedimentary rocks.

chemical weathering The wearing away of rock by chemical processes. These include solution, carbonation, hydrolysis, oxidation, and hydration.

chernozem soils Soils of the "black earth" belt produced by **calcification.** These are some of the most productive soils of the middle latitudes.

chinook winds Hot, dry adiabatic wind descending down the lee side of the Rocky Mountains onto the Great Plains of the western United States.

chlorosis An abnormal condition of plants in which the green parts lose their color or turn yellow as a result of disease. A lack of certain nutrients such as iron or nitrogen can also cause this condition.

chromosphere A transition region in the Sun between the denser, cooler **photosphere** and warm, less dense **corona.**

cinder cone A cone around a volcanic vent, composed exclusively of small fragmentary material, the result of highly explosive volcanic activity.

circumpolar vortex An upper air flow from west to east around a great polar low-pressure center. The flow consists of large undulations, called upper air waves or **Rossby waves.**

cirque A steep-walled amphitheater or basin, of glacial origin.

cirrocumulus (Cc) A high cloud consisting of small globular masses with a rippled appearance and blue sky. It is commonly called a "mackerel sky."

cirrostratus (Cs) A uniform, milky layer or veil of high sheetlike cloud above 6,000 meters (20,000 feet). It is often associated with a warm front.

cirrus (Ci) A high cloud above 6,000 meters (20,000 feet), consisting of ice crystals and delicate and wispy in appearance. It hardly interferes with sunlight and moonlight (often a ring will be present around the Sun and Moon).

clastic sedimentary rock Those rocks derived directly from sediments reduced by weathering of other rocks and transported and cemented together. Clastic sediments are derived from any one of the rock groups: igneous, sedimentary, and metamorphic.

clay-humus complex The cations held in solution in the root zone of clay minerals.

cleavage The tendency of a crystal to split along planes of weakness, determined by its molecular structure.

climate The long-term state of the atmosphere, consisting of the total complex of weather conditions. The term weather deals with the specific atmospheric conditions at a given point in time. Such elements as radiation, temperature, pressure, cloudiness, wind, precipitation, humidity, and evaporation are all important in describing the weather or climate of a place.

climatic controls Ocean currents, land masses, altitude, surface conditions, and air mass all are examples of important modifiers of the basic weather elements and serve as important controls of climate.

coal A mineral fossil fuel composed of carbon from the remains of plants.

coastal blowout A coastal dune with cuspate tips pointing upwind.

col A saddle-shaped notch in a ridge providing a pass over the range, commonly resulting from the development of back-to-back cirques.

cold front A frontal contact zone in which cold air is invading the warm air zone.

colloid A substance in a state of extremely fine subdivision, with particles from 10^{-5} to 10^{-7} centimeters in diameter, suspended in water. It plays a role in the adhesion of particles therefore affecting the soil structure. Colloids are also important in nutrient uptake in that they can attract and hold ions of dissolved substances.

color infrared A type of photographic film sensitive to green and red and invisible infrared light. When viewing this film, what is infrared shows as red, what is red shows as green, and what is green shows as blue. The blue is then eliminated by using a yellow filter. The advantage of color infrared film is its superior penetration through haze, being dependent entirely upon reflected sunlight. Healthy vegetation reflects more infrared and therefore is brighter red in the photograph.

columns Limestone or gypsum cave deposits in the form of a column, the product of a **stalactite** and a **stalagmite** meeting each other in a cavern environment.

community A group of plants and animals growing in a particular place, usually one common habitat. The term can be applied on a variety of scales. The biosphere is one community, a biome is a smaller subdivision, and the California Coastal Sage Scrub is yet another.

composite volcano A symmetrical cone built up over a long period of time as the result of a number of eruptions, consisting of a number of layers of ash, cinders, and lava. The lava is more acidic and explosive than the **shield volcano**. Fujiyama and Pico de Orizaba are good examples of composite volcanoes.

compound shoreline A shoreline exhibiting features of both emergence and submergence.

condensation The process by which a gas is reduced to a liquid. In such a process heat is liberated at the rate of 580 calories per gram of water at standard temperature and pressure.

condensation nuclei Minute aerial particles of dust, smoke, or salt with a marked affinity for water. The cloud formation process will not take place without these particles.

conduction Transfer of heat from one molecule to adjacent molecules by contact.

cone of depression A conical depression of the water table caused by pumping extraction of groundwater.

conglomerate A sedimentary rock composed of smooth rounded pebbles, cemented together by calcium carbonate, silica, or iron oxide.

contact metamorphism Metamorphism caused by a rise in temperature, usually as a result of the intrusion of a mass of molten igneous rock at a high temperature. It can produce a fusion or recrystallization of the minerals or grains in a rock. Coarse, grained sandstone changes into quartzite, limestone can be altered into marble.

continental arctic (cA) Very cold and very dry air, well developed in the winter, forming over the Arctic Ocean. It merges with the continental polar air of Canada.

continental glacier An ice sheet glacier that covers several thousand square miles, forming a thick ice cap thousands of feet thick. It spreads out from the zone of greatest thickness enveloping all landforms it encounters, ceasing its spread only when ablation rates at its outer edge balance the rate of spreading.

continental polar (cP) An air mass that originates over north central Canada, characterized by low temperature and low moisture content. This air mass is responsible for cold waves and low freezing temperatures as far south as Mexico and Florida.

continental rise The gently sloping area at the base of the continental slope.

continental shelf The gently sloping (1° or less) margins of a continent, submerged beneath the sea, extending from the coast to a point where the seaward slope increases markedly, usually at about 200 meters (600 feet) depth.

continental slope The marked slope (2–5°), from the edge of the continental shelf sloping to the deep-sea or abyssal plain.

continental subarctic climate (Dfc, Dwc) Similar to the **humid continental** climate with respect to the continental location and wide ranges in temperature. However, this climate is colder with lower precipitation and shorter summers resulting from its poleward location. This climate lies in the source region of **continental polar** air masses.

continental tropical (cT) A warm, dry air mass, well developed in the summer over the southwestern United States. It is best developed over the Sahara.

convection The mass movement of a fluid as a result of differences in temperature and therefore differences in densities within the fluid. Heating of air or water will cause convection.

convergence An atmospheric condition that exists when an air mass or winds have inflow into a specified region.

core The hot, dense inner portion of the Earth consisting of iron and nickel. The outer core is liquid and the inner core is solid.

Coriolis effect The effect resulting from the Earth's rotation, which tends to deflect the motion of objects. Any object or fluid moving horizontally in the Northern Hemisphere tends to be deflected to the right of its path of motion, regardless of its direction of motion. In the Southern Hemisphere, deflection is to the left. The effect is absent at the equator but increases progressively with latitude, reaching a maximum at the poles.

corona The outermost part of the Sun's atmosphere, extending beyond the orbit of the Earth. It is characterized by extremely low density, highly ionized atoms. Temperatures reach over 1 million degrees. During sunspot maximums the corona is at its greatest brightness and roughly circular. Near sunspot minimums the corona is flattened in the polar regions.

corrasion Mechanical erosion by sand, gravel, and boulders being carried by water or glaciers.

crescent dune See **Barchan dune.**

crevasse Stress crack in a glacier, occurring most commonly along the margins or near the terminal of the glacier in the zone of wastage.

crust The outer layer of the Earth, ranging in thickness from 10 kilometers in the ocean floor to over 40 kilometers on the continents.

crystal form The external shape of crystals.

crystalline rock A rock with constituent minerals of crystalline form, developed either through cooling from a molten form or as a result of metamorphism.

cuesta Asymmetrical ridge formed by erosion in alternating resistant and weak sedimentary beds.

cumulonimbus (Cb) A **cumulus** cloud which grows to as high as 10 kilometers (6–7 miles), often with its upper part spread out like an anvil. It is usually associated with thunderstorms and torrential rain as well as tropical hurricanes and tornadoes.

cumulus (Cu) A convection cloud which grows vertically from a flat base into a large white globular or domed summit.

Curie point Temperature above which a magnetic material loses its magnetization. Each material has a different Curie point.

curvature effect In cloud physics, as cloud droplets decrease in size, they exhibit a greater surface curvature that results in a more stable droplet and reduced droplet growth.

cuspate delta A **delta** form in which an arcuate delta has been modified by longshore currents, creating the cusps.

cyclone A low-pressure cell with decreasing pressure toward the center.

cyclonic precipitation Precipitation that develops as a result of lifting along a warm or cold front.

D

deciduous plant A plant that sheds its leaves and becomes dormant in an unfavorable season.

decomposers Organisms that feed on decaying plants and animals.

deflation Erosional process in which loose particles lying on the ground are picked up by the wind or rolled along the ground. Deflation produces blowouts and basins, occurring especially where the ground is dry and scattered with loose particles of sand, silt, and dust.

delta A depositional landform resulting from a stream deposit in a sea or lake.

dendritic drainage A stream pattern where tributary streams enter main streams at an acute angle resembling a tree and its branches.

density The number per unit area, such as people per square mile, or mass per unit volume, such as grams per cubic centimeter.

deposition The progressive accumulation of transported particles upon the stream bed and flood plain or floor of a lake or sea into which the stream empties.

desert biochore A vegetation pattern associated with extreme aridity. Plants are sparsely distributed and xerophytic.

desertification The rapid conversion of marginally habitable arid land into true desert. This process is enhanced by human action.

desert pavement (lag deposit) A rocky surface cemented together by evaporite salts left after wind has removed most small particles of material in an arid environment.

dew point The temperature at which air, being cooled, becomes saturated with water vapor.

diffusion The gradual dispersion of matter from an origin.

dike A mass of intrusive rock that has filled fractures in older rock. Shiprock, New Mexico, has dikes radiating outward from the volcanic neck of the old volcano, exposed by erosion.

dip The maximum slope of a stratum of sedimentary rock at a given point. The angle of inclination is given in degrees from the horizontal.

"dirty" moist air The atmosphere with hygroscopic nuclei.

discharge The quantity of water passing down a stream, depending on its volume and velocity.

dispersion The degree of compactness of the objects being studied in a given area. Trees, houses, populations, and mineral resources are distributed unevenly over the Earth. The degree of relative compactness of these elements is equated to dispersion.

doldrums An equatorial zone of light, variable winds and low pressure straddling the equator at latitude 3° to 10°.

drainage basin The area drained by a stream and its tributaries, separated from adjacent drainage basins by stream divides.

drizzle Small raindrops between 0.2 and 0.5 millimeters in diameter.

drowned river system A river system invaded by the sea, resulting from either subsidence of the land or rising sea level following the melting of the ice caps after the last Ice Age.

drumlin An elongate, asymmetrical hill of glacial materials deposited by glacial ice. The long axis parallels the direction of ice movement, and the steep end tends to face the direction from which the ice came.

dry adiabatic lapse rate The rate at which the temperature of a parcel of unsaturated air rises or falls. A rising parcel cools at the rate of 1° C per 100 meters (5.5° F per 1,000 feet). Air in a state of descent will heat at the same rate resulting from compression of the atmosphere. This occurs during a **Santa Ana** or **chinook** wind condition.

dry-farming Agriculture in a semiarid area, without the help of irrigation, by conserving soil moisture through mulching, and the utilization of two year's rain for one crop.

dune A mound of sand deposited by the wind.

dynes The force required to produce an acceleration of 1 centimeter per second every second on a 1-gram mass.

E

earthflow A type of **mass wasting** of water-saturated soil, overburden, or weathered weak bedrock. The material pulls away from the slope, leaving a steplike terrace and moves down as a plastic mass to form a bulging toe or lobes. **Earthflow** is common on hillsides after a period of long, continuous rainfall.

earthquake A rapid movement of rock masses along a fault. Earthquakes occur when faulting releases kinetic energy as shaking wavelike motion which is transmitted through the surface layer of the Earth. The shaking movement spreads in widening circles from the focal point of energy release, the **hypocenter.**

easterly wave A migratory wavelike disturbance in the tropical trade winds. Easterly waves can intensify into **tropical cyclones.**

ecosystem A **community** of plants and animals, viewed within its physical environment or habitat. The ecosystem is the result of interaction of soils, climate, vegetation, and animals.

edaphic Related to soils or the **regolith.**

elastic rebound theory The phenomenon whereby stressed rocks behave elastically before and after an earthquake, returning afterward to an undeformed, unstressed form.

electromagnetic radiation Energy propagated in the form of electromagnetic waves. These waves do not need molecules to propagate them, and in a vacuum they travel at approximately 300,000 kilometers per second or 186,000 miles per second.

electromagnetic spectrum The complete range of frequencies or wavelengths of electromagnetic energy including in order from the longest wave: radio, infrared, visible light, ultraviolet, x-rays, gamma rays, and cosmic rays.

elements The basic chemical units of matter, consisting of 92, from hydrogen to uranium, that occur naturally.

elevations and contour lines A contour line is an imaginary line showing the position of all those points on a map at some specific elevation above sea level. For example, the shoreline around an island would be a contour line, showing all points at zero elevation. Contour lines are always continuous and, if the map is large enough, will always form closed loops. Since contour lines represent only one elevation, they never intersect. The contour interval (C.I.) on a map is the vertical spacing between contour lines.

eluviation The removal and downward migration of soil components by water percolating through the soil. This term covers both leaching, a chemical solution process, and mechanical removal of finer particles.

entisol Soils with poor or no horizon development.

entropy The fraction of energy that is not used in doing work and is dissipated from the system.

environmental lapse rate The rate of decrease of temperature with altitude.

eolian Related to wind erosion or deposition.

epicenter The point on the ground surface above the **hypocenter,** where fault breaking starts in an earthquake.

equilibrium The situation in a stream where the average rate of supply of rock waste to the stream from all of its tributaries is equal to the rate at which the stream can transport the load.

equinox The position of the Earth's orbit where the Sun's rays are perpendicular at noon on the equator, on March 20 or 21 and September 22 or 23. On these dates, daylight and darkness are evenly divided into 12-hour periods at all latitudes. These dates are known as the autumnal equinox and vernal equinox, depending on the hemisphere.

erg A sandy desert with little vegetation.

erosion The progressive removal of material from the floor and sides of the stream channel.

erratic A boulder transported by glacial ice deposited on a surface of different bedrock, or surface material.

ERTS Earth Resources Technology Satellite.

esker Sinuous ridge of stream deposits, formed by a stream flowing beneath an ice sheet.

estuarine delta A delta deposit in a bay or drowned river mouth.

estuary The mouth of a river, stream, or other body of water (such as a bay) having a connection with the sea and within which salt water is measurably diluted with fresh water entering the system by way of streams and general surface runoff. They become nutrient traps for many life forms.

evaporation The process by which a liquid is changed to a gas. For water, heat is required at the rate of 580 calories per gram of water at standard temperature and pressure. The rate of evaporation of water is a function of air temperature, wind, and the kind of surface.

evapotranspiration The loss of moisture from the terrain by direct **evaporation** plus **transpiration** from plants.

evergreen Plants that retain green foliage year around.

exfoliation A weathering process where rock peels off in layers or concentric rings from the surface of exposed rocks.

exosphere The outer layer of the atmosphere where gases are highly ionized at very high temperature and low density.

extrusive rocks Molten rocks deposited on the Earth's surface from volcanic activity.

F

faulting The movement of blocks along cracks or joints in the Earth's crust, resulting from stress.

feldspar A family of minerals composed of aluminum silicates, a common constituent of igneous rock.

felsic rocks Igneous rocks that are rich in silica.

ferromagnesian Silicate containing significant iron and/or magnesium.

fiord A long, narrow arm of the sea, the result of the submergence of a deep glacial valley with steep walls. Fiords are found in Norway, Scotland, Greenland, Labrador, British Columbia, Maine, Alaska, southern Chile, and New Zealand.

firn Compact, granular snow partially converted to ice by melting and refreezing, induced by weight of overlying snow.

first-order stream The smallest tributary branch in a watershed.

fissure eruption An eruption of lava from a large crack rather than from a pipelike vent.

flood basalts Lavas resulting from fissure eruptions.

focus See **hypocenter.**

fog Suspended small water droplets in the air adjacent to the surface, creating a stratus cloud on the ground.

fog drip Moisture that condenses on plants in contact with a cloud or fog.

foliation A wavy laminated or banded fabric in such rocks as schist and gneiss, the result of metamorphism, recrystallization, and segregation of minerals into parallel layers.

food chain A dependent, interrelated food system where organisms, such as plankton, are at the base supporting higher more complex organisms, such as clams, oysters, worms, and shrimp. These organisms are then fed upon by higher forms, such as birds and fish and still larger predators.

footwall The fault block supporting the **hanging wall block.**

forest biochore A plant formation consisting of trees growing close together and forming a layer of foliage that largely shades the ground. Forests show stratification, with more than one layer producing distinct microclimates resulting from shading and transpiration of moisture. The forest biochore spans a great variety of climates, from the tropical rainforest climate to the subarctic.

fracturing The breaking or shattering of rock when it has been subjected to forces beyond its elastic limits.

friction The force of resistance to movement of one surface sliding over another.

frost wedging The repeated growth and melting of ice crystals in rock fractures. As the ice freezes, its volume increases. The expansion produces a mechanical breakup of the rock.

fumarole A vent in the Earth's surface that emits steam and hot gases.

fusion A nuclear process where hydrogen atoms are fused together to form helium and great quantities of energy are released.

G

galaxy A discrete aggregation of stars numbering in the billions. Our view of our galaxy is the Milky Way, one of the spiral arms.

gallery forests A border of trees along stream banks extending into adjacent grass or scrub land.

gamma rays The highest-energy and shortest waves of the **electromagnetic spectrum.**

geoid The terrestrial spheroid, an "Earth-shaped body," regarded as a mean sea-level surface projected through the continents.

geomorphic cycle A concept initiated by William Morris Davis, which states that landscapes proceed through a sequence from youth to maturity to old age.

geostrophic wind The ideal wind model occurring in the upper air, above 1,000 meters (3,000 feet), where the resultant wind is a balance between the pressure gradient force and the Coriolis effect. At this altitude friction is not important as a deflective force, and the air flows in the isobaric channels.

geosyncline A large linear depression or downfold in the Earth's crust. A geosyncline develops as a slow, continuous down-warping of sedimentary deposits thousands of feet thick.

geothermal energy A source of energy trapped beneath the Earth's crust in the form of superheated steam and/or hot water. This energy is then converted to electric power when the stream is released through drill holes and channeled to a steam driven generator.

geyser A spring that erupts hot water and steam intermittently at a vent in a volcanic region.

glacial valleys U-shaped valleys formed by alpine glacial erosion.

glacier A mass of ice, on land, that moves under its own weight driven by the force of gravity.

gleization A soil-forming process characteristic of poorly drained (non-saline) soils under cool, moist conditions.

gneiss A metamorphic coarse-grained crystalline rock of foliated texture and a streaked, wavy or banded appearance. Parent material is of many types such as granite, gabbro, or even sedimentary rocks.

graben A narrow trough between parallel faults where the blocks on each side have been uplifted. Death Valley and the Owens Valley in California are good examples of grabens.

graded bed The profile of a stream in **equilibrium.**

gradient See **Slope.**

granite A common plutonic igneous rock composed of feldspar, quartz, and mica.

grassland biochore The grassland is a region dominated by herbs, or nonwoody plants. It is common in regions of low precipitation from the arctic to the tropical steppe.

gravity Expressed in Newton's Law; each body in the universe attracts every other body with a force directly proportional to the product of their masses and inversely proportional to the square of the distance between them. The mass of a body is constant, while the force of attraction or weight of a body is determined by Newton's Law.

great circle The shortest distance between any two points on the surface of the globe.

great soil groups A Russian-American soil classification system. There are approximately 40 great soil groups.

greenhouse effect Warming of the atmosphere through the transmission of different wavelengths of light. Short-wave visual radiation can pass through the atmosphere to the surface of the Earth quite easily. However, because the Earth's reradiation is in the infrared, the atmosphere partially prevents outgoing radiation from rapidly escaping. Greenhouses also rely on this mechanism.

groundwater Water which is held in the pores and fractures of bedrock.

H

habitat The natural environment of a place, consisting of climate, soils, plants, and animals. The term is a synonym for environment.

Hadley cell A cycle of atmospheric circulation generated by rising air at the equator flowing poleward at elevation and sinking to the Earth's surface in the zone of the subtropical highs.

hail Ice pellets that fall from a cumulonimbus cloud. Its concentric layers indicate that the hailstone has ascended and descended several times in the cloud before falling out.

hailstones Transparent or partially opaque particles of ice that range in size from that of a small pea to that of golf balls or larger.

halite Rock salt, consisting of sodium and chlorine.

halomorphic soils Soils which are formed in the high latitudes of the tundra climate. They are very thin and may be considered an azonal type similar to podzols, but they can be poorly drained and take on an intrazonal form.

halophyte A plant capable of growing in highly saline or alkaline soils.

hanging valley An alpine glacial trough formed by a tributary glacier at an elevation higher than the valley floor cut by a principal valley glacier.

hanging wall The fault surface block that rests on the **footwall.**

hardness scale The Mohs scale of hardness of minerals, in which ten selected minerals have been arranged in an ascending scale:

1. Talc   6. Orthoclase
2. Gypsum   7. Quartz
3. Calcite   8. Topaz
4. Fluorite   9. Corundum
5. Apatite   10. Diamond

Hardness is a means of minerals identification. By scratching the unknown mineral with the above listed minerals, one can by elimination determine the hardness of the unknown mineral. Galena, for example, will scratch gypsum, hardness 2, but will be scratched by calcite, hardness 3. Therefore, galena's hardness is 2½.

hardpan A compact, hard layer lacking permeability consisting of clay minerals formed in the B horizon of a soil as the result of **illuviation.**

haystack Rounded hills characteristic of old age **karst topography.**

headland A promontory or point with a steep cliff-face projecting into the sea.

heat energy One of the many forms of energy a substance possesses, measured in calories or British thermal units (Btu).

herbivore An animal that survives by consuming primarily plant material.

heterosphere The outer layer of the atmosphere from 80 kilometers (50 miles) on up, consisting of a nonuniform composition of highly ionized gases arranged in spherical shells around the Earth.

hogback Narrow ridge formed in resistant rock strata tilted at a very steep angle.

homosphere The atmosphere from the Earth's surface upward to an altitude of about 80 kilometers (50 miles). Its chemical composition is highly uniform.

horizons Soil zones having distinctive physical and chemical characteristics.

horn A pyramidal peak formed when several cirques develop back to back.

hot springs Springs heated by shallow magma bodies or young, hot rocks, frequently found along fault lines.

humid continental climate (Dfa, Dfb) A climate found in the Northern Hemisphere in the upper latitudes. It is located in the interior and eastern sides of continents, far from the moderating marine influences of the west coast. The winters are very cold and snowy, being dominated by the continental polar air mass. This climate is characterized by wide ranges in annual temperature.

humidity The amount of water vapor present in air, measured in grams per volume of air or grams per kilogram of air.

humid subtropical climate (Cfa) A climate generally located on the southeastern side of the major continents in the middle latitudes. It is dominated by the western sides of oceanic high-pressure cells. In the summer, convective rainfall is dominant with warm, humid conditions. The winters are cool with frequent continental polar (cP) air mass invasions and frequent cyclonic storms.

humus Semidecayed organic plant and animal material, usually found near the surface of a soil in the **A horizon.**

hurricane (typhoon) An intense tropical cyclonic storm where sustained wind speeds exceed 120 kilometers per hour (75 mph). Torrential rain, thunder, and lightning are always associated with such storms. Hurricanes originate in latitudes 5° to 20° of the equator in all major oceans. Their tracks follow the trade winds and then move north into the belt of the westerlies.

hydraulic gradient A gravity slope determined by differences in elevation in a stream course.

hydrologic cycle The endless interchange of water between the oceans, air, and land. The cycle consists of evaporation and transpiration from the surface of the Earth; condensation and precipitation and fog drip in the atmosphere; surface runoff, infiltration, and groundwater movement to the streams, and finally back to the oceans.

hydrolysis A form of chemical weathering, involving a chemical reaction in water. Feldspar in granite is changed this way, producing a clay. The granite then loses its strength and begins to disintegrate.

hydrosphere The water sphere of the planet.

hygrophytes Plants which grow in very moist conditions.

hygroscopic nuclei See **condensation nuclei.**

hygrothermograph A recording instrument capable of recording temperature and relative humidity over various time spans. The humidity data are transmitted from a human hair sensing element to an inked pen on a rotating drum. Temperature data are transmitted from a bimetallic sensing element to the drum.

hypocenter The point where fault breaking starts.

I

icecap climate (EF) Source regions of Arctic and Antarctic air masses being situated on the Greenland and Antarctic icecaps. The annual range in temperature is low, and no month has an average temperature above freezing.

igneous rock A rock formed by the solidification of molten magma material.

illuviation The redeposition of materials in the B **horizon** through the process of **eluviation.**

infiltration The movement of water downward into the soil and groundwater.

infrared The portion of the **electromagnetic spectrum** just beyond the visible red end of the visible spectrum. These waves have a penetrating heating effect and are emitted by the sun, the surface of the Earth, as well as by clouds and the atmosphere.

infrared thermal scanner Instrumentation capable of sensing the environment in infrared light.

insolation Incoming solar energy at the Earth's surface. At any particular place on the Earth, the quantity of insolation will be determined by two factors: (1) the angle of the Sun's rays and (2) length of daylight. These two factors vary with latitude and season.

instability The term when applied to the atmosphere indicates that if a parcel of air is allowed to rise it will continue to rise in a spontaneous manner because it is lighter and warmer than the surrounding air. The air will continue to rise, cooling at the **dry adiabatic lapse rate** if unsaturated, until it has cooled to the same temperature as the adjacent air. In the case of absolute instability the environment has a lapse rate steeper than the dry adiabatic lapse rate.

internal coherence The degree of interaction of the internal components of the system determine its internal coherence. For example, a beehive is a system representing strong internal coherence, because of interaction of its members, and its members with the environment. Poor internal coherence is usually found where barriers and a lack of interaction exist. A lack of interaction between members of a region can produce weak internal coherence.

International Date Line Approximately 180° longitude, where each day begins and ends. As one crosses the line traveling west, the calendar is advanced one day; proceeding east, the calendar is set back one day.

intertropical convergence zone An area of low pressure on or near the equator (latitude 0° to 10°), where trade-wind air rises; also known as the intertropical front.

intrazonal soils Soils formed under conditions of very poor drainage such as bogs and meadows, or in playas or desert basins.

intrusive rock Igneous rock that solidified from magma beneath the surface of the Earth.

inversion An atmospheric condition where temperatures increase with height and produce a cooler layer of air at the surface. This creates a very stable condition because cooler air has a higher density and is heavier than the warmer air aloft. This produces air stagnation and air pollution buildup.

ion An electrically charged atom or molecule.

ionization The process when an atom gains or loses an electron to become an **ion.**

ionosphere See **heterosphere.**

isobar A line on a map that connects points of equal pressure. Where the isobars are closely spaced, wind velocities will be greater because of a greater difference in pressure in a given area.

J

jet stream An upper air wind system developed with waves known as **Rossby waves,** moving in an easterly pulsating action at altitudes between 9 and 12 kilometers (30,000 and 40,000 feet). The jet stream reaches velocities of 300 to 400 kilometers per hour (200 to 250 mph) and is often associated with cyclonic activity at the surface.

joints Cracks or fissures in bedrock.

jovian The giant gaseous plants: Jupiter, Saturn, Uranus, and Neptune.

K

kame A cone-shaped landform of assorted glaciofluvial debris, which apparently formed near ice margins from glacial meltwater.

Karst topography A landscape consisting of sinkholes, caverns, disappearing streams, and other solution features.

kettle A depression, often filled with water, which resulted from a buried ice block melting near the ice margin.

kinetic energy The energy that results from the motion of matter.

kinetic molecular theory The theory describing behavior of gases based on the motions of molecules of atomic particles. Only at extremely high temperature and high pressure do gases deviate from this theory.

L

laccolith A pool of intrusive magma forming a dome over strata and bulge in the Earth's crust. A term coined by G. K. Gilbert from two Greek words meaning "rock cistern."

lag deposits See **desert pavement.**

lagoon A shallow body of water along the coast bounded by an offshore bar or coral reef.

land breeze Wind blowing offshore, resulting from more rapid radiational cooling of land areas at night than over adjacent seas.

landslide Rapid sliding of large masses of rock with little or no flowage. It can take two basic forms: rockslide or slump.

langley Unit of solar energy equal to one gram-calorie per square centimeter.

lapse rate The rate of temperature change in the atmosphere with decreasing or increasing height. The environmental lapse rate is about 1° C for every 150 meters of ascent or 3.5° F per 1,000 feet. This lapse rate continues to the tropopause.

latent heat The amount of heat energy expended when a substance such as water changes states in a condensation or evaporation process. When condensation occurs, the heat released is latent heat of condensation. Evaporation indicates energy is required. Heat is absorbed to convert water to vapor. This serves as a mechanism for heat transfer from water to the atmosphere.

lateral moraine A **moraine** deposited at the side of a valley **glacier.**

laterite An iron- and aluminum-rich soil layer, which is the end product of weathering in the humid tropics.

laterization A soil-forming process occurring in humid, rainy, warm climates. A high mean temperature all year ensures sustained bacterial and chemical activity, which destroys dead vegetation as rapidly as it is produced. This accounts for low humus content. The high precipitation also accounts for heavy **eluviation** and a leached soil. Lateritic soils are very low in soil fertility.

latitude A measure of distance north or south of the equator, given in degrees; one degree equals approximately 110 kilometers (69 miles).

latosols Soils of the humid tropics are called latosols or laterites. The laterization process forms these soils.

leaching The removal of soluble salts and bases from the upper A horizon to the lower B horizon and on down into the groundwater.

limb or fold The rock strata on either side of the axis, or central line of a fold.

limestone Sedimentary rock composed of the elements calcium, carbon, and oxygen, resulting in the mineral calcium carbonate.

liquefaction Loss of soil strength during earthquake shaking. Seismic waves are amplified in areas of water-saturated soil where liquefaction is likely to occur.

lithosphere The Earth's rocky crustal zone, resting on the **asthenosphere** of the upper mantle.

load The total amount of material moved by a stream. It includes dissolved, suspended, and transported sediments of clay, silt, sand, and gravel.

loam A soil texture with approximately equal portions of sand, silt, and clay.

loess Silt deposits several feet thick that were deposited during the last glacial period. Loess is rich and fertile with excellent texture. The greatest deposits are in China.

longitude A measure of distance east or west of the prime meridian, given in degrees. Each hemisphere consists of 180 degrees of longitude.

longitudinal dunes (seif dunes) Dunes consisting of ridges that run parallel to the prevailing wind direction. They are usually located in regions of low sand supply and strong winds.

longshore current A current parallel to the shore in a direction away from the wind. This current is capable of moving sand along the sea bottom in a direction parallel to the beach. The movement of sand is commonly referred to as longshore transport.

longwave radiation Infrared, microwave, and radio waves.

lunar eclipse The eclipse of the Moon when it moves into the Earth's shadow.

M

macronutrients Soil nutrients required in large amounts by plants.

mafic rocks Igneous rocks that contain lower amounts of silica and high levels of ferromagnesium minerals.

magma Molten rock material under the surface of the Earth at very high temperature, charged with gas and under high pressure.

magmatic steam Steam from molten magma sources.

magnetosphere The area beyond the upper atmosphere where the Earth's magnetic field prevails over the Sun's magnetic field. The Earth's magnetic field resembles a simple bar magnet, the axis of which approximately corresponds with the geographic axis of the planet.

mantle That portion of the Earth above the core and below the crust. This portion represents the major mass of the Earth.

maquis French term for low-order shrub-form vegetation found in steppe climates near the Mediterranean.

marble A metamorphic rock composed of crystalline limestone or dolomite.

marine-built terrace A wave-cut bench or terrace resulting from wave action.

marine west coast climate (Cfb, Cfc) This climate is in the belt of the westerlies between 40° and 60° latitude on the west coasts of continents. Cyclonic storms with cool, moist maritime polar (mP) air masses are the dominant influences. Annual temperature range is small for this latitude.

maritime equatorial (mE) A warm, moist air mass originating over the oceans close to the equator.

maritime polar (mP) A cool, moist air mass originating over the North Pacific and North Atlantic at latitudes between 50° and 60° N. Most of the Pacific cyclonic storms that reach the Pacific Coast originate in this air mass source region.

maritime tropical (mT) An air mass originating in the subtropical high-pressure belts of the middle latitudes. It is generally warm and moist and unstable on its western side and stable with a strong inversion on the east. California and northwestern Mexico are dominated by the stable eastern side of the mT of the Pacific while the South and East are under the influence of the western side of the mT of the Atlantic.

mass wasting The various kinds of crustal downslope movements occurring under the pull of gravity.

mature stream system Said of a stream that has completed its phase of rapid downcutting and has developed a smoothly graded stream course. It is now in a state of **equilibrium.**

meanders Bends and curves in the stream course.

mean solar time Solar time corrected or averaged to take into account variations in position of the Sun at different places in Earth's orbit.

mechanical weathering (physical weathering) Weathering of a rock to produce fine particles from a massive rock without chemical changes occurring. However, in nature, chemical changes and mechanical weathering are both occurring simultaneously in the same rock mass.

Mediterranean climate (Csa, Csb) The unique feature of this climate is its winter maximum cyclonic rainfall pattern. This climate is located on the west sides of continents between 30° and 45° latitudes and is under the influences of winter cyclones and summer maritime tropical (mT) air.

Mediterranean woodland community Forest and scrub of the Mediterranean climate, consisting of evergreen hardwoods of drought-resistant nature.

megatherms Plants that grow in warm-temperature belts.

mercurial barometer A device that records rise and fall of atmospheric pressure in a column of mercury. The length of the column is equated to the force per unit area on the surface of the Earth at that location.

meridian A line of longitude, a **great circle** passing from pole to pole numbered from 0° to 180° east and west of the prime meridian.

mesa A large isolated tableland of resistant rock that is more extensive than a butte.

mesopause Temperature boundary between the **mesosphere** and **thermosphere**, where temperature no longer diminishes with increasing altitude.

mesophytes Plants that have intermediate water needs.

mesosphere A region between 50 and 80 kilometers (30 and 50 miles) height where temperatures fall to −85° C (−120° F) at the mesopause or beginning of the ionosphere.

mesotherms Plants that favor temperate climates of the middle latitudes.

metamorphic rock Rock formed by processes in which already consolidated rock undergoes changes in texture, composition, or chemical and physical structure. These changes may be brought about by heat and pressure.

meteoric steam Steam whose source is from precipitation.

meter A primary unit of measurement in the metric system, equaling a length of 39.37 inches. It was originally derived as 1/40 millionth of the meridian that passes through Dunkirk, France.

micron (micrometer) A unit of length equal to one-millionth of a meter.

micronutrients Soil nutrients required by plants in small amounts, but which are essential to plant health.

microtherms Plants that favor cool climates of the higher latitudes or elevations.

mid-oceanic ridge A rift zone extending 65,000 kilometers (40,000 miles) around the Earth. The ocean crust is in a process of being pulled apart along the mid-oceanic ridge, causing a rift valley to be along the center line of the ridge system.

middle-latitude cyclone (wave cyclone) The dominant type of weather disturbance of middle and high latitudes, a vortex that repeatedly forms, intensifies, and dissolves along the frontal zone between cold and warm air masses. The term "front" is a military term coming out of World War I.

middle-latitude desert climate (BWk) Found over an interior middle-latitude desert shut off by mountains and distance from maritime influences. The summers are dominated by continental tropical air while the winters are under the large continental polar air mass. It has great annual ranges in temperature with hot summers and cold winters.

middle-latitude steppe climate (BSk) A transitional climate between the continental climates and desert climates. The summers are hot with convective precipitation. Winters are cold under the influences of the continental polar air mass, with large annual ranges in temperature similar to the interior desert.

middle-latitude wave cyclone An extratropical cyclonic system that forms a wave pattern along the frontal boundary.

millibar A pressure unit of 1,000 dynes per square centimeter, used in recording atmospheric pressures as indicated by a barometer. Sea-level standard pressure is 1,013 millibars or 29.92 inches or 76 centimeters of mercury. High pressure can range as high as 1,040 millibars (30.7 inches or 78 centimeters) or higher. Low pressures range down to 982 millibars (29 inches or 74 centimeters) or lower.

mineral An inorganic substance with specific chemical composition and physical properties. Mixtures of minerals make up rocks. Nearly all minerals are crystalline. Some are simple in composition, consisting of a single element. Diamond contains only the element carbon. Pyrite consists of iron and sulfur. Minerals have several identifying properties: crystal form, hardness, specific gravity, color, luster and transparency, streak, cleavage, and fracture.

mist fogs Fogs that develop over water bodies in night and early morning hours as the result of radiational cooling.

mistral The term applied to a downslope katabatic wind in southern France. See **Santa Ana.**

Mohorovicic discontinuity (Moho) Contact zone between the Earth's crust and mantle. Seismic waves change velocity and direction at this boundary.

mollisols Soils with dark, rich organic topsoil and a high pH.

monocline A steplike fold resulting from erosion of an anticline in sedimentary or metamorphic rock.

monsoon The seasonal reversal of wind and pressure over land masses and neighboring oceans. Southern Asia is best known for this condition. It brings not only a change in wind and pressure but usually heavy rains and cloudy skies over the land mass during the summer.

moraine A body of glacial **till,** or sediment carried and deposited by a glacier. A moraine at the terminus of a glacier is a *terminal moraine.* As the end of the glacier wastes back, scattered debris is left behind. Successive halts in ice retreat produce *recessional moraines.*

mountain breeze Air movement upslope during the day and cool air drainage downslope at night in a mountainous region.

mudflow A mud stream of fluid consistency, which pours down a canyon like a river with considerable velocity. Mountains denuded of vegetation are especially vulnerable.

N

native elements Elements that are found naturally in pure forms. Gold is a well-known example.

neap tides Tides with minimal range between high and low tides, occurring when Earth, Sun, and Moon are in quadrature, forming a 90° angle.

negative vorticity Spin of the air around a low-pressure zone.

niche A place in the environment that is favorable to a particular plant or animal.

nimbostratus A low, thick **stratus** cloud with precipitation from which rain falls. "Nimbo" pertains to precipitation and indicates precipitation is coming from the cloud.

normal fault A fault resulting from tension forces and horizontal extension of the Earth's crust.

northeast trades Winds that blow from the subtropical high toward the equatorial low in the Northern Hemisphere.

nuclear energy Energy released when the atom is either split or fused.

nutrient cycle The movement of plant nutrients through the Earth's natural systems of atmosphere, biosphere, lithosphere, and hydrosphere.

O

oblate spheroid A flattened sphere, corresponding to the Earth's shape. The Earth's diameter of 12,742 kilometers (7,914 miles) is about 43 kilometers (27 miles) shorter through the poles than through the equator.

obsidian Volcanic glass formed from rapid cooling of magma during an eruption.

occluded front A region where the cold front has overtaken the warm front. The colder air of the fast-moving cold front remains next to the ground, forcing the warmer air aloft and producing heavy precipitation.

ocean currents Currents, either warm or cold, that are driven by the planet's global wind systems.

oceanic trenches Long, narrow sea-floor depressions whose bottoms reach depths of 9,000 meters (30,000 feet) or more. The Mindanao Trench is an example.

ore Minerals that have economic value.

organisms Living beings, ranging in size from whales to bacteria, deriving life from the environment and capable of reproduction.

orographic effect The forced ascent of air over a mountain barrier, often inducing precipitation or condensation on the windward side.

orographic precipitation Caused by the ascent (hence cooling) of moisture-laden air over a mountain range.

orographic uplift Vertical uplift of the air mass along a mountain barrier.

osmosis The tendency of a solvent to pass through a semipermeable membrane, such as cellophane or a living cell wall, toward a higher concentration tending to equalize the concentrations on both sides of the membrane. An opposing pressure will build up inside the cell or cells, referred to as osmotic pressure. At the root level this is responsible for nutrient uptake in the plant.

outwash Materials deposited beyond the forward limit of the glacier.

overthrust fault (thrust fault) A reverse fault of very low angle in which the upper block has been pushed very far forward over the lower block.

oxbow lakes Semicircular lakes formed in abandoned river channels as a result of a river meander that has been cut off.

oxidation Chemical weathering involving combination with oxygen. The red color in many rocks results from iron oxidizing to form a new substance, rust, consisting of an iron and oxygen compound.

oxides Compounds that have oxygen.

oxisols Thoroughly leached soils of humid tropics, often with a **hardpan.**

oxygen A colorless, tasteless gas forming about one-fifth of the Earth's lower atmosphere.

ozone layer A layer of ozone (O_3), largely occurring in the region between 19 and 64 kilometers (12 and 42 miles) in altitude. Ozone is produced by the action of ultraviolet waves on ordinary oxygen atoms. This layer serves as a shield, protecting the Earth's surface from most of the deadly ultraviolet radiation of the Sun.

P

pahoehoe A black, wrinkled, ropy, or corded surface of newly solidified mafic lava flow, typical of highly fluid lava.

parabolic dune A dune whose crest is bowed convexly downwind, opposite that of the barchan dune. A common example is the coastal blowout dune.

parallax The apparent change in position of an object resulting from the changed position from which it is viewed. If a nearby star is viewed from positions six months apart in the Earth's orbit, it will appear to move relative to far away stars in the background.

parallel of latitude A line on a map joining all points of the same angular distance north or south of the equator.

parent material Crustal material from which soil is derived.

particulates Dust-size particles in suspension in the atmosphere.

paternoster lakes A chain of lakes arranged in stairstep fashion in a glaciated valley.

pedalfers A leached soil from which cations such as calcium have been removed, leaving compounds of iron and aluminum. This soil is found in humid climates where precipitation exceeds 60 centimeters (24 inches).

pedestal rocks Rock formations formed by wind erosion with a narrow tapering neck and a larger cap.

pedocal A soil rich in calcium that has not been leached out. It is found in semiarid and desert regions.

pedologists Soil scientists.

peds Soil classes.

perched water table A water-saturated layer located above and separated from the normal water table level by an impervious layer of rock.

perennials Plants that live more than a single season.

perihelion The closest position of the Earth to the Sun, occurring on January 3.

permafrost Permanently frozen layer of soil in polar and subpolar environments of the high elevations and latitudes.

permeability The ability to allow the passage of water. Sand is permeable, in contrast to clay.

pH An indicator of the hydrogen ion concentration in a solution, used as a measure of soil acidity. A pH of 7 is neutral. Values below 7 are acid and values above 7 are basic. Most plants favor a soil pH between 5.5 and 8.5.

photochemical smog An air pollutant produced by reactions involving sunlight. Ozone is an example, being produced when sunlight acts upon nitrogen oxides.

photon The particle carrying electromagnetic energy. The amount of energy it carries determines its wavelength.

photosphere The visible surface layer of the Sun.

photosynthesis The biological synthesis or combining of carbon dioxide and water in green plant cells in the presence of light. Sugars are the product of this synthesis, resulting in radiant energy being converted to chemical energy.

phreatophytes Plants dependent on groundwater, which extend their roots deep into the soil.

phyllite A type of metamorphic rock, more strongly altered than slate, with a silky sheen.

pingo A circular mound with a frozen core, found in permafrost areas.

pioneer community The first plants to be established on new regolith.

plane of the ecliptic Plane of the Earth's orbit around the Sun, inclined 23½° to the plane of the equator.

plankton Minute floating organisms in seawater, used by marine life for food.

plateau A landform of limited local relief at high elevation.

playa A dry lake bed in a basin of interior drainage.

plugged dome A composite volcano subject to explosive eruption and characterized by rapidly congealing acidic lava in the central vent. Mount Lassen, California, is an example of a plugged dome.

plutonic rock Intrusive igneous rock that has cooled slowly at considerable depth in the Earth's crust, and having coarse texture. Granite is a good example.

pluvial A climate with relatively high precipitation.

podzolization A soil-forming process dominating climates that are humid but cool for part of the year, inhibiting bacteria action and high rates of chemical weathering. Such conditions exist in the middle and high latitudes of the marine west coast climate, humid continental, and subarctic climates. These soils have humus accumulation and high acidity with moderate to heavy leaching in the upper horizons.

podzols Soils of the cool humid climates, produced by **podzolization.**

point bar A sediment ridge formed on the inside curve of a meander. Formed in stream beds and exposed during low runoff.

polar easterlies Winds that blow from polar highs to subpolar lows in both hemispheres.

polar front A zone of intense interaction between unlike air masses in the upper latitudes.

polar icecap climate (EF) Climate of Antarctica and Greenland, where temperatures of the warmest month never average above 10° C (50° F).

polaris The north star, or pole star, used as a navigation star to determine latitude.

polar jet stream The **jet stream** associated with the polar front, located between 9 and 12 kilometers (30,000 and 40,000 feet).

porosity The ratio or percentage of pore space compared to the total volume of a substance. Porosity is not necessarily a function of particle size but of arrangement of particles.

potential energy The energy stored in a particle or body, capable of being released as kinetic energy or some other form.

prairie Tall grasslands of the middle latitudes where precipitation is generally above 50 centimeters (20 inches) annually.

precession A wobble in the Earth's axis resulting from gravitational forces of the Sun and Moon upon the equatorial bulge. This causes the Earth's axis to sweep out a complete circle in 26,000 years.

precipitation The fallout of moisture capable of being measured by weather instruments such as rain or snow gauges. This includes rain, hail, snow, sleet, and dew.

pressure gradient force The force created by differences in pressure between two points. The spacing of the **isobar** indicates the strength of the force. Closely spaced isobars indicate high wind and a strong pressure gradient force.

prime meridian (Greenwich meridian) The zero meridian, which is the reference line for measurement of longitude. This meridian passes through the North and South Poles and the Royal Observatory at Greenwich near London, England.

prominences Large ribbonlike loops of luminous gas extending outward from the Sun's **chromosphere,** and usually associated with flares and sunspots.

pygmy forest A climax forest of dwarfed and stunted plants.

pyroclastic rock Fragmental volcanic material (lava, cinders, ash, and dust), consolidated and compacted to form a solid mass.

Q

quadrangle A map bounded by parallels and meridians, which fits into a uniform topographic map series.

R

radar A system whereby radio waves are transmitted and then reflected by an object to a receiver. Electronic radar images can then be created showing the object or surface in that portion of the electromagnetic spectrum.

radial drainage pattern A drainage pattern where streams radiate out from the center of a high elevation, such as a volcanic peak or dome.

radiation Emission of electromagnetic energy waves, which transmit energy through space.

radiation fog Fog produced as the result of the radiation of heat from near the surface during night hours. It is likely to occur when the air temperature reaches the dewpoint temperature.

radiosonde A self-recording and radio-transmitting instrument carried by a helium balloon up to 30,000 meters (100,000 feet), sensing humidity, temperature, and pressure.

radio waves The longest wavelengths of electromagnetic energy (greater than 1.5 micron), representing 12 percent of the Sun's total energy.

rain Liquid precipitation.

rain shadow The drier side (leeward side) of a mountain area resulting from the drying of the air following the **orographic** effect.

rawinsonde Radio wind-sounding device. A self-recording and radio-transmitting instrument similar to the **radiosonde** but also sensing wind speed and direction.

reflection The process of returning or reflecting a portion of incident electromagnetic energy falling on a surface.

refraction The bending of a ray of light as it passes obliquely from one medium to another of different density in which the speed of light is different. Light is refracted as it enters the atmosphere because it is traveling through different densities of gas.

reg A rocky windblown desert resulting from deflation, containing **desert pavement.**

region A unit-area of the Earth's surface differentiated by its specific characteristics. A *formal* region can be described as an area with certain uniformity of characteristics, in contrast to a *functional* region characterized as a sphere of activity. The *study area* may constitute a formal region though it may also contain several functional regions, such as the habitat for deer. The unifying force in the functional region is one of movement and interdependence between the parts.

regional metamorphism (dynamic metamorphism) The alteration of rocks by pressure, associated with geologic movements, usually on a large scale. This can produce both physical and mineralogical changes; for example, shale is turned into slate, granite into gneiss.

regolith A mantle or layer of loose unconsolidated sandy silt and rocky waste, overlying bedrock.

rejuvenation In Davis's concept of **cycle of erosion,** the return of an older area to youth as the result of uplift or changes in base level of the system, for example, a drop in sea level.

relative humidity The water vapor present in a mass of air expressed as a percentage of the total amount that would be present were the air saturated at that temperature, or the ratio of the air's vapor pressure to the saturation vapor pressure. If the temperature of a given mass of air rises, the water holding capacity increases, yielding a lower relative humidity.

relief The physical configuration of the Earth's surface.

relief map A map which shows surface configuration by the following methods: contour lines, layer-tinting, hachures, hill shading.

remote sensing A method of collecting visual or electronic data which can later be converted to visual forms, using the various portions of the **electromagnetic spectrum** from x-rays to radio waves. A camera or sensing device capable of sensing the object in a specific range of frequencies is often used in connection with aircraft and satellite surveys. However, remote sensing in the broad sense can be a handheld camera recording visible light.

residual deposits The residue of rock material resulting from weathering.

respiration The process by which a living organism or cell takes in oxygen from the air or water, distributes and utilizes it, and gives off carbon dioxide.

reverse fault A fault caused by compression where the hanging wall has been raised relative to the footwall, resulting in crustal shortening. The San Fernando fault is a reverse fault. See **thrust fault.**

revolution The orbit of the Earth around the Sun, which takes approximately 365¼ days.

ria shoreline An irregular coastline resulting from the drowning of river mouths.

Richter scale A scale developed by Charles Richter used to measure the intensity of earthquakes. As the number on the Richter scale increases by 1, the tremor creates 10 times as great an amplitude of surface motion.

rift zone A zone that has undergone considerable crustal spreading, developing rift valleys and related volcanic activity and mountain building.

right-lateral fault A **strike-slip fault** in which the land on the opposite side of the fault moves to your right. The San Andreas Fault is an example.

Ring of Fire (circum-Pacific belt) A volcanic and earthquake-prone region ringing the Pacific Ocean along the western mountain borders of North and South America, Alaska, Siberia, and the Asian offshore island countries.

riptide A narrow current of seaward-moving water at a beach, which results from the piling up of water by incoming waves.

rock cycle The process in which igneous rock is eroded and converted to sedimentary and metamorphic rock, which in turn may be remelted along subduction zones to produce more igneous rock as rock returns to the magma state.

rockfall The most rapid of all **mass wasting** processes, involving the free-fall or rolling and bouncing of single masses of rock from a steep cliff, building up at the base a **talus slope.**

rock flour Finely ground rock debris resulting from abrasion within and beneath a glacier.

Rossby waves North-south oscillations of the **polar jet stream** in the prevailing westerlies, which shift storm tracks within the zone of the westerlies.

rotation The spinning of the Earth on its axis. One rotation requires approximately 24 hours.

runoff Flow of water off a land surface.

S

saddle A broad flat **col** in a ridge between two mountain peaks.

sag pond A pond formed in a fault zone.

salinization A soil-forming process where soluble salts accumulate in the upper horizons. This is the result of very high evaporation, which causes excessive capillary action. This process also is found in arid regions of poor drainage. Salts tend to accumulate because leaching is at a minimum.

salt Consisting of sodium chloride. It occurs in solution as brine or in solid sheets and crusts around lakes (such as the Dead Sea and Great Salt Lake).

saltation The bouncing, skipping process by which solid material moves along the bed of a stream or over a windblown surface.

sandstone A sedimentary rock, consisting mainly of sand grains of quartz, often with feldspar, mica, and a number of other minerals.

Santa Ana A hot, dry wind blowing from the northeast and descending the Sierra Nevada and San Gabriel Mountains into the low valleys of California, heating by compression at the downslope rate of 1° C per 100 meters (5.5° F per 1,000 feet). It is caused by a buildup of a mass of high pressure continental polar air mass over the Great Basin.

saturation Temperature at which the moisture capacity of air is reached; the state of the atmosphere when it can hold no more water vapor.

savanna biochore A plant formation consisting of a combination of trees and grassland in various proportions. It has a parklike appearance, with trees dotting a grassland. Water availability is less than in the forest biochore.

scale The proportion that a map, picture, or model bears to the thing it represents. The scale selected is determined by the questions raised about the area under study.

scarp The exposed surface where one block of rock has been displaced vertically relative to another along a fault line.

scattering The process by which small ice crystals, dust, or water particles in the atmosphere deflect radiation from its path in all directions.

schist A medium-grained rock which has been affected by regional metamorphism. It usually has a foliated, wavy texture. It is named for its major mineral; for example, garnet schist or mica schist.

sclerophyll Type of tough-leafed forest characteristic of Mediterranean climatic regimes.

scrub community A vegetation **community** in a semiarid climate characterized by small shrubs and trees.

sea arches Arches cut in rocky headlands by wave erosion.

sea breeze Wind blowing from sea to shore during the day because differential heating results in warmer temperatures over land. The rising air over the land is replaced by somewhat cooler and denser air from the sea.

second-order stream A stream formed by the joining of two **first-order streams.**

sedimentary rock Rock consisting of sediments or particles, laid down in layers, and cemented together.

seif dune See **longitudinal dune.**

seismic source Focus or hypocenter; the point in the Earth's crust where an earthquake originates.

seismologist A scientist who interprets earthquakes in the scientific endeavor of seismology.

shaking The wavelike motion of the Earth's surface induced by fault motion. In both the 1989 Loma Prieta and the 1971 San Fernando earthquake it has been estimated that 99 percent of the damage was solely the result of heavy shaking and only 1 percent of the monetary loss associated with the quake was directly related to faulting through structures.

shale A sedimentary rock formed from deposits of fine clay-sized particles compacted and cemented together.

shattering The disintegration of rock along new surfaces rather than old joints of a massive rock. This type of breakup produces highly angular pieces compared to the more rectangular blocks in **block separation.**

sheet wash The process by which soil is removed by runoff over a surface area without forming channels or streams.

shield volcano (dome volcano) A volcanic cone of basic lava with a small angle of slope (2° to 10°), and a large diameter at the base. Mauna Loa, Hawaii has a base 500 kilometers (300 miles) in diameter on the ocean floor. Its total height from the ocean floor to the summit is about 9,000 meters (30,000 feet); the height of the summit above sea level is 4,168 meters (13,675 feet). The lava is very fluid and less explosive than other types of volcanoes.

shifting cultivation A primitive form of agriculture practiced in the tropical rainy regions consisting of girdling the trees, burning, cutting, and clearing the forest. A crop is then planted until the soil is exhausted. This takes from 1 to 5 years. Then the population repeats the process on another tract of land. In 20 to 30 years the entire cycle can be repeated in the same place.

shoreline of emergence The set of coastal landforms that appear wherever a rising of the Earth's crust or a falling of sea level has occurred near the border of a continent.

shoreline of submergence The set of coastal landforms that appear wherever a sinking of the crust or a rising of sea level occurs near the border of a land area.

short-wave radiation Electromagnetic radiation including gamma, x-ray, ultraviolet, and light waves.

sial The granitic rocks of the continental crust, composed largely of silica and aluminum compounds.

sialic shield The heartlands of the continental crust, consisting of the very oldest rock. North America's geological heartland is the Canadian Shield.

silicates A large group of rock-forming minerals with silicon and oxygen. Quartz, for example, consists of SiO_2.

silt Particles from 2 to 50 microns in diameter.

siltstone A sedimentary rock consisting of silt-size particles laid down in water. Siltstone is coarser grained than **shale** and finer grained than **sandstone.**

sima Dense rocks of silica and ferromagnesium found underlying the less dense continents (**sial**) and forming the floors of much of the ocean.

sinkhole Surface depression resulting from the work of solution in areas of limestone or other soluble rock.

site Internal characteristics of a place. For example, if the place is your home, the internal characteristics would include a complete description of your house and the internal arrangement of rooms, living patterns, and even decorations and building materials. An example is our *study area.*

situation Relative location, the location of a place in relation to its wider surroundings; for example, the location of your house in relation to other houses, streets, stores, fire stations, or distance to the central city area.

sky coverage The fraction of the sky covered by clouds.

slate A gray or brick-red fine-grained rock that splits neatly into thin plates. Its parent material is **shale.**

sleet Small pellets of ice formed when rain falls through a zone of the atmosphere where the temperature is below freezing.

slippage A form of **mass wasting** in which a block separates from the main mass.

slope (gradient) A quantitative expression of steepness. A simple way to express and define slope is the change in elevation per unit of horizontal distance. It can also be computed in percent of grade where the horizontal is zero percent and 45° is 100 percent. To compute percent of slope, simply divide the vertical distance by the horizontal distance, using the same units, and multiply by 100.

slope orientation The compass direction of the slope face.

smog Mixed particles and chemical pollutants present in considerable densities in the air.

snow Precipitation in the form of hexagonal ice crystals.

snow line The lowest elevation or edge of the snow cover.

soil acidity see **pH.**

soil amendments Components added to the soil to improve its productivity.

soil creep Extremely slow downslope movement of soil.

soil profile The arrangement of the soil into **horizons** of different texture, structure, color, and consistency.

soil structure The way in which soil grains are grouped together into larger pieces.

soil texture The mixture of particle sizes composing the soil. Particles are classified into various grades from coarse gravel to fine clay.

solar constant The amount of energy received upon a unit area of surface perpendicular to the Sun's rays outside of the Earth's atmosphere. The value is 2 gram-calories per square centimeter per minute, or 2 langleys per minute, or 430 Btu per square foot per hour.

solar flares High energy releases of the Sun, sending bursts of ions toward the Earth that cause the auroras as well as serious disruption of radio communication.

solar wind A flow of charged particles (electrons and protons) emitted by the Sun.

solifluction The slow displacement and movement of water-saturated materials above the bedrock downslope due to gravity.

solum Horizons A and B, the "true soil."

solute effect The growth of a droplet that results from increased solute.

solution The chemical weathering process in which a mineral such as salt is dissolved by water. Rivers carry vast loads dissolved in solution.

source region Those land and ocean areas that strongly influence the temperature and moisture characteristics of the overlying **air mass.**

southeast trades Winds blowing from the Southern Hemisphere subtropical high to the equatorial low.

spatial distribution Configuration or arrangement of the pattern of the element under investigation. Spatial distribution includes three aspects: density, dispersion, and pattern.

spatial interaction The degree of movement and circulation within a given area. Isolation or central location plays a major role in determining how much interaction takes place. Australia, until the eighteenth century, remained almost unknown to the world and totally lacked spatial interaction within the world community because of its isolated position.

specific gravity The ratio of the mass of a body to the mass of an equal volume of water.

specific heat capacity The amount of energy necessary to change 1 gram of a substance by 1° C.

specific humidity The ratio of the weight of the water vapor in a parcel of the atmosphere to the total weight of the air (including the water vapor), stated in grams of water vapor per kilogram of air.

spit A wave depositional feature projecting from the mainland into an adjacent bay or sea.

spodosol Soil with a leached topsoil overlying subsoil that is more compact and is stained by aluminum and iron compounds.

spring Surface outflow of water where the water table is exposed.

spring tide Particularly high tide occurring twice monthly when Sun, Earth, and Moon are in line at the time of the new and full Moon.

stability The term when applied to air indicates that if a parcel of air is forced to rise it will tend to sink back to the ground because it is cooler and heavier than the surrounding air. See **instability.**

stack A steep pillar of rock rising from the sea, formerly part of a headland, but isolated by wave action, resulting in erosion.

stalactite A deposit of calcium carbonate extending downward from the ceiling of a cave. Dissolved minerals are redeposited as a result of precipitation of minerals under saturated conditions.

stalagmite A deposit of calcium carbonate extending upward from the floor of a cave, resulting from saturated water dripping from a stalactite directly above.

star dune A sand dune shaped like a pyramid up to 80 meters (300 feet) high.

stationary front A boundary surface between two dissimilar air masses where there is essentially no forward movement.

steam fog Fog created when cold air passes over a warmer body or surface.

steppe A short-grass **community** characteristic of semiarid areas.

stomata Microscopic openings in plant leaves through which carbon dioxide is absorbed and water vapor is lost through evapotranspiration.

stratification Layering of sedimentary beds.

stratocumulus (Sc) A low-lying cloud layer, consisting of dark gray globular masses within a continuous sheet. Sometimes open sky will appear between rolls of clouds, oriented at right angles to the wind.

stratosphere The layer above the **troposphere** where uniform temperatures prevail. However, upward through the stratosphere a slow rise in temperature occurs until 0° C (32° F) is reached at the stratopause.

stratovolcanoes See **composite volcanoes.**

stratus (St) A dense, low-lying cloud at any height up to roughly 2,500 meters (8,000 feet). It is often associated with fog and drizzle.

strike The compass direction along a rock stratum at right angles to the dip.

strike-slip fault (transcurrent fault) A fault where lateral motion dominates. The San Andreas fault system is a classic example.

subduction zone A region where an ocean plate is plunging beneath a continental plate. This is the case along much of the **Ring of Fire.**

sublimation The conversion of a solid to a vapor, or vice versa, without an intervening liquid stage. Snow crystals form in this manner as well as frost.

Snow and ice will also return in the vapor state to the atmosphere in this same manner without passing through the liquid state.

submarine canyon An undersea valley cutting into the **continental slope.**

subpolar lows Upper middle-latitude belts of low pressure lying between the westerlies and polar easterlies.

subsidence A form of mass movement where general settling of a large mass occurs. **Karst** regions are especially known for subsidence resulting from the carbonation process of weathering.

summer monsoon A wind system resulting from differential heating and cooling of land and adjacent oceanic areas. The summer monsoon of Asia occurs as the interior warms during the summer. This creates differences in pressure. The Asian summer monsoon brings rain from May through September.

summer solstice The position of the Earth in its orbit at which the Sun's rays are perpendicular at noon, on June 21 or 22, at the **Tropic of Cancer.** This position marks the northern limit for the perpendicular rays of the Sun. This date also represents the longest daylight period in the Northern Hemisphere.

sunspots Darker, cooler regions of the Sun consisting of magnetic fields with positive and negative polarity. Sunspots reach a maximum, on the average, every 11 years and are associated with solar disturbances, such as flares and prominences.

suspended load The fraction of the transported matter in a stream that is not dissolved or carried by **traction.**

swash The upsurge of water as a wave reaches shore, moving sand and gravel on the beach. See **Backwash.**

syncline A downfold in the Earth's crust with strata dipping inward toward a central axis, caused by compressive forces.

T

taiga A needle-leaf evergreen forest, extending across the northern continents. Trees commonly growing in this region are pines, spruce, fir, and birch.

talus slope Slopes of angular rock debris on a mountainside. The material is mainly formed as the result of frost action, and the slope is consistently around 35° with the horizontal.

tarn A small lake located on the floor of a **cirque.**

tectonic From a Greek word meaning "builder," applying to all internal forces which build up or form the features of the crust.

temperate forests Forests of the middle latitudes, consisting of either conifers or mixed forests.

temperature A measure of the mean velocity of the molecules, indicated in degrees.

terminal moraine Glacial debris deposited at the maximum point or terminus of ice movement.

terrestrial The Earth-like planets: Mercury, Venus, Earth, and Mars.

thermal infrared Invisible heat rays just beyond the red end of the visible spectrum; its waves are longer than those of the visible spectrum but shorter than radio waves. The Earth reradiates heat energy in this portion of the electromagnetic spectrum.

thermal springs Hot springs resulting from an interior heat source such as a magma pool or chamber.

thermosphere Uppermost of four primary layers of the atmosphere, where temperatures increase with increasing altitude reaching several thousand degrees.

third-order stream A stream formed from two **second-order streams.**

thrust fault A **reverse fault** of very low angle in which the upper beds have been pushed far forward over the lower beds.

thunderstorm An intense local storm associated with a large, dense **cumulonimbus** cloud in which there are very strong updrafts of air.

tidal flat An area of sand or mud uncovered at low tide. A habitat for marine organisms adapted to fluctuation in water level and salinity.

tide Periodic rise and fall of sea level in response to the gravitational attraction of the Sun and Moon.

till Unsorted rock debris deposited by glacial ice.

timberline The upper elevation limit of tree growth.

time zone A zone consisting of 15 degrees of longitude in which all clocks are synchronized. Twenty-four zones 15 degrees wide encompass the globe.

tombolo A long, narrow accumulation of sand, with one end attached to the land and the other end projecting into the sea to an offshore island or stack.

topography A description of the surface features of an area.

tornado A counterclockwise whirling storm (sometimes called twister) formed around an intensely low pressure system, with wind speeds greater than 320 kilometers per hour (200 miles per hour) and often a dark, funnel-shaped cloud. It occurs particularly in the Mississippi basin in spring and early summer, associated with a cold front commonly

along a squall line. It is very short-lived, lasting only an hour or two and usually only a few hundred yards or less in diameter. It can level all things in its path.

township A six-mile square known as a congressional township, referenced to an east-west base line, designated the *geographer's line.* Meridians and parallels laid off at six-mile intervals from the baseline form the boundaries of the township.

traction The slow movement of larger pieces of material along a stream bed as the result of the hydraulic force exerted by the water.

trade winds North and south of the **doldrums** are the trade winds covering a zone between latitudes 5° and 30°. The trades are a result of a pressure gradient between the subtropical belt of high pressure and the equatorial low of the doldrums. Deflection by the **Coriolis effect** influences the northeast and southeast trades.

transpiration The loss of water vapor from the leaves of plants through small pores known as **stomata.**

transportation The movement of the eroded particles by dragging along the bed, by suspension in the body of the stream, or in solution.

transverse dunes Wavelike ridges of sand separated by troughlike furrows. Their crests trend at right angles to the direction of the wind.

trellis stream pattern A rectangular pattern of drainage found in regions of anticlines and synclines that have eroded scarps where outcroppings of alternately more resistant and less resistant rocks occur at right angles to the slope.

Tropic of Cancer 23½° N latitude.

Tropic of Capricorn 23½° S latitude.

tropical A climatic zone bounded by the isotherm of 18° C (64.4° F), which represents mean temperature of the coolest month. This isotherm is the poleward boundary of the A climates.

tropical continental source The regions that give rise to tropical continental air masses, located in the deserts of the subtropics of North America, Africa, and Australia.

tropical cyclone The name for violent subtropical storms in south Asia, analogous to hurricane or typhoon. Winds are in excess of 120 kilometers per hour (74 mph).

tropical desert climate (BWh) A climate dominated by the continental tropical (ct) air masses and located along the Tropic of Cancer and Capricorn. Very high maximum temperatures and low precipitation make this climatic region home of the largest deserts in the world.

tropical monsoon climate (Am) A rainforest climate despite a dry season. The rainfall comes in the summer as the Asian land mass heats up, producing a thermal low pressure and pressure gradient toward the center of Asia. Maritime air moves onshore, becomes unstable, and produces very heavy precipitation in Southeast Asia. In the winter, high pressure develops over Asia and a reverse flow occurs, with dry cool air coming out of Asia. Northern Australia receives rainfall from the monsoon effect during this period (October–May).

tropical rainforest climate (Af) A wet equatorial climate with heavy convectional rainfall, very uniform in temperature throughout the year. This climate is located in the equatorial zone and extends into the **trade winds** of the low latitudes.

tropical savanna climate (Aw) A tropical climate type with seasonal variations of rainfall. The wet season occurs during the hemisphere's summer at high sun. The dry season is dominated by the continental tropical (ct) air mass. This climate is generally in the belt of the **trade winds** on both sides of the equator.

tropical steppe climate (BSh) A transition climate between the tropical savanna and tropical desert. The rainfall pattern is similar to the tropical savanna but generally much less.

tropics The zone between the **Tropic of Cancer** and **Tropic of Capricorn**. This zone is also referred to as the Torrid Zone.

tropopause The level at which the **troposphere** gives way to the **stratosphere,** where marked inversion begins.

tropophytes Plants that are adapted to seasonal variations of water and light and temperature.

troposphere The layer of the atmosphere in which the environment or normal lapse rate is consistent. This is the zone of weather and the living zone of the planet.

tsunami A series of waves generated in the ocean by earthquakes when a sudden shift in the ocean floor occurs. They travel outward from the area of disturbance in concentric expanding rings. When the waves reach a shoreline, the effect is to cause a rise in sea level as much as 30 meters (100 feet) above normal high tide, causing considerable destruction.

tundra climate (ET) A climate with no true warm season or summer, yet because of its coastal location, winters are not as cold as the continental climates. Located in the high latitudes as well as high altitudes above timberline. In the far north, the tundra climate forms a fringe along the Arctic coasts of North America and Asia.

tundra soils Soils that form in the high latitudes of the tundra climate. They are very thin and may be considered an azonal type similar to **podzols,** but they can be poorly drained and take on an intrazonal form.

turbidity Particulate air pollution or suspended sediment in water.

typhoon See **hurricane.**

U

ultisol Soils severely leached of bases that develop in hot, humid environments.

undertow The strong **backwash** creating muddy turbidity.

uniformitarianism The scientific assumption that the processes now operating to modify the landscape have operated in the same way in the geologic past.

unloading The removal of a load of regolith from a segment of the Earth's crust.

upslope fogs Fogs developing along lower slopes of mountain regions as the result of upslope movement of air.

upwelling The upward movement of cold ocean water, induced by surface winds and shoreline configuration; a very important factor in moderating tropical climates and creating low clouds and fog along the coast and over the cold ocean water.

V

vadose water Water that moves through permeable rock above the **water table.**

valley breeze Movement of air upslope during daylight warming of the mountain slope.

Van Allen radiation belt Two ring-shaped belts of charged particles (electrons and protons) within the **magnetosphere.** One exists 3,700 kilometers (2,300 miles) from the Earth's surface and the other more intense belt is at 13,000 to 19,000 kilometers (8,000 to 12,000 miles) out.

vapor pressure The pressure exerted by the water vapor present in the atmosphere.

ventifact A faceted rock created by the abrasive actions of windblown sand.

vernal equinox See **equinox.**

visibility The greatest horizontal distance one can see using a known standard reference feature such as a hill or building.

visible energy Electromagnetic energy with wavelengths ranging between 0.4 and 0.7 microns.

volcanic dome A volcanic feature with a dome form.

volcanic extrusive igneous rock Igneous rock formed by magma cooling on or near the surface of the earth.

W

warm front A zone where warm air is moving into a region of colder air.

watershed The surface area supplying runoff water to a master stream and its tributaries.

water table The upper limit of the subsurface area that is saturated with water.

wave-built terrace A gently sloping area seaward of the wave-cut bench, built up of materials eroded from the land.

wave-cut bench Gently sloping surface of sand or beach **regolith** cut at the base of a sea cliff by the erosive action of waves.

wave-cut cliff A steep escarpment eroded at the sea's edge by surf.

wave cyclone See **middle-latitude cycle.**

wave refraction A tendency for a wave front to be turned from its original direction as it approaches the coast. It may be retarded and turned by the shallowing of the water.

weathering The breakdown of rock materials by mechanical and chemical processes.

west coast desert climate (BWk) The climate located on the west coasts bordering the oceanic subtropical high-pressure cells, where upwelling and subsidence produce fogs and a pronounced inversion. This climate is very cool and moderate with extremely low precipitation.

westerlies A wind belt located between latitude 30° and 60°. The winds move out of the subtropical highs on the poleward side.

wet adiabatic lapse rate The rate of decrease in air temperature by dynamic expansion of air that is saturated, averaging about 0.5° C per 100 meters (2.7° F per 1,000 feet). This rate is less than the **dry adiabatic lapse rate** because of latent heat released. The actual rate varies with the amount of water vapor condensed. If the air is in a state of descent, it will heat at the same rate resulting from compression of the atmosphere.

wind The movement of air from areas of higher pressure to areas of lower pressure. The greater the pressure gradient, the greater the wind.

winter monsoon Continental offshore flow of air in winter, created by differences in heating, hence differences in pressure between land and sea.

winter solstice The position of the Earth in its orbit at which the Sun's rays are perpendicular at noon, on December 21, or 22, at the **Tropic of Capricorn.** This position marks the southern limit for the perpendicular rays of the Sun and the longest daylight period in the Southern Hemisphere and shortest in the Northern Hemisphere.

X

xerophytes Plants that are adaptable to arid habitats.

xerophytic plants Drought-resistant vegetation with very small leaves and other drought-resistant designs.

Y

yardang A sharp crest or ridge created by wind erosion.

yazoo tributary A tributary stream that parallels the course of a main stream for a considerable distance before entering it. An example is the Yazoo River, a tributary to the Mississippi River.

youthful stream valley The first stage in the cycle of landform development. Opening and deepening of the channel is the principal activity of a young stream, whose capacity for load exceeds the load available to it, hence erosion.

Z

zenith The point directly overhead, or 90° from the horizon.

zonal soils Soils formed under conditions of good drainage through the prolonged action of soil-forming processes with well-developed horizons.

zone of ablation The lower part of the glacier, where the rate of ice wastage is rapid and old ice is exposed to the glacial surface. It is identifiable by its "dirty" white surfaces.

zone of accumulation The upper end of the glacier, where snow is in the process of compaction and recrystallization. The surface is snow white.

Credits

Index

and mineral concentrations, 313
physical weathering, 371
slaking, 375
unloading, 374
Weather instruments, types of, 63
Wegener, Alfred L., 126, 316
Wegener/Bergeron/Findeisen theory, of ice
    crystal growth, 126
Westerlies, 98, 99
Widespread ascent, 122
Wind
    anticyclonic systems, 97–98
    and biosphere, 240
    bora, 105
    and buoyancy, 96
    centripetal force, 95–96
    chinook winds, 102, 470
    Coriolis effect, 95
    cyclonic systems, 97–98
    eolean landforms
        deflation basin, 470
        dunes, 473–477
        geographic significance of, 477
        lag deposits, 470–471
        loess, 471–472
        pedestal rocks, 470
        and vegetation, 468–470
        yardang, 470

and erosion, 467–468, 492
    processes in, 467–468
foehn winds, 102
and friction, 96
geostrophic wind, 95
global circulation, 98–99
gradient wind, 96
and gravity, 96
land breeze, 102–103
methods of transporting particles, 468
mistral, 105, 107
monsoons, 107–109
mountain breezes, 103, 108
polar easterlies, 98
polar jet stream, 99
Santa Ana winds, 102
sea breeze, 102
and slopes, 105–107
and soil formation, 206
upper air circulation, 99
valley breezes, 103, 108
vertical forces, 96
and waves, 450
westerlies, 98, 99
Winter monsoon, 107–109
Winter solstice, 40
Wisconsin Stage, 443
Wood, Harry O., 328

X

Xerophytic plants, 239, 242, 268, 271
X-rays, 72

Y

Yardang, 470
Yazoo streams, 403
Youthful stream valleys, 396

Z

Zenith, 36
Zonal soils, 515
Zone of aeration, groundwater, 408
Zone of saturation, groundwater, 409
Zone of wasting, 435